# Lecture Notes in Artificial Intelligence 12482

Subseries of Lecture Notes in Computer Science

More information about this series at http://www.springer.com/series/1244

Van-Nam Huynh · Tomoe Entani ·
Chawalit Jeenanunta · Masahiro Inuiguchi ·
Pisal Yenradee (Eds.)

# Integrated Uncertainty in Knowledge Modelling and Decision Making

## 8th International Symposium, IUKM 2020
## Phuket, Thailand, November 11–13, 2020
## Proceedings

*Editors*
Van-Nam Huynh
Japan Advanced Institute of Science and Technology
Nomi, Ishikawa, Japan

Chawalit Jeenanunta
Sirindhorn International Institute of Technology
Thammasat University
Pathum Thani, Thailand

Pisal Yenradee
Sirindhorn International Institute of Technology
Thammasat University
Pathum Thani, Thailand

Tomoe Entani
University of Hyogo
Kobe, Japan

Masahiro Inuiguchi
Graduate School of Engineering Science
Osaka University
Toyonaka, Osaka, Japan

ISSN 0302-9743 ISSN 1611-3349 (electronic)
Lecture Notes in Artificial Intelligence
ISBN 978-3-030-62508-5 ISBN 978-3-030-62509-2 (eBook)
https://doi.org/10.1007/978-3-030-62509-2

LNCS Sublibrary: SL7 – Artificial Intelligence



This Springer imprint is published by the registered company Springer Nature Switzerland AG
The registered company address is: Gewerbestrasse 11, 6330 Cham, Switzerland

# Preface

This volume contains the papers that were presented at the 8th International Symposium on Integrated Uncertainty in Knowledge Modelling and Decision Making (IUKM 2020) which took place during November 11–13, 2020. IUKM 2020 was originally planned to take place in Phuket, Thailand. However, it was moved online due to the COVID-19 pandemic and global travel restrictions.

The IUKM symposia aim to provide a forum for exchanges of research results and ideas, and experience of application among researchers and practitioners involved with all aspects of uncertainty modeling and management. Previous editions of the conference were held in Ishikawa, Japan (IUM 2010), Hangzhou, China (IUKM 2011), Beijing, China (IUKM 2013), Nha Trang, Vietnam (IUKM 2015), Da Nang, Vietnam (IUKM 2016), Hanoi, Vietnam (IUKM 2018), and Nara, Japan (IUKM 2019) with their proceedings being published by Springer in AISC 68, LNAI 7027, LNAI 8032, LNAI 9376, LNAI 9978, LNAI 10758, and LNAI 11471, respectively.

IUKM 2020 was jointly organized by Sirindhorn International Institute of Technology (SIIT), Thammasat University, Thailand's National Electronics and Computer Technology Center (NECTEC), and Japan Advanced Institute of Science and Technology (JAIST).

This year the conference received 55 submissions from 14 different countries. Each submission was peer-reviewed by at least two members of the Program Committee. After a thorough review process, 35 papers were accepted for presentation and inclusion in the LNAI proceedings.

We are very thankful to the local organizing team from SIIT and NECTEC for their hard working, efficient services, and wonderful local arrangements.

We would like to express our appreciation to the members of the Program Committee for their support and cooperation in this publication. We are also thankful to Alfred Hofmann, Anna Kramer, and their colleagues at Springer for providing a meticulous service for the timely production of this volume. Last, but certainly not least, our special thanks go to all the authors who submitted papers and all the attendees for their contributions and fruitful discussions that made this conference a great success.

November 2020

Van-Nam Huynh
Tomoe Entani
Chawalit Jeenanunta
Masahiro Inuiguchi
Pisal Yenradee

# Organization

## General Co-chairs

| | |
|---|---|
| Masahiro Inuiguchi | Osaka University, Japan |
| Pisal Yenradee | SIIT, Thammasat University, Thailand |

## Advisory Board

| | |
|---|---|
| Michio Sugeno | Tokyo Institute of Technology, Japan |
| Hung T. Nguyen | New Mexico State University, USA, and Chiang Mai University, Thailand |
| Sadaaki Miyamoto | University of Tsukuba, Japan |
| Akira Namatame | AOARD/AFRL, National Defense Academy of Japan, Japan |
| Thanaruk Theeramunkong | SIIT, Thammasat University, Thailand |
| Chai Wutiwiwatchai | National Electronics and Computer Technology Center, Thailand |

## Program Co-chairs

| | |
|---|---|
| Van-Nam Huynh | JAIST, Japan |
| Tomoe Entani | University of Hyogo, Japan |
| Chawalit Jeenanunta | SIIT, Thammasat University, Thailand |

## Local Arrangement Co-chairs

| | |
|---|---|
| Warut Pannakkong | SIIT, Thammasat University, Thailand |
| Sapa Chanyachatchawan | National Electronics and Computer Technology Center, Thailand |

## Publicity Chair

| | |
|---|---|
| Akkaranan Pongsathornwiwat | The National Institute of Development Administration, Thailand |

## Program Committee

| | |
|---|---|
| Tomoyuki Araki | Hiroshima Institute of Technology, Japan |
| Yaxin Bi | Ulster University, UK |
| Tru Cao | University of Texas Health Science Center at Houston, USA |
| Yong Deng | Xi'an Jiaotong University, China |

| | |
|---|---|
| Thierry Denoeux | University of Technology of Compiègne, France |
| Sebastien Destercke | University of Technology of Compiègne, France |
| Zied Elouedi | LARODEC, ISG de Tunis, Tunisia |
| Tomoe Entani | University of Hyogo, Japan |
| Katsushige Fujimoto | Fukushima University, Japan |
| Yukio Hamasuna | Kindai University, Japan |
| Ryo Haraguchi | University of Hyogo, Japan |
| Katsuhiro Honda | Osaka Prefecture University, Japan |
| Tzung-Pei Hong | National University of Kaohsiung, Taiwan |
| Jih Cheng Huang | Soochow University, Taiwan |
| Van-Nam Huynh | JAIST, Japan |
| Masahiro Inuiguchi | Osaka University, Japan |
| Radim Jirousek | University of Economics, Czech Republic |
| Yuchi Kanzawa | Shibaura Institute of Technology, Japan |
| Mayuka F. Kawaguchi | Hokkaido University, Japan |
| Vladik Kreinovich | The University of Texas at El Paso, USA |
| Yasuo Kudo | Muroran Institute of Technology, Japan |
| Yoshifumi Kusunoki | Osaka Prefecture University, Japan |
| Anh Cuong Le | Ton Duc Thang University, Vietnam |
| Bac Le | VNUHCM - University of Science, Vietnam |
| Churn-Jung Liau | Academia Sinica, Taiwan |
| Jun Liu | Ulster University, UK |
| Tieju Ma | East China University of Science and Technology, China |
| Luis Martinez | University of Jaen, Spain |
| Radko Mesiar | Slovak University of Technology in Bratislava, Slovakia |
| Tetsuya Murai | Chitose Institute of Science and Technology, Japan |
| Michinori Nakata | Josai International University, Japan |
| Canh Hao Nguyen | Kyoto University, Japan |
| Duy-Hung Nguyen | SIIT, Thammasat University, Thailand |
| Le Minh Nguyen | JAIST, Japan |
| Sa-Aat Niwitpong | King Mongkut's University of Technology North Bangkok, Thailand |
| Vilem Novak | University of Ostrava, Czech Republic |
| Warut Pannakkong | SIIT, Thammasat University, Thailand |
| Irina Perfilieva | University of Ostrava, Czech Republic |
| Zengchang Qin | Beihang University, China |
| Hiroshi Sakai | Kyushu Institute of Technology, Japan |
| Kao-Yi Shen | Chinese Culture University, Taiwan |
| Dominik Slezak | University of Warsaw, Infobright Inc., Poland |
| Roman Slowinski | Poznan University of Technology, Poland |
| Martin Stepnicka | University of Ostrava, Czech Republic |
| Kazuhiro Takeuchi | Osaka Electro-Communication University, Japan |
| Xijin Tang | Chinese Academy of Sciences, China |
| Yongchuan Tang | Zhejiang University, China |

| | |
|---|---|
| Roengchai Tansuchat | Chiang Mai University, Thailand |
| Phantipa Thipwiwatpotjana | Chulalongkorn University, Thailand |
| Araki Tomoyuki | Hiroshima Institute of Technology, Japan |
| Vicenc Torra | Umeå University, Sweden |
| Dang Hung Tran | Hanoi National University of Education, Vietnam |
| Seiki Ubukata | Osaka Prefecture University, Japan |
| Guoyin Wang | Chongqing University of Posts and Telecommunications, China |
| Koichi Yamada | Nagaoka University of Technology, Japan |
| Hong-Bin Yan | East China University of Science and Technology, China |
| Chunlai Zhou | Renmin University of China, China |

## Additional Referees

| | |
|---|---|
| Zhenzhong Gao | Osaka University, Japan |
| Shigeaki Innan | Osaka University, Japan |
| Ba-Hung Nguyen | JAIST, Japan |
| Duc-Vinh Vo | JAIST, Japan |

# Contents

## Machine Learning



## Machine Learning Applications

**Econometric Applications**

**Statistical Methods**

# Uncertainty Management and Decision Support

# Outcome Range Problem in Interval Linear Programming: An Exact Approach

Elif Garajová[1,2(✉)], Miroslav Rada[2,3], and Milan Hladík[1,2]

[1] Faculty of Mathematics and Physics, Department of Applied Mathematics, Charles University, Malostranské nám. 25, Prague, Czech Republic
{elif,hladik}@kam.mff.cuni.cz

[2] Department of Econometrics, University of Economics, nám. W. Churchilla 4, Prague, Czech Republic
miroslav.rada@vse.cz

[3] Department of Financial Accounting and Auditing, University of Economics, nám. W. Churchilla 4, Prague, Czech Republic

**Abstract.** Interval programming provides a mathematical model for uncertain optimization problems, in which the input data can be perturbed independently within the given lower and upper bounds. This paper discusses the recently proposed outcome range problem in the context of interval linear programming. The motivation for the outcome range problem is to assess further impacts and consequences of optimal decision making, modeled in the program by an additional linear outcome function. Specifically, the goal is to compute a lower and an upper bound on the value of the given outcome function over the optimal solution set of the interval program. In this paper, we focus mainly on programs with interval coefficients in the objective function and the right-hand-side vector. For this special class of interval programs, we design an algorithm for computing the outcome range exactly, based on complementary slackness and guided basis enumeration. Finally, we perform a series of computational experiments to evaluate the performance of the proposed method.

**Keywords:** Interval linear programming · Outcome range problem · Optimal solution set.

## 1 Introduction

Optimization under uncertainty has received a lot of attention in the literature throughout the past years. In this paper, we adopt the approach of *interval linear programming* [21], handling a class of uncertain optimization problems,

E. Garajová and M. Rada were supported by the Czech Science Foundation under Grant P403-20-17529S. M. Hladík was supported by the Czech Science Foundation under Grant P403-18-04735S. E. Garajová and M. Hladík were also supported by the Charles University project GA UK No. 180420.


V.-N. Huynh et al. (Eds.): IUKM 2020, LNAI 12482, pp. 3–14, 2020.
https://doi.org/10.1007/978-3-030-62509-2_1

in which the input data can be independently perturbed within a priori given intervals. Interval linear programming can be applied in modeling and solving various practical optimization problems, such as portfolio optimization [2] or transportation problems with interval data [4].

Analogously to classical linear programming, one of the main tasks solved in interval linear programming is to compute the best and the worst possible values optimal for some setting of the uncertain coefficients. This is known in the literature as the *optimal value range* problem. Methods for computing the optimal value range exactly have been derived for different types of interval linear programs [8,19]. Due to the NP-hardness of the problem, several algorithms for efficiently finding an approximation of the values were proposed (e.g. [17]). Since the original formulation of the problem admits infinite bounds of the range if there is an unbounded or infeasible program, a variant of the problem restricted only to finite values has also been introduced [10].

Recently, a related problem was proposed, referred to as the *outcome range problem* [18], dealing with the task of computing the maximal and the minimal possible values of a given linear function over the set of all optimal solutions of an interval linear program. The motivation for the outcome range problem is to evaluate further consequences and impacts of optimal decision making, which can be modeled by an additional outcome function. The authors prove NP-hardness of the problem and show two algorithms to approximate the bounds of the outcome range for programs with an interval right-hand side. In this paper, we derive an exact approach to computing the outcome range for interval linear programs with an interval objective and right-hand side. The proposed algorithm is based on guided basis enumeration, building on the framework introduced for computing an approximation of the optimal solutions via the interval simplex method [12,13] and a description of the optimal solution set in the special case [5]. Enumeration of possibly optimal solutions has also been previously applied in the context of fuzzy programming with an uncertain objective function [11].

## 2 Interval Linear Programming

### 2.1 Preliminaries and Notation

Throughout the paper, the following notation is used: For a given vector $v \in \mathbb{R}^n$, we denote by $\operatorname{diag}(v) \in \mathbb{R}^{n \times n}$ the diagonal matrix with entries $(\operatorname{diag}(v))_{ii} = v_i$ for $i \in \{1, \ldots, n\}$. The symbol $\mathbb{IR}$ denotes the set of all closed real intervals.

Given two real matrices $\underline{A}, \overline{A} \in \mathbb{R}^{m \times n}$ with $\underline{A} \leq \overline{A}$, we define an *interval matrix* $\mathbf{A} \in \mathbb{IR}^{m \times n}$ determined by the *lower bound* $\underline{A}$ and the *upper bound* $\overline{A}$ as

$$\mathbf{A} = [\underline{A}, \overline{A}] = \{A \in \mathbb{R}^{m \times n} : \underline{A} \leq A \leq \overline{A}\},$$

where the inequality $\leq$ is understood entrywise. An interval vector $\mathbf{a} \in \mathbb{IR}^n$ can be defined analogously. Alternatively, an interval matrix can also be determined by specifying its *center* $A^c = \frac{1}{2}(\overline{A} + \underline{A})$ and *radius* $A^\Delta = \frac{1}{2}(\overline{A} - \underline{A})$.

For interval vectors $\mathbf{a}, \mathbf{b} \in \mathbb{IR}^n$ and $\alpha \in \mathbb{R}$, we have the following interval-arithmetic operations:

$$\mathbf{a} \pm \mathbf{b} = [\underline{a} \pm \underline{b}, \overline{a} \pm \overline{b}], \quad \alpha \cdot \mathbf{b} = [\min\{\alpha\underline{b}, \alpha\overline{b}\}, \max\{\alpha\underline{b}, \alpha\overline{b}\}].$$

Finally, given a set $\mathcal{A} \subseteq \mathbb{R}^n$, we define its *interval enclosure* as an interval vector $\mathbf{v} \in \mathbb{IR}^n$ satisfying $\mathcal{A} \subseteq \mathbf{v}$. The tightest interval enclosure is known as the *interval hull* of the set.

### 2.2 Optimization with Interval Data

Interval programming deals with uncertain optimization problems, whose data can be perturbed within given interval ranges. For data given by an interval matrix $\mathbf{A} \in \mathbb{IR}^{m \times n}$ and interval vectors $\mathbf{b} \in \mathbb{IR}^m$, $\mathbf{c} \in \mathbb{IR}^n$, we define an *interval linear program* (ILP)

$$\min \mathbf{c}^T x \text{ subject to } \mathbf{A}x = \mathbf{b},\ x \geq 0 \tag{1}$$

as the set of all linear programs in the form

$$\min c^T x \text{ subject to } Ax = b,\ x \geq 0 \tag{2}$$

with $A \in \mathbf{A}$, $b \in \mathbf{b}$, $c \in \mathbf{c}$.

A particular linear program (2) in the ILP is called a *scenario*. Several different notions of feasible and optimal solutions in interval programming have been proposed in the literature. In this paper, we examine the solutions in a weak sense, meaning that feasibility or optimality is satisfied for at least one scenario. Namely, a given vector $x^* \in \mathbb{R}^n$ is

- a *(weakly) feasible solution* of ILP (1), if it is a feasible solution of the system $Ax = b$, $x \geq 0$ for some $A \in \mathbf{A}$, $b \in \mathbf{b}$,
- a *(weakly) optimal solution* of ILP (1), if it is an optimal solution for some scenario (2) with $A \in \mathbf{A}$, $b \in \mathbf{b}$, $c \in \mathbf{c}$.

Throughout the paper, we denote the set of all weakly optimal solutions by $\mathcal{S}$.

*Remark 1.* Note that the usual equivalent transformations used to rewrite a linear program into different forms (e.g. to obtain inequality constraints or free variables) are not always applicable in interval linear programming. For an ILP, applying the transformations may change the set of optimal solutions or the optimal values. However, for the special case of interval programs with a fixed (non-interval) constraint matrix, which is also addressed in this paper, the optimal set remains the same under these transformations (see [6] for details).

### 2.3 Optimal Value Range and Outcome Range

Apart from describing the optimal solutions, we are often also interested in the optimal objective values corresponding to these solutions. For interval programming, this usually means computing or approximating the optimal value range.

Given $A \in \mathbf{A}, b \in \mathbf{b}, c \in \mathbf{c}$, let $f(A,b,c)$ denote the optimal value of the corresponding scenario (2), with $f(A,b,c) = -\infty$ for an unbounded program and $f(A,b,c) = \infty$ for an infeasible program. Then, the *optimal value range* of ILP (1) refers to the interval $[\underline{f}, \overline{f}]$ bounded by the best optimal value $\underline{f}$ and the worst optimal value $\overline{f}$, where

$$\begin{aligned} \underline{f} &= \inf \{f(A,b,c) : A \in \mathbf{A}, b \in \mathbf{b}, c \in \mathbf{c}\}, \\ \overline{f} &= \sup \{f(A,b,c) : A \in \mathbf{A}, b \in \mathbf{b}, c \in \mathbf{c}\}. \end{aligned}$$

To avoid the possible infinite bounds, the *finite optimal value range* $[\underline{f}_{\text{fin}}, \overline{f}_{\text{fin}}]$ has been introduced, which is computed only over the bounded feasible scenarios.

Another related problem has been recently proposed, in which the goal is to compute the maximal and the minimal value of an additional linear function, called the *outcome function*, over the set $\mathcal{S}$ of all optimal solutions of an interval program. Given a linear function $g(x) = r^T x$ for some $r \in \mathbb{R}^n$, the *outcome range* of the ILP with respect to the outcome function $g$ is the interval $[\underline{g}, \overline{g}]$, where

$$\underline{g} = \inf \{r^T x : x \in \mathcal{S}\}, \qquad \overline{g} = \sup \{r^T x : x \in \mathcal{S}\}.$$

Note that the outcome range problem is not, in fact, a straight generalization of the optimal value range problem in the original formulation, since the optimal set does not reflect the possible infeasible or unbounded scenarios. However, if the objective vector is a fixed real vector, we can formulate the problem of computing the finite optimal value range as an outcome range problem, since

$$\underline{f}_{\text{fin}} = \inf \{c^T x : x \in \mathcal{S}\}, \qquad \overline{f}_{\text{fin}} = \sup \{c^T x : x \in \mathcal{S}\}.$$

Thus, it is also possible to adapt the algorithms for solving the outcome range problem to compute the finite optimal value range.

## 3 Outcome Range Problem

### 3.1 General Case

A straight-forward way to compute the outcome range with respect to some interval linear program is to find a description of the optimal solution set $\mathcal{S}$ and optimize the linear outcome function $g(x) = r^T x$ over $\mathcal{S}$. However, efficiently describing the set of all optimal solutions is a challenging problem.

Applying the theory of duality, the optimal set of (1) can be described as

$$Ax = b,\ x \ge 0,\ A^T y \le c,\ c^T x = b^T y,\ \underline{A} \le A \le \overline{A},\ \underline{b} \le b \le \overline{b},\ \underline{c} \le c \le \overline{c}. \quad (3)$$

However, note that this is not an interval program in sense of the definition introduced in Sect. 2, since there are dependencies between the occurrences of the interval-valued parameters $A$, $b$ and $c$. The outcome range over (3) can be computed using a solver for nonlinear programming (treating the interval parameters as variables), but it can be difficult to obtain an exact solution.

To efficiently compute an estimate of the outcome range, an (inner or outer) approximation of the optimal set $\mathcal{S}$ can be used to find some upper or lower bounds on the extremal outcomes. Several methods for approximating the possibly non-convex optimal set by a convex polyhedron or by an interval enclosure can be found in the literature (see e.g. [1,9,14,16] and references therein).

### 3.2 Special Case: Fixed Constraint Matrix

In [18], the authors addressed the outcome range problem for linear programs with an interval right-hand-side vector, examined its theoretical properties and designed two methods for approximating the bounds of the outcome range based on the McCormick relaxation of system (3) and on a local search algorithm. They also proved that the problem remains NP-hard even on this special class of ILPs.

In this section, we will focus on a slightly broader class of interval linear programs, for which we can derive a simplified description of the optimal set and propose a method to solve the outcome range problem exactly (albeit in exponential time). Namely, we will discuss the case of interval linear programs with a fixed (non-interval) constraint matrix, i.e. linear programs with an interval objective function and right-hand side.

**Problem Formulation.** Assume that an interval linear program in the form

$$\min \mathbf{c}^T x \text{ subject to } Ax = \mathbf{b},\ x \geq 0 \tag{4}$$

is given for $A \in \mathbb{R}^{m \times n}$, $\mathbf{b} \in \mathbb{IR}^m$ and $\mathbf{c} \in \mathbb{IR}^n$, satisfying $m \leq n$ and $\text{rank}(A) = m$, with a bounded optimal solution set $\mathcal{S}$ (unbounded optimal sets can also be treated with a small modification of the algorithm). Further, let a linear outcome function $r^T x$ be given for some $r \in \mathbb{R}^n$. The goal is to compute the exact bounds of the outcome range $[\underline{g}, \overline{g}]$ for $g(x) = r^T x$ over the optimal set $\mathcal{S}$ of ILP (4).

**Optimal Solution Set.** Since computing the outcome range amounts to optimizing $g(x)$ over the set of all optimal solutions $\mathcal{S}$, we employ a description of the optimal set derived in [5]. The result provides a decomposition of the optimal set described by system (3) into at most $2^n$ convex polyhedra based on complementary slackness, stating that the optimal solution set $\mathcal{S}$ can be described by ($x$-projection of) the union of feasible sets of all programs in the form

$$\begin{aligned} & Ax = b, & & \\ & x_i \geq 0, & (A^T y)_i = c_i, & \quad \text{for } i \in B, \\ & x_j = 0, & (A^T y)_j \leq c_j, & \quad \text{for } j \notin B, \\ & \underline{b} \leq b \leq \overline{b}, & \underline{c} \leq c \leq \overline{c}, & \end{aligned} \tag{5}$$

with $B \subseteq \{1, \ldots, n\}$. Note that for a fixed set $B$, system (5) is a linear system with constrained variables $b$ and $c$ (these can also be eliminated, see Remark 2). Moreover, we can also reduce the number of examined systems to index sets $B$ corresponding to a basis, if we are only interested in basic optimal solutions.

**Basis Enumeration.** Our goal is to compute the outcome range for a linear outcome function, thus we can restrict the enumeration in (5) to basic optimal solutions (and possible unbounded edges, if the optimal set is unbounded). Hereinafter, a *basis* is an index set $B \subseteq \{1, \ldots, n\}$ such that the matrix $A_B$ is non-singular ($A_B$ is the submatrix of $A$ formed by the columns indexed by $B$).

The main body of the proposed method is formed by a guided basis enumeration for finding all bases, which can be optimal for some scenario of the ILP. A similar approach was previously used for approximating the optimal solution set by the interval simplex method [12,13] or computing the worst finite optimal value [10]. Here, we adapt the idea to compute the exact outcome range. We explore the graph of all potential bases $G = (V, E)$, where

$$V = \{B \subseteq \{1, \ldots, n\} : |B| = m\}, \quad E = \{(B, B') \in V \times V : |B \cap B'| = m - 1\},$$

while exploiting the fact that the subgraph of all optimal bases is connected (see [10]). Thus, starting with an optimal basis, we will examine the neighboring nodes and further process only the proven optimal bases.

**The Algorithm.** The proposed method for computing the outcome range is presented in Algorithm 1. First, we use primal and dual feasibility to find a scenario $(A, b^*, c^*)$ that possesses an optimal solution and we find the corresponding optimal basis $B$ (Lines 1 and 2), which is used as the initial basis for the search. Here, we solve the linear program

$$Ax = b, \quad x \geq 0, \quad A^T y \leq c, \quad \underline{b} \leq b \leq \overline{b}, \quad \underline{c} \leq c \leq \overline{c}. \tag{6}$$

The initial outcome range is computed as the outcome range over all optimal solutions corresponding to the basis $B$ (Line 4). Then, the relevant neighboring nodes are considered and checked for optimality.

If the optimal solution set of the ILP can be unbounded, it is also possible to include a check for unbounded rays in the examined nodes.

Given the input data $(A, \mathbf{b}, \mathbf{c})$ of the ILP and a basis $B$, the function INTENCLOSURES called on Lines 7 and 12 computes interval enclosures $\mathbf{x}_B$ and $\mathbf{d}_N$ in the following way: First, $\mathbf{x}_B = [\underline{x}_B, \overline{x}_B]$ and $\mathbf{y} = [\underline{y}, \overline{y}]$ are interval enclosures of the feasible sets of the interval linear systems $A_B x_B = \mathbf{b}$ and $(A_B)^T y = \mathbf{c}_B$. These can be computed either exactly by interval multiplication as

$$\mathbf{x}_B = A_B^{-1}\mathbf{b}, \qquad \mathbf{y} = (A_B^{-1})^T \mathbf{c}_B, \tag{7}$$

or by utilizing some method of interval linear algebra for computing enclosures to interval linear systems of equations (see [3,20] and references therein). Then, we compute the interval vector $\mathbf{d}_N = [\underline{d}_N, \overline{d}_N]$ as

$$\mathbf{d}_N = \mathbf{c}_N - A_N^T \mathbf{y}, \tag{8}$$

evaluated in interval arithmetic (see Sect. 2.1).

**Algorithm 1.** Computing the Outcome Range

**Input:** ILP data $A \in \mathbb{R}^{m\times n}, \mathbf{b} \in \mathbb{IR}^m, \mathbf{c} \in \mathbb{IR}^n$, outcome function vector $r \in \mathbb{R}^n$
**Output:** Outcome range $[\underline{g}, \overline{g}]$ for $g(x) = r^T x$ over $\mathcal{S}$
1: $b^*, c^* \leftarrow$ solution of program (6)
2: $B \leftarrow$ optimal basis for the scenario $(A, b^*, c^*)$
3: $\mathcal{B} \leftarrow \{B\}$; Stack $\leftarrow \{B\}$
4: $[\underline{g}, \overline{g}] \leftarrow$ OUTRANGE$(B)$
5: **while** Stack $\neq \emptyset$ **do**
6: $\quad B \leftarrow$ Stack.POP()
7: $\quad \mathbf{x}_B, \mathbf{d}_N \leftarrow$ INTENCLOSURES$(B)$
8: $\quad$ Compute the neighborhood $N(B)$ as in (9) using $\mathbf{x}_B, \mathbf{d}_N$
9: $\quad$ **for** $B' \in N(B)$ **do**
10: $\quad\quad$ **if** $B' \notin \mathcal{B}$ **then**
11: $\quad\quad\quad \mathcal{B} \leftarrow \mathcal{B} \cup \{B'\}$
12: $\quad\quad\quad \mathbf{x}_B, \mathbf{d}_N \leftarrow$ INTENCLOSURES$(B')$
13: $\quad\quad\quad$ **if** $\exists j \in B\colon \overline{x}_j < 0$ **or** $\exists i \in N\colon \overline{d}_i < 0$ **then**
14: $\quad\quad\quad\quad$ **continue**
15: $\quad\quad\quad$ **if** rank$(B') = m$ **and** ISFEASIBLE$(B')$ **and** ISOPTIMAL$(B')$ **then**
16: $\quad\quad\quad\quad [\underline{h}, \overline{h}] \leftarrow$ OUTRANGE$(B')$
17: $\quad\quad\quad\quad [\underline{g}, \overline{g}] \leftarrow [\min\{\underline{g}, \underline{h}\}, \max\{\overline{g}, \overline{h}\}]$
18: $\quad\quad\quad\quad$ Stack.PUSH$(B')$

For a given optimal basis $B$ (and its corresponding node in the graph $G$), we then define the *neighborhood* $N(B)$ of $B$ computed on Line 8 as the set of all $B' = (B \setminus \{j\}) \cup \{i\}$ for some $i \in \{1, \dots, n\} \setminus B$ and $j \in B$ satisfying

$$\begin{aligned} &0 \in \mathbf{x}_j, S_{ji} < 0 \text{ and } \frac{\underline{d}_i}{S_{ji}} \geq \max\left\{\frac{\overline{d}_{i'}}{S_{ji'}} : S_{ji'} < 0, i' \notin B\right\}, \text{ or,} \\ &0 \in \mathbf{d}_i, S_{ji} > 0 \text{ and } \frac{\underline{x}_j}{S_{ji}} \leq \min\left\{\frac{\overline{x}_{j'}}{S_{j'i}} : S_{j'i} > 0, j' \in B\right\}, \end{aligned} \tag{9}$$

where $S_N = A_B^{-1} A_N$ (see [12,13] for details).

Moreover, we define the functions used in Algorithm 1 for checking optimality of a basis and computing the outcome range as follows:

- the function ISFEASIBLE$(B)$ returns `True` if the linear system

$$Ax = b, \quad \forall i \in B\colon x_i \geq 0, \quad \forall j \notin B\colon x_j = 0, \quad \underline{b} \leq b \leq \overline{b}, \tag{10}$$

is feasible for the given set $B$, otherwise it returns `False`. Assuming non-singularity of $A_B$, it is also possible to eliminate $x$ and rewrite the testing system as $A_B^{-1} b \geq 0,\ \underline{b} \leq b \leq \overline{b}$.
- the function ISOPTIMAL$(B)$ returns `True` if the linear system

$$\forall i \in B\colon (A^T y)_i = c_i, \quad \forall j \notin B\colon (A^T y)_j \leq c_j, \quad \underline{c} \leq c \leq \overline{c}, \tag{11}$$

is feasible for the given $B$, otherwise it returns `False`. Again, we can rewrite the system as $c_N - A_N^T (A_B^T)^{-1} c_B \geq 0,\ \underline{c} \leq c \leq \overline{c}$, with $N = \{1, \dots, n\} \setminus B$.

– the function OUTRANGE($B$) returns the restricted outcome range $[\underline{h}, \overline{h}]$ with respect to the optimal solutions corresponding to the basis $B$, i.e.

$$\underline{h} = \min r^T x \text{ subject to (10)}, \quad \overline{h} = \max r^T x \text{ subject to (10)}. \tag{12}$$

*Remark 2.* Note that the variables $b$ and $c$ in systems (10) and (10) can also be eliminated by reformulating the equations and inequalities using the corresponding bounds, such as $\underline{b} \leq Ax \leq \overline{b}$ instead of $Ax = b$ with $\underline{b} \leq b \leq \overline{b}$.

# 4 Computational Experiment

In this section, we examine the performance of Algorithm 1. We compare it to the nonlinear solver `fmincon` [15], a part of Matlab, optimizing on the optimal set described by system (3).

## 4.1 Instances

To compare Algorithm 1 with different approaches, we used the "class 2" instances from [18], which were kindly provided to us by authors. It is worth noting that "class 1" instances therein possess the property of basis stability and hence Algorithm 1 would terminate after one iteration.

The instances were (pseudo)randomly generated. Uncertainty in the data is only present in the right-hand-side vector, i.e. the objective function is crisp. Every original instance is determined by the number of constraints $m$, the number of variables $n'$ and the uncertainty parameter $\delta$. Coefficients of the programs are generated as integers from the following ranges: $A_{ij} \in [-10, 10]$, $\underline{c}_j \in [-20, -1]$, $\underline{b}_i \in [10, 20]$, $r_j \in [-20, 20]$. Also, to make the optimal set bounded, the last row of $A$ is restricted to be positive, i.e. to range $[1, 10]$. For coefficients of $\overline{b}$ it holds that $\overline{b} = \underline{b} + \delta e$, where $e = (1, \dots, 1)^T$ and $\overline{c} = \underline{c}$.

The instances were generated for parameters $\delta \in \{0.1, 0.25, 0.5, 0.75, 1\}$ and $(m, n') \in \{(10, 15), (30, 45), (50, 75), (80, 120), (100, 150), (200, 300), (300, 400), (400, 500), (500, 600)\}$, i.e. for 45 combinations of $(\delta, m, n')$ in total. For every combination, 30 instances were used, yielding 1350 instances in total.

The instances in [18] assume feasible region $Ax \leq b, x \geq 0$. Since Algorithm 1 is formulated for the form $Ax = b, x \geq 0$, we converted the instances by adding $m$ slack variables. For this technical reason, we have $n = n' + m$ variables.

## 4.2 Our Implementation

Algorithm 1 was implemented in Python with Gurobi [7] as the LP solver responsible for performing subprocedures OUTRANGE, ISFEASIBLE and ISOPTIMAL.

The implementation tries to reduce the number of linear programs to be solved and also their complexity in several ways. Firstly, programs in each of the subprocedures are solved by separate Gurobi LP objects. This usually means that Gurobi terminates after a small number of iterations (in particular for

ISFEASIBLE and ISOPTIMAL LPs), since the consecutive LPs differ only in the specification of the basis. Secondly, since OUTRANGE consists of quite simple programs (the feasible region is parallelotope intersected by the positive orthant), a lower bound for $\underline{h}$ and an upper bound for $\overline{h}$ of (12) can be easily obtained, which can be compared with the best values of outcome range found so far.

The enclosures for $\mathbf{x}$ and $\mathbf{d}$ computed by the INTENCLOSURES functions are implemented using interval arithmetic, as derived in formulas (7) and (8).

### 4.3 Results

**Comparison with `fmincon`.** Firstly, we compare the results obtained by Algorithm 1 with results of the Matlab nonlinear solver `fmincon`, which were kindly provided to us by the authors of [18]. `fmincon` tries to directly optimize $r^T x$ over the strong duality characterization (3) (with fixed $A$ and $c$). It was provided with 1800 s of computation time for each of the endpoints $\underline{g}, \overline{g}$ of outcome range. After 1800 s, the computation was terminated. Such computation time was sufficient for solving the problem sizes up to $(m, n') = (200, 300)$, however, it usually did not return the exact optimal solution.

Algorithm 1 was provided with 1800 s of computation time, as it computes both the endpoints simultaneously. This may seem as a disadvantage for Algorithm 1, however, it turns out that the OUTRANGE subprocedure generally takes about 20 % of the total time, i.e. each of the LPs in (12) takes only about 10 % of the total time.

The results of the experiment are summarized in Table 1. The table includes the results for all instances up to size (200, 300) and instances with smaller $\delta$ for the larger sizes (due to the time limit). The "opt found" column counts the number of instances that were solved to optimality by Algorithm 1; recall that there were 30 instances in total in each group. The most important factor for the computational time of Algorithm 1 is the number of optimal bases (the average number of optimal bases among the instances solved to optimality is denoted by $|B|$). This number is driven by the uncertainty factor $\delta$ and by the size of problems. This is a huge difference from `fmincon`, where only the size of instances matters. Hence Algorithm 1 is able to solve some of the larger instances for small $\delta$, where `fmincon` will surely timeout.

The performance of the methods within the time limit of 1800 s is shown in columns `BE` $\succ$ `fmin` and `fmin/BE` ratio. Note that both of the algorithms provide an inner approximation of the outcome range. The former column gives the number of instances for which the width of the computed outcome range is greater for Algorithm 1 (`BE` – basis enumeration). The latter column is the average ratio of widths of outcome ranges computed by `fmincon` and Algorithm 1. Note that for several instances of size (200, 500), `fmincon` returns wider outcome ranges, however, these cases are quite rare and the average ratio of widths of outcome ranges shows the superiority of Algorithm 1.

**Table 1.** Comparison of Algorithm 1 (denoted by BE, as basis enumeration, with the time limit of 30 min, and BE1m under the time limit of 1 min) with fmincon (fmin). Column $|B|$ is stands for the average number of optimal bases. The "time" columns show the average running time in seconds. Column "opt found" is the number of instances solved to optimality (out of 30). Columns BE $\succ$ fmin and BE $\succ$ BE1m is the number of instances for which BE computes a wider outcome range than fmincon or BE1m. The "ratio" columns stands for the average ratio of widths of outcome ranges. The symbol * denotes the instance sizes, for which fmincon did not return a solution within the given time limit.

| size $(m, n)$ | $\delta$ | $\|B\|$ | time BE | opt found | time fmin | BE $\succ$ fmin | fmin/BE ratio | BE $\succ$ BE1m | BE1m/BE ratio |
|---|---|---|---|---|---|---|---|---|---|
| (10, 25) | 0.1 | 1 | 0.01 | 30 | 1.54 | 30 | 0.71 | 0 | 1 |
| | 0.25 | 1.1 | 0.01 | 30 | 1.71 | 30 | 0.79 | 0 | 1 |
| | 0.5 | 1.3 | 0.01 | 30 | 3.14 | 30 | 0.83 | 0 | 1 |
| | 0.75 | 1.4 | 0.01 | 30 | 3.77 | 30 | 0.79 | 0 | 1 |
| | 1 | 1.3 | 0.01 | 30 | 1.24 | 30 | 0.76 | 0 | 1 |
| (30.75) | 0.1 | 1.4 | 0.02 | 30 | 6.52 | 30 | 0.84 | 0 | 1 |
| | 0.25 | 1.5 | 0.03 | 30 | 6.72 | 30 | 0.86 | 0 | 1 |
| | 0.5 | 2.6 | 0.03 | 30 | 5.3 | 30 | 0.88 | 0 | 1 |
| | 0.75 | 6.8 | 0.05 | 30 | 6.83 | 30 | 0.87 | 0 | 1 |
| | 1 | 9.1 | 0.06 | 30 | 6.88 | 30 | 0.88 | 0 | 1 |
| (50, 125) | 0.1 | 2 | 0.05 | 30 | 15.49 | 30 | 0.84 | 0 | 1 |
| | 0.25 | 2.3 | 0.05 | 30 | 16.25 | 30 | 0.86 | 0 | 1 |
| | 0.5 | 6.9 | 0.08 | 30 | 15.81 | 30 | 0.9 | 0 | 1 |
| | 0.75 | 24.6 | 0.17 | 30 | 19.96 | 30 | 0.87 | 0 | 1 |
| | 1 | 34 | 0.24 | 30 | 18.17 | 30 | 0.87 | 0 | 1 |
| (80, 200) | 0.1 | 3.2 | 0.1 | 30 | 49.24 | 30 | 0.73 | 0 | 1 |
| | 0.25 | 11.1 | 0.16 | 30 | 57.21 | 30 | 0.79 | 0 | 1 |
| | 0.5 | 57.1 | 0.65 | 30 | 52.06 | 30 | 0.79 | 0 | 1 |
| | 0.75 | 73.6 | 0.8 | 30 | 70.42 | 30 | 0.82 | 0 | 1 |
| | 1 | 2539.3 | 46.9 | 30 | 63.65 | 30 | 0.83 | 0 | 1 |
| (100, 250) | 0.1 | 4.6 | 0.14 | 30 | 101.56 | 30 | 0.66 | 0 | 1 |
| | 0.25 | 25.8 | 0.38 | 30 | 109.72 | 30 | 0.77 | 0 | 1 |
| | 0.5 | 101.4 | 1.7 | 30 | 111.26 | 30 | 0.79 | 0 | 1 |
| | 0.75 | 622.5 | 14.05 | 30 | 106.02 | 30 | 0.82 | 0 | 1 |
| | 1 | 2355.3 | 62.53 | 30 | 112.14 | 30 | 0.81 | 1 | 0.99 |
| (200, 500) | 0.1 | 28.9 | 1.7 | 30 | 783.18 | 30 | 0.57 | 0 | 1 |
| | 0.25 | 1019.1 | 149.85 | 29 | 856.51 | 30 | 0.65 | 3 | 0.99 |
| | 0.5 | 4699.7 | 797.22 | 24 | 800.73 | 29 | 0.74 | 12 | 0.97 |
| | 0.75 | 8219.4 | 1705.36 | 5 | 811.64 | 24 | 0.76 | 26 | 0.94 |
| | 1 | 3740 | 1761.74 | 1 | 745.46 | 27 | 0.73 | 27 | 0.93 |
| (300, 700) | 0.1 | 251.9 | 46.48 | 30 | * | * | * | 0 | 1 |
| (300, 700) | 0.25 | 2125.7 | 952.72 | 21 | * | * | * | 16 | 0.97 |
| (400, 900) | 0.1 | 500.5 | 380.83 | 27 | * | * | * | 6 | 0.98 |
| (500, 1100) | 0.1 | 543.5 | 1033.06 | 18 | * | * | * | 11 | 0.98 |

**One-Minute Perfomance.** Secondly, the last two columns of Table 1 show performance of Algorithm 1 with time limit of 1 min. It is worth noting that restricting the computational time reduces the number of bases to be examined for the larger instances. However, surprisingly, the optimal basis (or a nearly optimal one) is often found even in 1 min.

## 5 Conclusion

We have addressed the problem of computing the outcome range in interval linear programming, i.e. the problem of finding the maximal and the minimal value of an additional outcome function over the optimal solution set of an interval program. We designed an algorithm, based on guided basis enumeration and duality in linear programming, for computing the outcome range exactly on a special class of interval linear programs with a fixed constraint matrix. The conducted computational experiments show that the algorithm is competitive with the previously used nonlinear solver, working well especially for instances with narrow interval coefficients. The results also indicate that the algorithm is useful for providing a quick approximation of the outcome range by limiting its running time. However, further speedup could be achieved by a parallelized implementation of the algorithm or by deriving heuristics incorporating the outcome function in the basis search.

**Acknowledgements.** The authors would like to thank M. Mohammadi and M. Gentili for providing the test instances and results of the `fmincon` method.

## References

1. Allahdadi, M., Mishmast Nehi, H.: The optimal solution set of the interval linear programming problems. Optim. Lett. **7**(8), 1893–1911 (2012). https://doi.org/10.1007/s11590-012-0530-4
2. Chaiyakan, S., Thipwiwatpotjana, P.: Mean Absolute deviation portfolio frontiers with interval-valued returns. In: Seki, H., Nguyen, C.H., Huynh, V.-N., Inuiguchi, M. (eds.) IUKM 2019. LNCS (LNAI), vol. 11471, pp. 222–234. Springer, Cham (2019). https://doi.org/10.1007/978-3-030-14815-7_19
3. Corsaro, S., Marino, M.: Interval linear systems: the state of the art. Comput. Stat. **21**(2), 365–384 (2006). https://doi.org/10.1007/s00180-006-0268-5
4. D'Ambrosio, C., Gentili, M., Cerulli, R.: The optimal value range problem for the Interval (immune) transportation problem. Omega **95**, 102059 (2020). https://doi.org/10.1016/j.omega.2019.04.002
5. Garajová, E., Hladík, M.: On the optimal solution set in interval linear programming. Comput. Optim. Appl. **72**(1), 269–292 (2018). https://doi.org/10.1007/s10589-018-0029-8
6. Garajová, E., Hladík, M., Rada, M.: Interval linear programming under transformations: optimal solutions and optimal value range. CEJOR **27**(3), 601–614 (2018). https://doi.org/10.1007/s10100-018-0580-5
7. Gurobi Optimization, LLC: Gurobi optimizer reference manual (2020). http://www.gurobi.com
8. Hladík, M.: Optimal value range in interval linear programming. Fuzzy Optim. Decis. Making **8**(3), 283–294 (2009). https://doi.org/10.1007/s10700-009-9060-7
9. Hladík, M.: An interval linear programming contractor. In: Ramík, J., Stavárek, D. (eds.) Proceedings 30th International Conference on Mathematical Methods in Economics 2012, Karviná, Czech Republic, pp. 284–289 (Part I), Silesian University in Opava, School of Business Administration in Karviná, September 2012

10. Hladík, M.: The worst case finite optimal value in interval linear programming. Croatian Oper. Res. Rev. **9**(2), 245–254 (2018). https://doi.org/10.17535/crorr.2018.0019
11. Inuiguchi, M.: Enumeration of all possibly optimal vertices with possible optimality degrees in linear programming problems with a possibilistic objective function. Fuzzy Optim. Decis. Making **3**(4), 311–326 (2004). https://doi.org/10.1007/s10700-004-4201-5
12. Jansson, C., Rump, S.M.: Rigorous solution of linear programming problems with uncertain data. ZOR - Methods Models Oper. Res. **35**(2), 87–111 (1991). https://doi.org/10.1007/BF02331571
13. Jansson, C.: A self-validating method for solving linear programming problems with interval input data. In: Kulisch, U., Stetter, H.J. (eds.) Scientific Computation with Automatic Result Verification. Computing Supplementum, pp. 33–45. Springer, Vienna (1988). https://doi.org/10.1007/978-3-7091-6957-5_4
14. Lu, H.W., Cao, M.F., Wang, Y., Fan, X., He, L.: Numerical solutions comparison for interval linear programming problems based on coverage and validity rates. Appl. Math. Model. **38**(3), 1092–1100 (2014). https://doi.org/10.1016/j.apm.2013.07.030
15. MathWorks: MATLAB fmincon. https://www.mathworks.com/help/optim/ug/fmincon.html
16. Mishmast Nehi, H., Ashayerinasab, H.A., Allahdadi, M.: Solving methods for interval linear programming problem: a review and an improved method. Oper. Res. Int. J. **20**(3), 1205–1229 (2018). https://doi.org/10.1007/s12351-018-0383-4
17. Mohammadi, M., Gentili, M.: Bounds on the worst optimal value in interval linear programming. Soft. Comput. **23**(21), 11055–11061 (2018). https://doi.org/10.1007/s00500-018-3658-z
18. Mohammadi, M., Gentili, M.: The outcome range problem. arXiv:1910.05913 [math], April 2020. http://arxiv.org/abs/1910.05913
19. Mráz, F.: Calculating the exact bounds of optimal values in LP with interval coefficients. Ann. Oper. Res. **81**, 51–62 (1998). https://doi.org/10.1023/A:1018985914065
20. Neumaier, A.: Interval Methods for Systems of Equations. Encyclopedia of Mathematics and Its Applications. Cambridge University Press, Cambridge (1991). https://doi.org/10.1017/CBO9780511526473
21. Rohn, J.: Interval linear programming. In: Fiedler, M., Nedoma, J., Ramík, J., Rohn, J., Zimmermann, K. (eds.) Linear Optimization Problems with Inexact Data, Boston, MA, US, pp. 79–100. Springer (2006). https://doi.org/10.1007/0-387-32698-7_3

# Weak and Strong Consistency of an Interval Comparison Matrix

Milan Hladík[(✉)] and Martin Černý

Faculty of Mathematics and Physics, Department of Applied Mathematics, Charles University, Malostranské nám. 25, 118 00 Prague, Czech Republic
{hladik,cerny}@kam.mff.cuni.cz

**Abstract.** We consider interval-valued pairwise comparison matrices and two types of consistency – weak (consistency for at least one realization) and strong (acceptable consistency for all realizations). Regarding weak consistency, we comment on the paper [Y. Dong and E. Herrera-Viedma, Consistency-Driven Automatic Methodology to Set Interval Numerical Scales of 2-Tuple Linguistic Term Sets and Its Use in the Linguistic GDM With Preference Relation, IEEE Trans. Cybern., 45(4):780–792, 2015], where, among other results, a characterization of weak consistency was proposed. We show by a counterexample that in general the presented condition is not sufficient for weak consistency. It provides a full characterization only for matrices up to size of three. We also show that the problem of having a closed form expression for weak consistency is closely related with P-completeness theory and that an optimization version of the problem is indeed P-complete. Regarding strong consistency, we present a sufficient condition and a necessary condition, supplemented by a small numerical study on their efficiency. We leave a complete characterization as an open problem.

**Keywords:** Consistency · Decision making · Pairwise comparison matrix · Interval analysis.

## 1 Introduction

**Comparison Matrix.** A pairwise comparison matrix (or, multiplicative preference relation matrix) is an important tool in decision making, e.g., in the so called Analytic Hierarchy Process [16,18]. Matrix $A \in \mathbb{R}^{n\times n}$ is a pairwise comparison matrix if $0 < a_{ij} = 1/a_{ji}$ for each $i, j \in \{1,\dots,n\}$. Herein, the entry $a_{ij}$ expresses the preference of alternative $i$ over alternative $j$. The matrix is called *consistent* if

$$a_{ij} = a_{ik}a_{kj} \tag{1}$$

With a correction on paper DOI 10.1109/TCYB.2014.2336808.
Supported by the Czech Science Foundation Grant P403-18-04735S.
The counterexample and the correction provided by this paper were originally submitted to *IEEE Trans. Cybern.* as a technical note, but the journal editors resigned to accept this note which points to incorrect statements published there.


V.-N. Huynh et al. (Eds.): IUKM 2020, LNAI 12482, pp. 15–25, 2020.
https://doi.org/10.1007/978-3-030-62509-2_2

for each $i, k, j \in \{1, \ldots, n\}$. Consistency is equivalent to existence of a positive vector of weights $w \in \mathbb{R}^n$ such that $a_{ij} = \frac{w_i}{w_j}$. That is, the comparison matrix looks as follows

$$A = \begin{pmatrix} \frac{w_1}{w_1} & \frac{w_1}{w_2} & \cdots & \frac{w_1}{w_n} \\ \vdots & & & \vdots \\ \frac{w_n}{w_1} & \frac{w_n}{w_2} & \cdots & \frac{w_n}{w_n} \end{pmatrix}.$$

Since $w$ can be arbitrarily scaled, it is usually normalized such that $e^T w = 1$.

Consistency is rarely achieved in practice, so one has to estimate the weights from matrix $A$. Notice that a consistent matrix has the largest eigenvalue equal to $n$ (i.e., the size of $A$) and the corresponding Perron eigenvector is equal to the weight vector $w$. This motivated to compute the weight vector $w$ of an inconsistent matrix as its Perron vector. The corresponding eigenvalue $\lambda$ then satisfies $\lambda \geq n$. The proximity of $\lambda$ to $n$ measures the degree of consistency. If it is close in a specified sense, then $A$ is of acceptable consistency. More concretely, $A$ is of *acceptable consistency* if $\frac{1}{n-1}(\lambda - n) < \frac{1}{10} RI$, where $RI$ is the so called Random Index and its average value depending on $n$ is recorded in literature [16,18].

**Interval Comparison Matrix.** The entries of matrix $A$ are often subject to some kind of uncertainty, which can be due to vagueness of experts' assessments, incomplete data or aggregation of comparisons of a group of experts. We will model this uncertainty by the range of possible values, resulting in an interval valued matrix.

Let an interval comparison matrix $\mathbf{A} = [A^-, A^+]$ of size $n$ be given. That is, the $(i, j)$ entry of $\mathbf{A}$ is a positive compact interval $\mathbf{a}_{ij} = [a_{ij}^-, a_{ij}^+]$. The interval matrix $\mathbf{A}$ is an interval comparison matrix if the following two conditions are satisfied for each $i, j \in \{1, \ldots, n\}$

$$a_{ij}^+ = 1/a_{ji}^-, \quad a_{ii}^- = a_{ii}^+ = 1.$$

This condition can equivalently be expressed also as

$$[a_{ij}^-, a_{ij}^+] = \mathbf{a}_{ij} = \frac{1}{\mathbf{a}_{ji}} = \left[\frac{1}{a_{ji}^+}, \frac{1}{a_{ji}^-}\right], \quad \mathbf{a}_{ii} = 1.$$

Notice that due to inherent dependencies not every $A \in \mathbf{A}$ is a comparison matrix. That is why we introduce the set of comparison matrices in $\mathbf{A}$ as

$$\mathbf{A}^C := \{A \in \mathbf{A};\ a_{ij} = 1/a_{ji}\ \forall i, j\}.$$

**Interval Consistency.** One of the early works on interval comparison matrices is by [17], studying probability of rank reversal and also indicating possibly intractability of computing the exact range of weights by the eigenvector method. Consistency of an interval comparison matrix was often approached by proposing a kind of a compromise solution or other consistency indices. Compromise concepts of consistency were proposed, e.g., in [7,8,10,21–23], and some of them compared in [9,12].

In this paper, we address two concepts of consistency: weak and strong. *Weak* consistency means consistency of at least one matrix $A \in \mathbf{A}^{\mathcal{C}}$, while *strong* consistency means that every matrix $A \in \mathbf{A}^{\mathcal{C}}$ is of acceptable consistency. Notice that these two concepts are very natural from the perspective of interval community; for other interval matrix properties see, e.g., overviews [3,5,6,15].

## 2 Weak Consistency

We say that an interval matrix $\mathbf{A}$ is weakly consistent[1] if there is a consistent realization $A \in \mathbf{A}$.

Consistent matrices in $\mathbf{A}$ can be found by linear programming [1]. The problem is to find positive weights $w > 0$ such that $\frac{w_i}{w_j} \in \mathbf{a}_{ij}$, or in other words

$$a^-_{ij} w_j \leq w_i \leq a^+_{ij} w_j, \quad e^T w = 1, \quad w > 0.$$

This is a system of linear inequalities. Practically, the strict inequalities $w > 0$ can be replaced by $w \geq 0$, or by $w \geq \varepsilon$ for some sufficiently small $\varepsilon > 0$, or resolved by other means. So a solution $w$ can be computed by linear programming and the whole solution set of such weight vectors $w$ is a convex polyhedron. In contrast, the set $\mathbf{A}^{\mathcal{C}}$ of consistent matrices in $\mathbf{A}$ can be neither convex nor polyhedral.

Linear programming is an efficient tool for solving many optimization and optimization-related problems. However, it is still rather cumbersome (hard to parallelize, need of a solver, numerical issues, ...). So, if a problem can be solved directly, it is better to avoid calling a linear programming solver. In the rest of this section, we will be concerned with finding a closed form formula for weak consistency.

### 2.1 A Counterexample

To avoid solving linear programming problems, the following closed form formula was presented in [2] as Proposition 4.

**Proposition 1.** *An interval matrix* $\mathbf{A} = [A^-, A^+]$ *is weakly consistent if and only if*

$$a^+_{ik} a^+_{kj} \geq a^-_{ij}, \quad a^-_{ik} a^-_{kj} \leq a^+_{ij}$$

*for each* $i, k, j \in \{1, \ldots, n\}$.

Obviously, the condition presented above is necessary for weak consistency. However, it turns out that it is not a sufficient condition; to support it, we provide a counterexample.

[1] Sometimes the notion of "acceptable consistency" is used, but due to ambiguity we do not use it in this sense.

*Example 1.* Consider the interval matrix

$$\mathrm{A} = \begin{pmatrix} 1 & [2,4] & [4,8] & [4,8] \\ [0.25,0.5] & 1 & 4 & [4,8] \\ [0.125,0.25] & 0.25 & 1 & 2 \\ [0.125,0.25] & [0.125,0.25] & 0.5 & 1 \end{pmatrix}.$$

By an exhaustive enumeration, we can verify that the assumptions of Proposition 1 are satisfied.

Nevertheless, the interval matrix is not weakly consistent. Condition (1) for $(i,k,j) = (2,3,4)$ gives that the value of $a_{24}$ must be $a_{24} = 8$. Condition (1) for $(i,k,j) = (1,2,4)$ then proves inconsistency since there are no $a_{12} \in [2,4]$ and $a_{14} \in [4,8]$ such that $a_{12} \times 8 = a_{14}$.

The above example is the minimal counterexample. As we show below, Proposition 1 holds true as long as $n \leq 3$.

**Proposition 2.** *Proposition 1 holds true for every $n \leq 3$.*

*Proof.* It is sufficient to show validity of the statement for $n = 3$. For the sake of simplicity we consider a substitution $x_{ij} := \log_2(a_{ij})$. Condition (1) then reads $x_{ij} = x_{ik} + x_{kj}$ and the interval bound for $x_{ij}$ is $[x_{ij}^-, x_{ij}^+] = [\log_2(a_{ij}^-), \log_2(a_{ij}^+)]$. Consistency of $\mathrm{A} = [A^-, A^+]$ is now characterized by feasibility of the linear system

$$x_{12} + x_{23} = x_{13}, \quad x_{ij} \in [x_{ij}^-, x_{ij}^+]$$

in variables $x_{12}, x_{13}, x_{23}$. First, we eliminate variable $x_{12}$ by using the Fourier–Motzkin elimination technique. Since $x_{12}$ is involved in constraints $x_{12}^- \leq x_{12} \leq x_{12}^+$ and $x_{12} = x_{13} - x_{23}$, we obtain

$$x_{12}^- \leq x_{13} - x_{23} \leq x_{12}^+, \ x_{13}^- \leq x_{13} \leq x_{13}^+, \ x_{23}^- \leq x_{23} \leq x_{23}^+.$$

Since variable $x_{13}$ is involved in constraints $x_{13}^- \leq x_{13} \leq x_{13}^+$ and $x_{12}^- + x_{23} \leq x_{13} \leq x_{12}^+ + x_{23}$, its elimination leads to (omitting the redundant inequalities)

$$x_{13}^- \leq x_{12}^+ + x_{23}, \ \ x_{12}^- + x_{23} \leq x_{13}^+, \ \ x_{23}^- \leq x_{23} \leq x_{23}^+.$$

Now, elimination of $x_{23}$ results in system

$$x_{13}^- \leq x_{12}^+ + x_{23}^+, \ \ x_{12}^- + x_{23}^- \leq x_{13}^+,$$

which fully characterizes consistency of A.

*Remark 1.* The authors in [2] consider also (additive) preference relation matrices. Interval matrix $\mathrm{A} = [A^-, A^+]$ is then weakly (additive) consistent if there is $A \in \mathrm{A}$ such that $a_{ij} = a_{ik} + a_{kj} - 0.5$ for each $i,k,j \in \{1,\dots,n\}$. Then they provide a sufficient and necessary characterization (Proposition 2 in [2])

$$\begin{aligned} a_{ik}^+ + a_{kj}^+ &\geq a_{ij}^- + 0.5, \quad \forall i,k,j \in \{1,\dots,n\}, \\ a_{ik}^- + a_{kj}^- &\leq a_{ij}^+ + 0.5, \quad \forall i,k,j \in \{1,\dots,n\}. \end{aligned}$$

Again, it turns out that this statement is not correct. Basically, we can use the substitution $a_{ij} \mapsto \log_2(a_{ij}) + 0.5$, which transforms a multiplicative preference relation matrix to an additive one. In this way, the matrix from Example 1 transforms to

$$A = \begin{pmatrix} 0.5 & [1.5, 2.5] & [2.5, 3.5] & [2.5, 3.5] \\ [-1.5, -0.5] & 0.5 & 2.5 & [2.5, 3.5] \\ [-2.5, -1.5] & -1.5 & 0.5 & 1.5 \\ [-2.5, -1.5] & [-2.5, -1.5] & -0.5 & 0.5 \end{pmatrix},$$

which provides the resulting counterexample.

## 2.2 Extensions

Now, the question is whether we can provide a closed form characterization of weak consistency. Using the logarithmic transformation as in the proof of Proposition 2, we can formulate it as follows: Characterize when there is $X \in [X^-, X^+]$ such that $x_{ij} = x_{ik} + x_{kj}$ for each $i, k, j \in \{1, \dots, n\}$. There are many constraints redundant in the formulation. In view of $x_{ij} = -x_{1i} + x_{1j}$, we can reduce the problem using variables $x_{1i}$, $i \in \{2, \dots, n\}$ only:

$$\exists x_{12}, \dots, x_{1n} : x_{ij}^- \le -x_{1i} + x_{1j} \le x_{ij}^+. \tag{2}$$

This condition can be checked by means of linear programming, however, it does not provide a closed form characterization. Such a closed form expression characterizing weak consistency relates with P-completeness. Informally speaking, P-complete problems are the hardest problems having polynomial complexity and are hard to parallelize. More formally, recall that $P$ is the class of problems solvable in polynomial time and $NC$ is the class of problems that can be solved in polylogarithmic time with a polynomial number of processors. Then a problem is $P$-complete if it lies in $P$ and every problem in $P$ can be reduced to it by an $NC$-reduction.

The authors in [4,11] mentioned an open problem of checking solvability of an inequality system such that each constraint involves at most two variables. Our system (2) meets this condition, which indicates that deciding P-completeness for our problem need not be an easy task. However, we can at least show P-completeness for an optimization version of the problem.

**Proposition 3.** *The following problem is P-complete: Given $d \in \mathbb{Z}^{n-1}$, find a minimum value of $\sum_{i=2}^{n} d_{i-1} x_{1i}$ subject to constraints (2).*

*Proof.* By [11], it is a P-complete problem to find the optimum value of the transportation problem

$$\min \sum_{i=1}^{m} \sum_{j=1}^{\ell} c_{ij} y_{ij}$$

$$\text{subject to } \sum_{j=1}^{\ell} y_{ij} \le a_i, \ \forall i, \ \sum_{i=1}^{m} y_{ij} \ge b_j, \ \forall j, \ y_{ij} \ge 0,$$

where $a_i$ are integer supplies and $b_j$ integer demands. The dual problem reads

$$\max\ -a^T u + b^T v \text{ subject to } -u_i + v_j \leq c_{ij}, \forall i, j, \quad u, v \geq 0.$$

Let $M$ be sufficiently large value; see [19] for how to find such a value of polynomial size in similar problems. Now, we associate $n := m + \ell + 1$ and

$$x_{12} := u_1, \ldots, x_{1,m+1} := u_m,$$
$$x_{1,m+2} := v_1, \ldots, x_{1,m+\ell+1} := v_\ell.$$

The interval bounds are set as follows

$$x_{ij}^- := \begin{cases} 0 & i = 1,\ j \in \{2, \ldots, m + \ell + 1\}, \\ -M & \text{otherwise}, \end{cases}$$

$$x_{ij}^+ := \begin{cases} c_{ij} & i \in \{2, \ldots, m + 1\}, \\ & j \in \{m + 2, \ldots, m + \ell + 1\}, \\ M & \text{otherwise}, \end{cases}$$

and the objective coefficients

$$d_1 := -a_1, \ldots, d_m := -a_m,\ d_{m+1} := b_1, \ldots, d_{m+\ell} := b_\ell.$$

Now, to find the maximum value of $\sum_{i=2}^{n} d_{i-1} x_{1i}$ subject to constraints (2) means to solve the transportation problem.

## 3 Strong Consistency

It can hardly happen for a nontrivial interval matrix that every $A \in \mathbf{A}^C$ is consistent. That is why we define $\mathbf{A}$ to be *strongly consistent* if every $A \in \mathbf{A}^C$ is of acceptable consistency. There is not much known about strong consistency yet. We present a simple sufficient condition and a necessary condition, and then we perform a small numerical study. Below, $\lambda(A)$ denotes the largest eigenvalue of $A$ (in fact, in our case $\lambda(A)$ is equal to the spectral radius of $A$ since $A$ has positive entries).

**Sufficient Condition.** The sufficient condition is based on an upper bound obtain from matrix $A^+$. Notice that $A^+$ is usually not a comparison matrix, so the converse implication is not valid in general.

**Proposition 4.** *If* $\lambda(A^+) < n + \frac{n-1}{10} RI$, *then* $\mathbf{A}$ *is strongly consistent.*

*Proof.* For every $A \in \mathbf{A}$ we want $\frac{1}{n-1}(\lambda(A) - n) < \frac{1}{10} RI$, which is equivalent to $\lambda(A) < n + \frac{n-1}{10} RI$. By the Perron–Frobenius theory of nonnegative matrices [13], we have $\lambda(A) \leq \lambda(A^+)$, from which the statement follows.

**Necessary Condition.** In order to state a necessary condition (or a sufficient condition for not strong consistency), we suggest a heuristic for finding a least consistent matrix in $\mathbf{A}^{\mathcal{C}}$, that is, a matrix with largest $\lambda(A)$. Let $A \in \mathbf{A}^{\mathcal{C}}$ be an initial matrix and denote by $x, y \in \mathbb{R}^n$ its right and left eigenvectors, respectively, normalized such that $x^T y = 1$. By [14], the derivative of $\lambda(A)$ with respect to $a_{ij}$, $i \neq j$, reads $\lambda(A)' = y^T A' x = x_j y_i - a_{ij}^{-2} x_i y_j$. Thus the most (locally) convenient direction to adjust $A$ is to put

$$a_{ij} := \begin{cases} a_{ij}^{+} & \text{if } x_j y_i - a_{ij}^{-2} x_i y_j \geq 0, \\ a_{ij}^{-} & \text{if } x_j y_i - a_{ij}^{-2} x_i y_j < 0, \end{cases}$$

and $a_{ji} := 1/a_{ij}$. (Notice that for the case $x_j y_i - a_{ij}^{-2} x_i y_j = 0$ we can take another value from $\mathbf{a}_{ij}$ as well, but practically there is no difference since this case happens rarely.) We update the entries of $A$ for every $i < j$ and correspondingly for $i > j$. As a result, we obtain a new matrix $\tilde{A} \in \mathbf{A}^{\mathcal{C}}$ with possibly larger eigenvalue. We can iterate this process until we get stuck in a fixed point matrix or the eigenvalue will not increase any more. Eventually, we check if the resulting matrix is acceptably consistent.

*Example 2.* In order to assess efficiency of the sufficient condition and the necessary condition, we performed numerical experiments. They were carried out in MATLAB R2019b on a machine AMD Opteron(tm) Processor 6134 with 64431 MB RAM and 16 processors.

The problem instances were generated as follows. First, we generated a random interval vector $\mathbf{w}$ of length $n$. The midpoints of its entries were generated using the uniform distribution on $[1, 10]$ and the width of the intervals was set to $\delta$, which is a parameter. Then we constructed the interval matrix $\mathbf{A}$ by setting $\mathbf{a}_{ij} := \mathbf{w}_i/\mathbf{w}_j$ (using interval arithmetics), $i \neq j$, and $\mathbf{a}_{ii} := 1$, $i = 1, \ldots, n$. An interval matrix constructed in this way is always an interval comparison matrix.

The necessary condition was initiated by the matrix $A \in \mathbf{A}$ the entries of which are the interval midpoint values. Since this matrix need not be a comparison matrix, we correspondingly adjusted the entries under the diagonal.

We considered matrix size $n \in \{3, \ldots, 10\}$ since the comparison matrices in practice are not large. The value od parameter $\delta$ was chosen from the list $\{0.05, 0.1, 0.25, 0.5\}$. For each parameter setting, we run $10^5$ instances and calculated the following characteristics:

- `iterations`, the number of iterations for the necessary condition. We display its mean value and the standard deviation.
- `overestimation`, the overestimation of the sufficient condition over the necessary condition in terms of the maximum eigenvalue estimation. The overestimation is calculated as $\lambda(A^{+})/\lambda(\tilde{A}) - 1$, where $\tilde{A}$ is the matrix determined when evaluating the necessary condition. We display the mean value, the standard deviation, the minimum and maximum value of the overestimation.

Table 1 displays the results. We see that the local improvement heuristics utilized in the necessary conditions terminates always in the first iteration, so it

**Table 1.** (Example 2) Eigenvalue overestimation of the sufficient condition over the necessary condition.

| $n$ | $\delta$ | `iterations` | | `overestimation` | | | |
|---|---|---|---|---|---|---|---|
| | | mean | dev | mean | dev | min | max |
| 3 | 0.05 | 1 | 0 | 0.008446 | 0.003504 | 0.003375 | 0.03029 |
| 3 | 0.1 | 1 | 0 | 0.01674 | 0.00689 | 0.006701 | 0.05838 |
| 3 | 0.25 | 1 | 0 | 0.04082 | 0.01634 | 0.01659 | 0.1403 |
| 3 | 0.5 | 1 | 0 | 0.07851 | 0.02988 | 0.0328 | 0.2389 |
| 4 | 0.05 | 1 | 0 | 0.009532 | 0.00342 | 0.003851 | 0.03154 |
| 4 | 0.1 | 1 | 0 | 0.01881 | 0.006719 | 0.007567 | 0.06269 |
| 4 | 0.25 | 1 | 0 | 0.04616 | 0.01603 | 0.01909 | 0.1358 |
| 4 | 0.5 | 1 | 0 | 0.08887 | 0.02948 | 0.03763 | 0.268 |
| 5 | 0.05 | 1 | 0 | 0.01016 | 0.003262 | 0.004163 | 0.02833 |
| 5 | 0.1 | 1 | 0 | 0.02015 | 0.006423 | 0.008362 | 0.05886 |
| 5 | 0.25 | 1 | 0 | 0.04921 | 0.01527 | 0.02113 | 0.1363 |
| 5 | 0.5 | 1 | 0 | 0.09493 | 0.02832 | 0.04051 | 0.2687 |
| 6 | 0.05 | 1 | 0 | 0.01056 | 0.003098 | 0.00433 | 0.02853 |
| 6 | 0.1 | 1 | 0 | 0.02096 | 0.006081 | 0.008792 | 0.05539 |
| 6 | 0.25 | 1 | 0 | 0.05133 | 0.0146 | 0.02229 | 0.1336 |
| 6 | 0.5 | 1 | 0 | 0.09918 | 0.02708 | 0.04315 | 0.2611 |
| 7 | 0.05 | 1 | 0 | 0.01087 | 0.002968 | 0.004647 | 0.02729 |
| 7 | 0.1 | 1 | 0 | 0.02158 | 0.005807 | 0.008948 | 0.0536 |
| 7 | 0.25 | 1 | 0 | 0.05275 | 0.01395 | 0.02315 | 0.1364 |
| 7 | 0.5 | 1 | 0 | 0.1017 | 0.02568 | 0.04614 | 0.2492 |
| 8 | 0.05 | 1 | 0 | 0.01109 | 0.00282 | 0.004975 | 0.02562 |
| 8 | 0.1 | 1 | 0 | 0.02205 | 0.005563 | 0.009623 | 0.05195 |
| 8 | 0.25 | 1 | 0 | 0.05396 | 0.01328 | 0.0238 | 0.1239 |
| 8 | 0.5 | 1 | 0 | 0.104 | 0.02455 | 0.04823 | 0.2368 |
| 9 | 0.05 | 1 | 0 | 0.01128 | 0.002702 | 0.005125 | 0.02595 |
| 9 | 0.1 | 1 | 0 | 0.02242 | 0.005333 | 0.01013 | 0.05094 |
| 9 | 0.25 | 1 | 0 | 0.05479 | 0.0127 | 0.02521 | 0.1377 |
| 9 | 0.5 | 1 | 0 | 0.1058 | 0.02357 | 0.04782 | 0.2215 |
| 10 | 0.05 | 1 | 0 | 0.01144 | 0.002613 | 0.005434 | 0.02658 |
| 10 | 0.1 | 1 | 0 | 0.02267 | 0.005123 | 0.01043 | 0.05084 |
| 10 | 0.25 | 1 | 0 | 0.0555 | 0.01219 | 0.02527 | 0.1197 |
| 10 | 0.5 | 1 | 0 | 0.1071 | 0.02261 | 0.05048 | 0.2245 |

is very cheap to compute. The eigenvalue overestimation seems not to depend much on the matrix size (it grows very slowly with respect to the size), but in contrast it depends highly on the radii of input intervals. The wider intervals, the higher overestimation. For $\delta = 0.05$, the overestimation is about 1%, but for $\delta = 0.5$, the overestimation ranges from 8% to 11% on average. Anyway, we can conclude that the gap between the lower and the upper bound on the maximum eigenvalue is mild, and therefore the conditions for strong consistency can be considered to be efficient.

## 4 Conclusion

In this paper, we discussed weak and strong consistency of an interval pairwise comparison matrix. Weak consistency can be checked by linear programming, but attempts to find a closed form condition appeared in literature. We showed by counterexamples that two propositions in [2] (and a similar one in [20]) giving such conditions are not correct in general. We also showed that they remain valid for $n \leq 3$. Since the paper [2] was cited 173 times up to now, we feel that correcting it was necessary. Indeed, some of the citing articles build directly on those incorrect propositions; see, e.g., [24]. We also discussed existence of a closed form characterization of acceptable consistency. Since the problem is closely related to P-complete problems, it seems that finding such a characterization is hard if even possible.

Strong consistency has been less studied so far. We proposed a computationally cheap sufficient condition and a necessary condition (which is iterative, but practically terminates in one iteration). A small numerical study showed that the eigenvalue gap provided by these two conditions is moderate, and thus the conditions can be regarded to be efficient. A complete characterization of strong consistency is not known and remains as a challenging open problem.

## References

1. Arbel, A.: Approximate articulation of preference and priority derivation. Eur. J. Oper. Res. **43**(3), 317–326 (1989)
2. Dong, Y., Herrera-Viedma, E.: Consistency-driven automatic methodology to set interval numerical scales of 2-tuple linguistic term sets and its use in the linguistic GDM with preference relation. IEEE Trans. Cybern. **45**(4), 780–792 (2015)
3. Garloff, J., Adm, M., Titi, J.: A survey of classes of matrices possessing the interval property and related properties. Reliab. Comput. **22**, 1–10 (2016)
4. Greenlaw, R., Hoover, H.J., Ruzzo, W.L.: Limits to Parallel Computation: P-Completeness Theory. Oxford University Press, New York (1995)
5. Hladík, M.: An overview of polynomially computable characteristics of special interval matrices. In: Kosheleva, O., Shary, S.P., Xiang, G., Zapatrin, R. (eds.) Beyond Traditional Probabilistic Data Processing Techniques: Interval, Fuzzy etc. Methods and Their Applications. SCI, vol. 835, pp. 295–310. Springer, Cham (2020). https://doi.org/10.1007/978-3-030-31041-7_16

6. Horáček, J., Hladík, M., Černý, M.: Interval linear algebra and computational complexity. In: Bebiano, N. (ed.) MAT-TRIAD 2015. SPMS, vol. 192, pp. 37–66. Springer, Cham (2017). https://doi.org/10.1007/978-3-319-49984-0_3
7. Kuo, T.: Interval multiplicative pairwise comparison matrix: consistency, indeterminacy and normality. Inf. Sci. **517**, 244–253 (2020)
8. Liu, F.: Acceptable consistency analysis of interval reciprocal comparison matrices. Fuzzy Sets Syst. **160**(18), 2686–2700 (2009)
9. Liu, F., Huang, M.-J., Huang, C.-X., Pedrycz, W.: Measuring consistency of interval-valued preference relations: comments and comparison. Oper. Res. 1–29 (2020). https://doi.org/10.1007/s12351-020-00551-z
10. Liu, F., Yu, Q., Pedrycz, W., Zhang, W.G.: A group decision making model based on an inconsistency index of interval multiplicative reciprocal matrices. Knowl.-Based Syst. **145**, 67–76 (2018)
11. Lueker, G.S., Megiddo, N., Ramachandran, V.: Linear programming with two variables per inequality in poly-log time. In: Proceedings of the Eighteenth Annual ACM Symposium on Theory of Computing. (STOC 1986), pp. 196–205. ACM, New York (1986)
12. Meng, F., Tang, J.: A comparative study for consistency-based decision making with interval multiplicative preference relations. Int. J. Gen. Syst. **49**(4), 400–437 (2020)
13. Meyer, C.D.: Matrix Analysis and Applied Linear Algebra. SIAM, Philadelphia (2000)
14. Petersen, K.B., Pedersen, M.S.: The matrix cookbook, Technical University of Denmark (2012). http://www2.imm.dtu.dk/pubdb/p.php?3274, version 20121115
15. Rohn, J.: A handbook of results on interval linear problems. Technical report 1163, Institute of Computer Science, Academy of Sciences of the Czech Republic, Prague (2012). http://uivtx.cs.cas.cz/~rohn/publist/!aahandbook.pdf
16. Saaty, T.L.: Relative measurement and its generalization in decision making. Why pairwise comparisons are central in mathematics for the measurement of intangible factors. The analytic hierarchy/network process. RACSAM - Rev. R. Acad. Cien. Serie A. Mat. **102**(2), 251–318 (2008)
17. Saaty, T.L., Vargas, L.G.: Uncertainty and rank order in the analytic hierarchy process. Eur. J. Oper. Res. **32**(1), 107–117 (1987)
18. Saaty, T.L., Vargas, L.G.: Models, Methods, Concepts and Applications of the Analytic Hierarchy Process, International Series in Operations Research and Management Science, vol. 175, 2nd edn. Springer, New York (2012)
19. Schrijver, A.: Theory of Linear and Integer Programming. Wiley, Chichester (1998)
20. Wang, Y.M., Yang, J.B., Xu, D.L.: Interval weight generation approaches based on consistency test and interval comparison matrices. Appl. Math. Comput. **167**(1), 252–273 (2005)
21. Wang, Z.J.: A note on a goal programming model for incomplete interval multiplicative preference relations and its application in group decision-making. Eur. J. Oper. Res. **247**(3), 867–871 (2015)
22. Wang, Z.J., Lin, J.: Consistency and optimized priority weight analytical solutions of interval multiplicative preference relations. Inf. Sci. **482**, 105–122 (2019)

23. Wang, Z.J., Lin, J., Liu, F.: Axiomatic property based consistency analysis and decision making with interval multiplicative reciprocal preference relations. Inf. Sci. **491**, 109–137 (2019)
24. Zhang, Zhen., Yu, Wenyu, Guo, Chonghui: Group decision making based on acceptably consistent interval multiplicative preference relations. In: Chen, Jian, Nakamori, Yoshiteru, Yue, Wuyi, Tang, Xijin (eds.) KSS 2016. CCIS, vol. 660, pp. 165–174. Springer, Singapore (2016). https://doi.org/10.1007/978-981-10-2857-1_14

# The Facility Location Problem with a Joint Probabilistic Constraint

A. Suzuki(✉), T. Fukuba, and T. Shiina

Waseda University, Tokyo, Japan
aioizs@fuji.waseda.jp,tomohikari892@asa6gi.waseda.jp,tshiina@waseda.jp

**Abstract.** This study shows the effectiveness of the cutting plane method by applying it to the facility location problem with probabilistic constraints. Probabilistic constraints are those that should be satisfied at a certain probabilistic level and can consider the uncertainty of the parameters involved in the problem. Problems with such probabilistic constraints are generally difficult to solve. Therefore, based on previous research, we consider transforming a problem with probabilistic constraints into a 0–1 mixed integer programming problem under special conditions. Thereafter, we introduce the cutting plane method using a valid inequality of the feasible region.

**Keywords:** Cutting plane method · Facility location problem · Probabilistic constraints

## 1 Introduction

The facility location problem has been studied in various fields; however, for several actual situations, the assumption that the parameters involved in the problem are deterministic data does not apply. These data contain uncertainties and are thus represented as random variables. In this case, it is difficult to completely satisfy the constraints on demand. To address this problem, Charnes and Cooper [3] propose the introduction of a probabilistic constraint, which is to be satisfied at a certain probability level. However, T. Shiina [9] proposed that problems with a probabilistic constraint are generally difficult to solve for two reasons. First, the feasible region defined by a probabilistic constraint is usually non-convex. Second, even if the feasible region is convex, the objective function and probabilistic constraints generally include multidimensional integration. Thus, it is desirable to avoid these restrictions if possible. From previous studies [7], it is known that a linear programming problem with probabilistic constraints can be transformed into 0–1 mixed integer programming problem under special circumstances.

In this study, we first define the facility location problem with a joint probabilistic constraint and reformulate it to a stochastic mixed integer programming problem (SMIP). Thereafter, we introduce the cutting plane method using the


V.-N. Huynh et al. (Eds.): IUKM 2020, LNAI 12482, pp. 26–37, 2020.
https://doi.org/10.1007/978-3-030-62509-2_3

star inequality which is valid for SMIP. We present computational results that indicate that a large-scale problem can be solved efficiently.

In Sect. 2, we explain the problem outline and the research model, in addition well to clarifying the position of this research. In Sect. 3, we introduce a cutting plane method using strengthened star inequality. Section 4 describes the outline of the numerical experiment and the data generation method, and Section 4.2 describes the experimental results. Finally, Sect. 5 describes the conclusions of this study and the future tasks.

## 2 Problem Description and Formulation

Probabilistic constraints include a joint probabilistic constraint and a separate probabilistic constraint. In this study, we formulate facility location problem using a joint probabilistic constraint to handle the probability that the demand is simultaneously satisfied at all the demand points. Therefore, we model a facility location problem with a joint probabilistic constraint. This model aims to minimize the sum of the facility opening costs and transportation costs while satisfying demand at a certain rate $1 - \epsilon$ for the entire scenario, considering the uncertainty of the demand volume at each demand location. This model considers the following parameters and decision variables.

Sets

$I$: set of facilities
$J$: set of demand points
$S$: set of scenarios
$\epsilon$: confidence parameter(typically near zero)

Parameters

$f_i$: fixed cost of opening the facility $i$
$c_{ij}$: unit cost of the shipment from the facility $i$ to the demand point $j$
$Q_i$: capacity of the facility $i$
$\epsilon$: confidence parameter(typically near zero)

Variables

$x_{ij}$: amount of the product transported from the facility $i$ to the demand point $j$
$y_i$: 1 if facility is installed at site $i$, 0 otherwise
$\tilde{\xi}_j$: random vector indicating the demand from a demand point $j$

The model formulation is as follows:

$$\min \sum_{i \in I} f_i y_i + \sum_{i \in I} \sum_{j \in J} c_{ij} x_{ij} \tag{1}$$

$$\text{s.t.} \sum_{j \in J} x_{ij} \leq Q_i y_i \quad i \in I \tag{2}$$

$$\text{Prob}(\sum_{i \in I} x_{ij} \geq \tilde{\xi}_j \quad j \in J) \geq 1 - \epsilon \tag{3}$$

$$y_i \in \{0, 1\} \quad i \in I \tag{4}$$

$$x_{ij} \geq 0 \quad i \in I, j \in J \tag{5}$$

The objective in (1) is to minimize the total system costs including fixed costs of the facilities and costs of transportation from assigned facilities to each demand point. Constraint (2) takes care of the capacity restrictions at each facility. Constraint (3) includes a random variable $\tilde{\xi}$ and is a joint probabilistic constraint; this means that it is sufficient for the probability of satisfying the demand at all demand points to be $1 - \epsilon$ or more. Finally, constraint (4) defines the variable $y$ representing the installation of the facility as binary and constraint (5) limits the traffic $x$ to a non-negative value.

Here, the probabilistic constraint (3) of this model has the property that only the right-hand side variable $\tilde{\xi}$ is a random variable, which follows a finite discrete distribution. According to previous research [7], if such a condition holds, we can expand the probabilistic constraint of the problem and transform it into stochastic mixed integer programming problem (SMIP). Thus, the additional parameters and variables for SMIP are shown below:

Parameters

$\pi^s$: probability of scenario $s$

Variables

$v_j$: total amount of the product transported from each facility to the demand point $j$
$z^s$: binary variable indicating whether the demand is satisfied in scenario $s$

The SMIP formulation is as follows:

$$\min \sum_{i \in I} f_i y_i + \sum_{i \in I} \sum_{j \in J} c_{ij} x_{ij} \tag{6}$$

$$\text{s.t.} \ (2), (4), (5)$$

$$\sum_{i \in I} x_{ij} - v_j = 0 \quad j \in J \tag{7}$$

$$v_j + \xi_j^s z^s \geq \xi_j^s \quad j \in J, s \in S \tag{8}$$

$$\sum_{s \in S} \pi^s z^s \leq \epsilon \tag{9}$$

$$z^s \in \{0, 1\} \quad s \in S \tag{10}$$

Focusing on Eq. (8), if $z^s = 0$, it can be seen that the supply exceeds the demand as $v_j \geq \xi_j^s (j \in J, s \in S)$, so that the demand is reliably satisfied. Meanwhile, if $z^s = 1$, then $v_j \geq 0 (j \in J)$, and we cannot guarantee that the demand is satisfied. Constraint (9) is equivalent to $\sum_{s \in S} \pi^s (1 - z^s) \geq 1 - \epsilon$. The facility location problem dealt with in this study is shown in Fig. 1.

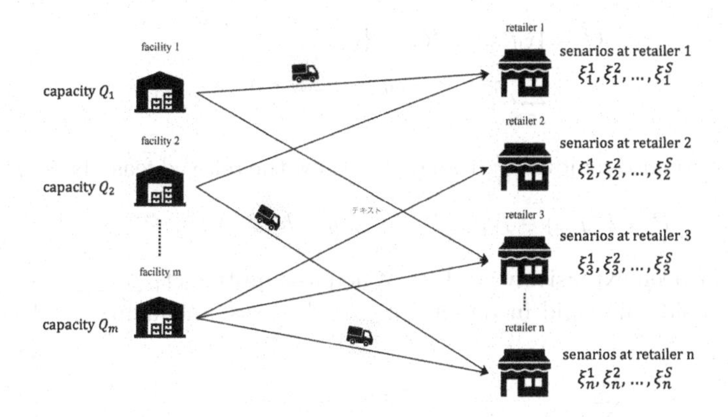


**Fig. 1.** The facility location problem with samples of each demand point

## 3 Methods

We apply the cutting plane method to the study model using a star inequality that is valid for SMIP shown by Luedtke et al. [7]. We will also explain how to create the cutting plane using the shortest path problem.

### 3.1 Star Inequality

We derive a star inequality that strengthens the SMIP formulation. First, we remove (7) from the formulation and show the relaxed feasible set $F$.

$$F := \{(v, z) \in R_+^n \times \{0,1\}^S : \sum_{s \in S} \pi^s z^s \leq \epsilon, v + \xi^s z^s \geq \xi^s, s \in S\} \tag{11}$$

Note that

$$F = \bigcap_{j=1}^{n} \{(v, z) : (v_j, z) \in G_j\} \tag{12}$$

$$G_j = \{(v_j, z) \in R_+ \times \{0,1\}^S : \sum_{s \in S} \pi^s z^s \leq \epsilon, v_j + \xi_j^s z^s \geq \xi_j^s, s \in S\} \tag{13}$$

Thus, a natural first step in developing a strong formulation for $F$ is to develop a strong formulation for each $G_j$. In particular, if an inequality is facet-defining for conv($G_j$), then the inequality is also facet-defining for conv($F$), which is the common set of $G_j$. We set $y = v_j, h^s = \xi_j^s$ for each s, and consider the general set $G$. Here, we assume without loss of generality that $h^1 \geq h^2 \geq \cdots \geq h^S$.

$$\begin{aligned} G = \{(y, z) \in R_+ \times \{0,1\}^S : \\ \sum_{s \in S} \pi^s z^s \leq \epsilon, y + h^s z^s \geq h^s, s \in S\} \end{aligned} \tag{14}$$

Thereafter, we also remove (9) from $G$ and show the relaxed feasible set $P$ below.

$$P = \{(y, z) \in R_+ \times \{0,1\}^S : y + h^s z^s \geq h^s, s \in S\} \tag{15}$$

This set has been extensively studied. When assumptions $t_1 \leq t_2 \leq \cdots \leq t_L$ and $h^{t_{L+1}} = 0$ hold, the valid inequality of [1] called star inequality, is obtained as follows.

$$y + \sum_{l=1}^{L} (h^{t_l} - h^{t_{l+1}}) z^{t_l} \geq h^{t_1}, \forall T = \{t_1, .., t_L\} \subseteq S \tag{16}$$

Furthermore, this star inequality can be strengthened by the knapsack constraint(9). Here, $p := \max\{k : \sum_{s=1}^{k} \pi^s \leq \epsilon\}$ is defined. From (9), $z^s = 1$ does not hold for all scenarios including $s = 1, \cdots, andp + 1$. Therefore, $y \geq h^{p+1}$ is determined. In addition to this, the mixed integer constraints of the feasible set $G$ can be expressed as $y + h^s z^s \geq h^s (s = 1, \cdots, p)$ because the constraints are redundant for $s = p+1, \cdots, S$. Summarizing these relationships, we can rewrite the mixed integer constraints as $y + (h^s - h^{p+1}) z^s \geq h^s (s = 1, \cdots, p)$. That is why we have the following feasible set G:

$$\begin{aligned} G = \{(y, z) \in R_+ \times \{0,1\}^S : \sum_{s \in S} \pi^s z^s \leq \epsilon, \\ y + (h^s - h^{p+1}) z^s \geq h^s, s = 1, ..., p\} \end{aligned} \tag{17}$$

Applying the star inequality (16) to this feasible set G gives the strengthened star inequality as follows.

$$\begin{aligned} y + \sum_{l=1}^{L} (h^{t_l} - h^{t_{l+1}}) z^{t_l} \geq h^{t_1} \\ \forall T = \{t_1, .., t_L\} \subseteq \{1, .., p\} \end{aligned} \tag{18}$$

Figure 2 below shows an example of how to set the scenario index set $T$ used in the strong star inequality. In the figure, the number of scenarios is $S = 50$, the confidence parameter $\epsilon = 0.1$, and by definition, the scenario occurrence probability is $\pi^s = 1/50 = 0.02$ and $p = 5$. This example shows a scenario index set $T = \{t_1, t_2, t_3\} = \{1, 2, 4\}$.

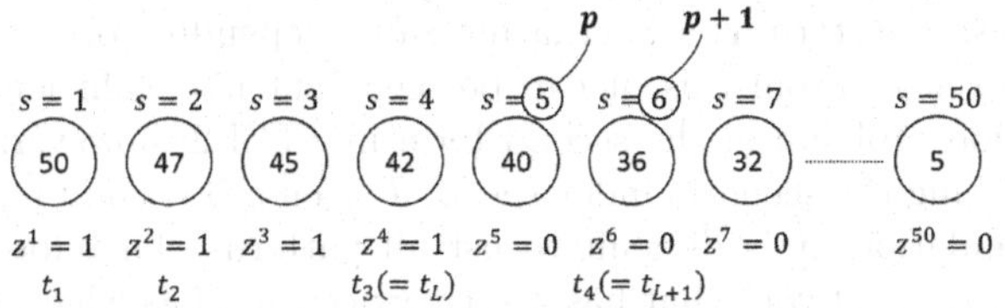

$p$ $p+1$

$s=1$ $s=2$ $s=3$ $s=4$ $s=(5)$ $s=(6)$ $s=7$ $s=50$

50 47 45 42 40 36 32 5

$z^1=1$ $z^2=1$ $z^3=1$ $z^4=1$ $z^5=0$ $z^6=0$ $z^7=0$ $z^{50}=0$

$t_1$ $t_2$ $t_3(=t_L)$ $t_4(=t_{L+1})$

**Fig. 2.** Example of scenario index set $T$

[7] proves the validity of the strengthened star inequality. Since When let $(y,z) \in G, l^* = \min\{l \in \{1,\ldots,L\} : z^{t_l} = 0\}$, $y \geq h^{t_{l^*}}$ holds, so the following relation can be shown.

$$y + \sum_{l=1}^{L} (h^{t_l} - h^{t_{l+1}}) z^{t_l} \geq h^{t_{l^*}} + \sum_{l=1}^{l^*-1} (h^{t_l} - h^{t_{l+1}}) = h^{t_1} \tag{19}$$

### 3.2 Introduction of Cutting Plane Method

Van Roy and Wolsey [10] show an algorithm of the cutting plane method, which is one of the solutions to the combinatorial optimization problem. Formulating the problem as an integer program and solving the linear relaxation problem generally yield a fractional solution. We sequentially add a plane that cuts off the fractional solution and leaves all integer solutions as a constraint to the problem. Finally, we obtain an integer solution. In this research model, the relaxed solutions $y^*$ and $z^*$ of SMIP may not satisfy the 0-1 integer condition. If the strengthened star inequality with such relaxed solutions does not hold the inequality relation, a valid inequality that removes the solutions is added. Thereafter, we illustrate the algorithm for introducing cutting plane in Fig. 3.

---

STEP0: Solve the relaxation problem of SMIP.
STEP1: The process ends if the relaxed solutions $y^*$ and $z^*$ satisfy the 0-1 integer condition.
STEP2: Substitute $z^*$ into the left-hand side of (18), and solve the maximizing deviation problem as the shortest path problem.
STEP3: If the result of STEP2 does not satisfy (18), a valid inequality that removes $z^*$ is added to the SMIP as a cutting plane, and if not possible, go to the branch and bound method. Back to STEP0.

---

**Fig. 3.** The cutting plane method using strengthened star inequality

Here, we explain how to make the cutting plane shown in STEP2 and STEP3 in Fig. 3. There are two terms on the left side of (18), the first term is a fixed value, and the second term is a fluctuating value depending on how to take the scenario index set $T$. From this, it can be seen that it is sufficient to consider the minimization problem of the second term in STEP2; however, because the method of selecting the scenario index set for $T$ is enormous and requires a large amount of calculation, it is difficult to list all patterns. Therefore, the second term is applied to a network that has $p+1$ vertices and has edges directed from the vertices in the direction of the larger vertex numbers; this term solves the network as the shortest path problem.

We define the network for the shortest path problem as a directed graph $G=(V,A)$.

Sets

$V$: set of vertexes($=p+1$)
$A$: set of edges (directed from vertex to vertex with larger number)
$V^+(i)=\{k:(i,k)\in A\}$: end-point set of edges exiting from vertex $i$
$V^-(i)=\{k:(k,i)\in A\}$: start-point set of edges connected to vertex $i$

Parameters

$s$: start-point of shortest path problem
$t$: end-point of shortest path problem
$c_{ij}$: weight of edge $ij(c_{ij}=(h^i-h^j)z^i)$

Variables

$w_{ij}$: 0-1 variable indicating whether the shortest path includes edge $ij$

The shortest path problem formulation is given as follows.

$$\min \sum_{(i,j)\in A} c_{ij}w_{ij} \tag{20}$$

$$\text{s.t.} \sum_{k\in V^-(i)} w_{ki} - \sum_{k\in V^+(i)} w_{ik} = -1, i=s \tag{21}$$

$$\sum_{k\in V^-(i)} w_{ki} - \sum_{k\in V^+(i)} w_{ik} = 0,$$
$$\forall i \in V\backslash\{s,t\} \tag{22}$$

$$\sum_{k\in V^-(i)} w_{ki} - \sum_{k\in V^+(i)} w_{ik} = 1, i=t \tag{23}$$

$$w_{ij}\in\{0,1\}, \forall(i,j)\in A \tag{24}$$

The objective function (20) is the minimization of the distance from the start point $s$ to the end point $t$, and (21)–(23) are the constraints on the flow conservation law at each vertex. Specifically, the network for the shortest path

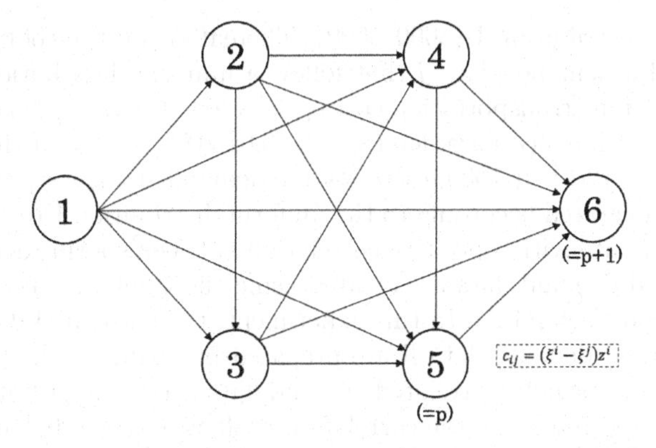


**Fig. 4.** Example of the shortest path problem network

problem when $p = 5$ is shown in Fig. 4. We solve this problem and define the vertex set of the shortest path excluding the end-point as the scenario index set $T$. In the example of Fig. 4, if the shortest path is $1 \rightarrow 3 \rightarrow 4 \rightarrow 6$, the scenario index set is $T = \{1, 3, 4\}$. Therefore, we present (25) that adapts this scenario index set $T$ to the strengthened star inequality (18).

$$y^* + (h^{t_1} - h^{t_3})z^{t_1^*} + (h^{t_3} - h^{t_4})z^{t_3^*} + (h^{t_4} - h^{t_6})z^{t_4^*} \geq h^{t_1} \tag{25}$$

If (25) is not satisfied, a cutting plane as in (26) can be added. However, if (25) is satisfied, the start-point s is shifted by one and the shortest path problem is solved again. If the cutting plane has not been added after all the patterns have been tried, the algorithm ends.

$$y + (h^{t_1} - h^{t_3})z^{t_1} + (h^{t_3} - h^{t_4})z^{t_3} + (h^{t_4} - h^{t_6})z^{t_4} \geq h^{t_1} \tag{26}$$

Finally, we show in (27), the general form of the cutting plane.

$$y + \sum_{l=1}^{L}(h^{t_l} - h^{t_{l+1}})z^{t_l} \geq h^{t_1}, \quad \forall T = \{t_1, .., t_L\} \tag{27}$$

## 4 Numerical Experiments

In this section, we describe how to generate the data used in the experiments based on Luedtke et al. [7].

### 4.1 Experimental Data

The facility $i$ is set at 40 places and the demand point $j$ is set at 100(or 200) places. The facility opening cost $f_i$ is generated according to the uniform distribution U(500000, 1000000), and the supply capacity $Q_i$ is generated according to the uniform distribution U(3000, 5000). Thereafter, on the coordinate plane, a facility $i$ and a demand place $j$ that follow a uniform distribution U(20, 200) are set, and the unit transportation cost $c_{ij}$ is calculated as $c_{ij} = \rho e_{ij}$ (constant $\rho := 0.5$). The confidence parameter is $\epsilon = 0.01, 0.05, 0.10$, and the number of scenarios is $S = 100, 500, 1000, 1500$. As for demand data, we prepare normal demand data generated according to the uniform distribution U(0, 500) and correlated demand data with a positive correlation between specific demand places. These correlated demand data are created using the Cholesky decomposition of Benninga [2] and Fushimi [5]. In this experiment, the correlated demand data I and the correlated demand data II are prepared by changing the way of giving the correlation coefficient. Correlated demand data I uses the distance $d$ between demand areas, and if $d < 20$, the correlation coefficient is $r = 0.8$, otherwise the correlation coefficient is $r = 0$. The distance $d$ is set on the coordinate plane according to the uniform distribution $U(10, 200)$. The correlated demand data I I is a data that gives a positive correlation ($0.5 \leq r < 1$) between all demand points regardless of the distance $d$.

CPLEX 12.6.2.0 was used as the MIP solver and all experiments were done on a computer with 3.60 GHz processor and 16.0 GB of memory.

### 4.2 Experimental Results

The facility location problem with a joint probabilistic constraint is solved using SMIP and SMIP that introduces the cutting plane method using strengthened star inequality. We compare the computation time of each solution. Using the normal demand data, the calculation results of the confidence parameter $\epsilon = 0.01, 0.05$ are shown in the following Tables 1 and 2.

**Table 1.** Calculation time for normal demand data ($\epsilon = 0.01$)

| $m$ | $n$ | $S$ | SMIP times($s$) | SMIP+strengthened star times($s$) |
|---|---|---|---|---|
| 40 | 100 | 100 | 2.0 | 3.0 |
| | | 500 | 2.0 | 23.0 |
| | | 1000 | 1001.0 | 68.0 |
| | 200 | 100 | 1.0 | 3.0 |
| | | 500 | 525.0 | 47.0 |
| | | 1000 | 3323.0 | 229.0 |

In Table 1, when the number of facilities is 40 and the number of demand points is 100, the calculation time of SMIP is $2.0, 2.0, 1001.0(s)$, and the SMIP+strengthened Star is $3.0, 23.0, 68.0(s)$. Similarly, for other conditions, the calculation time increases as the number of scenarios increases. From this, it can be seen that when the confident parameter $\epsilon$ is equal, the calculation time of the two solutions increases depending on the number of scenarios. In addition,

**Table 2.** Calculation time for normal demand data ($\epsilon = 0.05$)

| $m$ | $n$ | $S$ | SMIP times($s$) | SMIP+strengthened star times($s$) |
|---|---|---|---|---|
| 40 | 100 | 100 | 1.0 | 14.0 |
| | | 500 | 309.0 | 97.0 |
| | | 1000 | 30000.0 | 3974.0 |
| | 200 | 100 | 8.0 | 26.0 |
| | | 500 | 10000.0 | 1556.0 |

because the calculation time is longer when the number of demand points is 200 compared with when it is 100, it is understood that the calculation time also depends on the number of demand points. Next, we compare Table 1 and Table 2 while concentrating the confident parameter $\epsilon = 0.01, 0.05$. In the case of 40 facilities, 100 demand points, and 1000 scenarios, when $\epsilon = 0.01$, the calculation time of SMIP is 1001.0($s$), the SMIP+strengthened Star is 68.0($s$), and $\epsilon = 0.05$ The calculation time of SMIP is $> 30000.0(s)$, the SMIP+strengthened Star is 3974.0($s$). From this result, it can be seen that the larger $\epsilon$ is, the greater the amount of calculation. Meanwhile, in the case of a small-scale model with about 100 scenarios, the calculation time tends to be shorter if no cutting plane is added. In this regard, the computational complexity of the small-scale model is not so large and can be solved sufficiently by SMIP. Therefore, it may be considered that adding a cutting plane becomes redundant.

Next, we use the correlated demand data I to compare the computation time of the two solutions. The calculation results for each $\epsilon = 0.05, 0.10$ are shown in Tables 3 and 4 below.

**Table 3.** Calculation time for correlated demand data I ($\epsilon = 0.05$)

| $m$ | $n$ | $S$ | SMIP times($s$) | SMIP+strengthened star times($s$) |
|---|---|---|---|---|
| 40 | 100 | 100 | 2.0 | 4.0 |
| | | 500 | 4.0 | 6.0 |
| | | 1000 | 1773.0 | 30.0 |
| | | 1500 | 4793.0 | 741.0 |

From Tables 3 and 4, as in the case of normal demand data, it can be seen that when $\epsilon$ is equal, the calculation time for the two solutions increases depending on the number of scenarios. Furthermore, because it is possible to calculate a problem with $\epsilon = 0.10$ and 1500 scenarios, which is difficult when calculating with normal demand data, it is shown that the correlated demand data can handle a larger problem.

**Table 4.** Calculation time for the correlated demand data I ($\epsilon = 0.10$)

| $m$ | $n$ | $S$ | SMIP times($s$) | SMIP+strengthened star times($s$) |
|---|---|---|---|---|
| 40 | 100 | 100 | 3.0 | 5.0 |
| | | 500 | 32.0 | 15.0 |
| | | 1000 | 6597.0 | 355.0 |
| | | 1500 | 40000.0 | 2058.0 |

Finally, using the correlated demand data II, we compare the computation times of the two solutions. Furthermore, a comparison with correlation demand data I having different correlations is also shown.

**Table 5.** Calculation time for the correlated demand data II ($\epsilon = 0.05$)

| $m$ | $n$ | $S$ | SMIP times($s$) | SMIP+strengthened star times($s$) |
|---|---|---|---|---|
| 40 | 100 | 100 | 1.0 | 2.0 |
| | | 500 | 3.0 | 5.0 |
| | | 1000 | 3045.0 | 13.0 |
| | | 1500 | 4692.0 | 28.0 |

We compare the experimental results in Table 3 and Table 5 with the same confidence parameter $\epsilon$. In Table 3, when the number of scenarios is 1500, the calculation time of SMIP is 4793.0($s$), and the SMIP+strengthened Star is 741.0($s$). Meanwhile, in the case of 1500 scenarios in Table 5, the calculation time of SMIP is 4692.0($s$), and the SMIP+strengthened Star is 28.0($s$). From this, it can be seen that the correlated demand data II having a positive correlation between many demand points can be calculated faster by adding a cutting plane. In the correlated demand data II, the similarity of the demand that can be taken in each scenario is strong, and the range of the value is narrow. Thus, the cutting plane is considered to be strongly effective.

## 5 Conclusions

In this study, we introduced the cutting plane method for the facility location problem with a joint probabilistic constraint, and showed its effectiveness. From a comparison of these experimental results, it was found that the correlated demand data can be solved more efficiently compared with the normal demand data, and can handle even large-scale problems. Moreover, in the correlated demand data, it was found that giving a positive correlation between all the demand points can solve faster compared with giving a positive correlation between the specific demand points. Meanwhile, when the confidence parameter

$\epsilon$ is large, that is, a model with a large allowable probability that the constraint does not hold is still difficult to calculate. Therefore, as a future task, in addition to the solution in this study, we need to consider a strong formulation by adding other cuts. Furthermore, to approach the actual facility location problem, it is conceivable to consider indicators other than the distance between demand points when generating correlated demand data.

## References

1. Atamturk, A., Nemhauser, G., Savelsbergh, M.: The mixed vertex packing problem. Math. Program. **89**, 35–53 (2000)
2. Benninga, S.: Financial Modeling, 4th edn. The MIT Press, Cambridge (2014)
3. Charnes, A., Cooper, W.W.: Chance constrained programming problems using. Manag. Sci. **6**, 73–79 (1959)
4. Fujie, T.: Branch and cut method for mixed integer programming problems. Measur. Control **42**, 770–775 (2003)
5. Fushimi, M.: Stochastic Methods and Simulations. Iwanami publishers, Tokyo (1994)
6. Lejeune, M.A., Ruszczynski, A.: An efficient trajectory method for probabilistic production-inventory-distribution problems. Oper. Res. **55**, 378–394 (2007)
7. Luedtke, J., Ahmed, S., Nemhauser, G.: An integer programming approach for linear programs with probabilistic constraints. Math. Program. **122**, 247–272 (2010)
8. Murr, M.R., Prekopa, A.: Solution of a product substitution problem using stochastic programming. In: Uryasev, S.P. (ed.) Probabilistic Constrained Optimization. Nonconvex Optimization and Its Applications, vol. 49, pp. 252–271. Springer, Boston (2000). https://doi.org/10.1007/978-1-4757-3150-7_14
9. Shiina, T.: Stochastic Programming. Asakura publishers, Tokyo (2015)
10. Van Roy, T.J., Wolsey, L.A.: Solving mixed integer programming automatic reformulation. Oper. Res. **35**, 45–57 (1987)
11. Wolsey, L.A.: Integer Programming, pp. 133–166. A Wiley-Interscience Publication, Hoboken (1998)

# The Advantage of Interval Weight Estimation over the Conventional Weight Estimation in AHP in Ranking Alternatives

Masahiro Inuiguchi(✉) and Issei Torisu

Graduate School of Engineering Science/School of Engineering Science, Osaka University, Toyonaka, Osaka 560-8531, Japan
inuiguti@sys.es.osaka-u.ac.jp

**Abstract.** In this paper, we investigate the significance of interval weight estimation in the setting of Analytic Hierarchy Process (AHP). We consider several estimation methods for a normalized interval weight vector from a crisp pairwise comparison matrix. They have a desirable property. To avoid the non-uniqueness of the solution, we add an additional constraint, i.e., the sum of centers of interval weights is one. A few ranking methods under interval weights are considered. Numerical experiments are executed to compare the estimation accuracy of ranking alternatives under the assumption that the decision maker has a true interval weight vector. The advantage of interval weight estimation over crisp weight estimation is demonstrated.

**Keywords:** Interval weight estimation · AHP · Pairwise comparison matrix · Linear programming

## 1 Introduction

The Analytic Hierarchy Process (AHP) [1,2] is a useful tool for multiple criteria decision making. The decision maker is asked to provide pairwise comparison matrices (PCMs) showing the relative importance between alternatives and between criteria. A weight estimation method from a given PCM plays an important role for the evaluation of alternatives in AHP. The maximum eigenvalue method and the geometric mean method are popular weight estimation methods. Because human judgement is often imperfect, the obtained PCMs are often inconsistent. In AHP, the consistency index is defined to evaluate the consistency and if it is in the acceptable range, the estimated weights are accepted and used for the decision analysis.

On the other hand, from the viewpoint that the inconsistency of PCM is caused by the human vague evaluation on scores and weights, the interval AHP [3] was proposed. In the interval AHP, weights are estimated as intervals


V.-N. Huynh et al. (Eds.): IUKM 2020, LNAI 12482, pp. 38–49, 2020.
https://doi.org/10.1007/978-3-030-62509-2_4

reflecting the vagueness of human evaluation. Such interval weights are estimated by minimizing the sum of widths of interval weights under the adequacy to the given PCM and the normality of interval weights. The estimation problem is reduced to a linear programming problem. However, recently, it is shown that the interval weights estimated by the conventional interval weight estimation method are not very adequate. Then, several alternative interval weight estimation methods [4] have been proposed.

Numerical experiments [4] show that maximizing minimal range (MMR) method performs well. However, we revealed that the MMR method extended to a case with a given interval PCM does not satisfy a desirable property, i.e., the restorability of proper interval weights from a perfectly consistent interval PCM. We consider modified interval weight estimation methods satisfying the restorability, including the improved MMR method.

Moreover, the advantages of interval weights are not well demonstrated yet. We introduce a few methods for ranking alternatives based on the minimax regret and maximin principles. Then, we execute numerical experiments to examine the usefulness of estimated interval weights assuming a true normalized interval weight vector representing the decision maker's evaluations on criteria importance.

In next section, we introduce a general method for a normalized interval weight vector estimation satisfying the restorability of proper interval weights from a perfectly consistent interval PCM. In Sect. 3, two methods for ranking alternatives are introduced: one is based on the maximin rule and the other is based on the minimax regret rule. The numerical experiments for the examination of the significance of interval weight estimation is described in Sect. 4. Through numerical experiments, we evaluate the correctness of ranking alternatives and demonstrate the significance of interval weight estimation. Finally, in Sect. 5, the concluding remarks are described.

## 2 Estimation Methods of a Normalized Interval Weight Vector

In this paper, we treat a multiple criteria decision making (MCDM) problem. In the MCDM problem, we rank $m$ alternatives or select the best alternative considering $n$ criteria. Analytic Hierarchy Process (AHP) [1,2] is a useful tool for MCDM. It gives methods for evaluating the importance of criteria as well as the scores (utility values) of alternatives in view of each criterion based on PCMs $A = (a_{ij})$, where $a_{ij}$ shows the relative importance of the $i$-th criterion/alternative to the $j$-th criterion/alternative. It is assumed that a PCM $A$ satisfies the reciprocity, i.e., $a_{ij} = 1/a_{ji}$, $i \neq j$. Then a normalized weight vector $\boldsymbol{w} = (w_1, w_2, \dots, w_p)^{\mathrm{T}}$ ($p = m$ or $p = n$) is obtained by minimizing the deviations between $a_{ij}$ and $w_i/w_j$, $i \neq j$. The component $w_i$ of the normalized weight vector $\boldsymbol{w} = (w_1, w_2, \dots, w_p)^{\mathrm{T}}$ shows the importance of the $i$-criterion or the score (utility value) of $i$-th alternative in view of each criterion.

However, assuming that the decision maker has a vague evaluation on the importance of criteria and scores (utility values) of alternatives, we estimate normalized interval weight vectors, $\boldsymbol{W} = (W_1, W_2, \ldots, W_p)^{\mathrm{T}}$ with $W_i = [w_i^{\mathrm{L}}, w_i^{\mathrm{R}}]$, $i = 1, 2, \ldots, p$. Because the original estimation method based on minimizing sum of widths (MSW method) [3] for normalized interval vectors does not perform well, several interval weight estimation has been proposed. Among them, the maximizing minimal range (MMR) method [4] performed best in the estimation of normalized interval weights in the numerical experiments. However, it has been revealed that the MMR method does not satisfy the restorability of proper interval weights from a perfectly consistent interval PCM. In this section, we describe several modified methods satisfying the restorability including the improved MMR method. In this paper, we concentrate on the estimation of normalized interval weight vector showing importance of criteria from a given PCM $A$ assuming that scores of alternatives in view of each criterion are given.

The modified estimation methods are given in the following algorithm:

1. For each $k \in N = \{1, 2, \ldots, n\}$, Solve the following lexicographical linear minimization problem:

$$\begin{aligned}
&\text{lex-min} \quad \left(f_1^k(\boldsymbol{w}^{\mathrm{L}}, \boldsymbol{w}^{\mathrm{R}}, \Delta), f_2^k(\boldsymbol{w}^{\mathrm{L}}, \boldsymbol{w}^{\mathrm{R}}, \Delta)\right). \\
&\text{subject to} \\
&\quad \sqrt{a_{ij}} w_j^{\mathrm{L}} + \delta_{ij} = \sqrt{a_{ji}} w_i^{\mathrm{R}}, \ \delta_{ij} \geq 0, \ i, j \in N (i \neq j), \\
&\quad \sum_{i \in N \setminus j} w_i^{\mathrm{R}} + w_j^{\mathrm{L}} \geq 1, \ \sum_{i \in N \setminus j} w_i^{\mathrm{L}} + w_j^{\mathrm{R}} \leq 1, \ j \in N, \\
&\quad \sum_{i \in N} (w_i^{\mathrm{L}} + w_i^{\mathrm{R}}) = 2, \ w_i^{\mathrm{R}} \geq w_i^{\mathrm{L}} \geq \epsilon, \ \delta_{ij} \geq 0, \ i \in N,
\end{aligned} \tag{1}$$

where 'lex-min' stands for lexicographical minimization and $\boldsymbol{w}^{\mathrm{L}} = (w_1^{\mathrm{L}}, \ldots, w_n^{\mathrm{L}})^{\mathrm{T}}$, $\boldsymbol{w}^{\mathrm{R}} = (w_1^{\mathrm{R}}, \ldots, w_n^{\mathrm{R}})^{\mathrm{T}}$ and $\Delta = (\delta_{ij})$ ($n \times n$ matrix). For the pair of evaluation functions $(f_1^k, f_2^k)$, function pairs $(f^{\mathrm{W}}(\cdot \mid N \setminus \{k\}), f^{\mathrm{W}}(\cdot \mid \{k\}))$ and $(f_-^{\mathrm{D}}(\cdot \mid \{k\}), f_+^{\mathrm{D}}(\cdot \mid \{k\}))$ are considered. For $M \subseteq N$, functions $f^{\mathrm{W}}(\cdot \mid M)$, $f_-^{\mathrm{D}}(\cdot \mid M)$ and $f_+^{\mathrm{D}}(\cdot \mid M)$ are defined as follows:

$$f^{\mathrm{W}}(\boldsymbol{w}^{\mathrm{L}}, \boldsymbol{w}^{\mathrm{R}}, \Delta \mid M) = \sum_{i \in M} (w_i^{\mathrm{R}} - w_i^{\mathrm{L}}), \tag{2}$$

$$f_-^{\mathrm{D}}(\boldsymbol{w}^{\mathrm{L}}, \boldsymbol{w}^{\mathrm{R}}, \Delta \mid M) = \sum_{i \in N \setminus M} \sum_{j \in N \setminus (M \cup \{i\})} \delta_{ij}, \tag{3}$$

$$f_+^{\mathrm{D}}(\boldsymbol{w}^{\mathrm{L}}, \boldsymbol{w}^{\mathrm{R}}, \Delta \mid M) = \sum_{i \in M} \left( \sum_{j \in N \setminus M} (\delta_{ij} + \delta_{ji}) + \sum_{j \in M \setminus \{i\}} \delta_{ij} \right). \tag{4}$$

Let $\hat{f}_1^k$ and $\hat{f}_2^k$ be the optimal values of the first and second objective functions, respectively. We note that the adequacy of the given PCM $A$ is expressed by $a_{ij} \in [w_i^{\mathrm{L}}/w_j^{\mathrm{R}}, w_i^{\mathrm{R}}/w_j^{\mathrm{L}}]$, $i, j \in N$, $i < j$ and the first constraint guarantees this condition. The second constraint shows the normalized condition of interval weight vector $\boldsymbol{W}$. This guarantees the reachability [5] of all $w_i \in [w_i^{\mathrm{L}}, w_i^{\mathrm{R}}]$,

$i \in N$. Moreover, to avoid the non-uniqueness of the solution, third constraint, $\sum_{i \in N}(w_i^{\mathrm{L}} + w_i^{\mathrm{R}}) = 2$ is introduced.

2. For each $k$, we solve the following maximization and minimization problems (if Problem (1) has a unique solution $\tilde{w}_i^{\mathrm{L}}(k)$, $\tilde{w}_i^{\mathrm{R}}(k)$, $i \in N$, we can skip this step by defining $\hat{w}_i^{\mathrm{L}}(k) = \check{w}_i^{\mathrm{L}}(k) = \tilde{w}_i^{\mathrm{L}}(k)$ and $\hat{w}_i^{\mathrm{R}}(k) = \check{w}_i^{\mathrm{R}}(k) = \tilde{w}_i^{\mathrm{R}}(k)$, $i \in N$):

$$\begin{aligned}
&\text{maximize } w_k^{\mathrm{R}} \text{ / minimize } w_k^{\mathrm{L}}, \\
&\text{subject to} \\
&\quad \sqrt{a_{ij}}w_j^{\mathrm{L}} + \delta_{ij} = \sqrt{a_{ji}}w_i^{\mathrm{R}},\ \delta_{ij} \geq 0,\ i,j \in N(i \neq j), \\
&\quad \sum_{i \in N \setminus j} w_i^{\mathrm{R}} + w_j^{\mathrm{L}} \geq 1,\ \sum_{i \in N \setminus j} w_i^{\mathrm{L}} + w_j^{\mathrm{R}} \leq 1,\ j \in N, \\
&\quad \sum_{i \in N}(w_i^{\mathrm{L}} + w_i^{\mathrm{R}}) = 2,\ w_i^{\mathrm{R}} \geq w_i^{\mathrm{L}} \geq \epsilon,\ \delta_{ij} \geq 0,\ i \in N, \\
&\quad f_1(\boldsymbol{w}^{\mathrm{L}}, \boldsymbol{w}^{\mathrm{R}}, \Delta) \leq \hat{f}_1(k),\ f_2(\boldsymbol{w}^{\mathrm{L}}, \boldsymbol{w}^{\mathrm{R}}, \Delta) \leq \hat{f}_2(k).
\end{aligned} \tag{5}$$

Let $\hat{w}_i^{\mathrm{L}}(k)$ and $\hat{w}_i^{\mathrm{R}}(k)$, $i \in N$ be the values of $w_i^{\mathrm{L}}$ and $w_i^{\mathrm{R}}$, $i \in N$ at a solution maximizing $w_k^{\mathrm{R}}$. Similarly, let $\check{w}_i^{\mathrm{L}}(k)$ and $\check{w}_i^{\mathrm{R}}(k)$, $i \in N$ be the values of $w_i^{\mathrm{L}}$ and $w_i^{\mathrm{R}}$, $i \in N$ at a solution minimizing $w_k^{\mathrm{L}}$.

3. We calculate

$$\breve{w}_i^{\mathrm{L}} = \varphi_1\left(\hat{w}_i^{\mathrm{L}}(1), \hat{w}_i^{\mathrm{L}}(2), \ldots, \hat{w}_i^{\mathrm{L}}(n), \check{w}_i^{\mathrm{L}}(1), \check{w}_i^{\mathrm{L}}(2), \ldots, \check{w}_i^{\mathrm{L}}(n)\right),\ i \in N, \tag{6}$$

$$\breve{w}_i^{\mathrm{R}} = \varphi_2\left(\hat{w}_i^{\mathrm{R}}(1), \hat{w}_i^{\mathrm{R}}(2), \ldots, \hat{w}_i^{\mathrm{R}}(n), \check{w}_i^{\mathrm{R}}(1), \check{w}_i^{\mathrm{R}}(2), \ldots, \check{w}_i^{\mathrm{R}}(n)\right),\ i \in N, \tag{7}$$

where we consider (min, max) and (average, average) for the functions pair $(\varphi_1, \varphi_2)$ ('average' stands for the arithmetic mean).

4. $\breve{w}_i^{\mathrm{L}}$, $\breve{w}_i^{\mathrm{R}}$, $i \in N$ are transformed so that the sum of centers of interval weights becomes one. Namely, the estimated interval weight vector is composed of $\bar{W}_i = [\bar{w}_i^{\mathrm{L}}, \bar{w}_i^{\mathrm{R}}]$, $i \in N$, where

$$\bar{w}_i^{\mathrm{L}} = \frac{2\breve{w}_i^{\mathrm{L}}}{\sum_{i \in N}(\breve{w}_i^{\mathrm{L}} + \breve{w}_i^{\mathrm{R}})},\quad \bar{w}_i^{\mathrm{R}} = \frac{2\breve{w}_i^{\mathrm{R}}}{\sum_{i \in N}(\breve{w}_i^{\mathrm{L}} + \breve{w}_i^{\mathrm{R}})},\quad i \in N. \tag{8}$$

We note that if (average, average) is used for $(\varphi_1, \varphi_2)$, we have $\bar{w}_i^{\mathrm{L}} = \breve{w}_i^{\mathrm{L}}$ and $\bar{w}_i^{\mathrm{R}} = \breve{w}_i^{\mathrm{R}}$, $i \in N$.

In the procedure described above, by specifying $(f_1^k, f_2^k)$ at step 1 and $(\varphi_1, \varphi_2)$ at step 3, we obtain different estimation methods. In this paper, we consider only $(f^{\mathrm{W}}(\cdot \mid N \setminus \{k\}), f^{\mathrm{W}}(\cdot \mid \{k\}))$ and $(f_-^{\mathrm{D}}(\cdot \mid \{k\}), f_+^{\mathrm{D}}(\cdot \mid \{k\}))$ for $(f_1^k, f_2^k)$, and only (min, max) and (average, average) for $(\varphi_1, \varphi_2)$, Then, we have four different methods. They are called as follows:

**Maximizing the minimal range method with minimizing the sum of widths** (MMR$_{\mathrm{W}}$) is the estimation method specified by $f_1^k = f^{\mathrm{W}}(\cdot \mid N \setminus \{k\})$, $f_2^k = f^{\mathrm{W}}(\cdot \mid \{k\})$, $\varphi_1 = \min$ and $\varphi_2 = \max$.

**Maximizing the minimal range method with minimizing the sum of deviations** (MMR$_\mathrm{D}$): is the estimation method specified by $f_1^k = f_-^\mathrm{D}(\cdot \mid \{k\})$, $f_2^k = f_+^\mathrm{D}(\cdot \mid \{k\})$, $\varphi_1 = \min$ and $\varphi_2 = \max$.

**Averaging the minimal range method with minimizing the sum of widths** (AMR$_\mathrm{W}$) is the estimation method specified by $f_1^k = f^\mathrm{W}(\cdot \mid N \setminus \{k\})$, $f_2^k = f^\mathrm{W}(\cdot \mid \{k\})$, $\varphi_1 =$ average and $\varphi_2 =$ average.

**Averaging the minimal range method with minimizing the sum of deviations** (AMR$_\mathrm{D}$): is the estimation method specified by $f_1^k = f_-^\mathrm{D}(\cdot \mid \{k\})$, $f_2^k = f_+^\mathrm{D}(\cdot \mid \{k\})$, $\varphi_1 =$ average and $\varphi_2 =$ average.

The linear programming problems appeared in the above procedure can be solved sequentially by a technique of post optimality analysis. In those estimation methods resolve the non-uniqueness problem of the solution under a PCM $A$ by imposing an additional constraint, i.e., the sum of center values of interval weights is one. Namely, we assume that the decision maker has a consciousness on the normality so that the center values of interval weights representing her/his evaluation compose a normalized weight vector. These estimation methods satisfy the desirable property, i.e., the restorability of proper interval weights from a perfectly consistent interval PCM when the sum of center values of true interval weights is one although we use only crisp PCMs in the numerical experiments described in what follows.

We consider estimation methods by minimizing sum of widths (MSW) and minimizing sum of deviations (MSD) corresponding the original estimation method [3] and its modification [6]. Those methods solve linear programming problems (1) with objective functions $f^\mathrm{W}(\boldsymbol{w}^\mathrm{L}, \boldsymbol{w}^\mathrm{R}, \Delta \mid N)$ and $f_-^\mathrm{D}(\boldsymbol{w}^\mathrm{L}, \boldsymbol{w}^\mathrm{R}, \Delta \mid \emptyset) = f_+^\mathrm{D}(\boldsymbol{w}^\mathrm{L}, \boldsymbol{w}^\mathrm{R}, \Delta \mid N)$, respectively.

Moreover, we consider also MMR methods with center values of interval weights estimated by the conventional AHP (see [7]). In the conventional AHP, the maximum eigenvalue method and the geometric mean method are popular. Then we use those methods for the estimation of center values of interval weights. On the other hand, we have two MMR methods different in their evaluations functions, i.e., the sum of widths or the sum of deviations. Therefore, we have four methods by the combination, i.e., E-MMR$_\mathrm{W}$, E-MMR$_\mathrm{D}$, G-MMR$_\mathrm{W}$ and G-MMR$_\mathrm{D}$, where characters before hyphens, E and G, stand for 'the maximum eigenvalue method' and 'the geometric mean method', respectively while subscripts, W and D, stand for 'the sum of widths' and 'the sum of deviations'.

As a result, ten estimation methods for a normalized interval weight vector are considered in this paper.

## 3 Ranking Alternatives

In this paper, we compare ten estimation methods for a normalized interval weight vector as well as the estimation methods for a crisp weight vector used in the conventional AHP in ranking alternatives. For the estimation methods for a crisp weight vector, we use the maximum eigenvalue method and the geometric mean method. We assume that the scores of alternatives in each criterion are given. Then, we estimate the interval/crisp weights of criteria from a given PCM.

Under a crisp weight vector, the holistic score of an alternative is univocally defined by the weighted sum of scores. Then alternatives are ranked by the holistic scores. However, under a normalized interval weight vector, the holistic score of an alternative cannot be obtained univocally, because we do not know weights precisely but imprecisely by their ranges. Then, for ranking alternatives under a normalized interval weight vector, we adopt decision principles under strict uncertainty [8]. We utilize two principles, i.e., maximin rule and minimax regret rule, for ranking alternatives.

### 3.1 Maximin Rule

Let $u_i(o)$, $i \in N$ be the score of alternative $o$ in view of the $i$-th criterion. Let $W_i$, $i \in N$ be the interval weight of the $i$-th criterion. Following the maximin rule [8] the holistic score $MU(o)$ of an alternative is defined by the minimum weighted sum of scores, i.e., the worst score. Accordingly, $MU(o)$ is defined by

$$MU(o) = \min \left\{ \sum_{i \in N} w_i u_i(o) \;\middle|\; w_i \in W_i,\ i \in N \right\}. \tag{9}$$

The larger $MU(o)$ is, the more preferable alternative $o$ is. Then we obtain the following preference relation;

$$o_p \succsim_{\mathrm{MU}} o_q \Leftrightarrow MU(o_p) \geq MU(o_q),\ p, q \in L = \{1, 2, \dots, l\},\ p \neq q, \tag{10}$$

where $o_p$, $p \in L$ are alternatives. The preference relation $\succsim_{\mathrm{MU}}$ is a complete preorder (weak order).

### 3.2 Minimax Regret Rule

The maximum regret $r(o_p, o_q)$ of the selection of alternative $o_p$ to another alternative $o_q$ is defined by

$$r(o_p, o_q) = \max \left\{ \sum_{i \in N} w_i (u_i(o_q) - u_i(o_p)) \;\middle|\; w_i \in W_i,\ i \in N \right\}. \tag{11}$$

Then the maximum regret of the selection of alternative $o_p$ is defined by

$$\begin{aligned} R(o_p) &= \max_{q \in L} r(o_p, o_q) \\ &= \max \left\{ \sum_{i \in N} w_i (u_i(o_q) - u_i(o_p)) \;\middle|\; w_i \in W_i,\ i \in N,\ p \in L \right\}. \end{aligned} \tag{12}$$

The smaller $MU(o)$ is, the more preferable alternative $o$ is. Then we obtain the following preference relation;

$$o_p \succsim_{\mathrm{R}} o_q \Leftrightarrow R(o_p) \leq R(o_q),\ p, q \in L = \{1, 2, \dots, l\},\ p \neq q. \tag{13}$$

The preference relation $\succsim_{\mathrm{R}}$ is a complete preorder (weak order).

# 4 Numerical Experiments

## 4.1 Aim and Method for Experiments

In order to see the significance of the interval weight estimation, we conduct numerical experiments. In the numerical experiments, we suppose a multiple criteria decision problem where the scores (utility values) of alternatives in each criterion are given by experts. Moreover, as the alternatives are selected through a first-stage examination by experts, the holistic evaluation is assumed to be made by a weighted sum of scores. The importance of each criterion depends on the decision maker. We assume that the decision maker implicitly have an imprecise evaluation on the importance of each criterion. The importance of each criteria is assumed to be represented by an interval weight and these interval weights of all criteria are assumed to compose a normalized interval weight vector (true normalized interval weight vector). Because of the implicitness, the decision maker cannot provide the precise information about the weights of criteria, but we assume that s/he provides a crisp PCM. Therefore, the interval/crisp weights are estimated from the given PCM and the alternatives are ordered by using a ranking method with the estimated interval/crisp weights.

To simulate the multiple criteria decision problem described above, we assume a true interval weight $T_i$, $i \in N$. Then a PCM $A = (a_{ij})$ is generated by applying the following procedure to each pair $(i, j) \in N \times N$ such that $i < j$. Namely, $a_{ij}$ and $a_{ji}$ are generated by

(P1) Generate $r_i \in T_i$ and $r_j \in T_j$ independently, by random numbers obeying uniform distributions over $T_i$ and $T_j$.

(P2) Return to (P1), if there do not exist $t_k \in T_k$, $k \in N \setminus \{i, j\}$ such that

$$r_i + r_j + \sum_{k \in N \setminus \{i,j\}} t_k = 1. \tag{14}$$

(P3) Calculate $a'_{ij} = r_i / r_j$.

(P4) Obtain $a_{ij}$ by discretizing $a'_{ij}$ by

$$a_{ij} = \begin{cases} \dfrac{1}{9}, & \text{if } a'_{ij} \leq \dfrac{1}{9}, \\ \dfrac{1}{\text{round}(1/a'_{ij})}, & \text{if } \dfrac{1}{9} < a'_{ij} < 1, \\ \text{round}(a'_{ij}), & \text{if } 1 \leq a'_{ij} < 9, \\ 9, & \text{if } a'_{ij} \geq 9, \end{cases} \tag{15}$$

where round($\cdot$) is the round-off function.

(P5) Obtain $a_{ji} = 1/a_{ij}$. We note that we have $a_{ii} = 1$ from the definition of a PCM.

In the numerical experiments, we compare estimation methods of a normalized interval weight vector as well as the estimation methods of a crisp weight vector in ranking alternatives. Assuming a ranking method of alternatives, i.e.,

**Table 1.** Four cases for the true normalized interval weight vector

| | A | B | C | D |
|---|---|---|---|---|
| $X_1$ | [0.27, 0.33] | [0.23, 0.37] | [0.23, 0.37] | [0.25, 0.35] |
| $X_2$ | [0.21, 0.29] | [0.19, 0.31] | [0.20, 0.30] | [0.20, 0.30] |
| $X_3$ | [0.15, 0.25] | [0.15, 0.25] | [0.17, 0.23] | [0.15, 0.25] |
| $X_4$ | [0.09, 0.21] | [0.11, 0.19] | [0.10, 0.20] | [0.10, 0.20] |
| $X_5$ | [0.03, 0.17] | [0.07, 0.13] | [0.03, 0.17] | [0.05, 0.15] |

'minimax rule' or 'minimax regret rule, we compare the preference relations under the true normalized interval weight vector and the estimated normalized interval/crisp weight vector. If some of estimation methods of a normalized interval weight vector are better than the estimation methods of a crisp weight vector, we may find the significance of the estimation of a normalized interval weight vector.

### 4.2 Experimental Procedure

In the experiments, we consider five criteria and five alternatives, i.e., $n = 5$ and $l = 5$. For the true normalized interval weight vector, we considered four different cases as shown in Table 1.

For each true normalized interval weight vector, we generated 1,000 PCMs $A$ by the method described in the previous subsection. For each PCM, we estimated the normalized interval/crisp weight vector by the methods described in Sect. 2.

Let $t_i^{\mathrm{C}}$ be center values of intervals $T_i$ $(i \in N)$. The scores (utility values) of five alternatives in five criteria are generated randomly. Namely, for each $p \in L$, we apply the following procedure:

(U1) Generate randomly five numbers $s_i, i = 1, 2, \ldots, 5$ such that $\sum_{i \in N} s_i = 1$.
(U2) Calculate $s_i' = s_i / t_i^{\mathrm{C}}$.
(U3) Generate $v_i \in [0.95, 1.05]$, $i \in N$, independently by uniform random numbers.
(U4) Obtain $u_i(o_p) = v_i \cdot s_i'$, $i \in N$. We prepare 100 sets of scores (utility values) of five alternatives in five criteria.

By this score generation process, the center values of the intervals expressing the ranges of holistic scores (utility values) of alternatives locate around one. Therefore, the comparison among those alternatives does not become an easy task because of the closeness of their holistic scores.

Therefore, we have $1,000 \times 100 = 100,000$ samples for examination. For each examination, we compare the preference relations under the true normalized interval weight vector and the estimated normalized interval/crisp weight vector. For each examination, we count the number of alternative pairs such that true and estimated preference relations coincide.

**Table 2.** Frequency distribution about preference matched pairs in maximin rule

| True interval weights: C | | | | | | | | |
|---|---|---|---|---|---|---|---|---|
| Method | 0–3 pairs | 4 pairs | 5 pairs | 6 pairs | 7 pairs | 8 pairs | 9 pairs | 10 pairs |
| Eigenvalue | 25,889 | 13,228 | 14,896 | 15,287 | 13,663 | 9,712 | 5,517 | 1,808 |
| Geometric | 24,760 | 12,843 | 14,642 | 15,377 | 13,934 | 10,467 | 6,018 | 1,959 |
| MSW | 10,275 | 8,092 | 11,555 | 15,507 | 18,896 | 18,034 | 13,193 | 4,448 |
| MSD | 10,663 | 8,061 | 11,234 | 15,417 | 18,314 | 17,874 | 13,678 | 4,759 |
| $MMR_W$ | 986 | 1,818 | 3,221 | 7,173 | 15,769 | 23,660 | 28,780 | 18,593 |
| $MMR_D$ | 5,135 | 5,231 | 8,184 | 12,909 | 19,092 | 21,463 | 18,960 | 9,026 |
| $AMR_W$ | 1,388 | 2,205 | 4,344 | 9,269 | 16,762 | 25,249 | 26,130 | 14,653 |
| $AMR_D$ | 9,287 | 7,055 | 9,903 | 13,578 | 17,379 | 18,866 | 16,434 | 7,498 |
| E-$MMR_W$ | 1,357 | 1,874 | 3,558 | 7,600 | 15,227 | 24,528 | 27,610 | 18,246 |
| G-$MMR_W$ | 1,297 | 1,796 | 3,401 | 7,450 | 14,762 | 24,682 | 27,927 | 18,685 |
| E-$MMR_D$ | 5,940 | 5,679 | 8,365 | 12,259 | 17,703 | 21,334 | 19,230 | 9,490 |
| G-$MMR_D$ | 5,685 | 5,334 | 7,937 | 11,972 | 17,385 | 21,481 | 20,064 | 10,142 |

**Table 3.** Total numbers of preference matched pairs in maximin rule

| Method | A | B | C | D |
|---|---|---|---|---|
| Eigenvalue | 527,745 | 497,356 | 515,721 | 506,127 |
| Geometric | 536,619 | 499,618 | 524,989 | 511,959 |
| MSW | 679,226 | 594,680 | 646,765 | 640,582 |
| MSD | 678,082 | 603,846 | 647,303 | 645,096 |
| $MMR_W$ | 761,195 | 743,045 | 813,505** | 756,712 |
| $MMR_D$ | 731,962 | 734,007 | 718,071 | 733,390 |
| $AMR_W$ | 723,396 | 629,665 | 790705 | 681,313 |
| $AMR_D$ | 702,309 | 647,154 | 675,988 | 675,473 |
| E-$MMR_W$ | 815,560* | 791,886* | 811,106* | 851,568* |
| G-$MMR_W$ | 818,760** | 793,143** | 808,042 | 852,629** |
| E-$MMR_D$ | 716,724 | 722,531 | 714,973 | 728,937 |
| G-$MMR_D$ | 723,545 | 723,534 | 722,192 | 733,845 |

### 4.3 Results in the Maximin Rule

The results of the numerical experiments when the maximin rule is adopted are shown in Tables 2 and 3. In Table 2, for each estimation method, the frequency distribution of the number of alternative pairs such that true and estimated preference relations coincide under true normalized interval weight vector C. For other cases of true normalized interval weight vectors, we obtained similar results in general. Due to the space limitation, we show only the frequency distribution in case C. We compare preference relations obtained from true and estimated normalized interval weight vectors $1,000 \times 100 = 100,000$ times with 100 different scores and 1,000 different PCMs generated from the true normalized interval weight vector. As we have five alternatives, we obtain ${}_5C_2 = 10$ alternative pairs. Then if the preference relations perfectly coincide, we obtain 100,000 in the column of '10 pairs' and zeros in other columns. Moreover, the sum of numbers in each row becomes 10,000.

As the holistic scores of five alternatives are similar one another, the comparison of alternatives is not an easy task so that the numbers shown in the column of '10 pairs' are much smaller than 100,000 in Table 2. Similarly, the numbers shown in the column of '0–3 pairs' are not very small. However, we can see rather clearly the differences of accuracy in preference estimation by interval/crisp weight estimation. From Table 2, $\mathrm{MMR_W}$, G-$\mathrm{MMR_W}$ and E-$\mathrm{MMR_W}$ perform much better than the others because big numbers (more than 10,000) are located in columns of 7–10 pairs and small numbers (less than 10,000) are located in columns of 1–6 pairs. On the other hand, performances of the crisp weight estimations methods (eigenvalue and geometric) are much worse than the others. This implies that the estimation of interval weights is significant.

The similar observation is applicable in Table 3. In Table 3, the total number of alternative pairs such that true and estimated preference relations coincide in 10 (pairs) × 100,000 (examinations) = $1,00,000$ (pairwise comparisons) for all fours cases (A–D) of true normalized interval weight vectors. Asterisks ** and * show the best and the second best values in Table 3. As shown in Table 3, in all cases A–D, $\mathrm{MMR_W}$, G-$\mathrm{MMR_W}$ and E-$\mathrm{MMR_W}$ perform much better than the others. Among those three methods, G-$\mathrm{MMR_W}$ and E-$\mathrm{MMR_W}$ seem to perform a little better than $\mathrm{MMR_W}$ while $\mathrm{MMR_W}$ performs best in case D. Therefore, the combination of $\mathrm{MMR_W}$ with the conventional crisp weight estimation method works well.

It is unexpected that methods with minimization of sum of widths perform better than those with minimization of sum of deviations because it is known that the estimation of interval weights by minimization of sum of widths results in unbalanced width distribution (see [4]). The additional constraint, the sum of the center values of interval weights is one, may have good influence. MMR methods perform better than AMR methods.

From Table 3, we understand that G-$\mathrm{MMR_W}$ and E-$\mathrm{MMR_W}$ can estimate the preference relation about 80% of alternative pairs, although the given information is only a crisp PCM generated from true normalized interval weight vector.

### 4.4 Results in the Minimax Regret Rule

The results of the numerical experiments when the maximin rule is adopted are shown in Tables 4 and 5. Those correspond to Tables 2 and 3, respectively. We observe that values for estimation methods of normalized interval weight vector in Tables 4 and 5 are smaller than those in Tables 2 and 3 although we use the same $1,000 \times 100 = 100,000$ samples for examinations. On the other hand, values for estimations methods of normalized crisp weight vector in Tables 4

**Table 4.** Frequency distribution about preference matched pairs in minimax regret rule

| True interval weights: C | | | | | | | | |
|---|---|---|---|---|---|---|---|---|
| Method | 0–3 pairs | 4 pairs | 5 pairs | 6 pairs | 7 pairs | 8 pairs | 9 pairs | 10 pairs |
| Eigenvalue | 17,118 | 15,531 | 20,243 | 19,945 | 15,014 | 8,185 | 3,222 | 742 |
| Geometric | 16,782 | 15,560 | 20,002 | 20,097 | 15,288 | 8,178 | 3,272 | 821 |
| MSW | 10,776 | 12,108 | 18,328 | 20,967 | 18,942 | 11,299 | 5,996 | 1,584 |
| MSD | 11,068 | 12,380 | 18,370 | 21,070 | 18,427 | 11,375 | 5,734 | 1,576 |
| $MMR_W$ | 5,067 | 7,021 | 12,736 | 17,696 | 20,427 | 18,082 | 13,415 | 5,556 |
| $MMR_D$ | 6,737 | 8,482 | 15,678 | 20,281 | 20,014 | 16,426 | 9,345 | 3,037 |
| $AMR_W$ | 8,423 | 11,517 | 16,997 | 21,358 | 19,489 | 12,996 | 6,760 | 2,460 |
| $AMR_D$ | 8,887 | 10,601 | 17,598 | 20,500 | 19,256 | 13,982 | 7,116 | 2,060 |
| E-$MMR_W$ | 4,600 | 5,990 | 10,621 | 16,538 | 20,293 | 20,588 | 15,214 | 6,156 |
| G-$MMR_W$ | 4,662 | 5,922 | 10,593 | 16,495 | 20,061 | 20,618 | 15,389 | 6,260 |
| E-$MMR_D$ | 7,402 | 9,389 | 15,752 | 19,717 | 19,923 | 15,967 | 9,018 | 2,832 |
| G-$MMR_D$ | 7,197 | 9,290 | 15,430 | 19,632 | 20,055 | 16,205 | 9,199 | 2,992 |

**Table 5.** Total numbers of preference matched pairs in minimax regret rule

| Method | A | B | C | D |
|---|---|---|---|---|
| Eigenvalue | 529,947 | 546,032 | 531,269 | 535,007 |
| Geometric | 529,947 | 546,614 | 533,379 | 536,921 |
| MSW | 592,590 | 575,941 | 585,692 | 585,230 |
| MSD | 592.401 | 580,090 | 582,781 | 587,270 |
| $MMR_W$ | 639,413 | 627,348 | 674,510 | 654,713 |
| $MMR_D$ | 622,029 | 619,882 | 636,838 | 640,712 |
| $AMR_W$ | 628,495 | 603,904 | 606,526 | 623,472 |
| $AMR_D$ | 613,866 | 612,076 | 607,286 | 621,635 |
| E-$MMR_W$ | 692,770* | 701,643* | 693,368* | 734,391* |
| G-$MMR_W$ | 694,361** | 702,142** | 694,054** | 734,658** |
| E-$MMR_D$ | 621,566 | 653,816 | 630,182 | 648,362 |
| G-$MMR_D$ | 626,413 | 654,844 | 633,222 | 651,381 |

and 5 are larger than those in Tables 2 and 3. This may come from fact that the evaluation of maximum regret $R(o)$ depends on all alternatives while the evaluation of minimum score $MU(o)$ depends only on alternative $o$. Namely, the estimation of the minimum score $MU(o)$ depends only on the estimation quality of the holistic score interval of alternative $o$ although the estimation of the maximum regret $R(o)$ depend on the estimation quality of all interval weights.

Although values of 'eigenvalue' and 'geometric' are improved and values of others are decreases, Performances of the crisp weight estimations methods (eigenvalue and geometric) are worse than the others. This implies again that the estimation of interval weights is significant. Moreover, performances of G-$\mathrm{MMR_W}$ and E-$\mathrm{MMR_W}$ are better than others in all cases. In minimax regret rule, $\mathrm{MMR_W}$ does not outperform although it works still well. We confirm that the combination of $\mathrm{MMR_W}$ with the conventional crisp weight estimation method works well.

From Table 5, we understand that G-$\mathrm{MMR_W}$ and E-$\mathrm{MMR_W}$ can estimate the preference relation about 70% of alternative pairs.

## 5 Concluding Remarks

In this paper, we examined the significance of interval weight estimation instead of crisp weight estimation in the setting of AHP. We gave several estimation methods for a normalized interval weight vector. We considered the maximin rule and the minimax regret rule for ranking alternatives. We conducted numerical experiments to examine the significance of interval weigh estimation in ranking alternatives whose holistic scores are close one another. By a numerical experiment, we demonstrated the advantages of interval weight estimation over crisp weight estimation. The combination of $\mathrm{MMR_W}$ with the conventional crisp weight estimation performed best in the experiments.

Numerical experiments in other settings such that holistic scores of alternatives are not very close as well as the other interval weight estimation methods combined with the conventional crisp weight estimation are future research topics.

**Acknowledgement.** This work was supported by JSPS KAKENHI Grant Number JP17K18952.

## References

1. Saaty, T.L.: The Analytic Hierarchy Process. McGraw-Hill, New York (1980)
2. Saaty, T.L., Vargas, C.G.: Comparison of eigenvalue, logarithmic least squares and least squares methods in estimating ratios. Math. Model. **5**, 309–324 (1984)
3. Sugihara, K., Tanaka, H.: Interval evaluations in the analytic hierarchy process by possibilistic analysis. Comput. Intell. **17**(3), 567–579 (2001)
4. Inuiguchi, M., Innan, S.: Comparison among several parameter-free interval weight estimation methods from a crisp pairwise comparison matrix. CD-ROM Proc. MDAI **2017**, 61–76 (2017)
5. De Campos, L.M., Huete, J.F., Moral, S.: Probability intervals: a tool for uncertain reasoning. Int. J. Uncertain. Fuzziness Knowl. Based Syst. **2**(2), 167–196 (1994)
6. Torisu, I., Inuiguchi, M.: Increasing convergence of the quality of estimated interval weight vector in interval AHP. Proc. SCIS ISIS **2018**, 1400–1405 (2018)
7. Yamaguchi, M., Inuiguchi, M.: Estimation methods of interval weights centered at geometric mean from a pairwise comparison matrix. Proc. SCIS ISIS **2018**, 1394–1399 (2018)
8. French, S.: Decision Theory: An Introduction to the Mathematics of Rationality. Ellis Horwood Ltd., Hemel Hempstead (1986)

# Decision Support Framework for Composing of Different Questionnaires Based on Business Model with Optimization

Daniela Borissova[1](✉), Magdalena Garvanova[2], Zornitsa Dimitrova[1], Andrea Pandulis[2], and Ivan Garvanov[2]

[1] Institute of Information and Communication Technologies at the Bulgarian Academy of Sciences, 1113 Sofia, Bulgaria
dborissova@iit.bas.bg, zrn.dimitrova@gmail.com
[2] University of Library Studies and Information Technologies, 1784 Sofia, Bulgaria
{m.garvanova,a.pandulis,i.garvanov}@unibit.bg

**Abstract.** The digital transformation requires not only involving contemporary information technologies but also a business model that is capable to cope with specific challenges. These models should be integrated properly in a decision support system closely related to the business activities of the company. The current article describes a framework of a decision-support system for composing different questionnaires for specific survey taking into account several predefined target groups. A proper business model able to determine the most suitable questions for generating different target groups' questionnaires is formulated. This is realized by involving an expert that acts as a decision-maker to evaluate a predefined set of questions covering different sociological aspects toward predefined target groups. The distinguishing feature of this model is the ability to select a different number of questions for the questionnaires. Along with the formulated mathematical model, a software architecture framework for the implementation of this model is proposed. The numerical testing is done in case of the determination of four different questionnaires composed of an equal number of questions. Obtained results show the applicability of the formulated business model and the functionality of the proposed software framework.

**Keywords:** Information systems · Decision-making · Software framework · Questionnaires composing

## 1 Introduction

The organizations today are highly dependable on information technology not only to run their businesses but also to be competitive [1]. The role of top management is vital for planning and should provide motivated alternative decisions to cope with different challenges [2]. The digital transformation requires constantly updating the existing software systems involving new technologies. The decision support systems are designed to collect, process, store, retrieve, manipulating data to providing information


V.-N. Huynh et al. (Eds.): IUKM 2020, LNAI 12482, pp. 50–61, 2020.
https://doi.org/10.1007/978-3-030-62509-2_5

to support the business processes. As there is not a unified decision support system, each system is to be designed in such a way to be able to perform specific needs in accordance with the particular business model [3]. Whether using a web-based or desktop application, any information system must provide appropriate information to the user, for data collection, data processing and transmission of information to the user of the system [4, 5]. Different software systems are developed to support a variety of business activities. For example, systems to support the educational activities [6–8], a system for solving multiple-objective problems using secularizing and evolutionary approaches [9], business intelligence system for decision making for efficiency management [10], new generation software for digital entrepreneurship [11], etc. In addition, there exist a special kind of decision-making dealing with multi-criteria decision analysis (MCDA). The methods of MCDA allow making preference decisions by evaluating and prioritizing a limited set of alternatives based on multiple criteria [12]. When decision making is accomplished with uncertainty conditions or insufficient information some of the well-known principles of Wald, Savage Laplace, or Hurwicz could be used to determine the optimal alternative [13].

The growing companies' activities put a series of challenges related to cooperation between heterogeneous information in distributed networks [14]. To make this process easy, a meta-model of requirements about cooperative information systems using view-points could be used to decompose the system requirements to simplify the modeling [15]. Along with all requirements when developing the information system there is some portion of uncertainty that also should be considered [16]. The essence of any information system could be formalized as a representation of the knowledge that playing an important role in machine learning, data mining, and intelligent systems [17]. It is shown that by attribute reduction technique could be acquiring classification rules by using a minimal set of attributes. The attribute reduction techniques for massive and complex datasets are time-consuming and are a challenging research problem investigated by many authors. Some of them rely on attribute reduction considering incompletes with tolerance relation-based rough sets [18], or by composing rough approximations for distributed data [19], by using of multi-criterion strategy [20], etc. In contrast to these approaches, the aim of the paper is to propose a mathematically reasoned business model for the reduction of the given database to a number of subsets with known elements. In addition, these subsets should obey the main purpose and some additional requirements predefined for each subset. The subject of the investigations is related to extensive usage of smart technologies. The expanding of digital media, entertainment, advertising, retail, and transport to the digital economy are due to not only the transformation of business models to e-commerce [21] and software-as-a-service [22] but also due to the success of the developing modern mobile devices and applications. On the other hand, the e-media can be successful if a proper business model is implementing to meet the determined goal [23]. Some good practices related to the similar service-based management processes are to be considered including cyber resilience to protect our data [24, 25].

It should be mention and the growing widespread of ICT and new business challenges related to contemporary trends in digital cellular networks. There is no field to be affected by these technologies involving not only telecommunications but also different aspects of banking and different industry areas. With the massive usage of wireless technology, there are more and more disputes about their impact on the body and human health. A guide to evaluation and treatment the Internet addiction is given in [26]. The recent research shows that most of the power density and emitted electromagnetic waves are within the acceptable range, however, power density and emitted electromagnetic waves for some cells have hurtful effects in the long term for people who live near the base stations [27]. A review of the possible effects of radio-frequency electromagnetic field exposure on the central nervous system is presented in [28]. The investigation of the influence of mobile communications on public health is presented in [29]. The conducted investigations unequivocally show the influence of electromagnetic fields on human health [30, 31].

Among the various electronic devices, smartphones are used close to our bodies, and the time of usage is rapidly increasing. In addition, the use of smartphones is increased not only in adulthood but also among young people and the elderly, including young children. All of these motivated the authors to investigate the impacts of excessive use of smart technologies among different target groups. To do such an investigation it is needed to determine suitable questionnaires for different target groups to conduct the survey. The questionnaire should contain such questions that will reflect different aspects of the use of intelligent technologies, taking into account the specifics of the individual target groups. In this context, the determination of the most appropriate questions for the proper target group is of high importance. That is why in the current article a new mathematical business model is proposed to determine the sub-set of questions that are best suited to a particular target group. The rest of the article is organized as follows: Sect. 2 provides a detailed description of the problem. Section 3 contains the formulated mathematical business model and proper software architecture framework for its implementation. Section 4 contains the results of numerical testing. In Sect. 5 an analysis and discussion of obtained results are given. Section 6 represents conclusions concerning the proposed approach.

## 2 Problem Description

The business model is the core of the decision support system and is closely related to the purposes and tasks that should be performed. The main goal of the described software framework is to store and retrieve well-formulated questions from different domains to conduct a variety of surveys. It is supposed the existence of an already predefined set of questions (elements) stored in a database and that covers a variety of sociological aspects of a particular survey. The specific problem is related to the technique for the reduction of the questions (elements) by involving a DM in the evaluation process. The goal is to select a certain number of questions from a given domain (database) that is best suited for a given target group. For example, from particular domain database with questions has to extract only these questions that are the most adequate to the investigated target group. This particular problem for reducing a given set to sub-sets is illustrated in Fig. 1.

To cope with the formulated problem, a mathematical model is needed to determine the sub-sets of questions that will compose the particular questionnaire for the survey. This activity could be done by using methods of MCDA or by formulation a new one. Once the business model is formulated it should be implemented in a software system

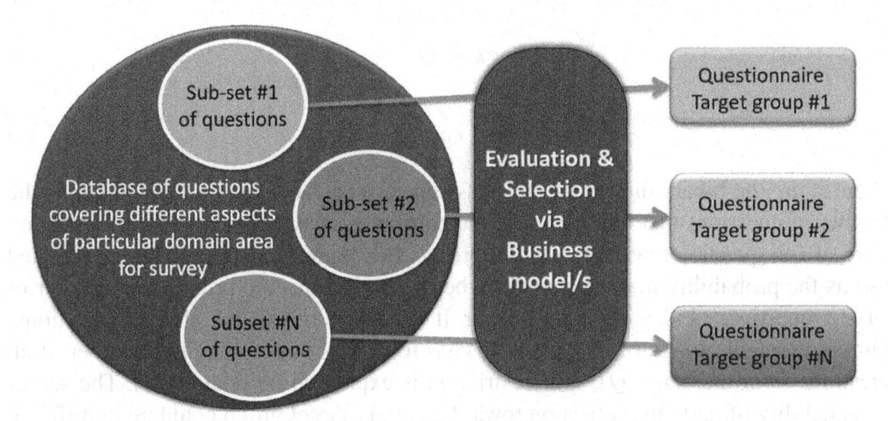


**Fig. 1.** Selection of a sub-set of questions to compose questionnaires for surveys when involving different target groups

capable to determine the best-suited questions for the groups to realize the requested survey.

# 3 Decision Support Framework for Selection of Questions to Compose Different Questionnaire Using Business Model for Thematic Survey

To be able to realize the described above decision problem it is needed to determine the main components of decision support system. These components could be roughly divided into two basic parts: a business model and a software framework. In accordance with the specifics of the formulated business model, a suitable software framework for model implementation should be proposed. These two basic parts of the information system for the determination of sub-sets with questions for different target groups are described below.

## 3.1 Business Model for Selection of Survey' Questions for Different Target Groups

The problem of the determination of sub-sets from an existing database could be approached by different MCDA techniques. In the current article, a business optimization model for the determination of a sub-set with questions that will compose the surveys for different target groups involving combinatorial optimization is formulated:

$$max \sum_{i=1}^{Q} \sum_{j=1}^{G} x_i e_{ij} \tag{1}$$

subject to

$$\sum_{i=1}^{Q} x_i = k \tag{2}$$

$$k < Q \tag{3}$$

$$0 \leq e_{ij} \leq 1 \tag{4}$$

where $x_i$ are the binary integer variables assigned to each question in the bank, $q$ is the number of questions that will compose the particular survey, and $e_{ij}$ are the evaluation score of $i$-th question toward $j$-th target group. The expression $e_{ij}$ could be interpreted also as the probability that the questions belong to the specific group. The number of survey' questions ($k$) has to be an integer. It should be noted also, that the questions' number that will be part of the survey ($k$) has to be less than the overall number of all questions from the bank ($Q$). This restriction is expressed by relation (3). The scores for suitability of particular question toward a certain target group could be in different ranges, but equal for all target groups. For example, a range between 0 and 1, or a range between 1 and 100 where the lower value means not appropriate and the highest value means the most appropriate. A corresponding optimization task based on the model (1)–(4) is to be solved as many times as the number of target groups is given.

The objective function (1) seeks to select these questions that have more profit for its maximization. This means that the questions with corresponding binary integer variables equal to 1 will be part of a particular survey. When determining the questions not suitable to be included in the survey the objective function (1) should seek the minimum.

It is possible also, instead of real values for evaluation score to use some fuzzy or linguistic expressions for evaluation of the questions to a given target group/s. In this situation, the objective function will have the following expression:

$$max \sum\nolimits_{i=1}^{Q} \sum\nolimits_{j=1}^{G} x_i(U, m)_{ij} \tag{5}$$

subject to the restrictions (2) and (3), and $U$ is a set and $m : U \rightarrow [0, 1]$ a membership function.

The distinguishing feature of the formulated both models is the absence of restriction about the number of generated surveys from the given database. In addition, the surveys for different groups can be composed of different numbers of questions.

Another suitable approach to evaluate the questions for different target groups is the Analytic Hierarchy Process (AHP). AHP is a well-known structured technique to analyze complex problems is proposed by Saaty [32]. This technique based on mathematics and psychology provides an accurate approach for quantifying both the weights of decision criteria and estimations obtained by pair-wise comparisons be-tween alternatives with respect to the evaluation criteria. AHP can be viewed as a form of rational decision-making.

### 3.2 Decision Support Framework for Implementation of Business Models for Composing Different Questionnaires

A software architecture framework for DSS that can implement the functionality of the proposed business models in the previous section is illustrated in Fig. 2.

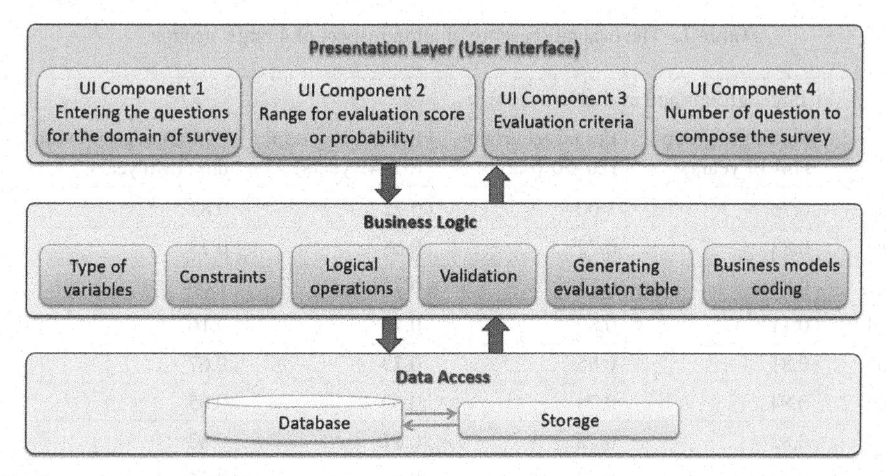


**Fig. 2.** Decision support framework for storage and processing the questions and composing different questionnaires

The three-tier architecture, composed of the presentation layer, business logic, and access to the database, provides an opportunity to use new technologies. The presentation tier communicates with the other tiers through the application program interface or user interface. That is why it has to be well done and contain all of the needed components. For the purpose of implementation of the business model (1)–(4), at least four components are to be present: 1) component for entering of questions; 2) component for a range of evaluation score; 3) component for questions' number to compose the survey; and 4) component for target groups' number. The second component of the three-tier architecture is the business logic or application layer, which refers to the logic that contains programming code, which represents the business rules and which supports the application's core functions. This is the most essential component closely related to the business model(s). It includes setting up the needed types of variables, constraints, logical operations for performing tasks like processing commands, validation, logical evaluations, decision making, database interactions, etc. Last but not least is the data access layer responsible to store and retrieve the information. This layer keeps the data independent of application servers or business logic, and its management and access are provided with programs like MongoDB, Oracle, MySQL, Microsoft SQL Server, etc.

## 4 Numerical Application

To demonstrate the applicability of the proposed model (1)–(4), an authorized expert considered as a person for decision-making is to evaluate a given excerpt of 25 questions in respect to four target groups. The questions aim to assess the psychological and physical impacts of excessive use of smart technologies. The evaluations of the given questions in respect to the suitability of four target groups distributed by ages are shown in Table 1.

The evaluation of all questions is done by using the range for a score between 0 and 1, where 0 mean not suitable question for the particular group and 1 means most

**Table 1.** The evaluation score of all in respect of 4 target groups

| | Target groups and evaluation score | | | |
|---|---|---|---|---|
| | For target group (14–19 years) | For target group (20–30 years) | For target group (31–49 years) | For target group with unknown years |
| q1 | 0.76 | 1.00 | 0.72 | 0.82 |
| q2 | 0.83 | 0.72 | 0.68 | 0.73 |
| q3 | 0.81 | 0.77 | 0.76 | 1.00 |
| q4 | 0.11 | 0.89 | 0.79 | 0.16 |
| q5 | 0.81 | 0.85 | 0.73 | 0.67 |
| q6 | 0.84 | 0.74 | 0.69 | 0.65 |
| q7 | 0.82 | 0.78 | 0.81 | 0.82 |
| q8 | 0.15 | 0.12 | 0.86 | 0.06 |
| q9 | 0.86 | 0.61 | 0.78 | 0.77 |
| q10 | 0.72 | 0.79 | 0.82 | 0.81 |
| q11 | 0.81 | 0.78 | 0.73 | 0.69 |
| q12 | 0.13 | 0.09 | 0.88 | 0.21 |
| q13 | 1.00 | 0.85 | 0.61 | 0.62 |
| q14 | 0.88 | 0.74 | 0.66 | 0.66 |
| q15 | 0.92 | 0.76 | 0.73 | 0.75 |
| q16 | 0.10 | 0.10 | 0.93 | 0.45 |
| q17 | 0.92 | 0.79 | 0.88 | 0.85 |
| q18 | 0.78 | 0.88 | 1.00 | 0.92 |
| q19 | 0.84 | 0.76 | 0.73 | 0.68 |
| q20 | 0.96 | 1.00 | 0.75 | 0.79 |
| q21 | 0.83 | 0.91 | 0.19 | 0.76 |
| q22 | 0.10 | 0.06 | 0.79 | 0.37 |
| q23 | 0.83 | 0.78 | 0.58 | 0.86 |
| q24 | 0.84 | 0.71 | 0.88 | 0.64 |
| q25 | 0.88 | 0.73 | 0.33 | 0.79 |

appropriate. These input data are used to compose four suitable questionnaires for four different groups considering the evaluation scores from the expert. It should be noted, that all of these questionnaires should contain the same number of questions to conduct the survey.

In order to realize the APH, some pair-wise comparisons are to be done between every two elements using a nine-point scale to express the degree of preference as shown in Table 2 [32].

**Table 2.** The fundamental scale for pair-wise comparisons

| Intensity of importance | Definition | Explanation |
|---|---|---|
| 1 | Same | Neither of the two alternatives is preferable over other |
| 3 | Weak | One alternative is preferred slightly over the other |
| 5 | Clear | One alternative is preferred clearly over the other |
| 7 | Strong | One alternative is preferred strongly over the other |
| 9 | Very strong | One alternative is preferred very strongly over other |
| 2, 4, 6, 8 | Compromise | Can be used for graduation between evaluation |

Instead of making the evaluation of questions using a given range or calculations for their probability, the AHP makes it possible to express DM preferences in a common way during the pair-wise alternative comparison.

## 5 Results Analysis and Discussion

The results for the formulated and solved four optimization tasks based on the proposed business model (1)–(4) are graphically visualized in Fig. 3.

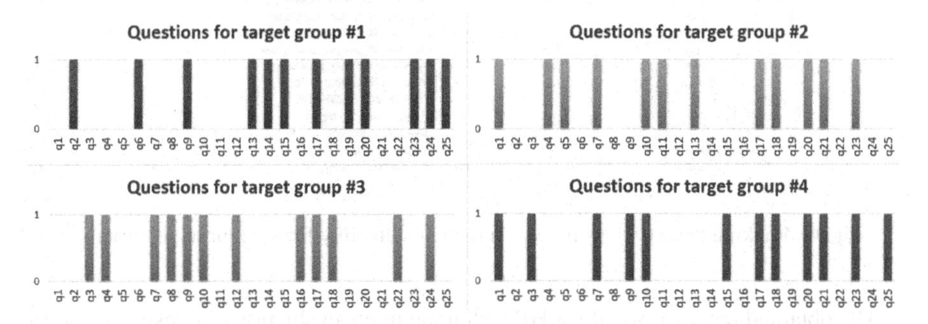


**Fig. 3.** Selected questions to compose questionnaires for different target groups

As could be seen from Fig. 3, there are some questions that are quite suitable and are selected to be part of all 4 questionnaires. For example, this is question #17, which can be considered as sufficiently universal for the purposes of the survey, regardless of the age limit of the target groups. Question #9 is selected for the questionnaires focused for target groups #1, #3 and #4, but it is not suitable for target group #2. The question #18 is appropriate for target groups #2, #3 and #4, question #13 best fits for target groups #1 and #2, etc. It is noticeable form Fig. 3 that the survey for the last target group #4 includes some questions that are not selected to compose the previous surveys for target groups #1, #2, and #3. This due the fact, that this questionnaire for target group #4 aims to investigate the consumers from mixed ages.

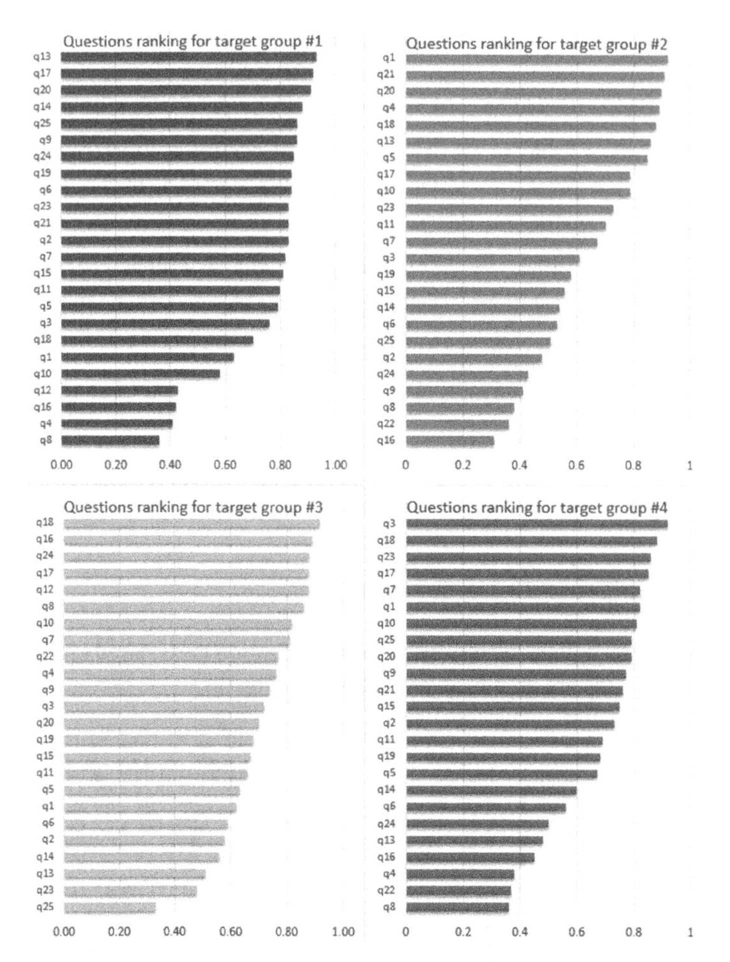


**Fig. 4.** Ranking the questions in respect to their suitability for the four target groups

The obtained results using the AHP technique in the evaluation of questions toward their suitability of the investigated target groups are shown in Fig. 4.

In contrast to the mixed-integer model (1)–(4) where tasks solutions determine 12 most suitable questions, the result from AHP are ranked lists of all questions with respect to criteria of suitability for a particular target group. Using the AHP techniques gives the performance of each question while the proposed optimization model only indicates if the question is selected or not. It should be noted the significant role of the expert or DM, which is responsible for the evaluation of all questions with respect to their suitability to the predefined target groups. There is no obstacle to use the evaluations of the given set of questions to create different questionnaires with different numbers of questions for the goal of the survey. If a different target group is needed to be involved within the survey, additional evaluation of the questions in respect to the new group is to be done. If a different target group is needed to be involved within the survey, additional evaluation of the questions with respect to the new group is to be done.

In case of evaluation of the questions by using some fuzzy or linguistic expressions for a given target group/s, the objective function (5) should be used. This scenario is planned as the future development of the described approach.

## 6 Conclusion

The current article describes a software framework for the selection of questions to compose different questionnaires to conduct a survey using a business model. The distinguishing feature of this approach is the possibility to determine the most appropriate questions for the questionnaires that best fit for the predefined target group. To realize a particular survey a business model based combinatorial optimization is proposed. This mathematical model is implemented via the business logic of the proposed software framework architecture. The essential part of the proposed business model respectively business logic is the presence of qualified DM involved to evaluate the set of questions in respect of its relevance for predefined groups.

The described approach is numerically tested in the case of the determination of questionnaires to compose surveys for four target groups. For the goal, an expert with a qualification in the area of smart technologies usage evaluated a set predefined number of questions. Using the proposed mathematical model along with these evaluations, four surveys for different target groups are determined. The obtained results show not only the correctness of the proposed business model and but also the applicability of the developed web-based software prototype with integrated business logic. These positive results motivated the authors to continue the investigation by using different business models that are able to incorporate the expertise of a group of DMs. In such a way the determined surveys will reflect the points of view of more experts and could be more objective.

**Acknowledgment.** This work is supported by the Bulgarian National Science Fund, Project title "Synthesis of a dynamic model for assessing the psychological and physical impacts of excessive use of smart technologies", KP-06-N 32/4/07.12.2019.

## References

1. Gerth, A.B., Peppard, J.: The dynamics of CIO derailment: How CIOs come undone and how to avoid. Bus. Horiz. **59**, 61–70 (2016)
2. Jarvelainen, J.: Understanding the stakeholder roles in business continuity management practices – A study in public sector. In: 53rd Hawaii International Conference on System Sciences, pp. 1966–1975. Maui, Hawaii (2020)
3. Borissova, D., Mustakerov, I.: An integrated framework of designing a decision support system for engineering predictive maintenance. Int. J. Inf. Technol. Knowl. **6**(4), 366–376 (2012)
4. Mustakerov, I., Borissova, D.: Software system for night vision devices design by reasonable combinatorial choice. In: Breaz, D., Breaz, N., Wainberg, D. (eds.) ICTAMI 2007, pp. 43–53 (2007)

5. Borissova, D., Mustakerov, I.: A framework for designing of optimization software tools by commercial API implementation. Int. J. Adv. Eng. Manage. Sci. **2**(10), 1790–1795 (2016)
6. Marinova, G., Guliashki, V., Chikov, O.: Concept of online assisted platform for technologies and management in communications – OPTIMEK. In: Hajrizi, E. (ed.) ICBTI'2014, pp. 136–145 (2014)
7. Mustakerov, I., Borissova, D.: A conceptual approach for development of educational Web-based e-testing system. Expert Syst. Appl. **38**(11), 14060–14064 (2011)
8. Borissova, D., Keremedchiev, D.: Intelligent System for Generation and Evaluation of e-Learning Tests Using Integer Programming. In: Simian, D., Stoica, L.F. (eds.) MDIS 2019. CCIS, vol. 1126, pp. 97–110. Springer, Cham (2020). https://doi.org/10.1007/978-3-030-39237-6_7
9. Kirilov, L., Guliashki, V., Genova, K., Zhivkov, P., Staykov, B., Vatov, D.: Interactive environment WebOptim for solving multiple-objective problems using scalarising and evolutionary approaches. Int. J. Reasoning-based Intell. Syst. **7**(1/2), 4–15 (2015)
10. Borissova, D., Cvetkova, P., Garvanov, I., Garvanova, M.: A framework of business intelligence system for decision making in efficiency management. In: Saeed, K., Dvorský, J. (eds.) CISIM 2020. LNCS, vol. 12133, pp. 111–121. Springer, Cham (2020). https://doi.org/10.1007/978-3-030-47679-3_10
11. Garvanova, M.: The digital entrepreneurship as a new generation software of the mind. In: Ivanova, S.V., Elkina, I.M. (eds.) icCSBs'2019, Future Academy, vol. LXXIV, pp. 294–300 (2019)
12. Belton, V., Stewart, T.: Multiple Criteria Decision Analysis: An Integrated Approach. Kluwer Academic Publishers, Massachusetts (2002)
13. Borissova, D., Mustakerov I.: A concept of intelligent e-maintenance decision making system. In: Innovations in Intelligent Systems and Applications (INISTA), pp. 1–6. IEEE, Albena Bulgaria, (2013)
14. Yahia, E., Aubry, A., Panetto, H.: Formal measures for semantic interoperability assessment in cooperative enterprise information systems. Comput. Ind. **63**(5), 443–457 (2012)
15. Kessi, K., Alimazighi, Z., Oussalah, M.: Requirement meta model of a cooperative information system oriented viewpoints. Procedia Comput. Sci. **64**, 474–482 (2015)
16. Taipalus, T., Seppanen, V., Pirhonen, M.: Uncertainty in information system development: Causes, effects, and coping mechanisms. J. Syst. Softw. **168**, 110655 (2020)
17. Liu, G., Feng. Y., Yang. J.: A common attribute reduction form for information systems. Knowl.-Based Syst. **193**, 105466 (2020)
18. Meng, Z., Shi, Z.: A fast approach to attribute reduction in incompletes with tolerance relation-based rough sets. Inf. Sci. **179**(16), 2774–2793 (2009)
19. Li, S., Hong, Z., Li, T.: Efficient composing rough approximations for distributed data. Knowl.-Based Syst. **182** (2019)
20. Gao, Y., Chen, X., Yang, X., Wang, P.: Neighborhood attribute reduction: A multicriterion strategy based on sample selection. Information **9**, 282 (2018). https://doi.org/10.3390/info9110282
21. Ballestar, M.T., Grau-Carles, P., Sainz, J.: Customer segmentation in e-commerce: Applications to the cashback business model. J. Bus. Res. **88**, 407–414 (2018)
22. Shalamanov, V.: Organizing for IT effectiveness, efficiency and cyber resilience in the academic sector: National and regional dimensions. Inform. Secur. Int. J. **42**, 49–66 (2019)
23. Borissova, D., Mustakerov, I. Korsemov D.: Business intelligence system via group decision making. Cybern. and Inform. Technol. **16**(3), 219–229 (2016)
24. Zaheer, H., Breyer, Y., Dumay, J.: Digital entrepreneurship: An interdisciplinary structured literature review and research agenda. Technol. Forecast. Soc. Chang. **148**, 119735 (2019). https://doi.org/10.1016/j.techfore.2019.119735

25. Andreev, R., Borissova, D., Shikalanov, A., Yorgova, T.: Chapter: Model-driven design of eMedia: Virtual technology transfer office, Book Title: Information Systems and Management in Media and Entertainment Industries, pp. 279–298. Springer (2016)
26. Young, K.S., Nabuco de Abreu, C. (eds.): Internet Addiction: A Handbook and Guide to Evaluation and Treatment Hoboken. John Wiley & Sons Inc., New york (2010)
27. Salih, Al-H. A., Saeed. A.T., Saber, Z. R.: A study on the effects of cellular mobile networks on people in Tikrit city based on power density measurements and calculations. Eng. Technol. Appl. Sci. Res. **9**(3), 4265–4270 (2019)
28. Kim, J.H., Lee, J.-K., Kim, H.-G., Kim, K.-B., Kim, H.R.: Possible effects of radiofrequency electromagnetic field exposure on central nerve system. Biomol. Ther. **27**(3), 265–275 (2019)
29. Markov, M. (ed.): Mobile Communications and Public Health. Taylor & Francis Group, Boca Raton (2019)
30. Garvanova, M., Garvanov, I., Borissova, D.: Influence of electromagnetic fields on human brain. In: 21st International Symposium on Electrical Apparatus and Technologies – SIELA 2020, pp. 1–4. Bourgas, Bulgaria (2020)
31. Garvanova, M., Shishkov, B., Vladimirov, S.: Mobile devices – effect on human health. In: Proceedings of 7th International Conference on Telecommunications and Remote Sensing, pp. 101–104, Barcelona, Spain (2018)
32. Schmoldt, D., Kangas, J., Mendoza, G.A., Pesonen, M. (eds.): The analytic hierarchy process in natural resource and environmental decision making. Managing For. Ecosyst. **3**, 15–35 (2001)

# Algorithms for Generating Sets of Gambles for Decision Making with Lower Previsions

Nawapon Nakharutai(✉)

Data Science Research Center, Department of Statistics, Faculty of Science, Chiang Mai University, Chiang Mai 50200, Thailand
nawapon.nakharutai@cmu.ac.th

**Abstract.** $\Gamma$-maximin, $\Gamma$-maximax, maximality and interval dominance are well-known criteria for decision making using lower previsions when precise probabilities are not available. This study proposes algorithms for generating a set of gambles that has a precise number of $\Gamma$-maximin (or $\Gamma$-maximax) gambles that can be used to generate random decision problems for benchmarking algorithms for finding $\Gamma$-maximin (or $\Gamma$-maximax) gambles. Since $\Gamma$-maximin and $\Gamma$-maximax imply maximality and interval dominance, the algorithms can also be used as an alternative algorithm for generating random decision problems with predetermined numbers of maximal and interval dominant gambles.

**Keywords:** Algorithm · Decision criteria · Lower prevision

## 1 Introduction

Consider a decision problem where we must choose a single option from a set of available options. Logically, we normally select the one that results in the best possible reward, i.e., an *optimal* option, according to some criterion. Suppose that each option induces an uncertain reward depending on the option and on the state of nature, for example, a lottery ticket. It is common to assume that rewards can be expressed on a utility scale. Thus, each option can be viewed as a bounded real-valued function on the set of states of nature which is called *gamble* [1]. In this way, we have framed a decision problem as choosing a gamble from a set of possible gambles with respect to some criteria.

If precise probabilities to all related events are given, then a simple criterion for choosing an optimal gamble is to maximise expected utility [2]. Whenever we cannot assign these precise probabilities due to incomplete information or disagreement between experts for providing probability of an event, it is recommended to handle this situation by using *lower previsions* which amount to probability bounds [3]. To evaluate the value of the lower prevision, one can solve

Supported by Data Science Research Center, Department of Statistics, Faculty of Science, Chiang Mai University, Chiang Mai 50200, Thailand.


V.-N. Huynh et al. (Eds.): IUKM 2020, LNAI 12482, pp. 62–71, 2020.
https://doi.org/10.1007/978-3-030-62509-2_6

linear programs as given in [4] and efficient algorithms for solving such linear programs can be found in [5,6].

$\Gamma$-maximin, $\Gamma$-maximax, maximality and interval dominance are four popular decision criteria associated with lower previsions [7]. To verify whether a gamble is $\Gamma$-maximin (or $\Gamma$-maximax) or not, the number of lower previsions that we have to evaluate is equal to the cardinality of the set of gambles. To determine whether a gamble is interval dominant, we have to evaluate the number of lower previsions for each gamble about twice of the cardinality of the set of gambles [4]. To check whether a gamble is maximal, we have to evaluate the number of lower previsions for each decision by solving either a single large linear program [9] or a sequence of smaller linear programs [4]. There are many proposed and developed algorithms for decision making with these criteria, for example, see [8–11].

Among these studies, there is an algorithm in [11] (i.e. Algorithm 6) for generating sets of gambles that have a precise number of maximal and interval dominant gambles. That algorithm can be used to generate decision problems for benchmarking algorithms for finding maximal and interval dominant gambles. However, as far as we know, there is no algorithm for generating sets of gambles that has a precise number of $\Gamma$-maximin or $\Gamma$-maximax gambles yet. Therefore, based on Algorithm 6 in [11], we would like to propose algorithms to generate sets of gambles that have a precise number of $\Gamma$-maximin or $\Gamma$-maximax gambles which will be useful for anyone who would like to benchmark their algorithms for $\Gamma$-maximin or $\Gamma$-maximax. Since both $\Gamma$-maximin and $\Gamma$-maximax imply maximality and interval dominance, the proposed algorithms can be used as an alternative algorithm for generating sets of gambles with pre-determined number of maximal and interval dominant gambles.

Also note that Algorithm 6 in [11] has severe computational issue because of the need to evaluate the values of a large number of lower previsions. In that case, the size of decision problems was limited. In fact, we might be able to extend the size of decision problems if we can reduce the number of lower previsions that we have to evaluate in that algorithm. We will show that the number of lower previsions we have to evaluate in these proposed algorithms is less than the one needed to be evaluated in Algorithm 6 in [11]. Therefore, combining the proposed algorithms, we could modify Algorithm 6 in [11] to reduce the number of lower previsions that needed to be evaluated for generating decision problems for benchmarking algorithms for maximality and interval dominance.

We structure this study as follows. In Sect. 2, we briefly review lower previsions and state four decision criteria. In Sect. 3, we propose algorithms for generating set of gambles with pre-specified number of gambles that are $\Gamma$-maximin and $\Gamma$-maximax. We also show how to combine these two algorithms for modifying Algorithm 6 in [11] to generate pre-determined number of gambles in the set that are maximal and interval dominant. Sect. 4 concludes this study.

## 2 Preliminaries

In this section, we briefly review the necessary preliminary definitions and notations used in this work.

### 2.1 Lower Previsions

Consider a finite possibility space $\Omega$, which indicates the set of all possible outcomes. A gamble $f$ on $\Omega$ is a bounded real-valued function which can be represented as an uncertain reward $f(\omega)$ if the true outcome is $\omega \in \Omega$. The set of all gambles on $\Omega$ is denoted by $\mathcal{L}$.

Following [1] and [3], we suppose that gambles can be modelled by a real-valued function $\underline{P}$ defined on dom$\underline{P}$, which is a subset of $\mathcal{L}$. Let $f \in$ dom$\underline{P}$, $\underline{P}(f)$ can be interpreted as the subject's supremum acceptable buying price for $f$, namely, for all $\alpha < \underline{P}(f)$, the gamble $f - \alpha$ is acceptable to the subject.

A basic condition for $\underline{P}$ to be consistent is given as follows.

**Definition 1.** *A lower prevision $\underline{P}$* avoids sure loss *if for all $n \in \mathbb{N}$, all non-negative $\lambda_1, \dots, \lambda_n \in \mathbb{R}$, all $f_1, \dots, f_n \in$ dom$\underline{P}$ and the following condition holds [12]:*

$$\max_{\omega \in \Omega} \left( \sum_{i=1}^{n} \lambda_i \left[ f_i(\omega) - \underline{P}(f_i) \right] \right) \geq 0. \tag{1}$$

Throughout this study, we assume that all $\underline{P}$ avoid sure loss. Otherwise, we can find that there are $n > 0$, $f_1, \dots, f_n \in$ dom$\underline{P}$ and non-negative $\lambda_1, \dots, \lambda_n \in \mathbb{R}$ such that

$$\max_{\omega \in \Omega} \left( \sum_{i=1}^{n} \lambda_i f_i(\omega) \right) < \sum_{i=1}^{n} \lambda_i \underline{P}(f_i). \tag{2}$$

In this case $\underline{P}$ does not avoid sure loss, meaning that the subject is willing to pay more than the maximum possible payoff which make no sense [12].

Given a lower prevision $\underline{P}$, we can define the so-called the *conjugate upper prevision* $\overline{P}$ on $-$dom $\underline{P} := \{-f : f \in \text{dom } \underline{P}\}$ by $\overline{P}(f) := -\underline{P}(-f)$ [12]. This can be seen as the subject's infimum acceptable selling price for $f$. In addition, to extend $\underline{P}$ to the set of all gambles $\mathcal{L}$, we can do this via its *natural extension* $\underline{E}$ [12]:

**Definition 2.** *The natural extension of $\underline{P}$ is given, for all $g \in \mathcal{L}$, by:*

$$\underline{E}_{\underline{P}}(g) := \sup \left\{ \alpha \in \mathbb{R} \colon g - \alpha \geq \sum_{i=1}^{n} \lambda_i (f_i - \underline{P}(f_i)), n \in \mathbb{N},\ f_i \in \text{dom } \underline{P},\ \lambda_i \geq 0 \right\}. \tag{3}$$

The natural extension $\underline{E}_{\underline{P}}(g)$ can be seen as the subject's supremum acceptable buying price for $g \in \mathcal{L}$ where prices $\underline{P}(f_i)$ for all $f_i \in$ dom $\underline{P}$ are taken into account. Note that $\underline{P}$ avoids sure loss if and only if $\underline{E}_{\underline{P}}$ is finite, and hence, $\underline{E}_{\underline{P}}$ is a lower prevision [12].

Likewise, the conjugate upper prevision $\overline{E}_{\underline{P}}$ is defined as [12]:

$$\overline{E}_{\underline{P}}(g) := -\underline{E}_{\underline{P}}(-g) \tag{4}$$

$$= \inf \left\{ \beta \in \mathbb{R}\colon \beta - g \geq \sum_{i=1}^{n} \lambda_i (f_i - \underline{P}(f_i)), n \in \mathbb{N}, f_i \in \text{dom } \underline{P}, \lambda_i \geq 0 \right\}. \tag{5}$$

Similarly, this can be viewed as the subject's infimum acceptable selling price for a gamble $g \in \mathcal{L}$ based on prices $\underline{P}(f_i)$ for all $f_i \in \text{dom } \underline{P}$ [12].

In this study, where there is no confusion, we simply write $\underline{E}_{\underline{P}}$ and $\overline{E}_{\underline{P}}$ as $\underline{E}$ and $\overline{E}$ respectively. Moreover, we assume that both $\Omega$ and dom $\underline{P}$ are finite so that lower previsions can be evaluated through solving linear programs given in [4–6].

## 2.2 Decision Criteria

We will review four decision criteria with respect to lower previsions: $\Gamma$-maximin, $\Gamma$-maximax, maximality and interval dominance. More detail and discussion of these criteria presented here can be found in [7].

Given a set of gambles $\mathcal{K}$, the first criterion simply selects gambles that maximise the lower natural extension [7]:

**Definition 3.** *The set of $\Gamma$-maximin gambles of $\mathcal{K}$ with respect to a natural extension $\underline{E}$ is*

$$\text{opt}_{\underline{E}}(\mathcal{K}) := \{f \in \mathcal{K} : \underline{E}(f) \geq \max_{g \in \mathcal{K}} \underline{E}(g)\}. \tag{6}$$

Along with the $\Gamma$-maximin, we also have another criterion that selects gambles that maximise the upper natural extension in $\mathcal{K}$ [7]:

**Definition 4.** *The set of $\Gamma$-maximax gambles of $\mathcal{K}$ with respect to a natural extension $\overline{E}$ is*

$$\text{opt}_{\overline{E}}(\mathcal{K}) := \{f \in \mathcal{K} : \overline{E}(f) \geq \max_{g \in \mathcal{K}} \overline{E}(g)\}. \tag{7}$$

Based on the strict partial preference orders [7], we have the two following criteria.

**Definition 5.** *The set of maximal gambles of $\mathcal{K}$ with respect to the given partial order*

$$f \succ g \text{ if } \underline{E}(f - g) > 0 \tag{8}$$

*is*

$$\text{opt}_{\succ}(\mathcal{K}) := \{f \in \mathcal{K}\colon (\forall g \in \mathcal{K})(g \not\succ f)\} \tag{9}$$

$$= \{f \in \mathcal{K} : (\forall g \in \mathcal{K})(\overline{E}(f - g) \geq 0)\}. \tag{10}$$

**Definition 6.** *The set of interval dominant gambles of* $\mathcal{K}$ *with respect to the given partial order*

$$f \sqsupset g \text{ if } \underline{E}(f) > \overline{E}(g) \tag{11}$$

*is*

$$\text{opt}_{\sqsupset}(\mathcal{K}) := \{f \in \mathcal{K}\colon (\forall g \in \mathcal{K})(g \not\sqsupset f)\} \tag{12}$$

$$= \{f \in \mathcal{K} : \overline{E}(f) \geq \max_{g \in \mathcal{K}} \underline{E}(g)\}. \tag{13}$$

Note that [7]:

$$\text{opt}_{\underline{E}}(\mathcal{K}) \cup \text{opt}_{\overline{E}}(\mathcal{K}) \subseteq \text{opt}_{\succ}(\mathcal{K}) \subseteq \text{opt}_{\sqsupset}(\mathcal{K}). \tag{14}$$

Now, we have seen four decision criteria with lower previsions. In the next section, we will discuss algorithms for generating random decision problems having a precise number of gambles that satisfy decision criteria as we want.

## 3 Algorithms for Generating Sets of Gambles

Suppose that we want to generate a set of $k$ gambles $\mathcal{K}$ such that exact $m$ gambles are $\Gamma$-maximin, i.e. $|\text{opt}_{\underline{E}}(\mathcal{K})| = m \leq k$. A straightforward idea to generate a set of gambles with a precise number of $\Gamma$-maximin gambles is to generate first $m$ $\Gamma$-maximin gambles and then generate $k - m$ non $\Gamma$-maximin gambles. To do so, we can start by generating $\mathcal{K} = \{f\}$, where $f$ is immediately $\Gamma$-maximin. Next, we generate a gamble $h$ for which

$$\text{opt}_{\underline{E}}(\mathcal{K} \cup \{h\}) = \text{opt}_{\underline{E}}(\mathcal{K}) \cup \{h\}, \tag{15}$$

and then we add $h$ to $\mathcal{K}$. We repeat this process until we have $|\text{opt}_{\underline{E}}(\mathcal{K})| = m \leq k$. After that we generate a gamble $h$ for which

$$\text{opt}_{\underline{E}}(\mathcal{K} \cup \{h\}) = \text{opt}_{\underline{E}}(\mathcal{K}), \tag{16}$$

and then we add $h$ to $\mathcal{K}$. We repeat this process until we have $|\mathcal{K}| = k$.

However, it may not be easy to randomly generate such gamble $h$ that satisfies Eq. (15) or Eq. (16). Therefore, we might have to sample many gambles until we obtain such set of gambles $\mathcal{K}$ that we desire.

Fortunately, after generating gamble $h$, it can be modified by shifting $h$ for some $\alpha \in \mathbb{R}$ to make a new gamble $h - \alpha$ satisfies either Eq. (15) or Eq. (16) as we want. The following theorem gives values of $\alpha$ for modifying $h$:

**Theorem 1.** *Let* $\mathcal{K}$ *be a set of gambles where a gamble* $f$ *in* $\mathcal{K}$ *is* $\Gamma$*-maximin. Let* $h$ *be another gamble and* $\alpha \in \mathbb{R}$*. Then*

1. $h - \alpha$ *and* $f$ *are both* $\Gamma$*-maximin if* $\alpha = \underline{E}(h) - \underline{E}(f)$*.*
2. $h - \alpha$ *is* $\Gamma$*-maximin but not* $f$ *if* $\alpha < \underline{E}(h) - \underline{E}(f)$*.*
3. $h - \alpha$ *is not* $\Gamma$*-maximin but* $f$ *is if* $\alpha > \underline{E}(h) - \underline{E}(f)$*.*

*Proof.* We know that $\underline{E}(f) = \max_{g\in\mathcal{K}} \underline{E}(g)$ since $f$ is $\Gamma$-maximin in $\mathcal{K}$. It is straightforward from the definition of $\Gamma$-maximin that

$$h - \alpha \text{ is also } \Gamma\text{-maximin in } \mathcal{K} \iff \underline{E}(h-\alpha) = \max_{g\in\mathcal{K}} \underline{E}(g) \quad (17)$$

$$\iff \underline{E}(h-\alpha) = \underline{E}(f) \quad (18)$$

$$\iff \alpha = \underline{E}(h) - \underline{E}(f). \quad (19)$$

Therefore, if $\alpha < \underline{E}(h) - \underline{E}(f)$, then $\underline{E}(f) < \underline{E}(h-\alpha)$. In this case, $f$ is no longer $\Gamma$-maximin and $h - \alpha$ is $\Gamma$-maximin in $\mathcal{K}$ as $\underline{E}(h-\alpha) = \max_{g\in\mathcal{K}} \underline{E}(g)$. On the other hand, if $\alpha > \underline{E}(h) - \underline{E}(f)$, then $f$ is still $\Gamma$-maximin while $h - \alpha$ is not $\Gamma$-maximin.

Using Theorem 1, we present an algorithm for generating a set of $k$ gambles $\mathcal{K}$ that has a precisely specified number of $m$ $\Gamma$-maximin gambles. The algorithm starts by generating $m$ $\Gamma$-maximin gambles and then adds $k - m$ gambles that are not $\Gamma$-maximin (see Algorithm 1).

**Algorithm 1.** Generate a set of $k$ gambles $\mathcal{K}$ such that $|\mathrm{opt}_{\underline{E}}(\mathcal{K})| = m \leq k$

**Output:** a set of $k$ gambles $\mathcal{K}$ such that exactly $m$ gambles are $\Gamma$-maximin.
Stage 1 Generate $\mathcal{K} = \{h_1\}$; ▷ $h_1$ is immediately $\Gamma$-maximin
Stage 2 Generating $m - 1$ $\Gamma$-maximin gambles
For $i = 2 : m$ do
2.1 Generate $h_i$
2.2 Choose $\alpha = \underline{E}(h_i) - \underline{E}(h_1)$
2.3 $\mathcal{K} \longleftarrow \mathcal{K} \cup \{h_i - \alpha\}$
Stage 3 Generating $k - m$ not $\Gamma$-maximin gambles
For $i = m + 1 : k$ do
3.1 Generate $h_i$
3.2 Choose $\alpha > \underline{E}(h_i) - \underline{E}(h_1)$
3.3 $\mathcal{K} \longleftarrow \mathcal{K} \cup \{h_i - \alpha\}$

Note that Algorithm 1 only needs to evaluate $k$ lower previsions, i.e. for each of gamble $h_i$. In addition, if we want those generated non-$\Gamma$-maximin in Stage 3 to be difficult to detect, then we can choose $\alpha$ to be slightly larger than $\underline{E}(h_i) - \underline{E}(h_1)$ for each $i = m + 1 : k$. For example, we can set $\alpha = \underline{E}(h_i) - \underline{E}(h_1) + \epsilon$, where $\epsilon$ is sampled from $(0, 1)$.

Let's see an example that illustrates Algorithm 1.

*Example 1.* We would like to generate a set of four gambles $\mathcal{K}$ that has two $\Gamma$-maximin gambles, where $\Omega = \{\omega_1, \omega_2, \omega_3\}$. To do so, for each $\omega_j$ and $i = 1, 2, 3, 4$, we first sample $h_i(\omega_j)$ uniformly from $(0, 1)$. Suppose that we can compute each $\underline{E}(h_i)$ and the result is displayed in Table 1:

We assign $\mathcal{K} = \{h_1\}$, where $h_1$ is immediately $\Gamma$-maximin. Next, for $h_2$, we choose $\alpha$ in stage 2.2 to be $\alpha_2 = \underline{E}(h_2) - \underline{E}(h_1) = 0.2243$, and obtain a gamble

**Table 1.** Generated gambles and their natural extensions.

| Gambles | Outcomes | | | $\underline{E}(h_i)$ |
|---|---|---|---|---|
| | $\omega_1$ | $\omega_2$ | $\omega_3$ | |
| $h_1$ | 0.2417 | 0.4039 | 0.0965 | 0.1373 |
| $h_2$ | 0.1320 | 0.9421 | 0.9561 | 0.3616 |
| $h_3$ | 0.5752 | 0.0598 | 0.2348 | 0.1279 |
| $h_4$ | 0.8212 | 0.0154 | 0.0430 | 0.0395 |

| Outcomes | $\omega_1$ | $\omega_2$ | $\omega_3$ |
|---|---|---|---|
| $h_2 - \alpha_2$ | −0.0924 | 0.7177 | 0.7318 |

$h_2 - \alpha_2$ which is another $\Gamma$-maximin gamble: Next, we construct two non-$\Gamma$-maximin gambles. For each $h_3$ and $h_4$, we find that $\underline{E}(h_3) - \underline{E}(h_1) = -0.0094$ and $\underline{E}(h_4) - \underline{E}(h_1) = -0.0978$. So, we can randomly choose a corresponding $\alpha$ in stage 3.2 to be $\alpha_3 = 0.1672 > -0.094$ and $\alpha_4 = -0.0133 > -0.0978$ respectively, and obtain $h_3 - \alpha_3$ and $h_4 - \alpha_4$ that are not $\Gamma$-maximin as follows:

| Outcomes | $\omega_1$ | $\omega_2$ | $\omega_3$ |
|---|---|---|---|
| $h_3 - \alpha_3$ | 0.4080 | −0.1074 | 0.0676 |
| $h_4 - \alpha_4$ | 0.8345 | 0.0287 | 0.0563 |

Eventually, we generate a set of four gambles $\mathcal{K} = \{h_1, h_2 - \alpha_2, h_3 - \alpha_3, h_4 - \alpha_4\}$ such that there are exactly two $\Gamma$-maximin gambles.

To generate a set of $k$ gambles $\mathcal{K}$ that has $m$ $\Gamma$-maximax gambles, i.e. $|\mathrm{opt}_{\overline{E}}(\mathcal{K})| = m \leq k$, this strategy can be applied as well. To do so, we first generate a gamble $\mathcal{K} = \{f\}$, where $f$ will be immediately $\Gamma$-maximax. Next, we generate a gamble $h$ and then $h$ is shifted by $\beta$ for some $\beta \in \mathbb{R}$ so that all $\Gamma$-maximax gambles in $\mathcal{K}$ and $h - \beta$ are $\Gamma$-maximax. We repeat this process until we obtain $m$ $\Gamma$-maximax gambles in $\mathcal{K}$. After that, we generate $k - m$ gambles that are not $\Gamma$-maximax. Specifically, we generate a gamble $h$ and then $h$ is shifted by $\beta$ for some $\beta \in \mathbb{R}$ so that all $m$ $\Gamma$-maximax gambles are still $\Gamma$-maximax but $h - \beta$ is not $\Gamma$-maximax. We repeat this process until we obtain $k - m$ gambles that are not $\Gamma$-maximax. Eventually, we end up with a set of $k$ gambles $\mathcal{K}$ such that $|\mathrm{opt}_{\overline{E}}(\mathcal{K})| = m \leq k$.

Ranges of $\beta$ to modify $h$ are given as follows:

**Corollary 1.** *Let $\mathcal{K}$ be a set of gambles where a gamble $f$ in $\mathcal{K}$ is $\Gamma$-maximax. Let $h$ be another gamble and $\beta \in \mathbb{R}$. Then*

1. *$h - \beta$ and $f$ are both $\Gamma$-maximax if $\beta = \overline{E}(h) - \overline{E}(f)$.*
2. *$h - \beta$ is $\Gamma$-maximax but not $f$ if $\beta < \overline{E}(h) - \overline{E}(f)$.*
3. *$h - \beta$ is not $\Gamma$-maximax but $f$ is if $\beta > \overline{E}(h) - \overline{E}(f)$.*

*Proof.* The proof is similar to the proof of Theorem 1, therefore the proof is omitted.

Similarly, using Corollary 1, we present an algorithm for generating a set of $k$ gambles $\mathcal{K}$ with a number of $m$ gambles that are $\Gamma$-maximax. The algorithm begins by generating $m$ $\Gamma$-maximax gambles and then adding $k-m$ gambles that are not $\Gamma$-maximax (see Algorithm 2).

**Algorithm 2.** Generate a set of $k$ gambles $\mathcal{K}$ such that $|\mathrm{opt}_{\overline{E}}(\mathcal{K})| = m \leq k$

**Output:** a set of $k$ gambles $\mathcal{K}$ such that exactly $m$ gambles are $\Gamma$-maximax.
Stage 1 Generate $\mathcal{K} = \{h_1\}$; ▷ $h_1$ is immediately $\Gamma$-maximax
Stage 2 Generating $m-1$ $\Gamma$-maximax gambles
For $i = 2 : m$ do
2.1 Generate $h_i$
2.2 Choose $\beta = \overline{E}(h_i) - \overline{E}(h_1)$
2.3 $\mathcal{K} \longleftarrow \mathcal{K} \cup \{h_i - \beta\}$
Stage 3 Generating $k-m$ not $\Gamma$-maximax gambles
For $i = m+1 : k$ do
3.1 Generate $h_i$
3.2 Choose $\beta > \overline{E}(h_i) - \overline{E}(h_1)$
3.3 $\mathcal{K} \longleftarrow \mathcal{K} \cup \{h_i - \beta\}$

Similar to Algorithm 1, we only need to evaluate $k$ upper previsions in Algorithm 2. Again, if we want those generated not $\Gamma$-maximax gambles in Stage 3 not to be easy to detect, we can set $\beta$ to be slightly larger than $\overline{E}(h_i) - \overline{E}(h_1)$ for each $i = m+1 : k$. For example, we can let $\beta = \overline{E}(h_i) - \overline{E}(h_1) + \epsilon$, where $\epsilon$ is sampled from $(0, 1)$.

Recall that both $\Gamma$-maximin and $\Gamma$-maximax gambles are maximal and interval dominant. So, both Algorithms 1 and 2 can be used to generate sets of gambles that have a precisely given number of maximal and interval dominant gambles as well. Since $\Gamma$-maximin and $\Gamma$-maximax are stronger criteria than maximality and interval dominance, readers must be aware when using Algorithms 1 and 2 to generate decision problems for benchmarking algorithms for finding maximal and interval dominant gambles.

An advantage of using Algorithms 1 and 2 to generate random decision problems for benchmarking algorithms with lower previsions is that these two algorithms need to evaluate only one lower prevision per gamble for choosing $\alpha$ and $\beta$, while Algorithm 6 in [11] has to evaluate a sequence of lower previsions per gambles in order to choose $\alpha$. Therefore, Algorithms 1 and 2 could be applied to reduce an effort to generate maximal and interval dominant gambles by Algorithm 6 in [11]. Specifically, suppose that we would like to generate a set of $k$ gambles $\mathcal{K}$ that has $m$ maximal gambles and $n$ interval dominant gambles with $\ell$ gambles are $\Gamma$-maximin (or $\Gamma$-maximax), where $\ell \leq m \leq n \leq k$. We can use Algorithm 1 (or 2) to generate $\ell$ gambles that are $\Gamma$-maximin (or $\Gamma$-maximax)

**Algorithm 3.** Generate a set of $k$ gambles $\mathcal{K}$ such that $|\mathrm{opt}_{\underline{E}}(\mathcal{K})| = \ell$ (or $|\mathrm{opt}_{\overline{E}}(\mathcal{K})| = \ell$), $|\mathrm{opt}_{\succ}(\mathcal{K})| = m$ and $|\mathrm{opt}_{\sqsupset}(\mathcal{K})| = n$ where $\ell \le m \le n \le k$.

**Input:** Numbers $\ell$, $m$, $n$, $k$ where $\ell \le m \le n \le k$.
**Output:** a set of $k$ gambles $\mathcal{K}$ such that $|\mathrm{opt}_{\underline{E}}(\mathcal{K})| = \ell$ (or $|\mathrm{opt}_{\overline{E}}(\mathcal{K})| = \ell$), $|\mathrm{opt}_{\succ}(\mathcal{K})| = m$ and $|\mathrm{opt}_{\sqsupset}(\mathcal{K})| = n$ where $\ell \le m \le n \le k$.

Stage 1 Generate $h_1$ and set $\mathcal{K} = \{h_1\}$; ▷ $h_1$ is immediately $\Gamma$-maximin (or $\Gamma$-maximax)
Stage 2 Generating $\ell - 1$ $\Gamma$-maximin (or $\Gamma$-maximax) gambles
For $i = 2 : \ell$ do
2.1 Generate $h_i$
2.2 Choose $\alpha = \underline{E}(h_i) - \underline{E}(h_1)$ ▷ to generate $\Gamma$-maximax gambles, replace $\alpha$ here by $\beta$ as chosen as in Algorithm 2
2.3 $\mathcal{K} \longleftarrow \mathcal{K} \cup \{h_i - \alpha\}$
Stage 3 Generate a sequence of $k - \ell$ gambles $h_{\ell+1}, \ldots, h_k$ such that $\overline{E}(h_i - h_j) < \overline{E}(h_i) - \underline{E}(h_j)$ for all $i, j \in \{\ell + 1, \ldots, k\}$
Stage 4 Generating $m - \ell$ maximal gambles
For $i = \ell + 1 : m$ do
4.1 Choose $\alpha$ such that $\max_{f \in \mathrm{opt}_{\succ}(\mathcal{K})} \underline{E}(h_i - f) \le \alpha \le \min_{f \in \mathrm{opt}_{\succ}(\mathcal{K})} \overline{E}(h_i - f)$
4.2 $\mathcal{K} \leftarrow \mathcal{K} \cup \{h_i - \alpha\}$
Stage 5 Generating $n - m$ interval dominant gambles
For $i = m + 1 : n$ do
5.1 Choose $\alpha$ such that $\min_{f \in \mathrm{opt}_{\succ}(\mathcal{K})} \overline{E}(h_i - f) < \alpha \le \overline{E}(h_i) - \max_{f \in \mathcal{K}} \underline{E}(f)$
5.2 $\mathcal{K} \leftarrow \mathcal{K} \cup \{h_i - \alpha\}$
Stage 6 Generating $k - n$ not interval dominant gambles
For $i = n + 1 : k$ do
6.1 Choose $\alpha$ such that $\alpha > \overline{E}(h_i) - \max_{f \in \mathcal{K}} \underline{E}(f)$
5.2 $\mathcal{K} \leftarrow \mathcal{K} \cup \{h_i - \alpha\}$

at the beginning and then we can generate the rest of gambles using Algorithm 6 in [11]. This approach is summarised in Algorithm 3.
Note that the condition in Stage 3 ensures that we can find $\alpha$ that satisfies Stage 5.1. and readers can also find more detail and discussion of the range of values $\alpha$ in Stages 4–6 in [11].

# 4 Conclusion

In conclusion, we proposed two algorithms (Algorithms 1 and 2) for generating random sets of gambles that have precise numbers of $\Gamma$-maximin gambles and $\Gamma$-maximax gambles, respectively. These proposed algorithms can also be useful to generate random sets of gambles that have an exact number of maximal and interval dominant gambles as both $\Gamma$-maximin and $\Gamma$-maximax imply maximality and interval dominance. This approach brings the number of evaluating lower previsions in Algorithms 3 to be less than the one in Algorithm 6 in [11]. Our proposed algorithms can be beneficial in generating randomly decision problems for the study of benchmarking algorithms with lower previsions.

**Acknowledgement.** This research is supported by Data Science Research Center, Department of Statistics, Faculty of Science, Chiang Mai University, Chiang Mai 50200, Thailand. The author would like to thank Prof. Matthias Troffaes and Assoc. Prof. Camila C.S. Caiado for their helps and suggestions.

## References

1. Walley, P.: Statistical Reasoning with Imprecise Probabilities. Chapman and Hall, London (1991)
2. Anscombe, F.J., Aumann, R.J.: A definition of subjective probability. Ann. Math. Stat. **34**, 199–205 (1963)
3. Williams, P.M.: Notes on conditional previsions. Int. J. Approximate Reasoning **44**(3), 366–383 (2007). https://doi.org/10.1016/j.ijar.2006.07.019
4. Troffaes, M.C.M., Hable, P.: Introduction to Imprecise Probabilities: Computation, pp. 329–337. Wiley (2014). https://doi.org/10.1002/9781118763117.ch16
5. Nakharutai, N., Troffaes, M.C.M., Caiado, C.: Efficient algorithms for checking avoiding sure loss. Proc. Mach. Learn. Res. **62**, 241–252 (2017)
6. Nakharutai, N., Troffaes, M.C.M., Caiado, C.: Improved linear programming methods for checking avoiding sure loss. Int. J. Approximate Reasoning **101**, 293–310 (2018). https://doi.org/10.1016/j.ijar.2018.07.013
7. Troffaes, M.C.M.: Decision making under uncertainty using imprecise probabilities. Int. J. Approximate Reasoning **45**, 17–29 (2007). https://doi.org/10.1016/j.ijar.2006.06.001
8. Kikuti, D., Cozman, F.G., Filho, R.S.: Sequential decision making with partially ordered preferences. Artif. Intell. **175**(7), 1346–1365 (2011). https://doi.org/10.1016/j.artint.2010.11.017
9. Jansen, C., Augustin, T., Schollmeyer, G.: Decision theory meets linear optimization beyond computation. In: Antonucci, A., Cholvy, L., Papini, O. (eds.) ECSQARU 2017. LNCS (LNAI), vol. 10369, pp. 329–339. Springer, Cham (2017). https://doi.org/10.1007/978-3-319-61581-3_30
10. Huntley, N., Hable, P., Troffaes, M.C.M.: Introduction to Imprecise Probabilities: Decision Making, pp. 190–206. Wiley (2014). https://doi.org/10.1002/9781118763117.ch8
11. Nakharutai, N., Troffaes, M.C.M., Caiado, C.: Improving and benchmarking of algorithms for decision making with lower previsions. Int. J. Approximate Reasoning **113**, 91–105 (2019). https://doi.org/10.1016/j.ijar.2019.06.008
12. Troffaes, M.C.M., de Cooman, G.: Lower Previsions: Wiley Series in Probability and Statistics. Wiley (2014). https://doi.org/10.1002/9781118762622

# PM 2.5 Problem in Chiang Mai, Thailand: The Application of Maximizing Expected Utility with Imbalanced Loss Functions

Sirapat Watakajaturaphon[1] and Parkpoom Phetpradap[1,2](✉)

[1] Department of Mathematics, Faculty of Science, Chiang Mai University, Chiang Mai 50200, Thailand
{sirapat_w,parkpoom.phetpradap}@cmu.ac.th
[2] Research Center in Mathematics and Applied Mathematics, Department of Mathematics, Faculty of Science, Chiang Mai University, Chiang Mai 50200, Thailand

**Abstract.** Over the past decade, PM 2.5 (particulate matters with diameters 2.5 μ or smaller) pollution has become a severe problem in Chiang Mai, Thailand. The problem occurs during the dry season from January to May. Undoubtedly, an efficient prediction model will significantly improve public safety and mitigate damage caused. Nonetheless, particular groups of people, especially ones who are vulnerable to the pollution, may prefer the prediction to be over-predicted rather than under-predicted. The aim of this research is to provide PM 2.5 density prediction models based on individual's preference. This will overcome the limit of classical prediction models where the over-prediction and under-prediction ratio are symmetric. The predictions are done via the maximizing expected utility technique with imbalanced loss functions. The study area is Chiang Mai province, Thailand. The study period is the dry season (January to May) from 2016 to 2018. The hourly data is provided by the Pollution Control Department, Ministry of Natural Resource and Environment, Thailand. The study results show that the predictions based on the maximizing expected utility technique with imbalanced loss functions improves the over prediction ratio of the prediction.

**Keywords:** Maximizing expected utility · Imbalanced loss function · PM 2.5 · Chiang Mai · Decision makings

## 1 Introduction

PM 2.5 (particulate matters with diameters 2.5 μ or smaller) pollution becomes a severe concern in Chiang Mai province, Thailand over the past decade [1]. This problem occurs every year during the dry season, from January to May,

Supported by Chiang Mai university.


V.-N. Huynh et al. (Eds.): IUKM 2020, LNAI 12482, pp. 72–83, 2020.
https://doi.org/10.1007/978-3-030-62509-2_7

and generally reached its peak in March. According to local environmental data sources such as Pollution Control Department [1], Climate Change Data Centre of Chiang Mai University [2], and Smoke Haze Integrated Research Unit [3], the averages of hourly PM 2.5 density during the dry season of 2016 to 2018 were above the safe level of 35.5 $\mu g/m^3$, and reached the peak at 80 $\mu g/m^3$ in April 2016. The main emission source is biomass open burning, such as forest fires, solid waste burning, and agricultural residue field burning [4,5]. Because of a tiny size of PM 2.5 particles, they can get deep into lungs and bloodstreams easily. The biggest impact of PM 2.5 pollution on public health is understood to be from long-term exposure to PM 2.5. It increases the mortality risk, particularly from cardiovascular causes. Exposure to high concentration of PM2.5 in a short-term can also exacerbate lung and heart conditions, and significantly affect quality of life. Children, the elderly and those with predisposed respiratory and cardiovascular diseases are known to be more vulnerable to the health impacts from the air pollution [6].

In 2019, a new safety criterion has been introduced to Chiang Mai province [7]. The danger levels, based on PM 2.5 density from 0 to 500 $\mu g/m^3$, are classified into six classes; Level 1 to 6 (Low, Moderate, Unhealthy for sensitive groups, Unhealthy, Very unhealthy and Hazardous) when the densities (in $\mu g/m^3$) are 0–12, 13–35, 36–55, 56–150, 151–250 and 251–500 respectively.

Since this pollution problem has a significant effect on human health, local traveling industry, and the economy, these results in an increase number of PM 2.5 measurement stations, both private and public, in Chiang Mai. The real-time PM 2.5 density information can be obtained through websites and mobile applications [1,2,7,8].

Undoubtedly, an efficient prediction model will significantly improve public safety and mitigate damage caused. As the problem occurs seasonally, the atmospheric parameters play parts of the problem. The Goddard Earth Observing System Model Version 5 (GEOS-5) is currently one of the widely used pollution prediction models developed by NASA's research team. However, it is believed that air pollution models may be different for each region due to many several factors [9,10]. The regional-developed models include a logistic regression model [11], a geographic information system-based model [12] and a fuzzy soft model [13].

However, it should be emphasized that the preference to obtain the predicted information of each individual is different. Some specific groups of people, especially the ones that are vulnerable to the pollution, may prefer the prediction to be *over-predicted* (the prediction value is higher than the actual one) rather than *under-predicted* (the prediction value is lower than the actual one). It can be seen that classical prediction models, such as time series models, will not fulfill this preference. Typically, the over-prediction and under-prediction ratio of prediction models are symmetric.

The aim of this article is to provide one-hour PM 2.5 density prediction models based on individual's preference. Note that, utility is a function that represents a personal preference [14]. For the health concern, we assume that over-predicted model is preferred. This will overcome the limit of classical prediction models where the over-prediction and under-prediction ratio are symmetric. The prediction is done via the maximizing expected utility technique. Note that a prediction model is still required for calculations as benchmarks. Therefore, we also construct a multivariate autoregressive model of PM 2.5 density based on four historical atmospheric parameters: PM2.5 density, relative humidity, wind speed and temperature. The study area is Chiang Mai province, Thailand. The study period is focused to the dry season(January to May) from 2016 to 2018. The hourly data is provided by the Pollution Control Department, Ministry of Natural Resource and Environment, Thailand [1].

The paper is structured as follows: In Sect. 2, we give backgrounds on the maximizing expected utility technique with imbalanced loss function. In Sect. 3.1, we give a brief information on the study area. A multivariate autoregressive model of PM 2.5 density is constructed and the prediction models based on the maximizing expected utility technique are developed in Sect. 3.2 and Sect. 3.3 respectively. The study results and discussions are included. Finally, the conclusion is given in Sect. 4.

## 2 Preliminaries

### 2.1 Maximizing Expected Utility Technique

When we are in a situation that we have to make a decision, we usually do not choose a decision arbitrarily from all available choices. Indeed, we want to make a decision that performs best based on some criterion. Such a decision is called an *optimal decision*. The classical approach for an optimal decision is the decision whose has the maximal expected utility. This type of decision making is called *maximizing expected utility* technique.

**Prediction Based on Maximizing Expected Utility.** Consider a discrete time process $(X_m)_{m \geq 0}$. Suppose that processes have been observed up to time $n$, that is $(X_1, X_2, \ldots, X_n) = (x_1, x_2, \ldots, x_n)$. In short, we write this as $\mathbf{X}_n = \mathbf{x}_n$. Let $X_{n+1}$ be the value that we need to predict based on observations $\mathbf{x}_n$. Define $\mathcal{X}_{n+1}$ as a set of all possible values of $X_{n+1}$ and let $x \in \mathcal{X}_{n+1}$. Let $\hat{X}_{n+1}$ be the best estimated of $X_{n+1}$, that is, the point estimator of $X_{n+1}$. In this case, a utility function is defined in terms of a loss function $L$ which quantifies the consequences of estimation residuals (the difference between the estimated and the actual values). There are two most commonly used loss functions:

- Square error loss (SEL) function is the square of the residual defined by

$$L_{SE}(x, X_{n+1}) = -(x - X_{n+1})^2 \tag{1}$$

- Absolute error loss (AEL) function is the absolute of the residual defined by:

$$L_{AE}(x, X_{n+1}) = -|x - X_{n+1}| \tag{2}$$

Note that the loss function is negative and our aim is to have as small loss as possible. In a sense of decision making, the point estimator $\hat{X}_{n+1}$ is equivalent to the optimal decision. In other words, $\hat{X}_{n+1}$ is the member in $\mathcal{X}_{n+1}$ that maximizes the expected value of the (negative) loss function. Even though the actual value $X_{n+1}$ is unknown, it may be predicted based on the observations $\mathbf{x}_n$. Therefore, we may consider the posterior expected value of the loss function given by $\mathbf{x}_n$ instead. Hence, by maximizing expected utility technique, the point estimator $\hat{X}_{n+1}$ can be obtained by:

$$\hat{X}_{n+1} = \arg \max_{x \in \mathcal{X}_{n+1}} E\big(L(x, X_{n+1})|\mathbf{X}_n = \mathbf{x}_n\big) \tag{3}$$

## 2.2 Imbalanced Loss Functions

It can be seen from (3) that the predicted values depend on the choice of loss function. Note that both SEL and AEL are balanced (symmetric) loss functions. In other words, the loss from positive and negative residuals contribute the same loss. However, for the PM 2.5 pollution problem the over prediction is preferred. Therefore, to make the prediction based on this preference, we should set the loss function to be imbalanced. For example, the imbalanced SE loss function may be defined as

$$L_{SE,\alpha}(x, X_{n+1}) := \begin{cases} -(x - X_{n+1})^2, & \text{if } x \geq X_{n+1} \\ -\alpha(x - X_{n+1})^2, & \text{if } x < X_{n+1} \end{cases}$$

Similarly, the imbalanced AE loss function may be defined as

$$L_{A,\alpha}(x, X_{n+1}) := \begin{cases} -|x - X_{n+1}|, & \text{if } x \geq X_{n+1} \\ -\alpha|x - X_{n+1}|, & \text{if } x < X_{n+1} \end{cases}$$

We will use $L_\alpha$ as a general imbalanced loss function. If the over prediction is preferred, we should set $\alpha > 1$. On the other hand, if the under prediction is preferred, we should set $0 < \alpha < 1$.

In summary, an algorithm to estimate $\hat{X}_{n+1}$ based on maximizing expected utility with an imbalanced loss function is $L_\alpha$ as follows:

---

**Algorithm 1.** *Prediction based on imbalanced loss function*

1. Define the imbalanced loss function $L_\alpha(x, X_{n+1})$
2. Calculate $E\big(L_\alpha(x, X_{n+1})|\mathbf{X}_n = \mathbf{x}_n\big)$
3. Find the optimal solution

$$\hat{X}_{n+1} = \arg \max_{x \in \mathcal{X}_{n+1}} E\big(L_\alpha(x, X_{n+1})|\mathbf{X}_n = \mathbf{x}_n\big)$$

---

*Remark 1.* By using the maximizing expected utility technique, finding the expected values of the imbalanced loss function is necessary. However, this may sometimes difficult to calculate explicitly. In this research, we approximate the expectation by Monte Carlo simulation. In other words, we may estimate the expectation by

$$E\big(L_\alpha(x, X_{n+1})|\mathbf{X}_n = \mathbf{x}_n\big) \approx \frac{1}{n}\sum_{i=1}^{k} L_\alpha(x, X_{n+1,i}). \tag{4}$$

where $X_{n+1,1}, X_{n+1,2}, \ldots, X_{n+1,k}$ are the simulated values of $X_{n+1}$. In should also be note that there are a number of ways to find the optimum values. For example, the problems could be solved by the quadratic programming for an imbalanced SEL and by the linear programming for an imbalanced AEL.

# 3 Main Results

## 3.1 The Study Area

The data of PM2.5 density, relative humidity, wind speed and temperature, are collected hourly from $1^{st}$ January 2016 to $31^{st}$ December 2018 at Yupparaj Wittayalai School measurement station, Chiang Mai, Thailand. The latitude and longitude coordinates of the station are 18.7917° N and 98.9888° E, respectively. All data used in this research is obtained from the Pollution Control Department, Ministry of Natural Resource and Environment, Thailand [1]. Note that, there are 130 missing data in January 2017 and there is no data for October 2018. The location of the station is shown on Fig. 1.

The plots of averages of hourly PM 2.5 density in each month of the year 2016 to 2018 are displayed on Fig. 2. It can be seen that the patterns of PM 2.5 density in each year are alike. Indeed, during the dry season (January to May), the averages of hourly PM 2.5 density are above 35.5 $\mu g/m^3$ (except January and May of 2017 and 2018) which are classified as danger level 3. In April 2016, the graph reaches the peak when the PM 2.5 reach danger level 4. In 2017 and 2018, they reach the peak in March. Compared with the other months of each year, the PM 2.5 averages are considered to be much lower. In this research, we focus on the situation when PM 2.5 density is high, for the sake of health concern. Therefore the period of study is from January to May (dry season) of each year.

## 3.2 Prediction Based on a Multivariate Autoregressive Model

In this section, we construct a multivariate autoregressive model of PM 2.5 density in the following hour, based on four atmospheric parameters including historical data of PM 2.5 density, relative humidity, wind speed and temperature.

First, for a discrete time $n$, we define the variables $X_n, H_n, W_n$ and $T_n$ as hourly average at time $n$ of PM 2.5 density, relative humidity, wind speed and

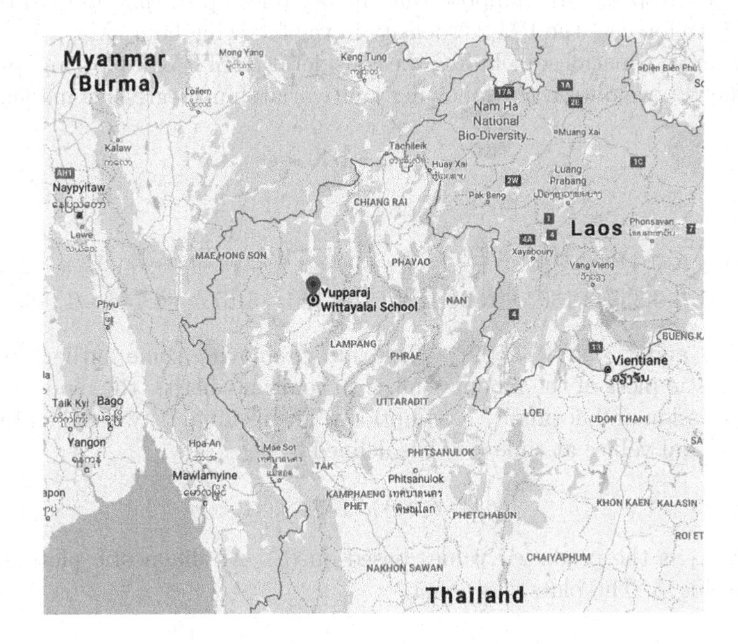


**Fig. 1.** Location of Yupparaj Wittayalai School station, Chiang Mai, Thailand

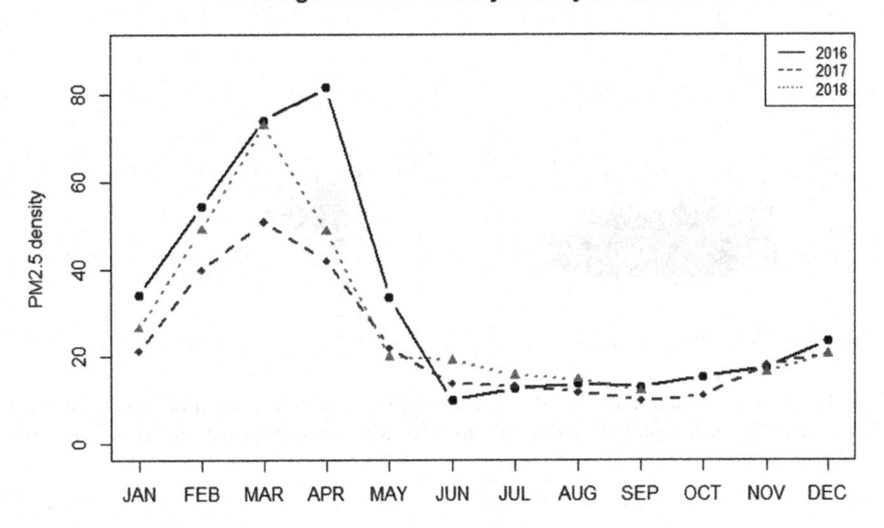


**Fig. 2.** Plot of average of hourly PM2.5 density in each month of 2016–2018

temperature respectively. Suppose that all variables up to time $n$ are observed, our aim is to forecast the PM 2.5 density in the following hour, $X_{n+1}$. By statistics tests and estimations, including the consideration of PACF plot, and p-values of parameters, we obtain the following multivariate autoregressive model:

$$X_{n+1} = a + \mathbf{b}\mathbf{X} + \varepsilon \tag{5}$$

where

$$\mathbf{X} = \left(X_n, X_{n-1}, X_{n-2}, H_{n-1}, H_{n-2}, H_{n-3}, W_n, W_{n-1}, T_{n-1}, T_{n-2}\right)^T,$$
$$\mathbf{b} = \left(1.167, -0.359, 0.130, -0.308, 0.381, -0.150, -1.677, -0.746, -1.003, 0.945\right),$$

$a = 10.003$ and $\varepsilon \sim N(0, 7.502^2)$. Next, after removing some outliers, we plot the diagnostic plots of the model. This is to confirm whether the model satisfies linear regression and normality assumptions. The residual between the predicted and the actual values at time $n+1$ is defined:

$$res_{n+1} = \tilde{X}_{n+1} - x_{n+1}, \tag{6}$$

where $\tilde{X}_{n+1}$ is the predicted values based on (5). All diagnostic plots are displayed in Fig. 3. The plots show that

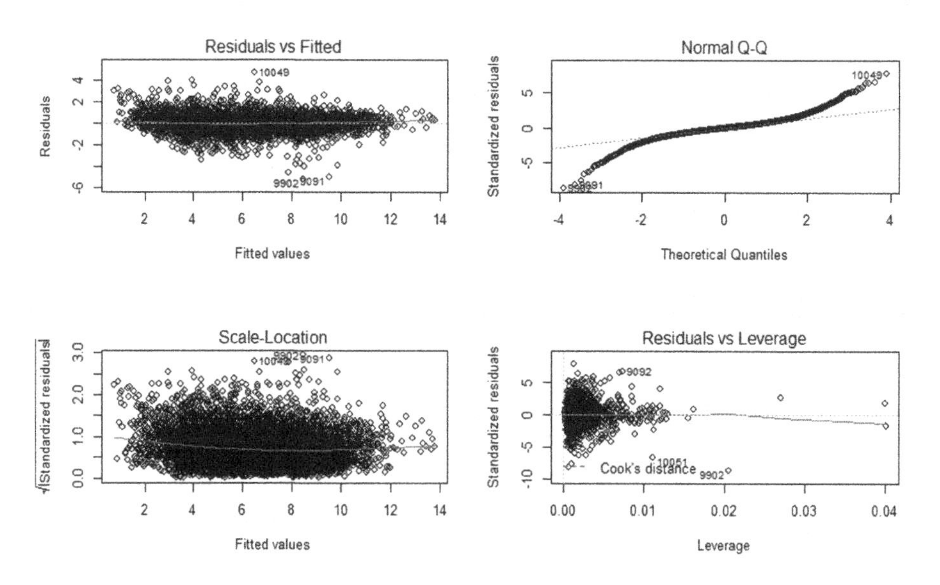


**Fig. 3.** Diagnostic plots: Residuals vs Fitted plot (top-left), Normal Q-Q plot (top-right), Scale-Location plot (bottom-left) and Residuals vs Leverage plot (bottom-right)

1. By Residuals vs Fitted plot (top-left), the pattern of points is indicated by the red line which lies horizontally on the zero line. Thus, the relationship of parameters is linear and the mean of residuals is very close to zero.

2. By Residuals vs Fitted plot and Scale-Location plot (bottom-left), the red lines are flat. Hence, the error terms have the same variance.
3. By Normal Q-Q plot (top-right), points are on the line in the middle of the graph but curve off near the ends. It indicates that the errors have a normal distribution with heavy tails.
4. By Residuals vs Leverage plot (bottom-right), neither extreme outliers nor influential points exist.

Hence, the multivariate autoregressive model defined in (5)is reliable.

In order to test the performance of the prediction model, the cross validation is applied. We use 80 % of the data as the train set and 20 % of the data as the test set. The statistics of the residuals are shown on Table 1.

**Table 1.** Residuals statistics of the multivariate autoregressive model

| | Mean | S.D. | Min | Median | Max | Proportion | |
|---|---|---|---|---|---|---|---|
| | | | | | | Under-predicted | Over-predicted |
| Residuals | 0.036 | 6.943 | −32.054 | 0.180 | 53.967 | 0.488 | 0.512 |

From Table 1, the under prediction ratio and the over prediction ratio are 0.488 and 0.512 respectively. We can see that the ratio is quite symmetric. Besides, 30% of the residuals are 2 $\mu g/m^3$ or less, and 60% of the residuals are 5 $\mu g/m^3$ or less. Thus, the multivariate autoregressive model performs effectively.

### 3.3 Prediction Based on Maximizing Expected Utilities

In this section, we make predictions based on the maximizing expected utility technique. The loss functions used are imbalanced SEL function and imbalanced AEL function. Since the over prediction is preferred in this research, the value of $\alpha$, the individual's preference, should be more than 1.

**Imbalanced Squared Error Loss.** In this section, we calculate the prediction values based on maximizing expected utility technique where the imbalanced SEL function is used. The values of $\alpha$ are set to be 1, 1.5, 2, 3, 5 and 10 respectively. The calculation is done via Monte Carlo simulation with 5,000 samples. The prediction performance is tested with the same random test set as in the multivariate autoregressive model in Sect. 3.2. The statistics of the residuals are displayed on Table 2. The histograms of residuals for each $\alpha$ are displayed on Fig. 4.

From Table 2, we can see that when $\alpha = 1$ the residuals' mean is 0.034 which is very close to zero. The over prediction and under prediction ratio are 0.511 and 0.489 respectively. This implies that the residuals are quite symmetric. This is unsurprising since $\alpha = 1$ implies the loss function is balanced. Besides, as the

**Table 2.** Residuals statistics of the prediction based on maximizing expected utility technique with Imbalanced-SEL function for $\alpha = 1, 1.5, 2, 3, 5, 10$.

| $\alpha$ | Residuals | | | | | Proportion | |
|---|---|---|---|---|---|---|---|
| | Mean | S.D. | Min | Median | Max | Under-predicted | Over-predicted |
| 1 | 0.034 | 6.944 | −32.005 | 0.132 | 53.756 | 0.489 | 0.511 |
| 1.5 | 1.261 | 6.946 | −31.080 | 1.361 | 55.234 | 0.388 | 0.612 |
| 2 | 2.127 | 6.946 | −29.662 | 2.243 | 56.468 | 0.332 | 0.668 |
| 3 | 3.334 | 6.950 | −28.842 | 3.516 | 56.974 | 0.259 | 0.741 |
| 5 | 4.845 | 6.957 | −27.115 | 4.916 | 58.808 | 0.186 | 0.814 |
| 10 | 6.852 | 6.958 | −25.233 | 7.007 | 60.711 | 0.118 | 0.882 |

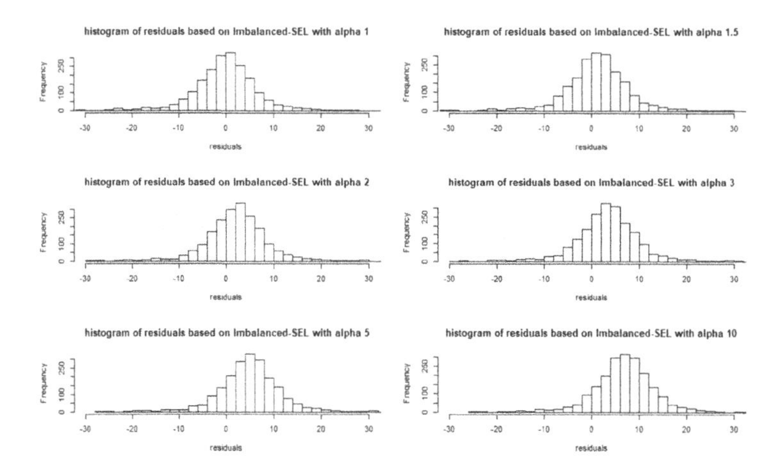


**Fig. 4.** Histogram of residuals based on maximizing expected utility technique with Imbalanced-SEL function for $\alpha = 1, 1.5, 2, 3, 5, 10$.

value of $\alpha$ increases, the over prediction ratio increases while the under prediction ratio decreases. From Fig. 4, it is clear that residuals increase as $\alpha$ increases. This implies that the model becomes less accurate as $\alpha$ increases.

**Imbalanced Absolute Error Loss.** In this section, we calculate the prediction values based on maximizing expected utility technique where the imbalanced AEL function is used. The values of $\alpha$ are set to be 1, 1.5, 2, 3, 5 and 10 respectively. The calculation is done via Monte Carlo simulation with 5,000 samples. The prediction performance is tested with the same random test set as in the multivariate autoregressive model in Sect. 3.2. The statistics of the residuals are displayed in Table 3. The histograms of residuals for each $\alpha$ are displayed on Fig. 5.

From Table 3, we can see that when $\alpha = 1$ the residuals' mean is 0.035 which is very close to zero. The over prediction and under prediction ratio are 0.513 and

**Table 3.** Residuals statistics of the prediction based on maximizing expected utility technique with Imbalanced-AEL function for $\alpha = 1, 1.5, 2, 3, 5, 10$

| $\alpha$ | Residuals | | | | | Proportion | |
|---|---|---|---|---|---|---|---|
| | Mean | S.D. | Min | Median | Max | Under-predicted | Over-predicted |
| 1 | 0.035 | 6.962 | −31.994 | 0.192 | 54.214 | 0.487 | 0.513 |
| 1.5 | 1.962 | 6.963 | −29.798 | 2.045 | 56.033 | 0.344 | 0.656 |
| 2 | 3.307 | 6.950 | −28.929 | 3.410 | 57.718 | 0.258 | 0.742 |
| 3 | 5.130 | 6.946 | −27.025 | 5.323 | 59.488 | 0.177 | 0.823 |
| 5 | 7.364 | 6.954 | −24.740 | 7.456 | 61.307 | 0.105 | 0.895 |
| 10 | 10.145 | 6.957 | −21.582 | 10.348 | 64.068 | 0.058 | 0.942 |

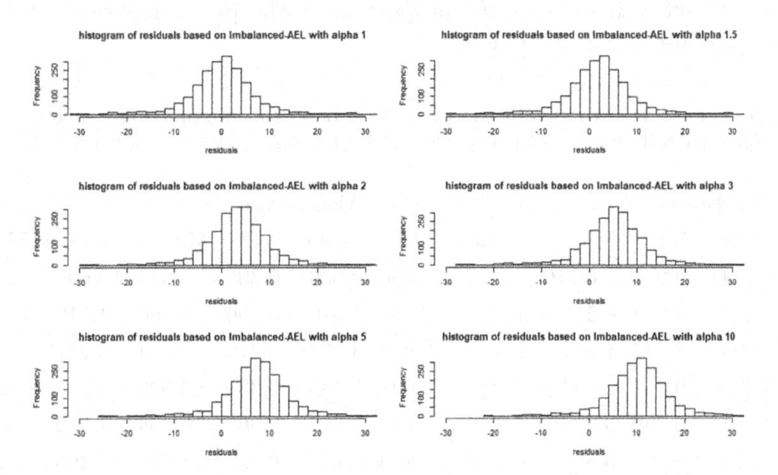


**Fig. 5.** Histograms of residuals based on maximizing expected utility technique with Imbalanced-AEL function for $\alpha = 1, 1.5, 2, 3, 5, 10$.

0.487 respectively. This implies that the residuals are quite symmetric. Similarly, as the value of $\alpha$ increases, the over prediction ratio increases, while the under prediction ratio decreases. From Fig. 5, it is clear that residuals increase as $\alpha$ increases. This implies that the model becomes less accurate as $\alpha$ increases.

**Discussion.** The prediction results from the imbalanced SEL functions agrees with the results from the imbalanced-AEL functions. When $\alpha = 1$, the residuals' mean is close to zero and the over prediction/under prediction ratio is quite symmetric. This is due to the fact that $\alpha = 1$ implies that the loss function is balanced. As $\alpha$ increases, the residuals' mean increase, the over-predicted proportion increases, while the under-predicted proportion decreases. Hence, as $\alpha$ increase, the prediction suits for individuals' that prefer over prediction. However, the drawback is that the prediction is less accurate. Therefore, the value of $\alpha$ must be chosen carefully in order to balance both accuracy and over-prediction ratio. For sensitive groups of people such as children, the elderly and those with predisposed respiratory and cardiovascular diseases, our suggestion is to choose the $\alpha$ value to be between 2 and 3. This would make both over-prediction ratio and accuracy of the model acceptable.

Note that the errors between the predictions based on maximizing expected utilities technique, $\hat{X}_{n+1}$, and the predictions based on the multivariate autoregressive model, $\tilde{X}_{n+1}$, are summarized in Table 4. As $\alpha$ increases, the residuals' mean increase, the over-predicted proportion increases, while the under-predicted proportion decreases. This agrees with the earlier results.

Finally, it should be note that this method provides better prediction ranges and over prediction/under prediction ratios compare to other naïve over-predicted methods such as straight addition or straight percentage addition (adding specific value or specific proportion to the prediction value from the multivariate autoregressive model).

**Table 4.** Residuals statistics of the prediction based on maximizing expected utility technique with Imbalanced-SEL function and AEL function for $\alpha = 1, 1.5, 2, 3, 5, 10$

| $\alpha$ | SEL Errors | | | | | AEL Errors | | | | |
|---|---|---|---|---|---|---|---|---|---|---|
| | Mean | S.D. | Min | Median | Max | Mean | S.D. | Min | Median | Max |
| 1 | −0.006 | 0.247 | −0.770 | −0.005 | 2.436 | 0.006 | 0.305 | −1.040 | 0.000 | 2.436 |
| 1.5 | 1.226 | 0.242 | 0.511 | 1.227 | 2.436 | 1.920 | 0.300 | 0.886 | 1.919 | 2.983 |
| 2 | 2.096 | 0.242 | 1.274 | 2.094 | 2.983 | 3.279 | 0.310 | 2.084 | 3.281 | 4.503 |
| 3 | 3.327 | 0.252 | 2.507 | 3.327 | 4.091 | 5.135 | 0.331 | 4.022 | 5.138 | 6.406 |
| 5 | 4.834 | 0.269 | 3.916 | 4.836 | 5.797 | 7.364 | 0.362 | 6.234 | 7.352 | 8.418 |
| 10 | 6.863 | 0.299 | 5.854 | 6.863 | 7.919 | 10.171 | 0.417 | 8.789 | 10.166 | 11.778 |

## 4 Conclusion

We construct PM 2.5 density prediction models based on individual's preference. The predictions are done via the maximizing expected utility technique with imbalanced loss functions. The squared error loss function and the absolute error loss function are used in this article. The multivariate autoregressive model of PM 2.5 density based on four historical atmospheric parameters: PM 2.5 density, relative humidity, wind speed and temperature is also constructed to be used as benchmarks. The study area is Chiang Mai province, Thailand. The study period is focused on the dry season (January to May) from 2016 to 2018. The hourly data is provided by the Pollution Control Department, Ministry of Natural Resource and Environment, Thailand.

The study results show that the over-prediction ratio increases when the imbalanced loss functions, $L_{\alpha}$, with $\alpha > 1$ are applied. This would extremely benefit some specific groups of people, especially ones who are vulnerable to air pollution, who prefer the prediction to be over-predicted rather than under-predicted.

However, the drawback of this technique is that the predictions become less accurate. Therefore, the value of $\alpha$ must be chosen carefully in order to balance both accuracy and over-prediction ratio.

## References

1. PCD (Pollution Control Department). http://www.pcd.go.th. Accessed 7 Jan 2020
2. Climate Change Data Centre of Chiang Mai University (CMU CCDC). http://www.cmuccdc.org. Accessed 15 May 2020
3. Smoke Haze Integrated Research Unit (SHIRU). http://www.shiru-cmu.org/. Accessed 15 May 2020
4. Chantara, S., Sillapapiromsuk, S., Wiriya, W.: Atmospheric pollutants in Chiang Mai (Thailand) over a five-year period (2005–2009), their possible sources and relation to air mass movement. Atmos. Environ. **60**, 88–98 (2012)
5. Wiriya, W., Prapamontol, T., Chantara, S.: PM10-bound polycyclic aromatic hydrocarbons in Chiang Mai (Thailand): seasonal variations, source identification, health risk assessment and their relationship to air-mass movement. Atmos. Res. **124**, 109–122 (2013)
6. Department for Environment Food and Rural Affairs. https://laqm.defra.gov.uk/public-health/pm25.html. Accessed 24 Mar 2020
7. Chiang Mai Air Quality Health Index (CMAQHI). https://www.cmaqhi.org/. Accessed 3 May 2020
8. IQAir. https://www.iqair.com/. Accessed 2 June 2020
9. Tan, J., Fu, J., Carmichael, G.: Why models perform differently on particulate matter over East Asia? A multi-model intercomparison study for MICS-Asia III. Atmos. Chem. Phys. Discuss. **20**(12), 7393–7410 (2019)
10. Bhakta, R., Khillare, P., Jyethi, D.: Atmospheric particulate matter variations and comparison of two forecasting models for two Indian megacities. Aerosol Sci. Eng. **3**(2), 54–62 (2019)
11. Pimpunchat, B., Junyapoon, S.: Modeling haze problems in the north of Thailand using logistic regression. J. Math. Fundam. Sci. **46**(2), 183–193 (2014)
12. Mitmark, B., Jinsart, W.: A GIS model for PM10 exposure from biomass burning in the north of Thailand. Appl. Environ. Res. **39**(2), 77–87 (2017)
13. Phetpradap, P.: A fuzzy soft model for haze pollution management in Northern Thailand. Article ID 6968705, 13 pages (2020)
14. Resnik, M.: An Introduction to Decision Theory. University of Minnesota Press, Minneapolis (2009)

# Identification of the Direct and Indirect Drivers of Deforestation and Forest Degradation in Cambodia

Sereyrotha Ken[1,2], Nophea Sasaki[3(✉)], Tomoe Entani[1], and Takuji W. Tsusaka[3]

[1] Graduate School of Applied Informatics, University of Hyogo, Kobe, Japan
{ab181201,entani}@ai.u-hyogo.ac.jp
[2] Wildlife Conservation Society, Phnom Penh, Cambodia
ken.rotha@gmail.com
[3] Natural Resources Management, Asian Institute of Technology, Pathum Thani, Thailand
nopheas@ait.ac.th

**Abstract.** Identification of the drivers and their agents of deforestation and forest degradation is important for effective implementation of the REDD+ activities. Here, we identified the direct and indirect drivers through the analysis of local perceptions. The mixed method was used for data collection and analysis. A survey of 215 families and four focus group discussions with 72 participants were conducted in seven community forests in Kampong Thom province, Cambodia. The Likert scale scoring was used to assess the level of acceptance by the locals. We found nine direct drivers and four main indirect drivers and six agents of forest deforestation and forest degradation.

**Keywords:** Illegal logging · Law enforcement · REDD+ · Likert scale · Multivariate ordered probit · Cambodia

## 1 Introduction

The REDD+ scheme or reducing emissions from deforestation and forest degradation, conversation, sustainable management of forests, and enhancement of forest carbon stocks is part of the United Nations Framework Convention on Climate (UNFCCC) to provide financial incentives for developing countries to reduce carbon emissions or increasing the removals. Deforestation and forest degradation (D & D hereafter) cause significant loss of carbon stocks, biodiversity, wildlife habitats, and other ecosystem services, on which billions of people depend for daily livelihoods. The outbreak of the COVID-19 (coronavirus disease 2019), which has infected 12.5 million and caused 560,000 deaths at the time of writing this paper, was possibly linked to the destruction of wildlife habitats [1, 2] as the world lost about 26 million hectares (ha) of forests annually between 2014 and 2018 – about 90% of this loss occurred in the tropics [3]. In addition, D & D are responsible for about 10% of global carbon emissions [4]. Although international efforts have been made to reduce D & D such as the REDD+ the UNFCCC, the 2014 New York Declaration on Forest, and the UN's Sustainable Development Goals in 2015, rate of


V.-N. Huynh et al. (Eds.): IUKM 2020, LNAI 12482, pp. 84–95, 2020.
https://doi.org/10.1007/978-3-030-62509-2_8

D & D does not slow down [3]. As D & D are complicated [5], understanding their drivers and agents of these drivers could provide the important basis for introducing the appropriate activities or interventions to reduce the drivers, which in turn can result in reducing D & D and related carbon emissions. Understanding the driver is also important for designing and implementing the REDD+ activities that involve multi-stakeholder participation [6, 7].

There are various methods used to identify the drivers of D & D. On the global scale, Curtis et al. [8] used satellite imagery and a forest loss classification model to identify the drivers of D & D. They found that clearing of forests for commodity production was the main driver, accounting for 27%, followed by logging activities (i.e. forestry, 26%), shifting agriculture (24%), and wildfire (23%). Using a questionnaire survey to landscape managers who managed 28 tropical landscapes around the tropics, Jayathilake et al. [5] was able to identify major drivers of D & D. Their findings indicate that commercial and subsistence agriculture were the main drivers of deforestation, followed by settlement expansion and infrastructure development. Specifically, land was cleared for rice, rubber, cassava and maize cultivation in these emblematic conservation landscapes. By conducting and analyzing data available in scientific literature and national and international reports, Kissinger et al. [6] identified five main drivers of deforestation in all tropical continents – the urban expansion, infrastructure, mining, agriculture for local subsistence and agriculture for commercial purposes. They further identified four main drivers of forest degradation – livestock grazing in forest, uncontrolled fires, fuelwood charcoal, and timber logging.

Although previous studies on identification of the drivers provide useful insights into the causes of tropical deforestation, they failed to focus on drivers that were actually perceived to be essential for the survival of the local people, especially for those whose daily livelihoods of many generations depended on forest ecosystem services for subsistence. In Vietnam, Van Khuc et al. [9] suggest the need to understand the drivers at the local level before introducing any interventions to reduce D & D. As the drivers involve many agents, any policy intervention is dooms to fail if such policies do not address the individual agents of the drivers [10]. Skutsch and Turnhout [10] analyzed the drivers of D & D in 12 countries across the tropics. They found that local agents were responsible for more than 70% of the drivers of D & D. The aim of this report was to identify the drivers and the agents of such drivers through a questionnaire survey, focus group discussions, and field observations in seven communities adjacent to the recently validated Tumring REDD+ project site located in the Prey Long Wildlife Sanctuary in Kompong Thom province, Cambodia. Local perceptions were collected using the Likert scale scoring method.

## 2 Study Methods and Materials

### 2.1 Description of the Study Site

The study site is in Kampong Thom province, Cambodia. The study site covers a total of 23 forestry communities with a total population of 5,267 families. Fieldworks were conducted in seven forestry communities of the 23 communities located inside the Prey Long Wildlife Sanctuary in Kompong Thom province, Cambodia (Fig. 1).

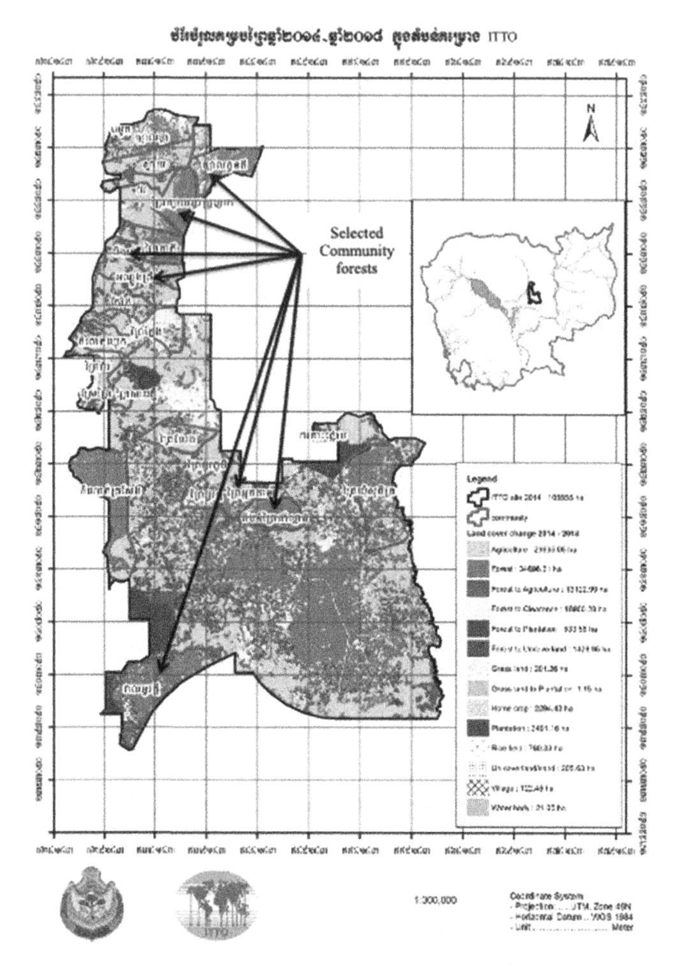


**Fig. 1.** Map of the study site showing the community forests selected for the survey
Source: FA [11]

## 2.2 Data Collection

Both the questionnaire survey and focus group discussions were conducted to understand the drivers and agents of D & D, and to discuss the consensus among the local communities on appropriate activities for addressing the drivers and their agents. The questionnaire was designed according to the Likert scale format to elicit the degree of agreement from local people toward the drivers of D & D. Moreover, focus group discussions were conducted to support and interpret the response to the survey in order to enhance the reliability of the results and their implications. In the Likert scale, the scores 1, 2, 3, 4 and 5 correspondingly referred to strongly disagree, disagree, neutral, agree and strongly agree to the questionnaire statement, respectively.

The drivers and agents of D & D with a score from 3 to 5 were considered as the perceived drivers of forest loss in study areas. Perception of nine direct drivers and eleven indirect drivers was elicited. For this study, direct drivers include illegal logging/unauthorized forest encroachment, commercial wood production, land clearing for commercial agriculture, charcoal production, land clearance for subsistence agriculture, new settlements/migration, natural disaster (flood, storm), human induced forest fire, and fuelwood (domestic usage or local consumption). Indirect drivers include road construction, population growth, public service, mining, hydropower development, shifting cultivation, livestock grazing, application of wood sector, fertile land availability limitation, land tenure and rights issue, limitation in law enforcement for illegal logging.

## 2.3 Sampling

The survey was implemented from 16 to 21 August 2018. There were 23 community forests with 5,267 households in the project area. Yet, due to the resource limitations, 7 community forests were included into the survey, namely, Veal O Khdey, Prey Cheam Smach, Prey Naktala, Prey Kbal Daun Tey, Prey Kbal Ou Kror Nhak, Beong Rolom, and Andoung Pring. These communities were purposely selected as they were located in or near the areas where the highest rate of forest cover change or vulnerability to future forest loss was observed. The sample size was calculated by using the following formula proposed by Yamane [12]:

$$n = \frac{N}{1 + Ne^2}$$

where: $n =$ minimum suggested sample size
$N =$ household population in the ITTO project site (5267 households)
$e =$ margin of error (7% or 0.07)

This equation suggested that a sample size was 197. In case of any errors in data collection which could affect the number of useable observations, around 10% additional households were interviewed. As a result, the actual sample size of the additional 10% became 29 but 4 of which were intentionally removed due to missing information in the entered database. Therefore, the total sample size was 215. In consultation with government officials and villager chiefs, four focus group discussions were conducted with local people in Prey Cheam Smach, Prey Naktala, Prey Kbal Ou Kror Nhak, and Prey Kbal Daun Tey on 28th and 29th August 2018. The number of participants was 72, including 39 females.

## 2.4 Data Analysis

The descriptive statistics for the five-level ordinal variables representing the extent of agreement with nine direct drivers, eleven indirect drivers, eleven types of agents, and eighteen types of REDD++ activities were presented using frequency distribution, mean, median, and standard deviation. These drivers, agents, and activities were ranked per respective category according to their mean scores. The quantitative results were discussed in conjunction with the insights from the focus group discussions. Furthermore, the multivariate ordered probit regression [13, 14] was employed to examine the association between the sociodemographic variables and the perception variables. It is ordered probit because the dependent variables were ordinal scale data [15], whilst it is multivariate because there were multiple dependent variables seemingly associated with each other through the unobserved error components [16]. Accordingly, four systems of equations were estimated: one with the nine direct drivers, another with the eleven indirect drivers, another with the eleven types of agents of D & D, and the other with the eighteen types of mitigation measures. The quantitative analyses were performed with STATA 16. In particular, the execution of the multivariate ordered probit model utilized the code package *cmp*.

# 3 Results and Discussions

## 3.1 Respondents' Profile

Table 1 summarizes the descriptive statistics of the respondents' profile. More than two thirds (68%) of the respondents were women and the rest (32%) were men. The higher number of female respondents was due to their availability during the time of the survey. Most women were at home, while men were at the field or far from home. Many of them were a labor worker in Thailand or Korea. The respondents' age ranged from 18 to 81 years old. The combined age segment from 18 to 50 accounted for 69.7% of the sample. Some of these adults worked on farm or used to go to the forest to collect NTFPs and participated in forest protection and management such as working as a ranger. Therefore, they witnessed how the forest in their community degraded or deforested in the past. On the other hand, 30.3% of the sample were older than 50. These people had experiences and knowledge about the changes in forest cover and conditions in their community and were thus able to provide their views on how and why the forest was lost or degraded in their region.

The majority (88.8%) of the respondents were married, while only 0.5% were single, 2.8% divorced, and the remaining 7.9% were unknown. Most (70.2%) of the households had 4 to 7 family members. Smaller households accounted for 21.9%, and larger ones 7.9%. Many (38.1%) of the respondents were educated up to primary school, 26.5% were illiterate, 21.4% finished secondary school, and only 7.0% completed high school. Most (70.2%) of the respondents had been living in the studied area for more than 20 years, followed by 15.4% with up to 10 years of residency, and 14.4% with 11–20 years of residency.

The dominant majority (80.5%) of the respondents were farmers, followed by labor workers (8.4%), business persons (5.1%), NTFPs collectors, and government officers (0.9% each). About a half (54.9%) of them had been with the same occupation for 10 years or less. The average household income was estimated to be around 2,000 US dollars per annum. Two thirds (68.8%) of the respondents were members of the community forests. But 43.7% of the interviewed respondents had ever participated in some activities for forest management or conservation, such as attending meetings and forest ranging.

## 3.2 Drivers of Deforestation and Forest Degradation

**Direct Drivers:** Table 2 presents the descriptive statistics of the response to the Likert-scale questions on direct drivers of D & D. Out of the nine items assessed, the top three

**Table 1.** Profile of the survey respondents ($n = 215$) in the study site

| Sociodemographic profile variable | Category or level | Frequency (%) | Mean (SD) |
|---|---|---|---|
| Gender of respondent* | Male | 32.6 | |
| | Female | 67.4 | |
| Age of respondent (years) | 18–30 | 23.7 | 42.3 (14.3) |
| | 31–40 | 28.8 | |
| | 41–50 | 17.2 | |
| | 51–60 | 17.7 | |
| | >60 | 12.6 | |
| Marital status of respondent | Single | 0.5 | NA |
| | Married | 88.8 | |
| | Divorced | 2.8 | |
| | Other | 7.9 | |
| Household size (headcount) | Less than 4 | 21.9 | 4.8 (1.6) |
| | 4–7 | 70.2 | |
| | More than 7 | 7.9 | |
| Level of education of respondent | No education | 26.5 | NA |
| | Informal education at pagoda | 1.9 | |
| | Literacy class | 2.8 | |
| | Primary school | 38.1 | |
| | Secondary school | 21.4 | |
| | High school | 7.0 | |
| | Diploma, vocational Education | 0.5 | |
| | College or higher | 1.9 | |
| Duration of residency of respondent (years) | 1–10 | 15.4 | 32.9 (18.4) |
| | 11–20 | 14.4 | |
| | 21–30 | 16.7 | |
| | >30 | 53.5 | |
| Primary occupation of respondent | Farmer | 80.5 | NA |
| | Labor worker | 8.4 | |
| | Businessperson | 5.1 | |

(*continued*)

direct drivers of D & D were illegal logging and unauthorized encroachment, commercial wood production, and land clearing for commercial agriculture. For these drivers, the average score exceeded 4.0. The other direct drives registered the average of below

**Table 1.** *(continued)*

| Sociodemographic profile variable | Category or level | Frequency (%) | Mean (SD) |
|---|---|---|---|
| | Government officer | 0.9 | |
| | NTFPs collector | 0.9 | |
| | Rancher | 0.5 | |
| | Other | 3.7 | |
| Duration with primary occupation (years) | 1–10 | 54.9 | 16.4 (14.0) |
| | 11–20 | 15.8 | |
| | 21–30 | 10.7 | |
| | >30 | 18.6 | |
| Household income from primary occupation (USD/year) | <500 | 28.4 | 1266.3 (1604.8) |
| | 500–1000 | 39.1 | |
| | 1001–2000 | 15.8 | |
| | >2000 | 16.7 | |
| Household income from other occupations (USD/year) | <500 | 61.4 | 790.7 (1181.2) |
| | 500–1000 | 9.3 | |
| | 1001–2000 | 21.9 | |
| | >2000 | 7.4 | |
| Community forest by respondent | Member | 68.8 | NA |
| | Non member | 31.2 | |
| Participation in forest management committee by respondent | Participant | 43.7 | NA |
| | Non participant | 56.3 | |

Note: *was adult person who was at home at the time of interviews.

4.0, though they were still above 3.0. Among them, charcoal production and land clearing for subsistence cultivation were perceived as relatively important direct drivers of D & D.

Collection of commercial wood products was the cause of rapid deforestation in the Philippines between 1970s and 1980s [17] and in the Baltistan Region of Pakistan [18]. Using high-resolution remote sensing data, a recent study also found that land clearing by the land economic concession caused rapid deforestation in Cambodia [19]. On the other hands, charcoal production has also caused forest degradation and deforestation in Mozambique [20], Brazil [21], and Angola [22].

**Indirect Drivers:** Table 3 presents the descriptive statistics of the response to the Likert-scale questions on indirect drivers of D & D. Out of the eleven items assessed, the top two indirect drivers of D & D were limitation in law enforcement for illegal logging and demand for timber. For these drivers, the average score exceeded 4.0. Two other

**Table 2.** Extent of agreement to the nine direct drivers in the study site ($n = 215$)

| Direct driver | | Average score (SD) | Median score |
|---|---|---|---|
| N1 | Illegal logging/unauthorized forest encroachment | 4.53 (0.60) | 5 |
| N2 | Commercial wood production | 4.20 (0.71) | 4 |
| N3 | Land clearing for commercial agriculture | 4.19 (1.15) | 5 |
| N4 | Charcoal production | 3.60 (1.12) | 4 |
| N5 | Land clearance for subsistence agriculture | 3.54 (0.75) | 4 |
| N6 | New settlements/migration | 3.43 (0.81) | 3 |
| N7 | Natural disaster (flood, storm) | 3.31 (0.91) | 3 |
| N8 | Human induced forest fire | 3.25 (0.96) | 3 |
| N9 | Fuelwood (domestic usage or local consumption) | 3.21 (0.77) | 3 |

indirect drives registered the average of below 4.0 but above 3.0, namely, land tenue and rights issue, and population growth. Hydropower development was perceived as the least significant indirect driver of D & D in the studied area. The top four indirect drivers are further explained in the following paragraphs.

**Table 3.** Extent of agreement to the eleven indirect drivers in the study site

| Indirect drivers | | Average score (SD) | Median score |
|---|---|---|---|
| P1 | Limitation in law enforcement | 4.33 (0.54) | 4 |
| P2 | Demand for timber | 4.15 (0.68) | 4 |
| P3 | Land tenure and rights issue | 3.72 (0.78) | 4 |
| P4 | Population growth | 3.47 (0.73) | 3 |
| P5 | Fertile land availability limitation | 2.94 (0.97) | 3 |
| P6 | Road construction | 2.79 (0.84) | 3 |
| P7 | Shifting cultivation | 2.73 (0.94) | 3 |
| P8 | Public service | 2.32 (0.89) | 2 |
| P9 | Mining | 1.95 (0.83) | 2 |
| P10 | Livestock grazing | 1.85 (0.69) | 2 |
| P11 | Hydropower development | 1.60 (0.63) | 2 |

### 3.3 Agents of the Drivers of Deforestation and Forest Degradation

Based on the focus group discussions, six agents of the drivers of D & D were identified. Detailed description of the agents is presented in Table 4.

**Table 4.** Agents' activities that were perceived to have contributed to deforestation and forest degradation in the study site

| Agents | Activities |
|---|---|
| Furniture makers | - Fell trees in huge amount for commercial wood products<br>- Trigger illegal logging from local people |
| Medium- and large-scale agricultural investors | - Convert forest land to agricultural land |
| Charcoal makers | - Produce charcoal |
| Land migrants | - Clear forest for land settlement<br>- Clear forest for agricultural land<br>- Trigger local people to clear forest land and sell to them |
| Firewood collectors | - Collect wood for domestic use |
| Subsistent farmers | - Clear forest land for growing crop |

### 3.4 Influences of Sociodemographic Factors on Respondents' Perception

Tables 5 show the results of the multivariate ordered probit analysis of how sociodemographic factors influenced the perceived direct and indirect drivers of D & D, perceived agents of D & D, and perceived activities to address the drivers. To keep the tables succinct, only the sign and statistical significance of the estimated coefficients are indicated. Although Tables 5 presents both the direct and indirect drivers, it comes from the two systems of regressions. It was found that women's agreement to the direct and indirect drivers was generally lower than men's. Eight out of the twenty drivers exhibited this tendency, while none of the other drivers showered any gender gap. Those who were farmers agreed less to the land fertility limitation as an indirect driver. Income level showed a mixed result though it presented significant association with a half of the drivers. With many of the drivers, community forest members had higher agreement with the drivers than non-members, while participants in community forest management had lower agreement than non-participants. Interestingly, this tendency was reversed for some indirect drivers that were considered less important. This result implies that general members of community forests and those engaged in management showed contrasting perception of the drivers, where the perception by the former was largely consistent with the ranking of the drivers while the perception by the latter was relatively neutral across the drivers. The other sociodemographic factors showed smaller or no relevance with the drivers of D & D in general.

In terms of perception of who were responsible for D & D, farmers tended to mention shifting cultivators, while non-farmers tended to cite investors in agriculture. However, those who were stably in the same job, including farming, were more likely to think firewood collectors were responsible for D & D. Similar to the case of the drivers, general members of community forests tended to be in line with the ranking of the agents, whereas management members tended to be relatively neutral cross the agents.

**Table 5.** Effects of sociodemographic factors on levels of agreement to direct and indirect drivers of D & D in the project area: multivariate ordered probit regressions ($n = 215$)

| Drivers | | Sociodemographic factors | | | | | | | | | |
|---|---|---|---|---|---|---|---|---|---|---|---|
| | | Women (1 if so) | Age (years) | Married (1 if so) | HH size (headcount) | Secondary School (1 if completed) | Residency Duration (years) | Farmer (1 if so) | Occupation Duration (years) | Ln income (Ln (USD/annum)) | CF member (1 if so) |
| Direct drivers | | | | | | | | | | | |
| 1 | Illegal logging/ unauthorized forest encroachment | | | | ++ | | | | | | +++ |
| 2 | Commercial wood production | – – – | | – | | | ++ | | | +++ | +++ |
| 3 | Land clearing for commercial agriculture | – – – | | | + | – | | – | | | +++ |
| 4 | Charcoal production | – | | | | – | | | | – – – | ++ |
| 5 | Land clearance for subsistence agriculture | – – – | | | | – – – | | | | | |
| 6 | New settlements/ Migration | – – – | | | | – – – | | | | | +++ |
| 7 | Natural disaster (flood, storm) | | | + | | | + | | – | | |
| 8 | Human induced forest fire | – – | | | | | | | | | +++ |
| 9 | Fuelwood (domestic usage or local consumption) | | | | | | | ++ | + | – – – | |
| Indirect drivers | | | | | | | | | | | |
| 1 | Limitation in law enforcement | – – | | | | | | | + | ++ | +++ |
| 2 | Demand for timber | | | | | | ++ | | | +++ | ++ |
| 3 | Land tenure and rights issue | | | | ++ | | | | | + | |
| 4 | Population growth | – | | | + | | | – | | | ++ |
| 5 | Fertile land availability limitation | | | ++ | | | + | – – – | | | |
| 6 | Road construction | | | | | | | | | | |
| 7 | Shifting cultivation | | | | | | | | – – – | – – – | – – – |
| 8 | Public service | | | | | | | | | – – – | |
| 9 | Mining | | | | | | | ++ | | +++ | – – – |
| 10 | Livestock grazing | | | | | | | ++ | | | – – – |
| 11 | Hydropower development | | + | | – | | | | | + | – – – |

Note: The two systems (the system of direct drivers and the system of indirect drives) were separately estimated but combined into one table. + and – indicate a positive coefficient and a negative coefficient, respectively. +++, ++, and + indicate $p < 0.01$, $< 0.05$, and $< 0.10$, respectively.

The other sociodemographic factors showed either mixed, smaller, or no association with the perception of which agents were more responsible for D & D.

## 4 Conclusion and Recommendations

Data from 215 respondents in seven communities in Kampong Thom province, Cambodia were analyzed to understand the local perceptions on the direct and indirect drivers of D & D, agents of such drivers, and appropriate activities to reduce these drivers according to five levels of Likert scale. Local people seem to accept that illegal logging and unauthorized forest encroachment (4.53), commercial wood production (4.20), and land clearing for commercial cultivation (large economic land concession) (4.19) were the main direct drivers of D & D. Local people considered charcoal production (3.60), land clearing for subsistence agriculture (3.54), new settlements (3.43), natural disasters such as drought and storm (3.31), human-induced forest fires (3.25), and fuelwood for domestic usage (3.21) as less relevant direct drivers.

They considered the limitation in law enforcement (4.33), high demand for timber (4.15), lack of land tenure and rights (3.72), and the growing population (3.47) as indirect drivers of D & D. Local people seemed to accept that these drivers can be reduced through the provision of farmland for households, financial incentives for agriculture, law enforcement on illegal logging, improvement in market access for agriculture products, community forest management, policy and governance reform, reforestation, environmental education on forest management, reform for land tenure and rights, agricultural intensification, and restoration of degraded forests.

Although the current study provides important information on local perceptions on direct and indirect drivers of D & D, seasonal field surveys may further increase "the accuracy of the results" and reduce the biases that would have arisen from the loss of memory the loss of memory about the exact events that happened in the study site.

**Acknowledgement.** We would like to thank Ms. Sokna Kry, a former student at the Natural Resources Management program at the Asian Institute of Technology in Thailand for assisting in data collection and drafting the related report. We also thank field staff of the Forestry Administration for assisting and facilitating the fieldwork.

## References

1. Afelt, A., Frutos, R., Devaux, C.: Bats, coronaviruses, and deforestation: Toward the emergence of novel infectious diseases? Front. Microbiol. **9**, 702 (2018). https://doi.org/10.3389/fmicb.2018.00702
2. Valitutto, M.T., Aung, O., Tun, K.Y.N., Vodzak, M.E., Zimmerman, D., Yu. J.H., et al.: Detection of novel coronaviruses in bats in Myanmar. PLoS ONE **15**(4), e0230802 (2020). https://doi.org/10.1371/journal.pone.0230802
3. NYDF Assessment Partners: Protecting and Restoring Forests: A Story of Large Commitments yet Limited Progress. New York Declaration on Forests Five-Year Assessment Report (2019). Climate Focus (coordinator and editor). Accessible at forestdeclaration.org on 14 May 2020
4. Houghton, R.A.: Carbon emissions and the drivers of deforestation and forest degradation in the tropics. Curr. Opin. Environ. Sustain **4**, 597–603 (2012)
5. Jayathilake, H.M., Prescott, G.W., Carrasco, L.R. et al.: Drivers of deforestation and degradation for 28 tropical conservation landscapes. Ambio (2020) https://doi.org/10.1007/s13280-020-01325-9

6. Kissinger, G., Herold, M., De Sy, V.: Drivers of Deforestation and Forest Degradation: A Synthesis Report for REDD+ Policymakers. Lexeme Consulting, Vancouver Canada (2012)
7. Salvini, G., Herold, M., De Sy, V., Kissinger, G., Brockhaus, M., Skutsch, M.: How countries link REDD+ interventions to drivers in their readiness plans: Implications for monitoring systems. Environ. Res. Lett. **9**(7), 074004 (2014). https://doi.org/10.1088/1748-9326/9/7/074004
8. Curtis, P.G., Slay, C.M., Harris, N.L., Tyukavina, A., Hansen, M.C.: Classifying drivers of global forest loss. Science **361**(6407), 1108–1111 (2018)
9. Van Khuc, Q., Tran, B.Q., Meyfroidt, P., Paschke, M.W.: Drivers of deforestation and forest degradation in Vietnam: An exploratory analysis at the national level. For. Policy Econ. **90**, 128–141 (2018)
10. Skutsch, M., Turnhout, E.: REDD+: If communities are the solution, what is the problem? World Dev. **130**, 104942 (2020). https://doi.org/10.1016/j.worlddev.2020.104942
11. FA: Sustainable Forest Management through REDD+ Mechanisms in Kampong Thom province. Progress Report submitted to the International Tropical Timber Organization. Forestry Administration, Phnom Penh (2019)
12. Yamane, T.: Statistics: An Introductory Analysis, 2nd edn. Harper and Row, New York (1967)
13. Bhat, C.R., Srinivasan, S.: A multidimensional mixed ordered-response model for analyzing weekend activity participation. Transp. Res. Part B Methodol. **39**, 255–278 (2005)
14. Ma, J., Ye, X., Shi, C.: Development of multivariate ordered probit model to understand household vehicle ownership behavior in Xiaoshan District of Hangzhou, China. Sustainability **10**, 3660 (2018). https://doi.org/10.3390/su10103660
15. Greene, W.H.: Econometric Analysis, 4th edn. Prentice-Hall, New Jersey (2000)
16. Salaisook, P., Faysse, N., Tsusaka, T.W.: Land Use Policy Reasons for adoption of sustainable land management practices in a changing context: A mixed approach in Thailand. Land Use Policy **96**, 104676 (2020). https://doi.org/10.1016/j.landusepol.2020.104676
17. Bensel, T.: Fuelwood, deforestation, and land degradation: 10 years of evidence from Cebu Province, The Philippines. Land. Degrad. Dev. **19**, 587–605 (2008)
18. Ali, J., Benjaminsen, T.A.: Fuelwood, timber and deforestation in the Himalayas: The case of Basho Valley, Baltistan region, Pakistan. Mt. Res. Dev. **24**(4), 312–318 (2004)
19. Davis, K.F., Yu, K., Rulli, M.C., Pichdara, L., D'Odorico, P.: Accelerated deforestation driven by large-scale land acquisitions in Cambodia. Nat. Geosci. **8**(10), 772–775 (2015). https://doi.org/10.1038/ngeo2540
20. Sedano, F., Silva, J.A., Machoco, R., Meque, C.H., Sitoe, A., Ribeiro, N., et al.: The impact of charcoal production on forest degradation: A case study in Tete, Mozambique. Environ. Res. Lett. **11**(9) (2016). https://doi.org/10.1088/1748-9326/11/9/094020
21. Sonter, L.J., Barrett, D.J., Moran, C.J., Soares-filho, B.S.: Carbon emissions due to deforestation for the production of charcoal used in Brazil' s steel industry. Nat. Climate Change **5**, 359–363 (2015). https://doi.org/10.1038/NCLIMATE2515
22. Chiteculo, V., Lojka, B., Surový, P., Verner, V., Panagiotidis, D., Woitsch, J.: Value chain of charcoal production and implications for forest degradation: Case study of Bié Province. Angola. Environ. **5**(11), 1–13 (2018). https://doi.org/10.3390/environments5110113

# Unit Commitment Problem in the Deregulated Market

R. Mikami[1(✉)], T. Fukuba[1], T. Shiina[1(✉)], and K. Tokoro[2]

[1] Waseda University, Tokyo, Japan
ryusei0614@toki.waseda.jp, tomohikari892@asagi.waseda.jp, tshiina@waseda.jp
[2] Central Research Institute of Electric Power Industry, Tokyo, Japan
tokoro@criepi.denken.or.jp

**Abstract.** In Japan, the electric power market has been fully deregulated since April 2016, and many Independent Power Producers have entered the electric power market. Companies participating in the market conduct transactions between market participants to maximize their profits. When companies consider maximization of their profit, it is necessary to optimize operation of generators in consideration of market transactions. However, it is not easy to consider trading in the market because the market contains many complex and uncertain factors. Even now, the number of participating companies is increasing, and research on operation of generators in consideration of market transactions is an important field. In the power market, there are various markets such as a day-ahead market and an adjustment market, and various transactions are performed between market participants. In this study, we discuss the day-ahead market trading. In the day-ahead market, power prices and demands vary greatly depending on the trends in power sell and purchase bidding. It is necessary for business operators to set operational schedules that takes into account fluctuations in power prices and demand. We consider an optimization model of generator operation considering market transactions and apply stochastic programming to solve the problem. In addition, we show that scheduling based on the stochastic programming method is better than conventional deterministic planning.

**Keywords:** Optimization · Stochastic programming

## 1 Introduction

In Japan, the electric power market has been fully deregulated since April 2016, and many Independent Power Producers have entered the electric power market. Companies participating in the market conduct transactions between market participants to maximize their profits. When companies consider maximization of their profit, it is necessary to optimize operation of generators in consideration of market transactions. However, it is not easy to consider trading in the market because it contains many complex and uncertain factors. Even now, the number of participating companies is increasing, and research on operation of generators


V.-N. Huynh et al. (Eds.): IUKM 2020, LNAI 12482, pp. 96–107, 2020.
https://doi.org/10.1007/978-3-030-62509-2_9

in consideration of market transactions is an important field. In the power market, there are various markets such as a day-ahead market and an adjustment market, and various transactions are performed between market participants. In this study, we assume the day-ahead market trading at JEPX (Japan Electric Power Exchange). In the day-ahead market, power prices and demands vary greatly depending on the trends in power sell and purchase bidding. It is necessary for business operators to set operational schedules that consider fluctuations in power prices and demand. In the day-ahead market, the market is divided into 30 min each day, and the supply side and the demand side bid at each time zone in a state where the market status of other market participants is unknown. A demand curve and supply curve are created by accumulating the bidding amount of each company in the price. The intersection of the demand and supply curves are the transactions actually performed in the market, the price is a contract price, and the volume at that time is the contract volume. Offer bids with a price lower than the contract price and a purchase bid with a price higher than the contract price are traded at the contract price. But, offer bids with a price higher than the contract price and purchase bids with a price lower than the contract price are rejected. Figure 1 shows how to determine the contract price and the contract volume using the demand curve and the supply curve.

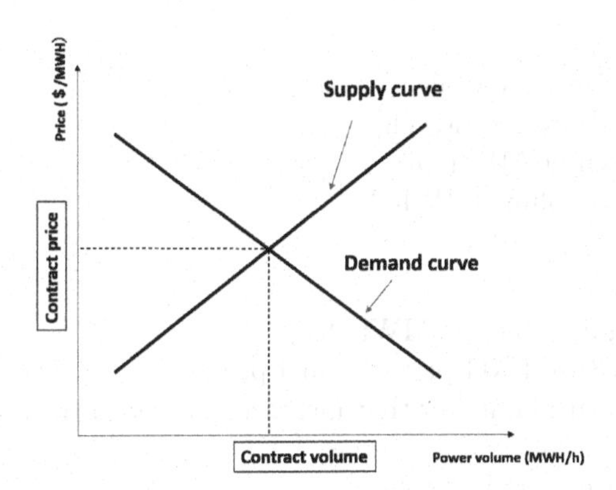


**Fig. 1.** Price and volume

In Fig. 1, the vertical axis represents the price and the horizontal axis the amount of electricity for bidding. In general, the higher the power price, the more the bidding volume increases. Hence, the supply curve increases to the right, while the volume of the bidding volume decreases. Fill prices and fill rates are affected by various factors in the market. In this study, we consider the total bidding volume on the demand side, the total bidding volume on the supply side, the fuel cost and the effect of the passage of time, and the fluctuations in the contract price and the quantitative amount due to the fluctuation of the total bidding volume in a short period of time. We consider an optimization model of generator operation taking into consideration market transactions and applying stochastic programming to solve the problem. In addition, we show that scheduling based on the stochastic programming method is better than conventional deterministic planning.

## 2 Demand Curve and Supply Curve

The contract price and the contract volume are determined from the demand curve formed by the buying bid on the demand side and the supply curve formed by the selling bid on the supplier side. However, JEPX discloses the total bid volume, the contract volume, and the contract price in each market every hour, but does not disclose the bidding volume at each bid price on the supply and demand sides. Therefore, actual demand/supply curve data cannot be obtained. The demand and supply curves are estimated based on the data released by JEPX [3], and the supply and demand curve model of JEPX obtained by the estimation is shown. This study uses this model [7] (Fig. 2).

Stochastic variable

$q_b^M$ Total purchase bidding volume[MWh/h]

Variable

$price_s$ Offer bid price[$ /MWh]
$price_b$ Purchase bid price[$ /MWh]
$q_s$ Offer bid volume[MWh/h]
$q_b$ Purchase bid volume[MWh/h]

Parameter

$q_s^M$ Total offer bidding volume[MWh/h]
$f$ Fuel price[$ /t](use LNG price for fuel price)
$t$ Trend term (natural number that increases by 1 with the passage of business days)
$a_0$ Constant term in supply curve
$a_1$ Coefficient of offer bidding volume in supply curve
$a_2$ Coefficient of total offer bidding volume in supply curve
$a_3$ Coefficient of fuel cost in supply curve
$b_0$ Coefficient of constant term in demand curve
$b_1$ Coefficient of bidding purchase volume in demand curve
$b_2$ Coefficient of total purchase bidding volume in demand curve
$b_3$ Coefficient of trend term in demand curve
$b_4$ Coefficient of fuel price in demand curve

Supply curve

$$price_s = a_0 + a_1 q_s - a_2 q_s^M + a_3 f \tag{1}$$

Demand curve

$$price_b = b_0 + b_1 b_s - b_2 q_b^M + b_3 t + b_4 f \tag{2}$$

Price and quality

$$price = \frac{b_1(a_2 q_s^M - a_0 - a_3 f) - a_1(b_2 q_b^M - b_0 - b_3 t - b_4 f)}{a_1 - b_1} \tag{3}$$

$$quality = \frac{(a_0 - a_2 q_s^M + a_3 f) - (b_0 - b_2 q_b^M + b_3 t + b_4 f)}{b_1 - a_1} \tag{4}$$

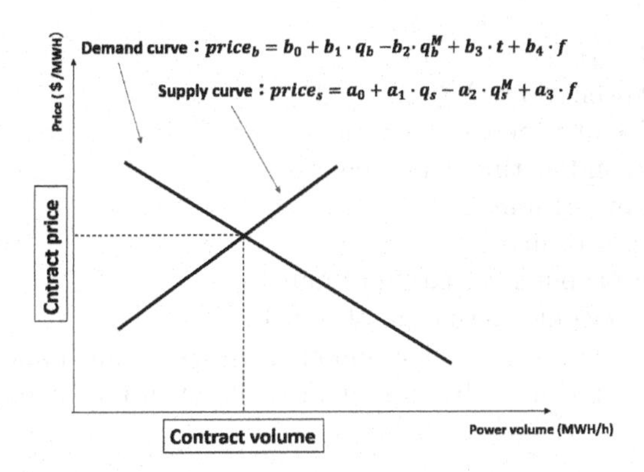


**Fig. 2.** Demand and supply curves

# 3 Model Formulation

The purpose of the generator operation plan in consideration of market transactions is to satisfy the operation constraints of the generator while taking into account fluctuations in electricity prices and demand. We calculate the costs necessary for the operation of the generator, such as fuel costs and start-up costs, from the revenues of the company, and maximizing (profit) excluding electricity purchase costs. The conventional unit commitment problem without considering the market requires that all the power demands be met by the own generator. In such a case, a new generator is often operated additionally to meet the demand, which leads to an increase in the total cost. When considering market transactions, you can purchase electricity from other businesses. Depending on the value of the demand, the total cost can be reduced by supplementing the power demand by purchasing power from another company rather than by starting a new generator. Therefore, in the formulation, when the total output of the generators operated by the utility is less than the demand, the shortage can be purchased from another electric utility. The generator commitment pattern is the same for all scenarios, but the generator output and power change every time. In the following, the formulation and the definition of symbols used will be described.

### 3.1 Definition

Variable

$x_{it}^s$ Output of unit i at time period t in scenario s
$y_t^s$ Power purchase amount at time t in scenario s
$u_{it}$ Variable indicating the commitment state of unit i at time t

Parameter

$I$ Number of units
$T$ Time
$S$ Number of scenario s
$p_s$ Probability of scenario s ($\sum_{s=1}^{S} p_s = 1$)
$K_t^s$ Electricity price at time t in scenario s
$d_t^s$ Electricity demand at time t in scenario s
$Q_i$ Maximum output of unit i
$q_i$ Minimum output of unit i
$R_i$ Upper limit of output fluctuation of unit i
$r_i$ Lower limit of output fluctuation of unit i
$L_i$ Minimum time that must be continuously on when unit i starts
$l_i$ Minimum time that must be continuously off when unit i stops

Function

$f_i(x_{it}^s)$ Fuel cost in unit i (quadratic function of $x_{it}^s$)
$g_i(u_{i,t-1}, u_{i,t})$ Startup cost for unit i (function of $u_{it}$)

### 3.2 Formulation

$$\min \sum_{s=1}^{S} p_s (\sum_{t=1}^{T} K_t^s (d_t^s - y_t^s) - \sum_{i=1}^{I} \sum_{t=1}^{T} f_i(x_{it}^s) u_{it}) - \sum_{i=1}^{I} \sum_{t=1}^{T} g_i(u_{i,t-1}, u_{i,t})$$

$$\text{s.t.} \tag{5}$$

$$\sum_{i=1}^{I} x_{it}^s + y_t^s \geq d_t^s, t = 1, \cdots T, s = 1, \cdots, S \tag{6}$$

$$y_t^s \geq 0, t = 1, \cdots, T, s = 1, \cdots, S \tag{7}$$

$$q_i u_{it} \leq x_{it}^s \leq Q_i u_{it}, i = 1, \cdots, I, t = 1, \cdots, T, s = 1, \cdots, S \tag{8}$$

$$r_i \leq x_{it}^s - x_{it-1}^s \leq R_i, i = 1, \cdots, I, t = 1, \cdots, T, s = 1, \cdots, S \tag{9}$$

$$u_{it} - u_{i,t-1} \leq u_{i\tau}, \tau = t + 1, \cdots, \min\{t + L_i - 1, T\},$$
$$i = 1, \cdots, I, t = 2, \cdots, T \tag{10}$$

$$u_{i,t-1} - u_{it} \leq 1 - u_{i\tau}, \tau = t + 1, \cdots, \min\{t + l_i - 1, T\},$$
$$i = 1, \cdots, I, t = 2, \cdots, T \tag{11}$$

$$u_{it} \in \{0, 1\}, i = 1, \cdots, I, t = 1, \cdots, T \tag{12}$$

(5) is the maximization of the expected value of the objective function(profit). Profit is calculated by multiplying the electricity price by the demand (income), and excluding the power purchase cost, fuel cost, and start-up cost (total cost). Since the commitment state does not change for each scenario, the start-up cost does not consider the occurrence probability of each scenario. On the contrary, fuel cost, and power purchase amount consider the occurrence probability of each scenario since these values change for each scenario. (6) is a constraint that the sum of the total output of the units and the power purchase amount at time t, in scenario s, satisfies the power demand, and (7) is a constraint that the power purchase amount at time t, in scenario s, is 0 or more. (8) are the upper and lower limits of generator output, (9) indicates the upper and lower limits of the output fluctuation of the generator. (10) indicates minimum time that must be continuously turned on when unit i starts, and (11) indicate minimum time that must be continuously turned off when unit i stops. (12) represents the 0–1 condition of the decision variable $u_{it}$. It is assumed that the power purchase fall short of the total output with respect to the demand at time t, in scenario s, and that power can be purchased at the power price at time t, in scenario s.

## 4 Solution

We used the L-shaped method to solve the UC problem. The L-shaped method is a method used to efficiently obtain a solution to a stochastic programming problem including second decision variables. The outline of the L-shaped method will be described. First, the L-shaped method solves the problem (first problem) of the first decision variable excluding the second decision variable, and obtains the first decision variable. Next, when the second problem is feasible, add an optimality cut. The optimality cut is a constraint used to approximate the value of second problem variable using a first decision variable. If it is not feasible, add a feasibility cut. Even if the first problem is feasible, the second problem is not always feasible. Feasibility cut is a constraint added to exclude such cases. Then, a new solution is obtained by solving the problem again based on the optimality cut and the feasibility cut to which the first problem has been added. By repeating this, add the cut to the first problem, the feasible region of the first decision variable is limited, and the optimal solution of the stochastic programming problem including the second decision variable can be obtained by repeating the iteration.

### 4.1 Formulation

Electricity price multiplied by demand (income) is not affected by decision variables. To maximize profit, solve the problem of minimizing the total (total cost) of power purchase cost, fuel cost, and start-up cost.

$$\min \sum_{s=1}^{S} p_s(\sum_{t=1}^{T} K_t^s y_t^s + \sum_{i=1}^{I}\sum_{t=1}^{T} f_i(x_{it}^s)u_{it}) + \sum_{i=1}^{I}\sum_{t=1}^{T} g_i(u_{i,t-1}, u_{i,t}) \tag{13}$$

Parameters to add

| | |
|---|---|
| $NCUT$ | Total number of optimality cuts |
| $\theta$ | Total fuel cost and electricity purchase cost |
| $\theta_s$ | Total fuel cost and power purchase cost in scenario s |

Master problem

$$\min \sum_{i=1}^{I} \sum_{t=1}^{T} f_i(0)u_{it} + \sum_{i=1}^{I} \sum_{t=1}^{T} g_i(u_{i,t-1}, u_{i,t}) + \theta$$

$$\text{s.t.} \tag{14}$$

$$u_{it} - u_{i,t-1} \leq u_{i\tau}, \tau = t+1, \cdots, \min\{t + L_i - 1, T\},$$
$$i = 1, \cdots, I, t = 2, \cdots, T \tag{15}$$

$$u_{i,t-1} - u_{it} \leq 1 - u_{i\tau}, \tau = t+1, \cdots, \min\{t + l_i - 1, T\},$$
$$i = 1, \cdots, I, t = 2, \cdots, T \tag{16}$$

$$u_{it} \in \{0, 1\}, i = 1, \cdots, I, t = 1, \cdots, T \tag{17}$$

$$\theta \geq \sum_{s=1}^{S} p_s \theta_s, s = 1, \cdots, S \tag{18}$$

$$\theta_s \geq \sum_{i=1}^{I} \sum_{t=1}^{T} \alpha_{it}^{s,ncut} u_{it} + \beta_s^{ncut}, ncut = 1, \cdots, NCUT, s = 1, \cdots, S \tag{19}$$

$$\theta, \theta_s \geq 0, s = 1, \cdots, S \tag{20}$$

Second problem

$$\min \sum_{t=1}^{T} K_t^s y_t^s + \sum_{i=1}^{I} \sum_{t=1}^{T} (f_i(x_{it}^s) - f_i(0))$$

$$\text{s.t.} \tag{21}$$

$$\sum_{i=1}^{I} x_{it}^s + y_t^s \geq d_t^s, t = 1, \cdots, T, s = 1, \cdots, S \tag{22}$$

$$y_t^s \geq 0, t = 1, \cdots, T, s = 1, \cdots, S \tag{23}$$

$$q_i u_{it} \leq x_{it}^s \leq Q_i u_{it}, i = 1, \cdots, I, t = 1, \cdots, T, s = 1, \cdots, S \tag{24}$$

$$r_i \leq x_{it}^s - x_{it-1}^s \leq R_i, i = 1, \cdots, I, t = 1, \cdots, T, s = 1, \cdots, S \tag{25}$$

## 4.2 Algorithm

Step 1: Set the provisional objective function value $\overline{z}$ to $\infty$ and the lower bound z of the objective function value $\underline{z}$ to 0.
Step 2: The optimal solution$\hat{u}_{it}$, $\hat{\theta}$ is obtained by solving the main problem.
Step 3: If $\sum_{i=1}^{I} \sum_{t=1}^{T} f_i(0)\hat{u}_{it} + \sum_{i=1}^{I} \sum_{t=1}^{T} g_i(\hat{u}_{i,t-1}, \hat{u}_{i,t}) + \hat{\theta}$ is greater than $\underline{z}$, set $\underline{z}$=$\sum_{i=1}^{I} \sum_{t=1}^{T} f_i(0)\hat{u}_{it} + \sum_{i=1}^{I} \sum_{t=1}^{T} g_i(\hat{u}_{i,t-1}, \hat{u}_{i,t}) + \hat{\theta}$.

If $\sum_{i=1}^{I}\sum_{t=1}^{T} f_i(0)\hat{u}_{it} + \sum_{i=1}^{I}\sum_{t=1}^{T} g_i(\hat{u}_{i,t-1}, \hat{u}_{i,t}) + \sum_{s=1}^{S} C_s(\hat{u})$ is lower than $\overline{z}$, set $\overline{z} = \sum_{i=1}^{I}\sum_{t=1}^{T} f_i(0)\hat{u}_{it} + \sum_{i=1}^{I}\sum_{t=1}^{T} g_i(\hat{u}_{i,t-1}, \hat{u}_{i,t}) + \sum_{s=1}^{S} C_s(\hat{u})$.
Step 4: If $\overline{z}$ is lower than $(1+\varepsilon)\underline{z}$, finish the algorithm.
Step 5: If $\hat{\theta}_s < C_s(\hat{u})$ for $\forall s \in S$, add the optimality cut to the main problem and go to step 2.

### 4.3 Optimality Cut

Find the first decision variable $\hat{u}_{it}$ by solving master problem, then solve the second problem for each scenario to find the second decision variable $\hat{x}^s_{it}$, $\hat{y}^s_t$. The optimality cut is obtained by using constraint (22), $x^s_{it} \leq Q_i u_{it}$, $q_i u_{it} \leq x^s_{it}$.Find the optimal solution of each dual problem,$\hat{\pi}^s_t$, $\hat{\lambda}^s_{it}$, $\hat{\mu}^s_{it}$. Using these, the following inequality becomes a valid inequality for the main problem.

$$\theta_s \geq \sum_{i=1}^{I}\sum_{t=1}^{T}(f_i(\hat{x}^s_{it}) - f_i(0)) + \sum_{i=1}^{I}\sum_{t=1}^{T}(\hat{\lambda}^s_{it} - \hat{\mu}^s_{it} - \hat{\pi}^s_t)\hat{x}^s_{it} +$$
$$\sum_{t=1}^{T}\hat{\pi}^s_t d^s_t + \sum_{t=1}^{T}(K^s_t - \hat{\pi}^s_t)\hat{y}^s_t + \sum_{i=1}^{I}\sum_{t=1}^{T}(\hat{\mu}^s_{it} q_i - \hat{\lambda}^s_{it} Q_i)u_{it} \tag{26}$$

$\alpha$, $\beta$ can be defined as follows.

$$\alpha^{s,ncut}_{it} = \hat{\mu}^s_{it} q_i - \hat{\lambda}^s_{it} Q_i$$
$$\beta^{ncut}_s = \sum_{i=1}^{I}\sum_{t=1}^{T}(f_i(\hat{x}^s_{it}) - f_i(0)) + \sum_{t=1}^{T}(K^s_t - \hat{\pi}^s_t)\hat{y}^s_t + \tag{27}$$
$$\sum_{i=1}^{I}\sum_{t=1}^{T}(\hat{\lambda}^s_{it} - \hat{\mu}^s_{it} - \hat{\pi}^s_t)\hat{x}^s_{it} + \sum_{t=1}^{T}\hat{\pi}^s_t d^s_t \tag{28}$$

## 5 Scenario Generation

The contract price and the fixed amount are affected by the total bid volume, the total bid volume, the fuel price, and the date. Make a five-day operation plan for the experiment. Due to the short period, fluctuations in fuel prices are not considered. In addition, the average value of each time zone is used for the total bidding volume $q^M_s$, and the total bidding volume $q^M_b$ is done. In this study, the scenario of electricity price and electricity demand is represented by a scenario tree that branches into two every day. The experimental period was 5 days. When the scenario is branched into two every day, the number of scenarios becomes 16, and the occurrence probability of each scenario $s$ is (1/16) (Fig. 3).

When considering market transactions, the electricity price and electricity demand fluctuate according to the tender trends of market participants. Since the day before the JEPX market is divided into 30 min each day, the number of hours per day is 48 with 30 min as one unit. Similarly, numbers are assigned

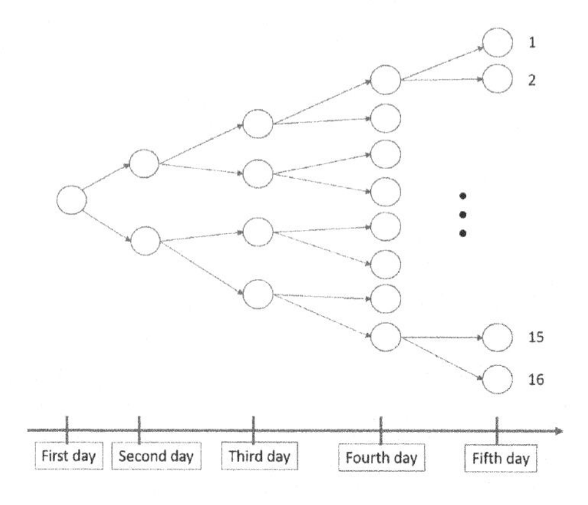


**Fig. 3.** Scenario tree

in order of time zone 1 from 0:00 to 0:30 and time zone 2 from 0:30 to 1:00. For the experiment, we refer to the transaction data for the month of December 2005 published by JEPX. Since the generator operation plan period is 5 days, the number of hours is 240. The electricity price and demand are created with reference to the average total bid amount and average total bid amount for each time period of the day in the month of December 2005. Create a scenario of contract price and contract quantitative change using demand curve and supply curve to consider the fluctuation of electricity price and demand in the market. Knowing the contract price and the contracted volume allows us to predict the demand for the individual business operator. In experiments, we assume that there will be a certain ratio of the demand for the personal business operator to the contracted amount. Perform the experiment (Figs. 4 and 5).

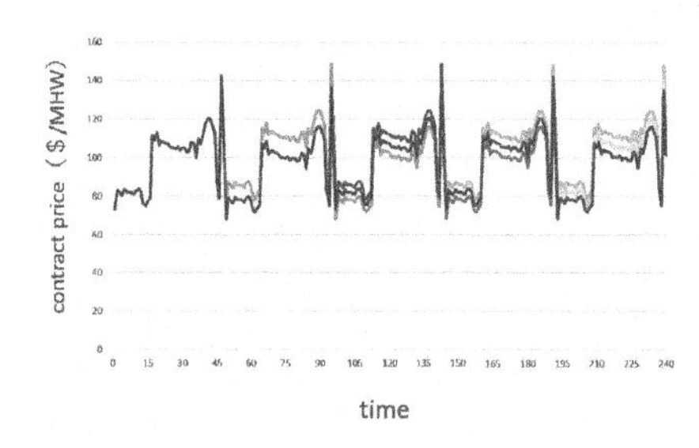


**Fig. 4.** Contract price

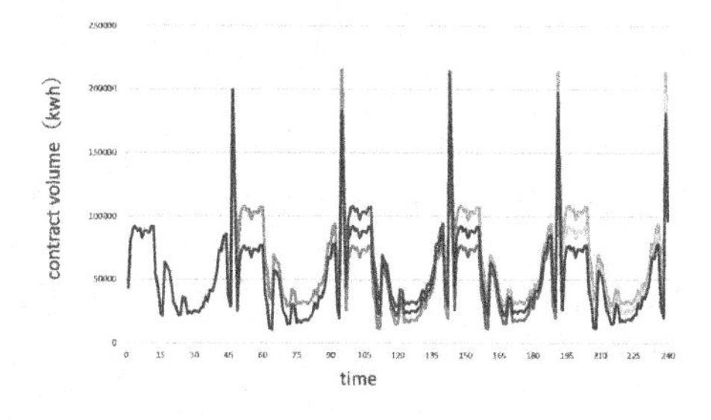


**Fig. 5.** Contract volume

# 6 Numerical Experiments

## 6.1 Value of the Stochastic Solution

A method for comparing a model based on stochastic programming with a deterministic model will be described. Birge and Louveaux [2] defines the evaluation of stochastic programming solutions. The value of the solution of the stochastic programming problem VSS: value of the stochastic solution is used. For the profit maximization problem, VSS is defined as VSS = RP − EEV. Recourse problem (RP) is the optimal objective function value of the stochastic programming problem. Expected result of using the expected value problem (EVV) solution is an optimal objective function value when a plan is made based on the average value of random variables. In the case of a start-stop problem, the EEV uses the start-stop state of each time zone of the generator obtained when solving the problem with the average value of the random variable. By fitting and determining the optimal output and power purchase for each scenario, the expected value of profit obtained based on that is the EEV. The starting and stopping states of the generator at each time zone obtained when solving the problem based on the average value of the random variables are not optimal. Therefore, from the relationship between RP and EEV, RP $\geq$ EEV, VSS $\geq 0$ holds. Assuming that the power demand that the operator must meet in each time zone will be about one or two times as much as the contract volume, the generator operation plans for each demand will be compared. The LNG price used for the fuel price, which is a parameter included in the demand curve and supply curve during the operation planning period, is assumed to be 400($)/t. In addition, we assume the company has 16 generators. We solved the problem with CPLEX 19.0.0 (Table 1).

**Table 1.** RP and EEV

| Assumption of demand to a contract volume | RP | EEV | VSS |
|---|---|---|---|
| Standard demand case | 743,627 | 732,881 | 1,746 |
| High demand case | 1,460,447 | 1,437,104 | 23,343 |

It can be said that a model using stochastic programming that considers power demand and price fluctuations has higher profits than a deterministic model, and is suitable for operation in an actual market where power demand and price fluctuations occur.

## 6.2 Comparision of the Model Considering Market Transactions

The model proposed in this study considers market transactions. In order to show its superiority, a comparison is made with a model used when market transactions used before the liberalization of electricity were not taken into account. A model taking market transactions into account is referred to as model A, and a model not taking market transactions as model B. The following shows the formulation of model B.

$$\max \sum_{s=1}^{S} p_s(\sum_{t=1}^{T} K_t^s d_t^s - \sum_{i=1}^{I}\sum_{t=1}^{T} f_i(x_{it}^s)u_{it}) - \sum_{i=1}^{I}\sum_{t=1}^{T} g_i(u_{i,t-1}, u_{i,t}) \tag{29}$$

$$\text{s.t.} \sum_{i=1}^{I} x_{it}^s \geq d_t^s, t = 1, \cdots, T, s = 1, \cdots, S \tag{30}$$

$$q_i u_{it} \leq x_{it}^s \leq Q_i u_{it}, i = 1, \cdots, I, t = 1, \cdots, T, s = 1, \cdots, S \tag{31}$$

$$r_i \leq x_{it}^s - x_{it-1}^s \leq R_i, i = 1, \cdots, I, t = 1, \cdots, T, s = 1, \cdots, S \tag{32}$$

$$u_{it} - u_{i,t-1} \leq u_{i\tau}, \tau = t+1, \cdots, \min\{t + L_i - 1, T\}, \quad i = 1, \cdots, I, t = 2, \cdots, T \tag{33}$$

$$u_{i,t-1} - u_{it} \leq 1 - u_{i\tau}, \tau = t+1, \cdots, \min\{t + l_i - 1, T\}, \quad i = 1, \cdots, I, t = 2, \cdots, T \tag{34}$$

$$u_{it} \in \{0, 1\}, i = 1, \cdots, I, t = 1, \cdots, T \tag{35}$$

Compare the case where model A is operated using commitment pattern obtained based on the EEV of model B (Table 2).

**Table 2.** Model A and model B

| Assumption of demand to a contract volume | Model A | Model B | Difference |
|---|---|---|---|
| Standard demand case | 743,627 | 736,382 | 7,244 |
| High demand case | 1,460,447 | 1,447,496 | 12,951 |

### 6.3 L-Shaped Solution

Business operators that aim to maximize profits with a generator operation planning model that considers market transactions has higher profits than a generator operation plan model that does not consider market transactions that was used before the liberalization of electricity. From the result, this model is excellent for them. We compared the problem solved using the L-shaped method with the one solved directly. It can be seen that solving the problem using the L-shaped method leads to a significant reduction in calculation time. Further, the objective function value can be calculated with a certain degree of accuracy (Tables 3 and 4).

**Table 3.** L-shaped solution

| Assumption of demand to a contract volume | Profit | Solution time (s) |
|---|---|---|
| Standard demand case | 677,609 | 617 |
| High demand case | 1,307,210 | 493 |

**Table 4.** Direct solution

| Assumption of demand to a contract volume | Profit | Solution time (s) |
|---|---|---|
| Standard demand case | 743,627 | 2185 |
| High demand case | 1,460,447 | 5442 |

## 7 Conclusion

In this study, we proposed a generator operation planning model that considers the market transactions required after electricity liberalization. The generator operation planning model that takes market transactions into account was used before the power liberalization. Profit was larger than the generator operation planning model that did not consider market transactions, indicating that it was an excellent model for operators aiming to maximize profit. Additionally, the model using the stochastic programming method, considering the power demand and price fluctuation, has higher profits than the deterministic model. In the actual market where the power demand and price fluctuates, it can be said that it is suitable for operation. The L-shaped method leads to a significant reduction in calculation time. Further, the objective function value can be calculated with a certain degree of accuracy.

## References

1. Cerisola, S., et al.: Stochastic power generation unit commitment in electricity market: a novel formulation and a comparison of solution methods. Oper. Res. **57**(1), 32–46 (2009)
2. Birge, J.R., Louveaux, F.: Introduction to Stochastic Programming, pp. 137–152. Springer, New York (1997). https://doi.org/10.1007/978-1-4614-0237-4
3. JEPX (Japan Electric Power Exchange). jepx.org
4. Shiina, T.: Stochastic programming. Asakura Shoten, Tokyo (2015)
5. Sakawa, M., Nisizaki, I.: Introduction to Mathematical Programming. Morikita Publishing, Tokyo (2014)
6. Watanabe, I., et al.: Research Report of Central Research Institute of Electric Power Industry, Simulation of Electricity Market-Development of Basic Model Considering Power Supply Start/Stop Plan (2004). (Project Number: R03016)
7. Yamaguchi, N.: Research report of Central Research Institute of Electric Power Industry, Empirical research on JEPX electricity trading trends using simultaneous equation model (2007). (Project number: Y06006)

# Risk Impact Assessment and Prioritization in Multimodal Freight Transportation Systems

Kwanjira Kaewfak[1,2](✉), Veeris Ammarapala[2], and Nantaporn Ratisoontorn[3]

[1] School of Knowledge Science, Japan Advanced Institute of Science and Technology, Ishikawa 923-121, Japan
kwanjira@jaist.ac.jp
[2] School of Management Technology, Sirindhorn International Institute of Technology, Thammasat University, Pathum Thani 12000, Thailand
[3] NECTEC, National Science and Technology Development Agency, Pathum Thani 12120, Thailand

**Abstract.** The major challenge in the development of multimodal freight transportation involves numerous uncertainties and inherent risks from the perspective of stakeholders. Risks are potential threats with impacting critical business operations. Nonetheless, identifying and evaluating risks are more difficult because of the uncertainties and subjectivities of relevant data. This study proposes an integration of fuzzy set theory with failure mode and effects analysis (FMEA) to model multimodal transportation risks quantitatively. The major aim of this study is to identify and prioritize quantitative risks associated with multimodal freight transportation. A qualitative expert-panel interview of operators of multimodal freight transportation in Thailand is conducted. The result helps identify, understand, and prioritize risks related to multimodal freight transportation and also raises user's attention to the high priority risks. The process is useful for industries in minimizing the consequences of risks in multimodal freight transportation.

**Keywords:** Multimodal freight transportation · Risk management · Risk analysis · FMEA · Fuzzy logic · Fuzzy FMEA

## 1 Introduction

Freight transportation is a vital element in logistics system, as it ensures the timely availability and efficient movement of raw materials and finished products. With the dynamic requirements of global supply chains, multimodal transportation is currently a key element of modern transportation systems. Furthermore, the EU Transport Commission reported that road freight transport continues

Supported by SIIT-JAIST-NECTEC Dual Doctoral Degree Program Scholarship.


V.-N. Huynh et al. (Eds.): IUKM 2020, LNAI 12482, pp. 108–121, 2020.
https://doi.org/10.1007/978-3-030-62509-2_10

to grow, in particular, is projected to increase by around 40% by 2030 and by little over 80% by 2050. Therefore, the EU transport policy aims to reduce road transport to less polluting and more energy efficient modes of transport. Over the years, multimodal transportation has become an area of focus because of the problems of road safety, environment concerns, and traffic congestion. However, the major challenges in the development of multimodal freight transportation systems involve numerous uncertainties and inherent transportation risks. Risks are a potential threat that directly impact transportation systems [7]. Especially, in multimodal freight transportation, risks can lead to accidents which directly result in cost, time, and quality of logistics systems. Therefore, to manage and reduce the consequences of risks, it is necessary to discuss the background of the risks related to multimodal freight transportation.

Spadoni *et al.* [13] evaluated individual and societal risk connected with the road transport of dangerous materials between factories. Zaman *et al.* [16] presented the risk assessment of ship collisions. Risk identification and risk evaluation are used as steps of a formal safety assessment for ship collisions. However, very few studies focused on multimodal transportation that risk is an important factor and impacts core transportation operations [7]. Therefore, to minimize these potential risks, this study proposes a novel framework for analyzing and synthesizing risks in multimodal freight transportation on the basis of the fuzzy set theory, using FMEA approach.

In this study, we identified 62 current risks in the literature by presenting a holistic view of risk in multimodal freight transportation. Risk analysis is used to determine the priorities of risks. The procedure combines classifying and assessing transportation risk incidents and the measures to minimize risks with appropriate controls. Following the steps to minimize these potential failures, this study adapts the fuzzy set theory using FMEA method. It examines the potential risks described in prior studies and identifies new risks. A Risk Priority Number (RPN) is used for prioritizing the failure modes. Then, fuzzy rule base is utilized to connect the variables and obtains fuzzy RPN. The objective of this study is to identify and analyze multimodal transportation risks for practical understanding and implementation in business.

This paper is organized as follows. Related work is briefly presented in Sect. 2. Sections 3 presents the application of fuzzy FMEA. Section 4 provides the result and analysis. Section 5 contains the conclusion, limitation and further study.

## 2 Related Work

Multimodal freight transportation is a key element in the freight transportation industry. The main advantage of multimodal transportation solutions is its relatively low external costs. The external cost of a multimodal train per tonne-km is only 28% of the external cost of a general freight truck [4]. Moreover, multimodal freight transportation uses less energy than road transport and makes significant contribution in achieving a sustainable EU transport sector.

Risk has a major impact on multimodal transport and logistics. Past incidents provide an important perspective on the risk factors in transportation systems. An example is the ship collisions in the Malacca Strait [16]. It was defined as a high risk area for navigation, which severely affects the safety of marine traffic. Another serious incident in Italy in 2009, the Viareggio accident was a very serious event involving the transportation of dangerous substances. The resulting flash fire caused severe damage and affected houses and vehicles near the railway [8]. Therefore, risk management and its analysis become a significant concern for transportation activities.

Many studies have examined risk management [5,6], especially in the context of risk in logistics. Generally, The risk management process composes of risk identification and risk assessment. Risk identification is the fundamental phase for identifying risk events and risk sources. Wu and Blackhurst [15] stated that the most significant step in the risk assessment is the determination and classification of the categories of risk. Subsequently, risk can be weighted, compared and quantified. The diversity in risk assessment ensures appropriate techniques for various circumstance. Failure mode and effects analysis (FMEA) is one of proper applications in analyzing and preventing the failures in logistics and transportation.

Failure mode and effects analysis (FMEA) was originated in the 1950s by the aeroplane industry and has been utilized in several applications involving the assessment and prevention of potential risks. In case of logistics and transportation, it is utilized to examine the potential failures and their causes and effects in the freight transportation process. Numerous studies use FMEA in the logistics and transportation sector [7,16]. Kengpol and Tuammee [7] proposed FMEA for analyzing the nature of multimodal transportation risks. FMEA was used to assess the risk score of occurrence, severity, as well as detection of failure modes. Each risk score is calculated based on a scale of 1–10 and computed a RPN to select risk criteria in multimodal transportation. Therefore, the calculation of FMEA can be given as follows.

Firstly, the potential effect of each failure mode is determined. The risks are evaluated by expert opinion and will subsequently be calculated as RPN for each failure mode. The RPN described the overall risk scores is calculated by multiplying the three input factors: severity ($S$): the fact of being severe, occurrence ($O$): possibility of the failures and detection ($D$): likelihood of controls to detect. All input factors are normally determined by weighting on a scale of 1 to 10, as follows:

$$PRN = S \times O \times D \tag{1}$$

However, it is ambitious to accurately estimate the possibility of failure events or risks using FMEA calculation because of the RPN limitations in the assessment. Generally FMEA operation determines scores between 1 to 10 that is allowed to determine the probability of $S$, $O$, and $D$ [2]. These scores probably present in the same RPN value however the impact of these risks that were concealed may be different. Moreover, RPN scores are assessed from different expert knowledge and experience; thus results might be less accurate [2].

The fuzzy logic was initially presented by Zadeh in 1965, which was oriented to the rationality of uncertainty order to imprecision or vagueness. A significant contribution of fuzzy logic is its capability of representing vague data. A fuzzy logic is a class of objects with a continuum of grades of membership. The membership of an element to a fuzzy set is a single value between zero and one. Various works now combine fuzzy concepts with other scientific disciplines as well as modern technologies. Triangular fuzzy numbers (TFNs) are used in this applications due to their computational simplicity and useful fuzzy information processing. The definitions and algebraic operations are described as follows.

TFNs can be defined by a triplet $(l, m, u)$ and its membership function $\mu_A(x)$ can be defined by Eq. (2). Equation (2), $l$ and $u$ are the lower and upper bounds of the fuzzy number $A$, respectively and $m$ is the modal value for $A$.

$$\mu_A(x) = \begin{cases} \dfrac{x-l}{m-l} & ,l \leq x \leq m \\ \dfrac{u-x}{u-m} & ,m \leq x \leq u \\ 0 & ,otherwise \end{cases} \tag{2}$$

(where $\mu_A(x)$ is a real number of the interval [0,1] and $l$, $m$, $u$ are non-negative real numbers)

Define two TFNs A and B by the triplets A $= (l_1, m_1, u_1)$ and B $= (l_2, m_2, u_2)$
Then: $Addition: A(+)B = (l_1, m_1, u_1)(+)(l_2, m_2, u_2) = (l_1+l_2, m_1+m_2, u_1+u_2)$
$Multiplication: AB = (l_1, m_1, u_1)(l_2, m_2, u_2) = (l_1 l_2, m_1 m_2, u_1 u_2)$,
$Inverse = (l_1, m_1, u_1)^{-1} \approx (\frac{1}{u_1}, \frac{1}{m_1}, \frac{1}{l_1})$

In general, FMEA is used to determine the risk priority of failure modes, which is divided into two phases: (i) define the different failure modes and their impacts, (ii) identify failure modes by the probability of occurrence and their severity [1,11]. Many researchers mention the deficiency of conventional RPN used in FMEA. Chin *et al.* [3] noted that the assessment grade of FMEA is often subjective and qualitatively described in natural language. As a result, it is relatively hard to calculate the possibility of failure mode. Lin *et al.* [10] indicated that a combination of occurrence, severity and detection presenting exactly the same RPN value may have different risk implications. Furthermore, the mathematical formula for calculating RPN is still questionable and debatable.

Fuzzy logic is derived from fuzzy set theory and intended to model logical reasoning with vague or imprecise statements [11]. Fuzzy logic that uses linguistics variables can be a useful model. Linguistics variables are easier to use than numerical assignments for describing failure modes. Therefore, fuzzy logic method integrated FMEA approach was proposed. Fuzzy FMEA used the linguistic variables to assess the risks related with failure directly. Furthermore, fuzzy FMEA can deal with vagueness and qualitative data. Consequently the structure of the combination $S$, $O$ and $D$ was more flexible.

The literature contains the applications of fuzzy FMEA. Renjith *et al.* [11] developed a fuzzy FMEA for prioritizing failure modes in the case of LNG storage facilities. Li *et al.* [9] used fuzzy FMEA to evaluate the potential risks of road transportation of hazardous materials.

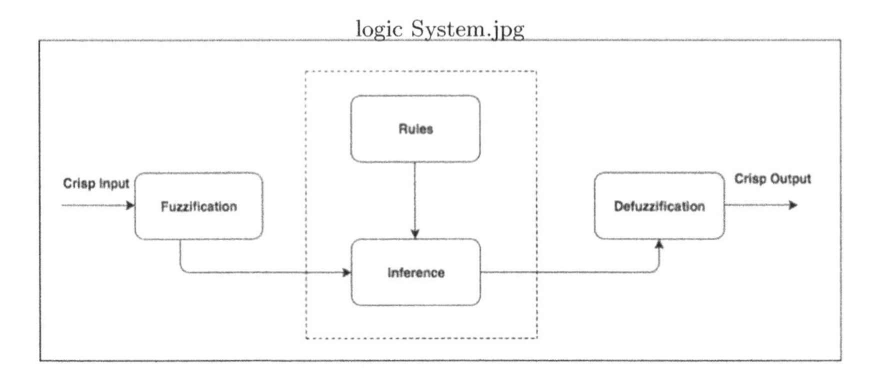


**Fig. 1.** Diagram of the fuzzy rule base system.

The fuzzy logic system has been used to classify and evaluated risks according to the rules of fuzzy logic. Figure 1 represents the basic diagram of the fuzzy rule base system [2].

**The algorithm of fuzzy rule base** has the following steps:

(i) Initialization: identify the linguistic variables, create the membership functions, and set the rule base.
(ii) Fuzzification: convert crisp input data into fuzzy values using the membership functions.
(iii) Inference: assess the rules in the rule base and combine the results of each rule.
(iv) Defuzzification: Convert the output data into non-fuzzy values.

Primarily, the use of fuzzy logic in FMEA can provide a detailed explanation of component failures and accident scenarios. The overall data of linguistic variable was utilized to express the $S$, $O$, and $D$. These data will then be the input of fuzzification to define the degree of membership in each input [2]. The input will be analyzed by the fuzzy rule base. Then the output in defuzzification step will be converted into crisp output to prioritize the failure modes. To achieve the outcome, the determination of each rule will be combined at the fuzzy inference stage. The technique for evaluation called the Mamdani (max–min) inference is used to obtain the result. Then, output interface transmits results of fuzzy inference into output.

After the defuzzification process, the result are modified to be crisp output. A popular technique used in this process was the center of gravity (COG) method [2]:

$$COG = \frac{\int_a^b \mu_A(X)\, x dx}{\int_a^b \mu_A(X)\, dx} \tag{3}$$

## 3 Fuzzy FMEA Application

The proposed Fuzzy FMEA model has the following steps:

**Step I:** Define risks in multimodal transportation.

This step determines the various potential risks involving multimodal transportation in Thailand. The study utilizes a qualitative expert opinion system to collect and manage risk data in multimodal freight transportation. The decision-making environment relies on the opinions of experts.

**Step II:** Determine all potential failure modes and effects in multimodal transportation risks.

The second step studies the risks involved in multimodal transportation to implement and creates the questionnaires used for data collection. Then, we estimate all failure modes, impact, and current control of each failure mode [2]. In this stage, the study identified 62 current transportation risks by presenting a holistic view of risks in multimodal freight transportation.

**Step III:** Estimate the $S$, $O$, and $D$ ratings of each failure mode by FMEA users.

Generally, during the process of risk analysis, risk scores would be assessed using three parameters: occurrence, severity, and detection. In traditional risk assessment, occurrence, severity and detection are described as crisp values. The classical FMEA considered 10 classes, as shown in Table 1, 2 and 3. The scores of $S$, $O$ and $D$ are obtained by the assessment, according to Eq. (1). These scores rely on the knowledge and experience of experts. The overall scores are computed regarding to the Geometric Mean (GM) method. The GM method is employed to calculate each score and prioritize the failures. The problems will be presented with FRPN higher than or equal to average FRPN. They must be solved and improved accordingly [2].

**Table 1.** Occurrence scales used in FMEA [11].

| Rank | Probability of failures | Linguistic variable |
|---|---|---|
| 1 | < 1:20000 | Unlikely |
| 2 | 1:20000 | Very remote |
| 3 | 1:10000 | Remote |
| 4 | 1:2000 | Very low |
| 5 | 1:1000 | Low |
| 6 | 1:200 | Moderate |
| 7 | 1:100 | Moderate high |
| 8 | 1:20 | High |
| 9 | 1:10 | Very high |
| 10 | 1:2 | Almost certain |

**Step IV:** Compute the RPN scores using algorithm of fuzzy set theory.

The original computation of RPN results in a value scale from 1 to 1000. The RPN value obtained from the experts shows the problem which does not affect the corrective actions too much [2]. Therefore, the RPN scores are modified into fuzzy values using membership functions and linguistic expression. Table 4 shows the failure classification on the calculation of RPN.

Due to several uncertainties, these values are non-crisp in nature [11]. Therefore, fuzzy logic can be used for determining the risks involving uncertainties in the parameters of risk analysis. The crisp input values of $S$, $O$, and $D$ are normally measured on a 10-point scale [11]. These values are turned into fuzzy values using the membership function and linguistic expression. In classical FMEA, it is difficult to determine whether measures should be taken in regard to those risks. The 5-scale fuzzy FMEA is applied throughout all classes; therefore, this risk analysis would be more precise. The membership function values are generated by the fuzzy FMEA scale. For example, the following fuzzy scale of $S$, $O$, and $D$ are defined as {Very High, High, Medium, Low and Very low} = {(7,9,10),(5,7,9),(3,5,7),(1,3,5),(0,1,3)} respectively.

**Table 2.** Severity scales used in FMEA [11].

| Rank | Severity of each effect of failure or error | Effect |
|---|---|---|
| 1 | No reason to expect failure to have any effect on safety | None |
| 2 | Very minor effect on system performance to have any effect on safety | Minor |
| 3 | Minor effect on system performance to have any effect on safety | Minor |
| 4 | A failure is not serious enough to cause system damage, but can result in unscheduled maintenance | Low |
| 5 | The system requires repair. A failure that may cause moderate property damage | Moderate |
| 6 | System performance is degraded. A failure causes system damage | Significant |
| 7 | System performance is severely affected but functions. The system may not operate | Major |
| 8 | System is inoperable with loss of primary function. Failure can involve hazardous outcomes | Extreme |
| 9 | Failure involves hazardous outcomes. Failure will occur with warning | Serious |
| 10 | Failure is hazardous and occurs without warning. A failure is serious enough to cause system damage | Hazardous |

**Step V:** Ranking FRPN.

In this study, potential risks in multimodal freight transportation are calculated using 62 current transportation risks on the basis of the corresponding RPN value, as shown in Table 4.

In the overview of the fuzzy logic process, five classes of the fuzzy FMEA scale are reformed from the classical FMEA scale for multimodal freight transportation risks. After that they are coded using the Fuzzy Logic Designer Tool in Matlab. This study develops the fuzzy design (Mamdai min-max) using the input and output. Subsequently, the $S$, $O$ and $D$ parameters and if-then rules (based on the expert knowledge) are generated with linguistic variables as inputs

**Table 3.** Detection scales used in FMEA [11].

| Rank | Likelihood of detection of failure or error | Degree of importance |
|---|---|---|
| 1 | Current control almost certainly will detect a potential failure mode | Almost certain |
| 2 | Very high likelihood that current control will detect failure modes | Very high |
| 3 | High chance that the design control will almost certainly detect a potential failure mode | High |
| 4 | Moderately high likelihood current control will detect failure modes | Moderately high |
| 5 | Moderate chance that the design control will detect a potential failure mode | Moderately |
| 6 | Low likelihood that current control will detect failure modes | Low |
| 7 | Very low likelihood that current control will detect failure modes | Very low |
| 8 | Remote chance that the design control will detect a potential failure mode | Remote |
| 9 | Defect most likely remains undetected | Very remote |
| 10 | System failures are not detected | Almost impossible |

to evaluate the risks [11]. Finally, the FRPN score is the output of the fuzzy inference system. Similarly, the membership function of the output (FRPN) is also generated. The process of fuzzification uses the triangular membership function to present the input $S$, $O$, and $D$ (Fig. 2a) and the output membership function corresponding to FRPN (Fig. 2b).

The 125 rules (5·5·5) are created to control the output value by the rule constructed using the MATLAB Fuzzy Logic Toolbox (Fig. 3). These rules are created to include combinations of $S$, $O$, and $D$. The rules defined by logistics and transportation experts, for example:

Rule# 1: **If** Occurrence = very low **and** Severity = very low **and** Detection = very low **then** Risk = very low.
Rule# 2: **If** Occurrence = very low **and** Severity = very low **and** Detection = low **then** Risk = very low.
Rule# 3: **If** Occurrence = very low **and** Severity = very low **and** Detection = medium **then** Risk = low.

The result of the defuzzification process using the center of gravity method (Eq. 3) are evaluated based on the conclusions of the fuzzy values. The hazards or risks considered in this study are ranked according to risk analysis experts in Table 5. Thus, the corrective actions of multimodal transportation area are prioritized according to FRPN scores.

**Step VI:** Corrective actions of multimodal transportation areas with high FRPN.

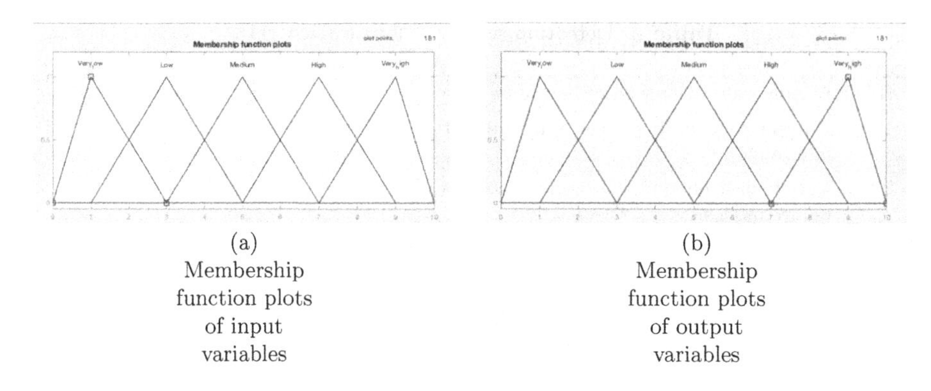


(a) Membership function plots of input variables

(b) Membership function plots of output variables

**Fig. 2.** Membership function plots.

fuzzy.png

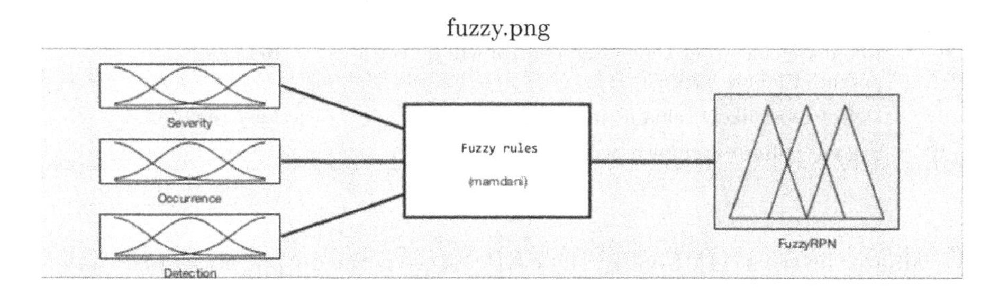


**Fig. 3.** MATLAB fuzzy logic toolbox.

**Table 4.** Failure classification on basis of RPN.

| Rank | Range | Linguistic variable |
|---|---|---|
| 1 | RPN $<$ 5 | None |
| 2 | $5 \leq$ RPN $\leq 10$ | Very low |
| 3 | $10 \leq$ RPN $\leq 20$ | Low |
| 4 | $20 \leq$ RPN $\leq 50$ | High low |
| 5 | $50 \leq$ RPN $\leq 100$ | Low medium |
| 6 | $100 \leq$ RPN $\leq 150$ | Medium |
| 7 | $150 \leq$ RPN $\leq 250$ | High medium |
| 8 | $250 \leq$ RPN $\leq 350$ | Low high |
| 9 | $300 \leq$ RPN $\leq 500$ | High |
| 10 | RPN $>$ 500 | Very high |

This study has the advantage of eliminating failures that would likely occur in the working process of transportation. This prioritization of risk could help users to select the action wisely and reduce time consumption. The result shows that the risk with the highest score is in freight damaged transportation.

## 4 Results and Analysis

The selected multimodal freight transportation and case organizations were chosen in collaboration. All the organizations interviewed are part of the multimodal freight transportation process and act in various roles, representing different parts of the logistics and transportation sector. This was considered essential for an accurate analysis. The interviews revealed that the risk scores were identified varied between the experts. Therefore, reliable prior literature and expert knowledge are used to identify risks in multimodal freight transportation.

The risk assessment based on the traditional FMEA method is calculated using the RPN calculation method. The $S$, $O$, and $D$ are computed using multiplication according to the traditional FMEA method by which results are presented in different RPN values. It is evident that there is only one failure with RPN $> 500$. This means that the failure is very small (less than half), and therefore, corrective actions are not required [2]. The main drawback of the expert interview is that they cannot express their opinions against the team members. Therefore, bias in the risk assessment is possible in this process. The assessment of FRPNs in continuous value of data presents scores on a scale of 1 to 10, which is simple and prominent for users. The evaluation format of Fuzzy FMEA with individual scores is unique to users. The Fuzzy FMEA can represent the concept and evaluation without bias in assessing the scores from expert team. In summary, we found 62 failures by using the Fuzzy FMEA methodology, with 30 failures requiring high priority because the FRPN is greater than 5.51, on average.

Table 5 provides a break-down of the risk analysis results. Of the 62 risks identified in the interviews and literature, critical failure or risks have been identified on the basis of corresponding RPN and fuzzy RPN value. We observed that many critical failures give identical RPN values, which makes it difficult to prioritize these risks. After the defuzzification process, new FRPN values are obtained. The ranking in Table 5, we can prioritize these risks based on their fuzzy RPN scores, for example, the first top three risks are damages in transportation, hazardous materials, and fuel price with fuzzy RPN scores of 7.93, 7.50, and 7.11, respectively.

**Table 5.** RPN, FRPN and prioritization of failures.

| Risk No. | Failure mode | S | O | D | RPN | RPN priority | FRPN | FRPN priority |
|---|---|---|---|---|---|---|---|---|
| 1 | Damages in transportation [6,7] | 8 | 9 | 6 | 502 | 1 | 7.93 | **1** |
| 2 | Fuel price [12,14] | 5 | 9 | 8 | 414 | 2 | 7.11 | **3** |
| 3 | Unpredictability of regulation and law [12] | 4 | 7 | 8 | 246 | 8 | 6.57 | 10 |
| 4 | Strikes [6,14] | 7 | 6 | 7 | 265 | 7 | 6.81 | 6 |
| 5 | Border policies [14] | 4 | 6 | 8 | 231 | 11 | 6.37 | 14 |
| 6 | Fierce competition [14] | 5 | 6 | 7 | 216 | 15 | 6.57 | 10 |
| 7 | Bottlenecks in routes [7] | 5 | 8 | 5 | 220 | 14 | 5.6 | 27 |
| 8 | Economic crisis [14] | 5 | 5 | 9 | 221 | 13 | 6.65 | 8 |
| 9 | Operating in heavy traffic [7] | 6 | 8 | 7 | 326 | 5 | 6.93 | 5 |
| 10 | Difficult navigability [14] | 5 | 4 | 4 | 99 | 49 | 4.39 | 53 |
| 11 | Storms [7] | 4 | 3 | 4 | 49 | 61 | 3.87 | 60 |
| 12 | Internet connection [14] | 5 | 6 | 4 | 129 | 37 | 5.5 | 31 |
| 13 | Railway network capacity [14] | 5 | 5 | 6 | 161 | 27 | 5.78 | 21 |
| 14 | IT vulnerability [6,12] | 7 | 8 | 3 | 191 | 22 | 6.47 | 12 |
| 15 | Fire [14] | 9 | 5 | 4 | 180 | 24 | 4.67 | 49 |
| 16 | Route restrictions [14] | 5 | 7 | 5 | 203 | 18 | 5.56 | 29 |
| 17 | Inland traffic accident [6,14] | 6 | 9 | 4 | 236 | 10 | 5.97 | 19 |
| 18 | Maritime accidents [6] | 5 | 5 | 4 | 110 | 44 | 5.6 | 27 |
| 19 | Capacity of road network [7,14] | 5 | 8 | 5 | 207 | 16 | 5.67 | 26 |
| 20 | Capacity problem in railroad traffic [14] | 5 | 5 | 5 | 128 | 38 | 5.49 | 32 |
| 21 | Terrorism [6] | 9 | 4 | 5 | 152 | 30 | 3.72 | 61 |
| 22 | Cargo being stolen or tampered [14] | 9 | 4 | 5 | 159 | 28 | 4.32 | 55 |
| 23 | Unpredictable traffic congestion [7,14] | 6 | 8 | 6 | 291 | 6 | 6.17 | 17 |
| 24 | Explosion of gas line [8,14] | 9 | 3 | 5 | 143 | 33 | 3.29 | 62 |
| 25 | Infrastructure limitations [12,14] | 4 | 3 | 4 | 48 | 62 | 3.97 | 59 |
| 26 | Customs formalities and ambiguities [6,14] | 4 | 5 | 5 | 113 | 43 | 5.06 | 41 |
| 27 | Wind [14] | 5 | 3 | 5 | 86 | 54 | 5.34 | 36 |
| 28 | Traffic route breakdowns [12,14] | 6 | 6 | 4 | 147 | 32 | 5.78 | 21 |
| 29 | Heavy rains [7,14] | 5 | 7 | 5 | 196 | 21 | 5.56 | 29 |
| 30 | Driving time legislation [14] | 5 | 7 | 4 | 148 | 31 | 5 | 42 |
| 31 | Intersecting rail and road traffic [7,14] | 4 | 4 | 5 | 86 | 55 | 4.25 | 57 |
| 32 | Floods [7,14] | 5 | 4 | 4 | 91 | 53 | 4.59 | 51 |
| 33 | Climate change [7,14] | 4 | 6 | 4 | 107 | 46 | 5 | 42 |
| 34 | On-time/on-budget delivery [6] | 8 | 8 | 4 | 241 | 9 | 6 | 18 |
| 35 | Port capacity and congestion [6] | 4 | 6 | 5 | 109 | 45 | 5 | 42 |
| 36 | Demand Volatility [12] | 5 | 4 | 3 | 77 | 58 | 4.6 | 50 |
| 37 | Shippers breaking the contract [14] | 5 | 4 | 4 | 74 | 59 | 4.28 | 56 |
| 38 | Excessive handling due to border crossing [6] | 6 | 7 | 4 | 161 | 25 | 6.34 | 15 |
| 39 | Port congestion [14] | 6 | 6 | 4 | 156 | 29 | 5.73 | 23 |
| 40 | Ship collisions [14] | 5 | 5 | 5 | 125 | 40 | 5.2 | 40 |

*(continued)*

**Table 5.** (*continued*)

| Risk No. | Failure mode | S | O | D | RPN | RPN priority | FRPN | FRPN priority |
|---|---|---|---|---|---|---|---|---|
| 41 | Changes in schedules [14] | 5 | 5 | 3 | 80 | 56 | 5 | 42 |
| 42 | Railway operator's low service [14] | 5 | 5 | 6 | 131 | 36 | 5.71 | 24 |
| 43 | Insufficient personnel [14] | 5 | 4 | 3 | 67 | 60 | 4.15 | 58 |
| 44 | Fault in cargo/traffic control systems [14] | 6 | 7 | 5 | 199 | 20 | 5.97 | 19 |
| 45 | Chemicals [8,14] | 8 | 7 | 7 | 352 | 4 | 6.62 | 9 |
| 46 | Lack of skilled workers [7,12,14] | 8 | 6 | 4 | 187 | 23 | 4.99 | 47 |
| 47 | Stoppages made with cargo onboard [14] | 7 | 5 | 6 | 222 | 12 | 6.67 | 7 |
| 48 | Shipment quality information [14] | 5 | 3 | 5 | 94 | 51 | 5.47 | 33 |
| 49 | Railway operator's monopoly [14] | 5 | 3 | 6 | 91 | 52 | 5.24 | 38 |
| 50 | Shipment capacity [14] | 4 | 5 | 4 | 77 | 57 | 4.38 | 54 |
| 51 | Document interpretation problems [14] | 8 | 6 | 4 | 205 | 17 | 5.44 | 34 |
| 52 | Shipping company's monopoly [14] | 5 | 5 | 5 | 121 | 41 | 5.23 | 39 |
| 53 | Cargo handling equipment condition [14] | 4 | 4 | 5 | 96 | 50 | 4.38 | 54 |
| 54 | Insufficient railroad wagons [14] | 4 | 5 | 6 | 120 | 42 | 5.36 | 35 |
| 55 | Insufficient documentation [7,14] | 6 | 4 | 5 | 140 | 34 | 6.21 | 16 |
| 56 | Organizational borders [14] | 4 | 5 | 6 | 127 | 39 | 5.71 | 24 |
| 57 | Terminal's monopoly [14] | 5 | 4 | 5 | 101 | 48 | 4.58 | 52 |
| 58 | Lack of multimodal equipment [14] | 7 | 6 | 5 | 202 | 19 | 6.94 | 4 |
| 59 | Other hazardous materials [8,14] | 8 | 8 | 6 | 369 | 3 | 7.5 | **2** |
| 60 | Transport providers' fragmentation [6] | 5 | 5 | 5 | 131 | 35 | 5.25 | 37 |
| 61 | Escalating rate of pollution [7,12] | 7 | 5 | 4 | 161 | 26 | 6.39 | 13 |
| 62 | Unexpected port/cargo handling productivity [14] | 4 | 5 | 5 | 101 | 47 | 5 | 42 |

## 5 Conclusion, Limitation and Further Study

Multimodal freight transportation is increasingly complex and vulnerable to various risks. It is difficult to predict the process, as it involves risk in all activities. From the managerial perspective, risks are potential threats that can have undesirable impacts on normal activities or prevent action. Especially, multimodal freight transportation is the integration of two or more modes of transport to move goods from the source to the destination. Accidents can occur at any point during the transportation and logistics processes. The main objective of this study was to identify and assess the potential risks effecting multimodal freight transportation in Thailand. We investigate the major visibility risks involved with complex multimodal freight transportation. With the help of prior literature and expert knowledge, 62 potential risks were identified. We found the fuzzy FMEA approach to be an effective tool for prioritizing the critical failure of components in complex systems. The results of our study provide several implications for managerial practice in the realm of multimodal freight transportation and risk management. The analysis divided the risks and the nature of their impacts of severity, occurrence and detection. Furthermore, the results suggest that a logistics and transportation manager should consider the source and nature of risk impacts to minimize the consequences of risks in multimodal freight transportation.

However, this empirical study has limitations. Risk assessment cannot be evaluated solely from total scores and these factors should be adjusted from the expert opinions. In addition, the interviewees are mostly the experts in logistics and transportation in Thailand, which might introduce various perspectives on factors affecting risk assessment. The majority of data acquired in this study is also subjective to the economic environment of Thailand. Therefore, the factors must be adjusted before they can be applied to other case studies.

The main contributions of this study are a novel model and reliable approach for prioritizing risks. This risk analysis helps managers focus on the most relevant risk in multimodal freight transportation and take corrective actions. For further research, we plan to develop a new platform to analyze risk systematically and extend to other complex and critical installations.

## References


1. Bowles, J.B., Peláez, C.: Fuzzy logic prioritization of failures in a system failure mode, effects and criticality analysis. Reliab. Eng. Syst. Saf. **50**(2), 203–213 (1995)
2. Chanamool, N., Naenna, T.: Fuzzy FMEA application to improve decision-making process in an emergency department. Appl. Soft Comput. **43**, 441–453 (2016)
3. Chin, K.S., Chan, A., Yang, J.B.: Development of a fuzzy FMEA based product design system. Int. J. Adv. Manuf. Technol. **36**(7–8), 633–649 (2008)
4. Forkenbrock, D.J.: Comparison of external costs of rail and truck freight transportation. Transp. Res. Part A Policy Pract. **35**(4), 321–337 (2001)
5. Hallikas, J., Lintukangas, K.: Purchasing and supply: an investigation of risk management performance. Int. J. Prod. Econ. **171**, 487–494 (2016)
6. Ho, W., Zheng, T., Yildiz, H., Talluri, S.: Supply chain risk management: a literature review. Int. J. Prod. Res. **53**(16), 5031–5069 (2015)
7. Kengpol, A., Tuammee, S.: The development of a decision support framework for a quantitative risk assessment in multimodal green logistics: an empirical study. Int. J. Prod. Res. **54**(4), 1020–1038 (2016)
8. Landucci, G., Antonioni, G., Tugnoli, A., Bonvicini, S., Molag, M., Cozzani, V.: HazMat transportation risk assessment: a revisitation in the perspective of the Viareggio LPG accident. J. Loss Prev. Process Ind. **49**, 36–46 (2017)
9. Li, Y.L., Yang, Q., Chin, K.S.: A decision support model for risk management of hazardous materials road transportation based on quality function deployment. Transp. Res. Part D Transp. Environ. **74**, 154–173 (2019)
10. Lin, Q.L., Wang, D.J., Lin, W.G., Liu, H.C.: Human reliability assessment for medical devices based on failure mode and effects analysis and fuzzy linguistic theory. Saf. Sci. **62**, 248–256 (2014)
11. Renjith, V., Jose kalathil, M., Kumar, P.H., Madhavan, D.: Fuzzy FMECA (failure mode effect and criticality analysis) of LNG storage facility. J. Loss Prev. Process Ind. **56**, 537–547 (2018)
12. Shankar, R., Choudhary, D., Jharkharia, S.: An integrated risk assessment model: a case of sustainable freight transportation systems. Transp. Res. Part D Transp. Environ. **63**, 662–676 (2018)

13. Spadoni, G., Leonelli, P., Verlicchi, P., Fiore, R.: A numerical procedure for assessing risks from road transport of dangerous substances. J. Loss Prev. Process Ind. **8**(4), 245–252 (1995)
14. Vilko, J., Ritala, P., Hallikas, J.: Risk management abilities in multimodal maritime supply chains: visibility and control perspectives. Accid. Anal. Prev. **123**, 469–481 (2019)
15. Wu, T., Blackhurst, J. (eds.): Managing Supply Chain Risk and Vulnerability: Tools and Methods for Supply Chain Decision Makers. Springer, London (2009). https://doi.org/10.1007/978-1-84882-634-2
16. Zaman, M.B., Kobayashi, E., Wakabayashi, N., Khanfir, S., Pitana, T., Maimun, A.: Fuzzy FMEA model for risk evaluation of ship collisions in the Malacca Strait: based on AIS data. J. Simul. **8**(1), 91–104 (2014)

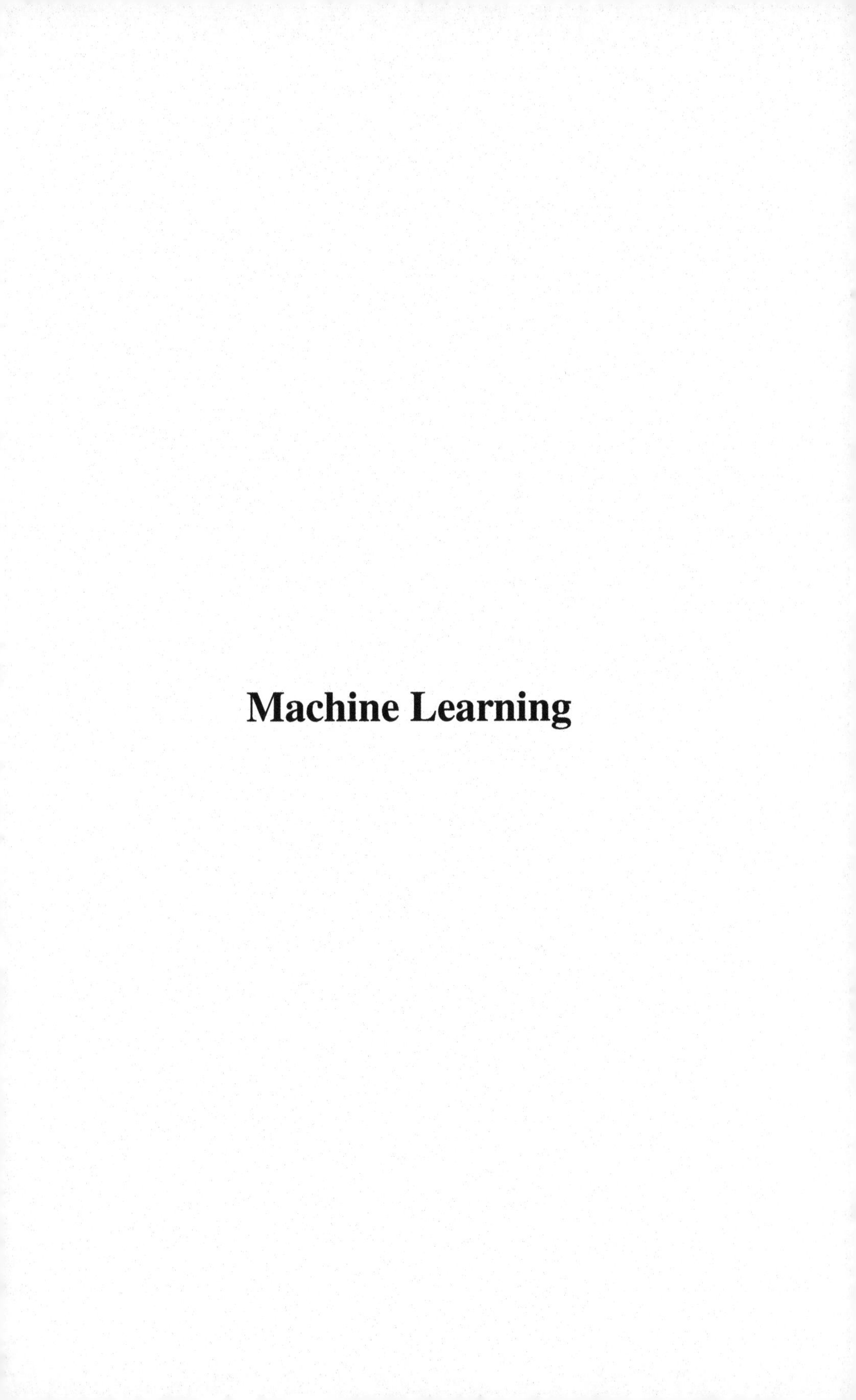

# Machine Learning

# Dynamic Features Spaces and Machine Learning: Open Problems and Synthetic Data Sets

Sema Kayapinar Kaya[1], Guillermo Navarro-Arribas[2], and Vicenç Torra[3](✉)

[1] Department of Industrial Engineering, Munzur University, Tunceli, Turkey
[2] Department of Information and Communications Engineering, CYBERCAT-Center for Cybersecurity Research of Catalonia, Universitat Autònoma de Barcelona, Bellaterra, Spain
guillermo.navarro@uab.cat
[3] Department Computing Sciences, Umeå University, Umeå, Sweden
vtorra@ieee.org

**Abstract.** Dynamic feature spaces appear when different records or instances in databases are defined in terms of different features. This is in contrast with usual (static) feature spaces in standard databases, where the schema of the database is known and fixed. Then, all records in the database have the same set of variables, attributes or features. Dynamic feature mining algorithms are to extract knowledge from data on dynamic feature spaces. As an example, spam detection methods have been developed from a dynamic feature space perspective. Words are taken as features and new words appearing in new emails are, therefore, considered new features. In this case, the problem of spam detection is represented as a classification problem (a supervised machine learning problem).

The relevance of dynamic feature spaces is increasing. The large amounts of data currently available or received by systems are not necessarily described using the same feature spaces. This is the case of distributed databases with data about customers, providers, etc. Industry 4.0, Internet of Things, and RFIDs are and will be a source of data in dynamic feature spaces. New sensors added in an industrial environment, new devices connected into a smart home, new types of analysis and new types of sensors in healthcare, all are examples of dynamic feature spaces. Machine learning algorithms are needed to deal with these type of scenarios.

In this paper we motivate the interest for dynamic feature mining, we give some examples of scenarios where these techniques are needed, we review some of the existing solutions and its relationship with other areas of machine learning and data mining (e.g., incremental learning, concept drift, topic modeling), we discuss some open problems, and we discuss synthetic data generation for this type of problem.

**Keywords:** Dynamic feature space · Dynamic feature mining · Clustering · Classification · Association rule mining


V.-N. Huynh et al. (Eds.): IUKM 2020, LNAI 12482, pp. 125–136, 2020.
https://doi.org/10.1007/978-3-030-62509-2_11

# 1 Introduction

A dynamic feature space corresponds to the case in which the description of the data changes over time. That is, the set of variables, attributes, or features that describe each of the objects of our concern change as time passes. This is in contrast with standard databases in which we have a static feature space with all variables well defined from the start.

Dynamic feature space is more general than other concepts that appear in the literature of data mining as data streams and concept drift. The main characteristics of data streams is that information (e.g., records) arrive continuously and need to be processed in real time. That is, we cannot wait to make decisions until the *last* piece of information arrives. In addition, data streams tend to be very large. In most cases, the feature space of the data is fixed a priori (e.g., information from a set of fixed sensors).

Concept drift is also usually processed with static feature spaces. Concept drift is when high level information (the concepts) extracted from low level information (e.g., the records) change with respect to time. This can be due to changes in the sources generating this low level information. For example, in a city, squares hold different type of activities (street market, parades, festivals, holiday traffic, night traffic, daily busy traffic, etc.). An intelligent system monitoring public spaces by means of a set of sensors and video cameras may need to take into account concept drifts due to the different type of activities that may take place, as e.g. we need to take into account that *normal activity* will change its pattern. Another example is when monitoring people health over large periods of time (i.e., years). In this case, we also need to take into account that concepts as e.g. *good health* will need to be adapted because of age.

Concept drift techniques have been extensively applied to spam detection. Classifiers need to change over time their model of what a spam message is. This is because the sources (spammers) generating spam messages also change the way they produce them.

Other areas where concept drift can be applied include the Internet of Things (IoT) and Industry 4.0. The Internet of Things is defined as the network of communication –via internet– of physical objects or *things* that are embedded with e.g. sensors, internet connectivity, actuators, Radio-Frequency Identification (RFID) tags, and mobile phones (see [6,12]).

The term Internet of Things (IoT), which emerged in the early 21st century, is the most important technological component of the basic philosophy of Industry 4.0. Kevin Ashton, co-founder of MIT's Auto-ID Center, introduced the term of Internet of Things in 1999 [6]. The implementation of IoT, more particularly within the industry, is supported by the integration of several effective technologies such as RFID and wireless sensor networks (WSN). RFIDs are crucial components as they can monitor goods and materials at a certain distance without touching them. WSN are constituted of sensor nodes where each node is equipped with sensors to monitor in a cooperative manner physical conditions such as heat, pressure, vibration, pH level, humidity, and noise level [6]. WSN typically contain hundreds or even thousands of sensor nodes that are able to automatically sense and exchange information among different devices. See Fig. 1. Reliability, accuracy, flexibility, cost-effectiveness, and ease of installation are the top features of WSNs. Since there is no need for infrastructure or human intervention, WSN is easily installed. They perform their tasks by perceiving, calculating and taking action in the environment. They can organize themselves (self-organization) and adapt to support different practices. They are a useful component of Industry 4.0.

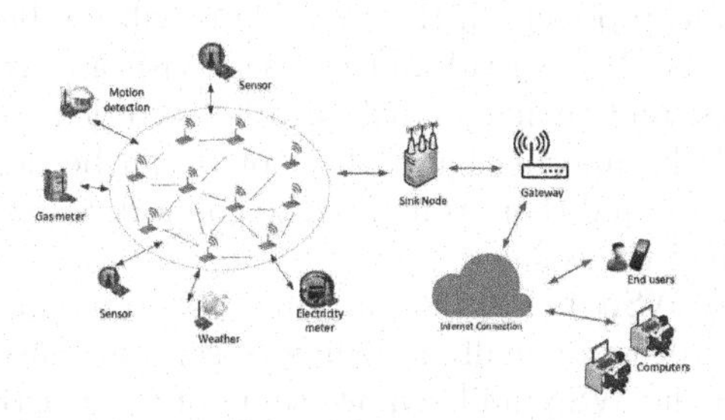


**Fig. 1.** The basic architecture of a Wireless Sensor Network (WSN).

It is usual that in Industry 4.0 and Internet of Things, all processes change over time (in some cases imperceptible at first, but noticeable at last). In order that predictions are still valid after a while and avoid machine learning degradation [23] (model decay), concept drift needs to be taken into account. In these areas we have a static feature space when e.g. the set of sensors do not change over time.

While concept drift is relevant in areas as Industry 4.0 and IoT, it is not enough for systems that are expected to have an extended life cycle. In this case, it is necessary to deal with dynamic feature spaces. Pervasiveness of systems for sensing and storing needs to take into account that the features stored in the past may be different from those stored later. Some features will disappear and new features added. Old features may become obsolete and be removed, sensors retrieving data may become faulty and stop working. On the other hand, new sensors will be added and new variables will be defined.

This is the case in Industry 4.0 and IoT where features correspond to sensors in industrial environments, or values from devices connected into the cloud. In medical databases with data about patients, new features may correspond to new types of analysis and results from new exploration procedures. At the same time, old exploration procedures and analysis will not be used anymore.

### 1.1 Some Scenarios

We can consider the following scenarios that can be modeled with a dynamic feature space.

- **Monitoring customers and visitors in a shop.** Let us consider the case of a shop where RFID readers are located at the entrance of the shop so that all RFIDs that are worn by customers are read. This can be modeled as a dynamic feature space. This is so because it is not possible to know a priori the set of all existing RFIDs and, even if this were possible, we cannot be sure on the RFIDs available in the future. A database is populated with all the information read. We can even consider that the database is tagged after the clients and visitors leave the shop with information on whether they have or have not bought products in the shop. This will be done linking pairs of records based on the RFIDs read. In the case of visitors having bought, which types of products, and their cost will also be recorded. In this scenario we consider that all the information to be stored can be deduced from RFID readings themselves, and there is no need to connect to external databases as e.g. the database of the cashier.
- **Industry 4.0.** A WSN is installed in a factory to monitor the production. Analysis of the process or malfunctioning of sensors cause engineers to add new sensors into the WSN and remove some of the existing ones. Replacement, installation and removal of sensors causes that some new records in the database have a different structure than previous ones. Data mining techniques are to be applied into this database to monitor and predict different components of the factory.

### 1.2 Structure of the Paper

In Sect. 2 we review existing approaches in the literature that relate to dynamic feature spaces. This is a relatively new area and the literature is limited. We discuss topics and research solutions that are relevant to dynamic feature spaces, and underline the differences and specific requirements.

In Sect. 3 we discuss the need to consider new features and their relationships with previous ones. We consider that this is a very relevant problem in dynamic feature space.

We will also argue that there is a lack of publicly available data for research in dynamic feature space. This makes analysis and test of algorithms difficult. Because of that, we discuss in Sect. 4 our approach to appropriate synthetic data generation.

The paper finishes with some conclusions and lines for future work.

## 2 Related Work

Incremental learning [17] is the area of machine learning where new training examples are considered once a model is already learned. These additional training examples are used to retrain the model or to adapt it. Incremental learning is to permit adaptability of the models.

We have already discussed in the introduction the relevance of concept drift research. The problem of concept drift is considered in [16]. It is based on considering an incremental learning framework where at a given time $t_0$ we have

$X^H = (X_1, \ldots, X_t)$ instances and then we consider a new instance $X_{t+1}$ to classify. Information on the right class of $X_{t+1}$ may be available later. This approach is used to define an approach for model testing. Instances $X_i$ are defined in a feature space $\mathcal{F}$.

The mathematical formulation of the problem in [16] is based on assuming that each instance $X_t$ is generated by a source $S_t$ and that sources may be change on time. That is, source $S_i$ may be equal or different to source $S_j$ for $i \neq j$. Each source may have different processes to generate the instances, and these processes can be noisy, or even include periodicity.

Different types of machine learning models can be built for concept drift. As an example, Law and Zaniolo consider in [11] the problem of nearest neighbor classification for data streams. In their solution all data is assumed to be in the same feature space (with a fixed number of variables known a priori). Nevertheless, classification may need to change with time due to concept drift. That is, the system may need to reconsider over time its classification process. Their solution is implemented by means of an updating of the classifiers involved in the system. Classification is based on $k$-nearest neighbors where the value of $k$ is considered dynamically.

In dynamic feature space problems, features may change over time. We can model this situation by means of considering a set of features that depend on time. That is, we define $\mathcal{F}_t$ as the set of features at time $t$. We consider that $\mathcal{F}_t$ is increasing with respect to time as the system does not unlearn previous seen features (i.e., $t < t'$ implies that $\mathcal{F}_t \subseteq \mathcal{F}_{t'}$).

Instances are not necessarily equally spaced in time. Because of that we consider that the first feature is time itself. We model this establishing that $\mathcal{F}_0$ is just a set that represents time (i.e., a time stamp).

The case of concept drift with dynamic features is discussed in e.g. [15]. The authors use the terms descriptional concept drift for the case in which the features do not change (i.e., only the distribution of the classes change). In contrast, the authors use the term contextural concept drift for the case in which the features change. They propose the Feature Adaptive Ensemble (FAE) algorithm for this type of concept drift.

Topic modeling is an active area of research (see e.g. [4,19]) that is also related to concept drift. More specifically, the detection of bursty topics. A synthetic dataset generation method has been recently proposed for generating data for topic modeling that permits to include concept drifts. See for [3] for details.

Research in concept drift as well as most reviewed literature related to dynamic feature mining focus on supervised machine learning and more particularly to classification problems. That is, when we have a set of records and there is a distinguished attribute that is the attribute to be learned. This attribute is categorical (the class). The literature on regression problems (when the attribute to be learned is numerical) and on unsupervised methods (e.g., clustering, association rule mining, methods to find latent variables as principal components and SVD) is scarce.

Algorithms for supervised learning with concept drift with dynamic feature spaces include the algorithms by Katakis et al. [9], StreamMiner, the Concept Drift Committee algorithm, and the FAE [15].

The algorithm by Katakis et al. [9] uses models that can add features dynamically. In particular, $k$-Nearest Neighbors and Naive Bayes. StreamMiner retrains classifiers after a set of training data is available, the Concept Drift Committee algorithm retrains decision trees after each training instance. In contrast FAE uses Naive Bayes, and new features may be added at any time.

Gomes et al. [5] focus on the problem of finding recurrent concepts. That is, detecting concepts that appear again in the data (instead of focusing on new concepts). Their approach is based in considering incremental feature selection to reduce the size of the model when new features are needed.

Most of the research on dynamic feature space deals with problems related to classification of text [17,18]. In this context, following the standard approach for text classification, the feature space is the set of words. Then, new features correspond to new words when they appear in the new instances being considered. More particularly, research focuses on systems for spam detection and for processing of posts in social networks. Mails and posts are modeled as a stream of texts, and they are used to feed the algorithms. New words in the documents are considered when they appear instead of considering all possible words from the start. This is, as e.g. [5] explains, because "using a very large vocabulary of words is clearly inefficient".

The literature consider a variety of data sets for testing algorithms on concept drift and dynamic feature spaces. Katakis et al. [9,10] use Spam Assassin [22]. FAE is evaluated using Spam Assassin [22], the KDD Cup 2000 [24], and the BYU Bookstore datasets [20]. The system in [17] uses the ENRON, the ECML, and the PU datasets. All of them are defined in terms of sets of emails.

Another problem of interest for dynamic feature spaces found in the literature is related to market basket analysis. Market basket analysis is usually processed using association rule mining algorithms and frequent itemset mining. These algorithms process databases where each record represents a shopping cart. Algorithms help to extract rules that relate the sets of objects bought. That is, rules of the form $A, B, C \Rightarrow D, E$ that indicate that people that buy $A, B$ and $C$ also buy $D$ and $E$.

Examples of algorithms for dynamic databases in the area of market basket analysis include [1,2,14]. Recall that in dynamic databases records are received as a stream. In [1], the authors consider the case of databases in which there are additions of new items and deletions. Naturally, in a database where records represent shopping carts new items can be seen as new features. Therefore this problem can be also classified as a dynamic feature space.

Methods to deal with dynamic feature spaces seems to lack the analysis of new features in relation to existing ones. That is, when a new feature is added, it can be related to (e.g., substitute) a previous one.

As a summary of what we have discussed so far, we underline the following:

- Dynamic feature space problems extend naturally concept drift problems, and in most of the applications we expect both elements to be taken into account.
- Literature on dynamic feature space focuses on classification problems. Algorithms focus on text and systems are mainly for Spam detection. No much literature on other types of problems, no literature on methods for numerical data (typical IoT and Industry 4.0 scenarios). Literature on unsupervised methods and regression methods for dynamic feature space is scarce.
- Data sets for testing dynamic feature spaces focus on textual data (related to Spam detection problems). Market basket analysis problems can also be seen from a dynamic feature space perspective, but data is also categorical.
- Solutions on dynamic feature space lack analysis of new features (i.e., on how these features relate to, substitute, complement a previous one).

We discuss in more detail the last item in the next section.

## 3 On the Relationships Between Features

When streaming data arrives, new data may incorporate new features. Incremental learning systems consider these new features and may revisit the models on the basis of these new features. This implies that new models are available in which the new features are taken into consideration.

Existing systems do not try to establish any relationship between new features and previous ones. That is, most systems do not evaluate if the new feature is redundant with existing ones, complementary, or whether it is replacing a feature that was previously present and it is no longer available.

We consider that this is a relevant question in dynamic feature problems. In some applications relationships need to be established between new features and previous ones to inform users about the novelty of the new variables. Relationships can be described by means of statistical (including imputation) and machine learning models. Accuracy of the models and (co)relation of the new variable in terms of the other ones can be used as metrics of this relationships. Approaches based on association rule mining can also help to establish relationships between features.

The relationship between variables links with the example we have given in the introduction where an RFID reader detects all RFIDs of a given customer (and reads all relevant information) at the entrance of a shop. The system also detects and reads all RFIDs of the customer at the exit of the shop. This information is stored and can then be used to infer at the entrance e.g. the probability of a client to buy a product or customize dynamic advertisements in the shop.

It is also relevant to mention in this context that the presence of a new product implies that we add the corresponding feature in the database. Nevertheless, the absence of a RFID reading from a customer does not mean that this product has not been bought or no longer exists. It only means that is not worn by the customer in the shop. In other words, we can consider that a feature can be in three different states: present (occasionally with some value), absent, and not-known. In some scenarios, we may consider that for an old feature (after an appropriate period of time, say $t_f$), the only possible values are: present and absent. In contrast, for new features, we may consider either the three values or only present and not-known.

# 4 On Data for Dynamic Features Space

As stated in previous sections, one of the standard application of dynamic feature mining is spam detection. There are quite a few algorithms that are defined and evaluated for data sets based on text. We have discussed them in Sect. 2. Nevertheless, available data for the numerical case is more limited. We propose three different approaches to generate synthetic numerical data with different levels of complexity. The data we have generated is available from [25] (see section *Data*).

## 4.1 Data Generation for Dynamic Features Space with No Concept Drift

We consider the case that data is completely synthetic and the case in which partially synthetic data is generated from a real data set.

For completely synthetic data, we start building $k$ $n$-dimensional clusters, and then hide some of the variables using some arbitrary noise procedure. The process is as follows.

- First, for given $k$ and $n$, define $k$ clusters in an $n$-dimensional space to obtain $r$ records (where the number of records is known). This is done randomly building $k$ cluster centers and populating each cluster with $r/k$ records around the center. This generates a total of $r$ records. When all variables are considered as independent, we use a Normal distribution for each variable centered in the corresponding variable of the cluster center.
  Here we can also create variables that depend on other variables that can be later used to study the problems discussed in Sect. 3.
- Second, for each dimension or feature $i$, we define a probability of the variable being absent. The probability of record $j$ being removed is $t_q(a, b, c, d)(j)$ where $t_q(a, b, c, d)$ defines a trapezoidal distribution as given in Fig. 2. In this definition, $0 \leq q \leq 1$ is the probability in the interval $[b, c]$ In this way, different records include different dimensions according to the probabilities established by the trapezoidal distribution. Note that this definition corresponds to a window where values are not present, which permits to implement different windows for different variables of the file (a kind of sliding window that moves across the variables). We have defined a similar procedure where the sliding window are the values to be published.

We proceed in the same way to generate the synthetic data for a given real data set. In this case, we apply the second step of the process to the given data

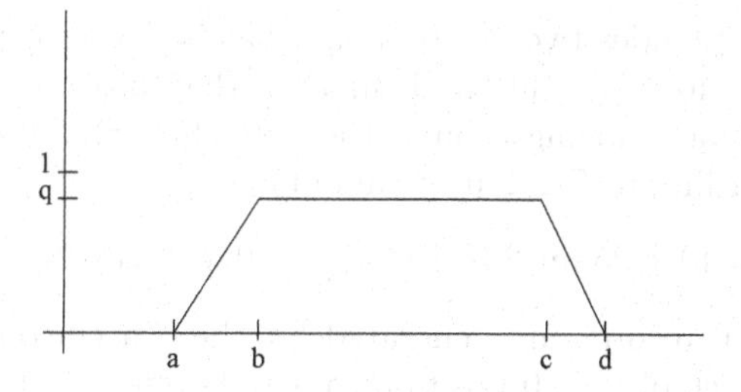


**Fig. 2.** Graphical representation of the trapezoidal distribution $t_q(a, b, c, d)(j)$.

file. We have used in this case the data sets in [7] (available at [21]). There are two datasets. They are the very large Census (with 124,998 records described by 13 features) and Forest (with 581,009 records described by 10 features). Given these datasets we consider that first features appear first while the last ones appear latter. More specifically, the $j$th feature (over a total of $n$ features) in a file with $m$ records is expected to appear in record $m*(j-1)/n$th and disappear in record $m*j/n$. Then, to build a dynamic feature data, we build a probability distribution for each pair *(record, feature)* and select its presence randomly according to this probability.

In this way, the first feature will mainly disappear when the record number increases. In contrast the last feature will not be present (with high probability) for first records, and then appear later. In any case, for any record we may have the presence or absence of a feature is based on some probability.

### 4.2 Data Generation for Dynamic Features Space with Concept Drift

We can proceed similarly to generate completely synthetic data with concept drift. In this case, we define $n$-dimensional clusters with centers that change over time. That is, the position of each cluster center depends on time (is a function of time or, in fact, on the record number).

This is similar to the synthetic data used by Ibrahim et al. [8]. Their data sets S1 and S2 are built as two dimensional datasets with centers changing over time. Data set S1 consists on records centered in a single cluster that changes on time. This single cluster has 11 consecutive positions and 100 records are generated for each position. The 11 positions are (according to Fig. 2 in [8] about):

$$v_{1i}^1 = 8.1i, v_{1i}^2 = 8.1i$$

for $i = 0, \ldots, 10$. We use here $v_{1i}^a$ to denote the $a$th dimension of the only cluster (1st cluster) at $i$th position.

Data set S2 [8] includes two clusters one that is always centered in position (40,40) (i.e., $v_0^1 = 40, v_0^2 = 40$) and another that has 10 different positions that change over time (moving around the first cluster). These 10 consecutive positions are (according to Fig. 2 in [8] about):

$$v_{1i}^1 = 40 + 25\sin(2\pi i/10), v_{1i}^2 = 40 + 25\cos(2\pi i/10)$$

for $i = 0, \ldots, 9$. 100 records are generated for the 0th cluster and 100 records are generated for each of the 10 positions of the 1st cluster. This process results into 110 records. We use here $v_{ci}^a$ to denote the $a$th dimension of the $c$th cluster at $i$th position.

Data of the 0th cluster is interleaved with data from the other cluster. More particularly, we have 100 records from cluster center $(v_{10}^1, v_{10}^2)$, then 10 records from cluster center $(v_0^1, v_0^2)$, then 100 records from cluster center $(v_{11}^1, v_{11}^2)$, then 10 more records from cluster center $(v_0^1, v_0^2)$, and so on.

Moshtaghi et al. [13] describe data sets that are similar to S1 and S2 above but differ on some of the parameters (as e.g., the number of records in each cluster) and, probably, on the exact position of the clusters. They also state that cluster $(v_0^1, v_0^2)$ corresponds to noise.

In order to generate data that include dynamic features space, we proceed, as described in Subsect. 4.1, removing some of the variables with a given probability. That is, we establish a probability of a variable being absent or present using the trapezoidal distribution described above.

### 4.3 More Complex Models via Markov Chains

In order to generate synthetic data for dynamic features space with more complex concept drifts, as well as more complex structure on the variables present in the records, we propose the use of Markov chains, or, alternatively, probabilistic automata.

We consider a set of states, where each state represents encodes a set of cluster centers. Then, we have transition matrices with probabilities for moving from one state to the other as usual in Markov chains.

The difference between the Markov chain and the probabilistic automata, is that the automata uses a transition function $\delta : Q \times \Sigma \to P(Q)$ that uses an input symbol (from the set of symbols $\Sigma$) in addition to the current state (from the state space $Q$) to give a probability into the space of states. In our setting, we use the number of records already generated as the input symbol. In this way, we can control the number of records generated in each state. Therefore, $\Sigma$ is the set of integers and the input parameter of $\delta$ is the number of records that have been already generated.

As a simple example, we would represent data from S2 as a 10 state model. Let $q_0, q_1, \ldots, q_9$ be the 10 states. Each state needs to generate 110 records. So, the transition function only allows to transit from $q_0$ to $q_1$ when 110 records have been generated. That is, we define $\delta(q_0, 110)(q_1) = 1$, and then $\delta(q_0, a)(q_0) = 1$ for $a < 110$. We have $\delta(q_0, a)(q) = 0$ otherwise. Similarly, we transit from $q_1$ to $q_2$ when 220 records have been generated and so on.

Then, each state $q_i$ generates 110 as follows: about 10 records from cluster center $(v_0^1, v_0^2)$ and 100 records from cluster center $(v_{1i}^1, v_{1i}^2)$.

## 5 Summary

This paper discusses the use of data mining and machine learning for dynamic feature spaces. We discuss some areas for applying these techniques. Among others, we give some examples of their application in Internet of Things, and Industry 4.0.

We have underlined some open problems and a specific research agenda for dynamic feature spaces. We have also discussed that the available data for algorithm developing and testing focus on textual data (for Spam detection), and proposed ways to synthetic data generation.

**Acknowledgements.** This work was partially supported by the Wallenberg AI, Autonomous Systems and Software Program (WASP) funded by the Knut and Alice Wallenberg Foundation.

## References

1. Abuzayed, N., Ergenç, B.: Dynamic itemset mining under multiple support thresholds. In: Proceedings of the FSDM 2016, pp. 141–148 (2016)
2. Abuzayed, N.N., Ergenç, B.: Comparison of dynamic itemset mining algorithms for multiple support thresholds. In: Proceedings of the IDEAS 2017 (2017)
3. Belford, M., Mac Namee, B., Greene, D.: Synthetic dataset generation for online topic modeling. In: Proceedings of the AICS 2017, pp. 7–8 (2017)
4. Blei, D.M., Ng, A.Y., Jordan, M.I.: Latent Drichlet allocation. J. Mach. Learn. Res. **3**, 993–1022 (2003)
5. Gomes, J.B., Gaber, M., Sousa, P.A.C., Menasalvas, E.: Mining recurring concepts in a dynamic feature space. IEEE Trans. Neural Networks Learn. Syst. **25**(1), 95–110 (2014)
6. Gubbi, J., Buyya, R., Marusic, S., Palaniswami, M.: Internet of Things (IoT): a vision, architectural elements, and future directions. Future Gener. Comput. Syst. **29**, 1645–1660 (2013)
7. Herranz, J., Nin, J., Solé, M.: Kd-trees and the real disclosure risks of large statistical databases. Inf. Fusion **13**(4), 260–270 (2012)
8. Ibrahim, O.A., Keller, J.M., Bezdek, J.C.: Evaluating evolving structure in streaming data with modified Dunn's indices. IEEE Trans. Emerg. Top. Comput. Intell. (2020, in press). https://doi.org/10.1109/TETCI.2019.2909521
9. Katakis, I., Tsoumakas, G., Vlahavas, I.: On the utility of incremental feature selection for the classification of textual data streams. In: Bozanis, P., Houstis, E.N. (eds.) PCI 2005. LNCS, vol. 3746, pp. 338–348. Springer, Heidelberg (2005). https://doi.org/10.1007/11573036_32
10. Katakis, I., Tsoumakas, G., Vlahavas, I.: Tracking recurring contexts using ensemble classifiers: an application to email filtering. In: Proceedings of the KAIS (2009)

11. Law, Y.-N., Zaniolo, C.: An adaptive nearest neighbor classification algorithm for data streams. In: Jorge, A.M., Torgo, L., Brazdil, P., Camacho, R., Gama, J. (eds.) PKDD 2005. LNCS (LNAI), vol. 3721, pp. 108–120. Springer, Heidelberg (2005). https://doi.org/10.1007/11564126_15
12. Lee, I., Lee, K.: The Internet of Things (IoT): applications, investments, and challenges for enterprises. Bus. Horiz. **58**, 431–440 (2015)
13. Moshtaghi, M., Bezdek, J.C., Erfani, S.M., Leckie, C., Bailey, J.: Online cluster validity indices for performance monitoring of streaming data clustering. Int. J. Intell. Syst. **34**, 541–563 (2019)
14. Otey, M.E., Wang, C., Parthasarathy, S., Veloso, A., Meira, W.: Mining frequent itemsets in distributed and dynamic database. In: Proceedings of the ICDM 2003 (2003)
15. Wenerstrom, B., Giraud-Carrier, C.: Temporal data mining in dynamic feature spaces. In: Proceedings of the ICDM 2006 (2006)
16. Zliobaite, I.: Learning under concept drift: an overview. Arxiv:1010.4784v1 (2010). https://arxiv.org/pdf/1010.4784.pdf
17. Sanghani, G., Kotecha, K.: Incremental personalized E-mail spam filter using novel TFDCR feature selection with dynamic feature update. Expert Syst. Appl. **115**, 287–299 (2019)
18. Song, G., Ye, Y., Zhang, H., Xu, X., Lau, R.Y.K., Liu, F.: Dynamic clustering forest: an ensemble framework to efficiently classify textual data stream with concept drift. Inf. Sci. **357**, 125–143 (2016)
19. Steinhauer, H.J., Helldin, T., Mathiason, G., Karlsson, A.: Topic modeling for anomaly detection in telecommunication networks. J. Ambient Intell. Humanized Comput. (2019, in press)
20. http://byubookstore.com
21. http://www.ppdm.cat/gransDades.php
22. http://spamassassin.apache.org/publiccorpus/
23. https://towardsdatascience.com/why-machine-learning-models-degrade-in-production-d0f2108e9214
24. http://www.ecn.purdue.edu/KDDCUP/
25. http://www.ppdm.cat/links.php

# An Evidential Reasoning Framework for User Profiling Using Short Texts

Duc-Vinh Vo(✉), Anh Hoang, and Van-Nam Huynh

Japan Advanced Institute of Science and Technology, Nomi, Japan
vinhvo@jaist.ac.jp

**Abstract.** Short texts-based user profiling is a challenging task despite its benefits to many applications such as job offering, item recommendation, and targeted advertisement. This work proposes an evidential reasoning framework for user profiling problem, which is capable of not only working with short texts but also dealing with uncertainty inherently in user corpora. The framework consists of three phases: representation learning, inference and combination, and user profile extraction. Particularly, in the first phase a word embedding technique is used to convert words into semantic vectors, then $k$-means clustering is utilized for learning abstract concepts that hide inside user plain texts, and then organizing the learned concepts into a hierarchical structure. In the second phase, by considering each document in user corpus as an evidential source that partially contributing to user profiles, and taking the concept hierarchy into account, we infer a mass function associated with each user document by maximum a posterior estimation, then apply Dempster's rule of combination for fusing all documents' mass functions into an overall one for the user corpus. In the third phase, we apply the so-called pignistic probability principle to extract top-$n$ keywords from user's overall mass function to define the user profile. Thanks to the ability of combining pieces of information from many documents, the proposed framework is flexible enough to be scaled when input data coming from not only multiple modes but different sources on web environments. Empirical results on datasets crawled from Twitter and Facebook validate the effectiveness of our proposed framework.

**Keywords:** User profiling · Short texts · Mass functions · Dempster-shafer theory

## 1 Introduction

With the dramatic spread of social networks such as Twitter, Facebook, Instagram, WeChat, etc., inferring user profiles from the information they shared has attracted much attention from communities because of its potential applications in job hunting, targeted advertising, and recommendation systems. For example, given a queried topic an application is required to identify a list of experts/users who are knowledgeable about that topic. In this context, the task is referred to


V.-N. Huynh et al. (Eds.): IUKM 2020, LNAI 12482, pp. 137–150, 2020.
https://doi.org/10.1007/978-3-030-62509-2_12

as expert finding or expert profiles [1]. In general, user profiling is the process of automatically converting user information into a predefined, interpretable format that reflects the most important aspects of the user's profile which is useful for further decision making in practical applications. In this work, we address the problem of *identifying user profiling primarily based on short texts created by users on social networks* defined as follows:

**Problem Statement.** Given a set of short documents created by users/experts on micro-blogging platforms (e.g., Twitter, Facebook, Instagram), the problem is to extract, per each user, a semantic profile that partially reflects his or her preferences toward predefined fields such as entertainments, technologies, politics, educations, etc. The resulting profile is represented by a list of keywords extracted from user's corpus. Mathematically, the target is to construct a function $f$ such that: $f : \langle \boldsymbol{\mathcal{U}}, \boldsymbol{\mathcal{D}} \rangle \mapsto \boldsymbol{\mathcal{W}}$, where $\boldsymbol{\mathcal{U}} = \{u_1, u_2, ..., u_m\}$ represents a set of users; $\boldsymbol{\mathcal{D}} = \{D_1, D_2, ..., D_m\}$ is a set of corpora, each corpus $D_i$ consists of all short documents created by the corresponding user $u_i$; and $\boldsymbol{\mathcal{W}} = \{w_1, w_2, ..., w_m\}$ is the set of users' profiles with $w_i = \{w_{i,1}, w_{i,2}, ..., w_{i,n}\}$ being the profiling result for the $i^{th}$ user, i.e., the top-$n$ keywords extracted from $u_i$'s dictionary.

When working with *short texts*, the nearly uniform distribution over terms in user's vocabulary causes challenges for previous approaches, especially models which is, in essence, based on statistical properties in users' corpora. Furthermore, users usually create a lot of short documents, each of which, under the semantic aspect, hides an amount of abstract concepts partially contributing to the user profile where plain texts do not clearly show.

Taking these observations into consideration, our idea aims at learning hidden, abstract concepts from words in user corpus and represent these concepts at multiple levels of abstractions, each of which semantically reflects partial information for inferring user profiles. Next, we build a map to transform plain-texted documents into abstract concept-based documents and consider each of those as an evidential source of information for inferring the user profile. After that, we construct a mechanism to quantitatively evaluate the critical level of each source, and then, inspired by Dempster-Shafer theory, we combine all of these quantitative information piece into an overall one which is useful for inferring the user profile. Finally, we utilize the so-called pignistic probability principle to extract top-$n$ keywords for obtaining the desired profile.

To summarize, our work makes the following key contributions: (1) inspired by Dempster-Shapfer theory and maximum a posterior estimation, a way of inferring a mass function for short texts document in the user corpus is proposed, and a general combination paradigm is derived for fusing all documents' mass functions into an overall one where the resulting profile is defined from; (2) overall, we propose an evidential reasoning framework for automatically inferring user profiles based on short texts created by the social networks' users. The advantage of the proposed framework is that it is flexible enough to be scaled to work well when the data coming from not only multiple modes but different sources on web environments due to the ability of combining pieces of information from many documents.

## 2 Related Works

Key topics related to our work are *user profiling* and *topic models.*

**User profiling**, also known as expert profiling, basing on user behavior has gained much attention from researchers for years. The approaches based on texts has increasingly attracted much attention, especially after Craswell et al. introduced the expert search task at Text REtrieval Conference Enterprise Track 2005 [7]. Later, Balog and de Rijke [3] proposed a method to model expert profiles as a vector where each component reflects someone's skills expressed by a score. In addition, Balog et al. [2] also introduced a generative language modeling algorithm for the problem. Although these tasks were inspired from the web track, experiments are conducted on the internal enterprise data for mining the relationships between entities within an organization. Recently, along with the increasing trend on the amount of texts created by users, the need of advanced methods based on texts therefore appears. In [13], Estival et al. introduce a way of finding author profiles based on texts from English emails. However, this approach is infeasible because of privacy issues in user texts. The problem turns out to be more challenging when working with short texts, e.g., a tweet on Twitter. There are some works coping with short texts like applying a neural network to discriminate gender on manually labeled tweets [9], or identifying the most effective feature set for author identification by focusing on messages retrieved from Twitter in [14]. Similarly, [4] aims at finding demographic information by social media texts, or [19] identifies a user's occupation thanks to his or her tweets. Although these methods is designed for short texts, they limited user profiling for a particular application, and do not take into account the uncertainty associated with each document in user corpus.

**Topic Models.** A topic model is a type of statistical model for inferring the distribution of *abstract topics* appear in a collection of text corpus. Latent Dirichlet Allocation (LDA) introduced by Blei et al. [5] is one of the state-of-the-art for topic model. LDA is a generative statistical model. It assumes that each document is a mixture of topics and each topic is a mixture of words. The primary objective is to infer topic distribution specific to each document and word distribution specific to each topic in the entire corpus. Two commonly used techniques are Variational Inference [5] and Gibb Sampling [15]. Conjugate distribution pairs play a vital role in the learning process of LDA model, especially the Dirichlet-Multinomial conjugacy because it allows the integral in the denominator of Bayesian theorem to be integrated out, making the inference process computationally feasible One possible approach is to apply topic model for inferring user profiles, e.g., LDA with collapsed Gibb sampling [5,15]. However, this method does not perform well on short texts because it lacks of capturing statistical properties of words' semantics in such kind of data.

To our knowledge, although there are works focus on short texts, the desired profiles are primarily toward a particular application, such as detecting basic demographic information [4,19], inducing the geographical location [16,20], inferring user preferences in politics and their intentions on voting [6,26], or learning semantic user profiles that are primarily used to improve the problem of sentiment classification [25]. Furthermore, previous works do not take into account the degree of belief that associates with each document in user's corpus while inferring user profiles. This is especially important in practical applications when data coming from a variety of sources. For example, a user may own many social network accounts simultaneously like LinkedIn, Twitter, Facebook, Instagram, and so on. Our approach, on the other hand, aims at capturing user profiles for general purpose applications, e.g., the resulting profiles in somehow should reflect user preferences at multiple fields like entertainments, technologies, politics, educations, etc. In addition, thanks to the mechanism of fusing all pieces of information in each document into an overall one where we base on to extract the user profile, our model is flexible, easy to be scaled when working when data coming from not only multiple modes, but different sources on web environments. Technical details will be explained in Sect. 4.

## 3 Background: Dempster-Shafer Theory

This section provides a brief introduction of basic concepts in Dempster-Shafer theory [10] that are necessary for our work. Then, in Sect. 4, these concepts, especially the D-S mass functions, will be applied for building and combining mass functions associated with a set of hypotheses for text data.

### 3.1 D-S Mass Functions

Let $\Omega = \{\omega_1, \omega_2, ..., \omega_M\}$ be defined as a finite set of $M$ hypotheses in which all elements are mutually pairwise exclusive and exhaustive. These hypotheses are called ***frame of discernment*** [23] which is interpreted as all possible answers to a concern question (e.g., which word in the dictionary should be selected to be a keyword appearing user profiles). A ***mass function*** $m$, also called the ***basis probability assignment*** (BPA) in the language of Dempster-Shafter theory, is defined as a mapping rule $m : 2^{\Omega} \mapsto [0, 1]$, satisfying $m(\emptyset) = 0$, and $\sum_{S \subseteq \Omega} m(S) = 1$. Each mass value associated with $m(S)$, where $S \subseteq \Omega$, is a degree of belief exactly committed to the proposition $x \in S$ and *nothing more* (i.e., not a specific subset of $S$). This proportion $m(S)$ considered as a certain piece of evidence which is useful for fusion information from multiple sources [12]. The situation of *open-world* and *total ignorance* are modeled as $m(\emptyset) > 0$ and $m(\Omega) = 1$, respectively. The set of all subsets $S \subseteq \Omega$ such that $m(S) > 0$ is called the **focal set** of $m$. $m$ is a generalization of categorical, Bayesian, and consonant mass functions [11].

### 3.2 Dempster's Rule of Combination and Decision Making

Consider the situation that there are two distinct, independent sources of information over the same frame of discernment $\Omega$, represented as $m_1$ and $m_2$, respectively. It is necessary to define a way for combining these two beliefs into a single one. Dempster's rule of combination is a power tool for fusing such pooling of evidences and is defined as follows.

$$(m_1 \oplus m_2)(S) = \begin{cases} 0 & \text{if } S = \emptyset \\ \frac{1}{1-\kappa} \sum_{S' \cap S'' = S} \left[ m_1(S') \times m_2(S'') \right] & \text{if } S \neq \emptyset \end{cases} \tag{1}$$

where $\kappa = \sum_{S' \cap S'' = \emptyset} m_1(S') \times m_2(S'')$ - degree of conflict between two sources, (2)

Assume that knowledge about a random variable $\mathcal{X}$ is presented by a corresponding mass function $m$. And our task is to select one element $\omega_i \in \Omega$ as the predicted value for $\mathcal{X}$. According to Smets [24], in order to make such a decision, D-S mass functions represented by $m$ should be transformed into the so-called *pignistic probability distribution* $BetPm : \Omega \mapsto [0, 1]$ defined as follows:

$$BetPm(\omega_i) = \sum_{S \subseteq \Omega,\ \omega_i \in S} \frac{m(S)}{|S|} \tag{3}$$

Then, the element $\omega_i$ with highest pignistic probability is selected as the predicted value for $X$, i.e., $\mathcal{X} = \omega^* = \text{argmax}_{\omega_i \in \Omega} pl(\omega_i)$.

## 4 The Proposed Framework

Our proposed framework is summarized in Fig. 1. Some key components will be described in this section.

### 4.1 Learning Abstract Concepts at Multiple Levels of Abstractions

There are two steps at this component: (1) word embedding for converting words into their corresponding semantic vectors; and (2) $k$-means clustering for learning a hierarchy of abstract concepts. Details are described as follows.

***Word Embedding.*** Two common approaches for word vectors in literature are the global matrix factorization [8] and local context window model [17]. Each approach has its own advantages and disadvantages in capturing optimal structure and statistical information stored inside the corpus. In this work, we applied a word embedding technique that utilize advantages of both model families, called Global Vector which is preferred to as GloVe [18]. An additional reason is that word vectors are trained on Twitter which consists of 2 billion

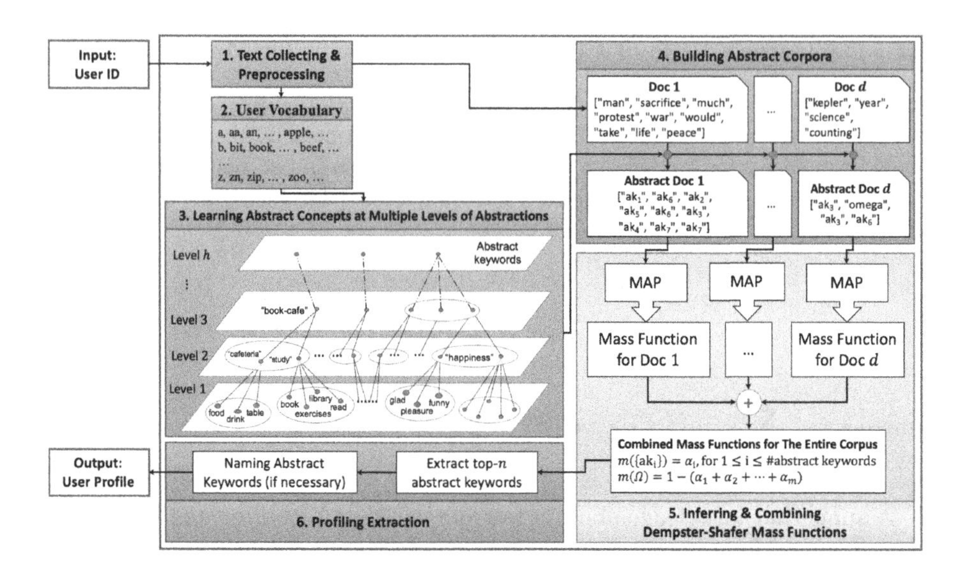


**Fig. 1.** The proposed framework for user profiling using short texts.

tweets, comprised by around 27 billion tokens. The lengths of resulting vectors are 25, 50, 100, 200, respectively[1]. This pretrained model matches the datasets we are working on.

***Learning a Hierarchy of Abstract Concepts by $k$-means Clustering.*** The idea here is to learn a set of concepts and conceptualizations that shared between the terms in user vocabulary, and to represent these abstract concepts at multiple levels of abstractions. Plain texts can be seen as the lowest abstraction level. We called the words at a higher conceptual level abstract keywords. Each abstract keyword is represented by a centroid returned by $k$-means clustering with the meaning that these abstract keywords capture some virtual features that are shared between the words *within* a cluster. For example, three words *kitty*, *puppy*, and *sparrow* can be represented by an abstract keyword $ak_i$, called "*pet*". This step ends up at a hierarchical structure. Each level contains a lists of abstract keywords. Each abstract keyword is represented by an abstract name and its corresponding word vector in $\mathbb{R}^n$ space, where $n \in \mathbb{Z}^+$.

## 4.2 Building Abstract Corpora

At this step, basing on refined texts and abstract keywords learned from previous steps, we build an abstract corpus per user as following: for each document $d \in \mathcal{U}$, a word $w_i$ is replaced by its nearest abstract keywords according to a distant function, e.g., Euclidean distance or cosine similarity. If the similarities between word $w_i$ and abstract keywords $ak_i, \forall i$ are greater than a predefined threshold

[1] The model is available at https://nlp.stanford.edu/projects/glove/.

$\epsilon$, then $w_i$ is replaced by an extra keyword called 'omega' with the meaning that $w_i$ is similar to all abstract keywords and this is interpreted as total ignorance in Dempster-Shafer theory.

### 4.3 D-S Mass Functions: Inference and Combination

From now we call *documents (corpus)* instead of *abstract documents (abstract corpus)* for brevity. In order to find the answer for user profiling, we now solve two subtasks: (1) infer a mass function for each document; and (2) combine all documents' mass functions into an overall one for the entire corpus.

***Inferring D-S Mass Functions.*** Consider a given set $\mathcal{W} = \{w_1, w_2, ..., w_N\}$ of $N$ independent, identically distributed (i.i.d.) draws from a multinomial distribution $\mathcal{V}$ of size $V$. In our case, $\mathcal{W}$ is a document created by one user, e.g., a tweet on Twitter or a status on Facebook, which contains a list of abstract keywords $w_i$ $(1 \leq i \leq N)$. $\mathcal{V}$ is the set of all possible abstract keywords $\mathcal{V} = \{ak_1, ak_2, ..., ak_V,$ 'omega'$\}$ at a specific abstraction level in the hierarchical structure of user vocabulary, each abstract keyword is represented by an abstract name and its corresponding word vector. The likelihood of these drawings in a given document is computed by (4).

$$L(\boldsymbol{p}|\boldsymbol{w}) = p(\mathcal{W}|\boldsymbol{p}) = \prod_{i=1}^{N}\prod_{t=1}^{V} p_t^{[w_i = ak_t]} = \prod_{t=1}^{V} p_t^{n_t} \tag{4}$$

where $n_t$ is the number of times abstract keyword $ak_t$ was observed as a word[2] in the document $\mathcal{W}$, $\sum_{t=1}^{V} n_t = N$, and $\sum_{t=1}^{V} p_t = 1$. Here we assume that abstract keywords in vocabulary $\mathcal{V}$ follow a multinomial distribution, denoted as $Mult(ak_t \in \mathcal{V}|\boldsymbol{p})$, where $\boldsymbol{p}$ is the probability that an abstract keyword $ak_t$ is observed as a word $w_i$ in a given document. Inspired by maximum a posterior distribution when applying Bayes rule with the Dirichlet distribution on the prior parameter $\boldsymbol{p}$: $\boldsymbol{p} \sim Dir(\boldsymbol{p}|\boldsymbol{\alpha})$, where $\boldsymbol{\alpha}$ is a concentration parameter vector which each element $\alpha_i$ corresponds to $p_i$ in $\boldsymbol{p}$, the probability for each term $t$ in a document $d$ is determined as follow:

$$p_t = \frac{n_t + \alpha_t - 1}{\sum_{t'=1}^{V}(n_{t'} + \alpha_{t'} - 1)}, \; \forall t \in [1, V] \tag{5}$$

Applying (5) for inferring the mass function associated with each document $d$ gives the answers shown in (6), (7)

[2] Abstract keyword $ak_t$ refers to one element of $\mathcal{V}$, and word $w$ refers to a particular observation in a given document, respectively. We refer to abstract keyword if the category in a multinomial is meant and to words if a particular observation or count in a document is meant. Thus several words in a text corpus can be classed as the same abstract keyword in the vocabulary.

$$m(\{ak_t\}) = \frac{(\#\text{times } ak_t \text{ appears in } d) + \alpha_t - 1}{(\#\text{words in } d) + \sum_{t=1}^{V} \alpha_t - V}, \tag{6}$$

$$m(\Omega) = \frac{(\#\text{times 'omega' appears in } d) + \alpha_{omega} - 1}{(\#\text{words in } d) + \sum_{t=1}^{V} \alpha_t - V}.$$
$$\text{where } \{ak_t\} \subseteq \Omega = \{ak_1, ak_2, ..., ak_V\}, \forall t \in [1, V] \tag{7}$$

***Combining all Masses Functions.*** Consider a set $\mathcal{M} = \{m_1, m_2, ..., m_M\}$, each $m_i$ is a mass function defined for the $i^{th}$ document over a frame of discernment $\Omega$ where $\Omega = \{ak_1, ak_2, ..., ak_V\}$ is a set of all possible abstract keywords, $|\Omega| = V$. Each mass $m_i$ can be calculated via (6), (7). We derive a general formula for combining $M$ mass functions into an overall one for the entire corpus as shown in (8) and (9).

$$m^{(1,2,..,M)}(\{ak_t\}) = \frac{1}{\mathcal{K}} \left( \prod_{i=1}^{M} m_i(\{ak_t\}) + \sum_{k=1}^{M-1} \sum_{\substack{S_j \subseteq \{1,...,M\} \\ |S_j|=k}} \prod_{u \in S_j} m_u(\{ak_t\}) \prod_{\substack{v=1 \\ v \notin S_j}}^{V} m_v(\Omega) \right), (1 \leq t \leq V) \tag{8}$$

$$m^{(1,2,...,M)}(\Omega) = \frac{1}{\mathcal{K}} \sum_{i=1}^{M} m_i(\Omega) \tag{9}$$

$$\mathcal{K} = \sum_{t=1}^{V} \left[ \prod_{i=1}^{M} m_i(\{ak_t\}) + \sum_{k=1}^{M-1} \sum_{\substack{S_j \subseteq \{1,...,M\} \\ |S_j|=k}} \prod_{u \in S_j} m_u(\{ak_t\}) \prod_{\substack{v=1 \\ v \notin S_j}}^{V} m_v(\Omega) + \prod_{i=1}^{M} m_i(\Omega) \right] \tag{10}$$

where $m^{(1,2,...,M)}(\{ak_t\})$ is a mass value assigned for a particular subset $\{ak_t\} \subseteq \Omega$, $\forall t \in [1, V]$, and $\mathcal{K}$ is the normalization factor. These formula could be proven by induction. The combined mass function is the fundamental building block for extracting top-$n$ keywords to define the user profiles.

### 4.4 Extracting User Profile and Naming the Abstract Keywords

After inducing the mass function for the entire corpus, the final step is to extract top-$n$ abstract keywords and naming these as actual keywords if necessary. This can be implemented by firstly applying the so-called pignistic probability principle via (3) [24] to calculate mass for all singleton in $\Omega$, then select the top-$n$ $w_i$ words such that $BetPm(w_i)$ are highest.

## 5 Empirical Results

### 5.1 Settings

**Datasets.** We work with two datasets collected from social networks which are Twitter[3] and Facebook[4]. Twitter dataset consists of **1189** users. For each user,

[3] Crawled from https://dev.twitter.com.

[4] Crawled by a python tool developed by our team with respected to Facebook Policies.

we collect all tweets from the beginning of registration up to May 31, 2015. Similarly, Facebook dataset consists of **1259** users. For each user, we collected all posts on his/her timeline from the date of registration to April $3^{rd}$, 2020.

**Ground Truth Keywords.** In order to get the ground true keywords per user, we carried out two processes which are automatically and manually. For the manual process, the annotators (totally 6 annotators) manually inspect the content of all posts and list the top 100 keywords for one user profile. For automatic process, we extract the hashtags from user posts and rank these hashtags by their frequency in all posts. For instance, the original hashtag #EqualityCannotWait is replaced by *equalitycannotwait* and we leave it as the ground truth keyword. We found that the maximum cosine similarity between the word embeddings of the keywords listed by annotators and the ones extracted automatically is high, ranging from 0.72 to 0.86. Finally, the annotators are asked to compare the keywords extracted by the two processes, ranked and created the final top 50 keywords as the profile for one user.

**Research Questions and Evaluation Metrics.** The experiments are for primarily answering these research questions: **RQ1.** How is the performance of the proposed framework and baseline models in inferring actual keywords using short texts for user profiling problem? **RQ2.** How is the performance of our proposed framework and baseline models in capturing the conceptual abstractions shared between words that reflect users' preferences in their profiles? **RQ3.** What is the impact of word vector's length on the overall performance of profiles derived by different methods?

To answer the **RQ1** we use a common, standard precision metric coupled with the ground true labels collected as introduced above. This measure is good enough for evaluating the ability of different competitors in inferring top $n$ keywords for user profile with $n$ ranging from 5 to 50 depending on applications. The standard precision score is defined as 1 iff $w_*$ and $w_{gt}$ are identical, 0 otherwise. Turning to the **RQ2**, the performance of competitors was evaluated by the semantic precision, denoted as $s$-precision. This metric is a common measure for evaluating the relevance score between a retrieved keyword $w_*$ and the ground truth keyword $w_{gt}$, defined by the cosine similarity between $e(w_*)$ and $e(w_{gt})$, where $e(w_i)$ is the word embedding vector of the word $w_i$. For the **RQ3**, we conduct the experiment at different configurations to learn user profiles by changing the length of the word vectors to different values that are commonly used in literature like 25, 50, 100, and 200 [18].

**Baselines.** The proposed framework is compared with two approaches for text-based user profiling. The first one is based on terms' frequency in user corpus to define the profile such as Term Frequency-Inverse Document Frequency method (**TFIDF**) - a numerical statistics based-estimator that reflects how important a word is to a document in a collection or corpus, word count method, and

Rapid Automatic Keyword Extraction method (**RAKE**) [21]. We conducted experiments[5] on these methods and found that TFIDF gives the best results according to the standard precision and semantic precision as introduced above.

The second approach is one of the state-of-the-art method for text-based user profiling problem. This is primarily based on topic model. Some methods on this direction includes the Latent Dirichlet Allocation [5], the Author-Topic Model for Authors and Documents (ATM) [22], and the Gibbs Sampling algorithm for the Dirichlet Multinomial Mixture model (GSDMM) [27]. Basing on these models, we derived user profile based on LDA with Gibb Sampling technique (LDA) as followings: (1) Build LDA model with $n$ topics[6] ($n$ is also the size of user profile); (2) For each topic-term distribution: extract the term $t$ with highest probability and remove it from the topic-term, then add this term to a set $S$; (3) Return $S$ as the user profile. We then conducted experiments between ATM, GSDMM, and LDA-UP and found that LDA-UP gives best performances.

Our methods called **Dempster-Shafer theory-based user profile.** For each user, our proposed framework extracts two types of profiles. The first one is represented by a list of abstract keywords, called *Abstract-DST profile (ab_dst)*. The second one is represented by actual keywords, called *Actual-DST profile (ac_dst)*, obtained by naming all abstract keywords to their nearest words in user vocabulary according cosine similarity.

To summary, thanks to sub-experiments between models, we now just need to compare these following methods: (1) TFIDF-based models (TFIDF), (2) Gibb-Sampling-LDA-based models (LDA), and (3) Dempster-Shafer-based models (ab_dst and ac_dst).

### 5.2 Results

In all experiments, we set the value for hyper-parameters as follows: *abstract_level = 2*, and *distance_function = cosine_similarity*. The number of keywords in profiles vary from 5 to 50 with *stepsize = 5*. Furthermore, in order to tune the value for threshold $\epsilon$ and the number of abstract keywords $c$, we split our datasets (all users' corpora) into training and validation sets with ratio 7/3 respectively. We use the train set for choosing the best value of $\epsilon = 0.8$, and the number of clusters in each abstraction level is $k = 50\%$ of vocabulary size[7]. We report the average results from 10 times of conducting the same experiments on test sets.

Figure 3 and Fig. 2 respectively shown the performance of user profiling on Twitter and Facebook datasets. These figures show the performance of profiles extracted by four different methods including TFIDF, LDA, and our proposed

[5] These experiments are carried out by directly counting on preprocessed texts without applying word embedding. The stopwords are also eliminated.

[6] In our datasets, the coherence and perplexity score of the LDA topic model are both optimal and remain stable when the number of topics is greater than or equal 12.

[7] This value seems reasonable for English. Let take WordNet dataset as an example, the total number of English words are 147249. There are 316244 of synonyms. This implies that the average number of synonyms is around 2.15. In our experiment, the value of 2 gives best performance on training set.

ones (*ac_dst* and *ak_dst*). As illustrated in the figures, *ac_dst* and *ak_dst* outperform all the baseline models in both metrics (standard precision and semantic precision). These results give the answers for the first two research questions **RQ1** and **RQ2**. From these experiments, an overall conclusion can be drawn is that our proposed framework is stable and efficient for user profiling problem using short texts (for most values of $n$-the size of user profiles).

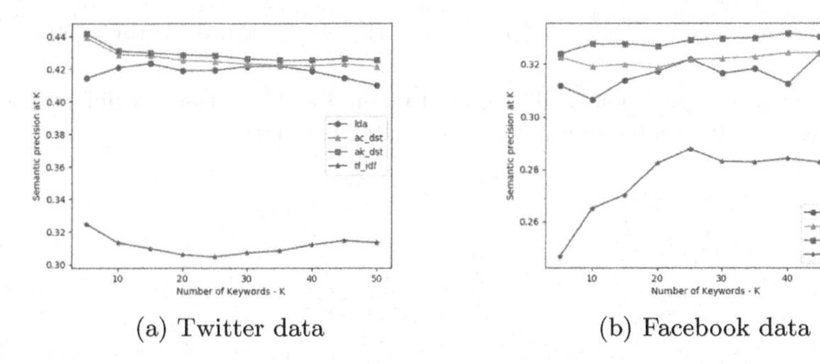


(a) Twitter data (b) Facebook data

**Fig. 2.** Semantic precision performance of User Profiles derived by 4 different methods: tf_idf, lda, ak_dst, and ac_dst

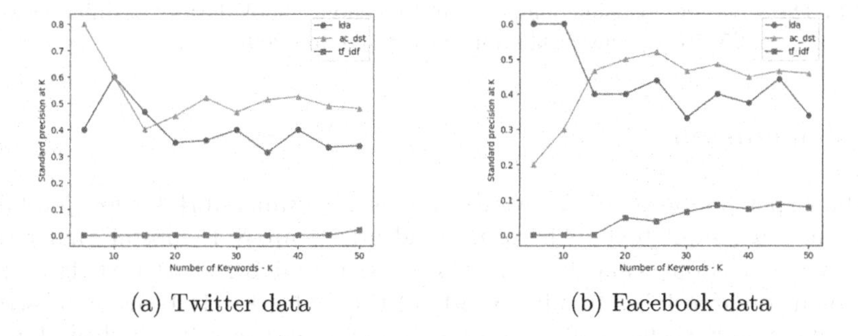


(a) Twitter data (b) Facebook data

**Fig. 3.** Standard precision performance of User Profiles derived by 4 different methods: tf_idf, lda, ak_dst, and ac_dst

***The Impact of Word Vectors' Length.*** Finally, we turn to the final experiment that check how does the length of word vectors affect the overall performance of different methods. For inspection, we measure the semantic precision of all competitors at different word vectors' sizes like 25, 50, 100, and 200. Figure 4 and Fig. 5 show experimental results on both datasets. We observe that word vector of size 25 gives best performance. Furthermore, the performance ranking between all competitors are identical, which are $ak\_dst > ac\_dst > lda > tf\_idf$. This observation again validates the stable of our proposed framework when the size of word vectors are changed.

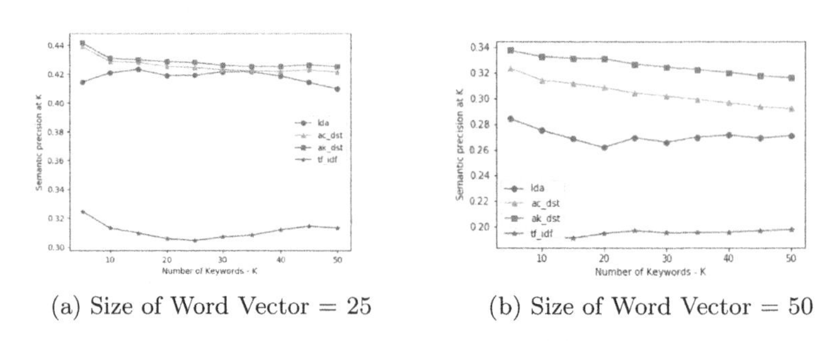


(a) Size of Word Vector = 25 (b) Size of Word Vector = 50

**Fig. 4.** The semantic precision of all competitors on Twitter dataset at different word vectors' sizes: 25, 50 (similar results for sizes of 100 and 200).

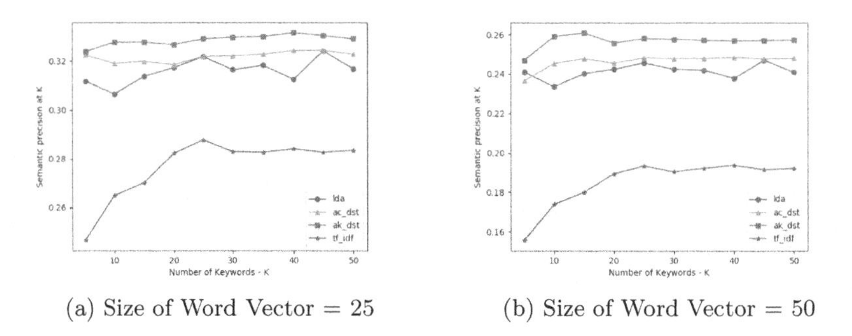


(a) Size of Word Vector = 25 (b) Size of Word Vector = 50

**Fig. 5.** The semantic precision of all competitors on Facebook dataset at different word vectors' sizes: 25, 50 (similar results for sizes of 100 and 200).

# 6 Conclusion

In this paper, we proposed an evidential reasoning framework for user profiling problem using short texts. The primary idea is to apply Dempster-Shafer theory, word embedding, and $k$-means clustering for mining short text data. The proposed framework essentially consists of three phases: (1) Learning abstract concepts at multiple levels of abstraction from user corpora; (2) Evidential inference and combination for user modelling; and (3) User profile extraction. In the empirical section, we conducted experiments to answer three research questions for validating the effectiveness and stability of our proposed framework.

It is reasonably that user profiles should be updated through times. So that the topic of dynamic user profiling which captures the change in user profile over time still attracts attention from researchers. And we would leave this problem as for future works.

**Acknowledgment.** This research was supported in part by the US Office of Naval Research Global under Grant no. N62909-19-1-2031.

## References


1. Balog, K.: Expertise retrieval. Found. Trends® Inf. Retrieval **6**(2–3), 127–256 (2012). https://doi.org/10.1561/1500000024
2. Balog, K., Bogers, T., Azzopardi, L., de Rijke, M., van den Bosch, A.: Broad expertise retrieval in sparse data environments. In: Proceedings of the 30th Annual International ACM SIGIR conference on Research and Development in Information Retrieval, SIGIR '07, pp. 551–558. ACM Press, New York (2007)
3. Balog, K., de Rijke, M.: Determining expert profiles (with an application to expert finding). In: Proceedings of the 20th International Joint Conference on Artifical Intelligence, IJCAI'07, pp. 2657–2662. Morgan Kaufmann Publishers Inc., San Francisco (2007)
4. Bergsma, S., Durme, B.V.: Using conceptual class attributes to characterize social media users. In: Proceedings of the 51st Annual Meeting of the Association for Computational Linguistics, vol. 1: Long Papers, pp. 710–720. Association for Computational Linguistics, Sofia (2013)
5. Blei, D.M., Ng, A.Y., Jordan, M.I.: Latent dirichlet allocation. J. Mach. Learn. Res. **3**, 993–1022 (2003). https://doi.org/10.1162/jmlr.2003.3.4-5.993
6. Cohen, R., Ruths, D.: Classifying political orientation on twitter: it's not easy! In: Proceedings of the 7th International Conference on Weblogs and Social Media, ICWSM 2013, pp. 91–99. The AAAI Press (2013)
7. Craswell, N., de Vries, A.P., Soboroff, I.: Overview of the TREC 2005 enterprise track. TREC **5**, 199–205 (2005)
8. Deerwester, S., Dumais, S.T., Furnas, G.W., Landauer, T.K., Harshman, R.: Indexing by latent semantic analysis. J. Am. Soc. Inf. Sci. **41**(6), 391–407 (1990). https://doi.org/10.1002/(sici)1097-4571(199009)41:6⟨391::aid-asi1⟩3.0.co;2-9
9. Deitrick, W., Miller, Z., Valyou, B., Dickinson, B., Munson, T., Hu, W.: Gender identification on twitter using the modified balanced winnow. Commun. Netw. **04**(03), 189–195 (2012). https://doi.org/10.4236/cn.2012.43023
10. Dempster, A.P.: Upper and lower probabilities induced by a multivalued mapping. Ann. Math. Stat. **38**(2), 325–339 (1967). https://doi.org/10.1214/aoms/1177698950
11. Dempster, A.P.: A generalization of bayesian inference. J. Roy. Stat. Soc. Ser. B (Methodol.) **30**(2), 205–232 (1968). https://doi.org/10.1111/j.2517-6161.1968.tb00722.x
12. Denoeux, T.: A k-nearest neighbor classification rule based on dempster-shafer theory. IEEE Trans. Syst. Man Cybern. **25**(5), 804–813 (1995). https://doi.org/10.1109/21.376493
13. Estival, D., Gaustad, T., Hutchinson, B., Pham, S.B., Radford, W.: Author profiling for english emails. In: Proceedings of the 10th Conference of the Pacific Association for Computational Linguistics, pp. 263–272 (2007)
14. Green, R., Sheppard, J.: Comparing frequency- and style-based features for twitter author identification. In: FLAIRS 2013 - Proceedings of the 26th International Florida Artificial Intelligence Research Society Conference, pp. 64–69 (01 2013)
15. Griffiths, T.L., Steyvers, M.: Finding scientific topics. Proc. Natl. Acad. Sci. **101**(Supplement 1), 5228–5235 (2004). https://doi.org/10.1073/pnas.0307752101
16. Han, B., Cook, P., Baldwin, T.: A stacking-based approach to twitter user geolocation prediction. In: Proceedings of the 51st Annual Meeting of the Association for Computational Linguistics: System Demonstrations, pp. 7–12. Association for Computational Linguistics (2013)

17. Mikolov, T., Chen, K., Corrado, G., Dean, J.: Efficient estimation of word representations in vector space (2013). http://arxiv.org/abs/1301.3781
18. Pennington, J., Socher, R., Manning, C.: Glove: global vectors for word representation. In: Proceedings of the 2014 Conference on Empirical Methods in Natural Language Processing (EMNLP), pp. 1532–1543. Association for Computational Linguistics, Doha (2014). https://doi.org/10.3115/v1/D14-1162
19. Preoţiuc-Pietro, D., Lampos, V., Aletras, N.: An analysis of the user occupational class through twitter content. In: Proceedings of the 53rd Annual Meeting of the Association for Computational Linguistics and the 7th International Joint Conference on Natural Language Processing, vol. 1: Long Papers, pp. 1754–1764. Association for Computational Linguistics, Beijing (2015). https://doi.org/10.3115/v1/P15-1169
20. Rahimi, A., Cohn, T., Baldwin, T.: Twitter user geolocation using a unified text and network prediction model. In: Proceedings of the 53rd Annual Meeting of the Association for Computational Linguistics and the 7th International Joint Conference on Natural Language Processing, vol. 2: Short Papers, pp. 630–636. Association for Computational Linguistics, Beijing (2015). https://doi.org/10.3115/v1/P15-2104
21. Rose, S., Engel, D., Cramer, N., Cowley, W.: Automatic keyword extraction from individual documents. In: Text Mining, pp. 1–20. John Wiley & Sons, Ltd. (2010). https://doi.org/10.1002/9780470689646.ch1
22. Rosen-Zvi, M., Griffiths, T., Steyvers, M., Smyth, P.: The author-topic model for authors and documents. In: Proceedings of the 20th Conference on Uncertainty in Artificial Intelligence (UAI), pp. 487–494 (2004)
23. Shafer, G.: A Mathematical Theory of Evidence. Princeton University Press, Princeton (1976)
24. Smets, P.: Data fusion in the transferable belief model. In: Proceedings of the Third International Conference on Information Fusion, vol. 1, pp. PS21–PS33. IEEE (2000). https://doi.org/10.1109/ific.2000.862713
25. Tang, D., Qin, B., Liu, T.: Learning semantic representations of users and products for document level sentiment classification. In: Proceedings of the 53rd Annual Meeting of the Association for Computational Linguistics and the 7th International Joint Conference on Natural Language Processing, vol. 1: Long Papers, pp. 1014–1023. Association for Computational Linguistics, Beijing (2015). https://doi.org/10.3115/v1/P15-1098
26. Volkova, S., Coppersmith, G., Durme, B.V.: Inferring user political preferences from streaming communications. In: Proceedings of the 52nd Annual Meeting of the Association for Computational Linguistics, vol. 1: Long Papers, pp. 186–196. Association for Computational Linguistics, Baltimore (2014). https://doi.org/10.3115/v1/P14-1018
27. Yin, J., Wang, J.: A dirichlet multinomial mixture model-based approach for short text clustering. In: Proceedings of the 20th ACM SIGKDD International Conference on Knowledge Discovery and Data Mining - KDD'14. ACM Press (2014). https://doi.org/10.1145/2623330.2623715

# Basic Consideration of Co-Clustering Based on Rough Set Theory

Seiki Ubukata(✉), Narihira Nodake, Akira Notsu, and Katsuhiro Honda

Osaka Prefecture University, 1-1 Gakuen-cho, Naka-ku, Sakai, Osaka 599-8531, Japan
{subukata,notsu,honda}@cs.osakafu-u.ac.jp

**Abstract.** In the field of clustering, rough clustering, which is clustering based on rough set theory, is a promising approach for dealing with the certainty, possibility, and uncertainty of belonging of object to clusters. Generalized rough C-means (GRCM), which is a rough set-based extension of hard C-means (HCM; k-means), can extract the overlapped cluster structure by assigning objects to the upper areas of their relatively near clusters. Co-clustering is a useful technique for summarizing co-occurrence information between objects and items such as the frequency of keywords in documents and the purchase history of users. Fuzzy co-clustering induced by multinomial mixture models (FCCMM) is a statistical model-based co-clustering method and introduces a mechanism for adjusting the fuzziness degrees of both objects and items. In this paper, we propose a novel rough co-clustering method, rough co-clustering induced by multinomial mixture models (RCCMM), with reference to GRCM and FCCMM. RCCMM aims to appropriately extract the overlapped co-cluster structure inherent in co-occurrence information by considering the certainty, possibility, and uncertainty. Through numerical experiments, we verified whether the proposed method can appropriately extract the overlapped co-cluster structure.

**Keywords:** Clustering · Co-clustering · Rough set theory · Rough clustering · Rough co-clustering

## 1 Introduction

In recent years, the amount of data used during data analysis tasks is increasing, and the demand for clustering, which is a technique for automatically classifying and summarizing data, is also increasing. Hard $C$-means (HCM; $k$-means) [1] is one of the most widely used partitive clustering methods. In HCM, the membership of object to cluster is represented by a binary value $\{0, 1\}$ and each object is assigned to one and only one cluster. Therefore, HCM cannot deal with situations such that an object belongs to multiple clusters at the same time. Thus, there are disadvantages such as a risk of misclassifying objects located near the cluster boundary.

In order to solve such problems, many extensions of HCM have been proposed using soft computing approaches such as fuzzy theory and rough set theory.


V.-N. Huynh et al. (Eds.): IUKM 2020, LNAI 12482, pp. 151–161, 2020.
https://doi.org/10.1007/978-3-030-62509-2_13

| Item / Object | $a_1$ | $a_2$ | $a_3$ | $a_4$ | $a_5$ |
|---|---|---|---|---|---|
| $x_1$ | 1 | 1 | 0 | 0 | 0 |
| $x_2$ | 0 | 1 | 1 | 0 | 0 |
| $x_3$ | 1 | 1 | 1 | 0 | 1 |
| $x_4$ | 0 | 0 | 1 | 1 | 1 |
| $x_5$ | 0 | 0 | 0 | 1 | 1 |

(a) Co-occurrence information matrix

Co-cluster 1
Co-cluster 2

| Item / Object | $a_1$ | $a_2$ | $a_3$ | $a_4$ | $a_5$ |
|---|---|---|---|---|---|
| $x_1$ | 1 | 1 | 0 | 0 | 0 |
| $x_2$ | 0 | 1 | 1 | 0 | 0 |
| $x_3$ | 1 | 1 | 1 | 0 | 1 |
| $x_4$ | 0 | 0 | 1 | 1 | 1 |
| $x_5$ | 0 | 0 | 0 | 1 | 1 |

(b) Co-clustering

Co-cluster 1
Co-cluster 2

| Item / Object | $a_1$ | $a_2$ | $a_3$ | $a_4$ | $a_5$ |
|---|---|---|---|---|---|
| $x_1$ | 1 | 1 | 0 | 0 | 0 |
| $x_2$ | 0 | 1 | 1 | 0 | 0 |
| $x_3$ | 1 | 1 | 1 | 0 | 1 |
| $x_4$ | 0 | 0 | 1 | 1 | 1 |
| $x_5$ | 0 | 0 | 0 | 1 | 1 |

(c) Rough co-clustering

**Fig. 1.** Examples of co-occurrence information matrix, co-clustering, and rough co-clustering.

Fuzzy $C$-means (FCM) [2] proposed by Bezdek is a widely used fuzzy extension of HCM. In fuzzy approaches, the domain of the membership of object to cluster is relaxed to the unit interval $[0, 1]$ to represent ambiguity. In FCM, nonlinear regularization is introduced by adding an exponential weight to the objective function of HCM to realize fuzzification.

Rough $C$-means (RCM) is a rough set-based extension of HCM. RCM deals with the certainty, possibility, and uncertainty of belonging of object to clusters by introducing the lower, upper, and boundary areas of the clusters, respectively. These areas are analogous to the lower and upper approximations, and boundary region in rough set theory. RCM can extract the overlapped cluster structure by assigning objects to the upper areas of their relatively near clusters. RCM is roughly classified into two methods, LRCM [3] proposed by Lingras and West, and PRCM [4] proposed by Peters. Ubukata et al. proposed generalized rough $C$-means (GRCM) [5] by unifying LRCM and PRCM. GRCM is further generalized to linear function threshold-based $C$-means (LiFTCM) [6].

Co-clustering is a useful technique for summarizing co-occurrence information between objects and items such as the frequency of keywords in documents and the purchase history of users. For example, in document analysis, each document (object) is considered to be characterized by the frequency of various keywords (items), and the goal is to extract the co-cluster structure composed of highly related documents-keywords. Since co-occurrence data have characteristics different from vector data on Euclidean space, analysis from a viewpoint different from the conventional clustering methods for vector data such as co-clustering is necessary. Figures 1(a) and 1(b) show examples of co-occurrence information matrix and co-clustering, respectively.

Fuzzy clustering for categorical multivariate data (FCCM) [7] is an FCM-type fuzzy co-clustering, in which the aggregation degree of objects and items is defined as the criterion of clustering. In FCCM, the co-cluster structure is estimated by the fuzzy memberships of object and item, and these memberships are fuzzified by the entropy-based regularization terms. However, FCCM has no corresponding statistical model. Multinomial mixture models (MMMs) [8] is a probabilistic model for co-clustering and can estimate multinomial distributions that generate data using the EM framework. Honda et al. proposed fuzzy co-clustering induced by multinomial mixture models (FCCMM) [9] by focusing on the similarity between FCCM and MMMs. FCCMM is based on the objective function composed of the log-likelihood function of MMMs and the K-L information-based regularization term, and introduces a mechanism for adjusting the fuzziness degrees of both objects and items.

Although fuzzy co-clustering approaches are useful, it is also useful to consider co-clustering based on rough set theory, which deals with uncertainty from a different viewpoint from fuzzy theory. In this paper, we propose a novel rough co-clustering method, rough co-clustering induced by multinomial mixture models (RCCMM), with reference to GRCM and FCCMM. RCCMM aims to appropriately extract the overlapped co-cluster structure inherent in co-occurrence information by considering the certainty, possibility, and uncertainty. Figure 1(c) shows an example of rough co-clustering. Through numerical experiments, we verified whether the proposed method can appropriately extract the overlapped co-cluster structure.

The remainder of the paper is organized as follows. Section 2 describes the preliminaries for our study. Rough co-clustering induced by multinomial mixture models is proposed in Sect. 3. In Sect. 4, our numerical experiments are discussed. Finally, the conclusions are presented in Sect. 5.

## 2 Preliminaries

### 2.1 $C$-Means-Type Clustering

This section explains $C$-means-type clustering including hard $C$-means (HCM) and generalized rough $C$-means (GRCM).

Let $U$ be the set of $n$ objects:

$$U = \{\boldsymbol{x}_1, ..., \boldsymbol{x}_i, ..., \boldsymbol{x}_n\}, \tag{1}$$

where each object $\boldsymbol{x}_i$ is a point in the $m$-dimensional real vector space:

$$\boldsymbol{x}_i = (x_{i1}, ..., x_{ij}, ..., x_{im})^\top \in \mathbb{R}^m. \tag{2}$$

Let $C$ be the number of clusters. $C$-means-type clustering extracts $C$ clusters from $U$. Let $u_{ci} \in \{0, 1\}$ be the membership of object $i$ to cluster $c$. Let $\boldsymbol{b}_c$ be the representative point of cluster $c$, which is called the cluster center:

$$\boldsymbol{b}_c = (b_{c1}, ..., b_{cj}, ..., b_{cm})^\top \in \mathbb{R}^m. \tag{3}$$

Let $d_{ci}$ be the Euclidean distance between cluster center $\boldsymbol{b}_c$ and object $i$:

$$d_{ci} = \|\boldsymbol{x}_i - \boldsymbol{b}_c\| = \left( \sum_{j=1}^{m} (x_{ij} - b_{cj})^2 \right)^{\frac{1}{2}}. \tag{4}$$

**HCM.** The optimization problem of HCM is given as follows:

$$\text{min.}\ J_{HCM} = \sum_{c=1}^{C}\sum_{i=1}^{n} u_{ci} d_{ci}^2, \tag{5}$$

$$\text{s.t.}\ u_{ci} \in \{0,1\}, \forall c, i, \tag{6}$$

$$\sum_{c=1}^{C} u_{ci} = 1, \forall i. \tag{7}$$

A sample algorithm of HCM is shown below.
[**Algorithm: HCM**]

**Step 1** Set the number of clusters, $C$.
**Step 2** Initialize $\boldsymbol{b}_c$.
**Step 3** Calculate $u_{ci}$ by nearest assignment:

$$u_{ci} = \begin{cases} 1 & \left(c = \arg\min_{1 \le l \le C} d_{li}\right), \\ 0 & \text{(otherwize)}. \end{cases} \tag{8}$$

**Step 4** Calculate $\boldsymbol{b}_c$ by the center of the cluster:

$$\boldsymbol{b}_c = \frac{\sum_{i=1}^{n} u_{ci} \boldsymbol{x}_i}{\sum_{i=1}^{n} u_{ci}}. \tag{9}$$

**Step 5** Repeat **Step 3–4** until $u_{ci}$ does not change.

In Eq. (8), if there are multiple clusters with the minimum distance, the sum-to-one constraints (Eq. (7)) are violated. Examples of manners to satisfy the constraints include assigning object to the cluster with a small index, randomly assigning it to one of the clusters, and equally distributing its membership value to the clusters, in which the Boolean domain of Eq. (6) is violated. Thus, the constraints of hard clustering inhibit cluster overlap.

**GRCM.** RCM can deal with the certainty, possibility, and uncertainty of belonging of object to clusters by considering the lower, upper, and boundary areas of clusters. The lower, upper, and boundary areas are analogous to the lower and upper approximation, and boundary region in rough set theory. The lower and upper areas of the cluster are constructed so as to satisfy the following basic theorems of the lower and upper approximations.

**Theorem 1:** Each object belongs to the lower approximation of at most one cluster.
**Theorem 2:** Each object belonging to the lower approximation of a cluster belongs to the upper approximation of the same cluster.
**Theorem 3:** Each object not belonging to any lower approximations belongs to the upper approximations of at least two clusters.

The boundary area of the cluster is obtained by the difference of the memberships to the upper area and to the lower area. RCM does not have an explicit objective function but is constructed by modifying the HCM algorithm.

Let $\underline{u}_{ci}, \overline{u}_{ci}, \hat{u}_{ci} \in \{0,1\}$ be the membership of object $i$ to the lower, upper, and boundary areas of cluster $c$, respectively. Let $d_i^{min}$ be the minimum distance between object $i$ and all the cluster centers:

$$d_i^{min} = \min_{1 \leq l \leq C} d_{li}. \tag{10}$$

In each iteration, the assignment of object to the upper area of cluster is first calculated. Object $i$ is assigned to the upper areas of its relatively near clusters by increasing the threshold based on a linear function of $d_i^{min}$:

$$\overline{u}_{ci} = \begin{cases} 1 & (d_{ci} \leq \alpha d_i^{min} + \beta), \\ 0 & (\text{otherwize}), \end{cases} \tag{11}$$

where $\alpha$ $(\alpha \geq 1), \beta$ $(\beta \geq 0)$ are parameters that adjust the roughness of clustering. If $\alpha = 1, \beta = 0$, an HCM-like assignment is obtained. The larger the parameters are, the rougher the partition is, and the larger the cluster overlap is.

The memberships $\underline{u}_{ci}$ and $\hat{u}_{ci}$ of the lower and boundary areas, respectively, are calculated by using $\overline{u}_{ci}$:

$$\underline{u}_{ci} = \begin{cases} 1 & \left(\overline{u}_{ci} = 1 \wedge \sum_{l=1}^{C} \overline{u}_{ci} = 1\right), \\ 0 & (\text{otherwize}), \end{cases} \tag{12}$$

$$\hat{u}_{ci} = \begin{cases} 1 & \left(\overline{u}_{ci} = 1 \wedge \sum_{l=1}^{C} \overline{u}_{ci} \geq 2\right), \\ 0 & (\text{otherwize}). \end{cases} \tag{13}$$

New cluster center $\boldsymbol{b}_c$ is calculated by the convex combination of the center of the lower, upper, and boundary areas of cluster $c$:

$$\boldsymbol{b}_c = \underline{t}\frac{\sum_{i=1}^{n} \underline{u}_{ci}\boldsymbol{x}_i}{\sum_{i=1}^{n} \underline{u}_{ci}} + \overline{t}\frac{\sum_{i=1}^{n} \overline{u}_{ci}\boldsymbol{x}_i}{\sum_{i=1}^{n} \overline{u}_{ci}} + \hat{t}\frac{\sum_{i=1}^{n} \hat{u}_{ci}\boldsymbol{x}_i}{\sum_{i=1}^{n} \hat{u}_{ci}}, \tag{14}$$

where $\underline{t}, \overline{t}, \hat{t} \geq 0$ s.t. $\underline{t} + \overline{t} + \hat{t} = 1$ are priority weights of the lower, upper, and boundary areas, respectively. If the lower or boundary areas are empty, we temporarily set $\overline{t} = 1$.

A sample algorithm of GRCM is shown below.

**[Algorithm: GRCM]**

**Step 1** Set the number of clusters, $C$, the roughness parameters, $\alpha \geq 1, \beta \geq 0$, and the priority weights, $\underline{t}, \overline{t}, \hat{t} \geq 0$ s.t. $\underline{t} + \overline{t} + \hat{t} = 1$.
**Step 2** Initialize $\boldsymbol{b}_c$.
**Step 3** Calculate $\overline{u}_{ci}$ by Eqs. (10) and (11).
**Step 4** Calculate $\underline{u}_{ci}$ and $\hat{u}_{ci}$ by Eqs. (12) and (13), respectively.
**Step 5** Calculate $\boldsymbol{b}_c$ by Eq. (14).
**Step 6** Repeat **Step 3–5** until $\overline{u}_{ci}$ does not change.

### 2.2 Fuzzy Co-Clustering

Co-clustering deals with co-occurrence information of objects and items, and extracts a co-cluster structure by simultaneously grouping both objects and items. Let $R = \{r_{ij}\}$ be the $n \times m$ co-occurrence information matrix between $n$ objects and $m$ items. In the context of co-clustering, co-occurrence degree $r_{ij}$ can take a real number, whereas in the context of multinomial mixture models, it takes a natural number. Some normalizations such as $r_{ij} \in \{0, 1\}$ or $r_{ij} \in [0, 1]$ make it easier to interpret co-occurrence degree. The goal of fuzzy co-clustering is to simultaneously estimate the fuzzy memberships $u_{ci}$ and $w_{cj}$ of object $i$ and item $j$ to co-cluster $c$, respectively. Objects and items that are highly related have high fuzzy memberships to the same co-cluster.

**Simple FCCMM.** In this paper, we introduce a simple version of FCCMM with the same setting of the item fuzziness degree as MMMs for simplicity. The optimization problem of simple FCCMM is given by:

$$\max. \quad J_{FCCMM}^{simple} = \sum_{c=1}^{C}\sum_{i=1}^{n}\sum_{j=1}^{m} u_{ci} r_{ij} \log w_{cj} + \lambda_u \sum_{c=1}^{C}\sum_{i=1}^{n} u_{ci} \log \frac{\alpha_c}{u_{ci}}, \tag{15}$$

$$\text{s.t.} \quad u_{ci}, w_{cj}, \alpha_c \in (0, 1], \forall c, i, j, \tag{16}$$

$$\sum_{c=1}^{C} u_{ci} = 1, \forall i, \sum_{j=1}^{m} w_{cj} = 1, \forall c, \sum_{c=1}^{C} \alpha_c = 1, \tag{17}$$

where $\lambda_u$ ($\lambda_u \geq 0$) is the object fuzziness degree. If $\lambda_u = 0$, a crisp object partition is obtained. The larger the $\lambda_u$ is, the fuzzier the object partition is. This model matches MMMs when $\lambda_u = 1$. Object and item memberships are fuzzified by the K-L information-based regularization term and the logarithm, respectively. These fuzzification manners are induced by MMMs. Original FCCMM has a parameter for adjusting the fuzziness degree of item memberships. Crisp item memberships are not usually considered because they produce co-clusters containing one and only one item due to the sum-to-one constraint of Eq. (17).

## 3 Proposed Method

In this paper, we propose a novel rough co-clustering method, rough co-clustering induced by multinomial mixture models (RCCMM), with reference to GRCM and FCCMM.

In preparation for formulating RCCMM, we first consider a hard version of FCCMM, which is called HCCMM, by setting $\lambda_u = 0$ in Eq. (15). Then, the optimization problem of HCCMM is described as follows:

$$\text{max.}\quad J_{HCCMM}=\sum_{c=1}^{C}\sum_{i=1}^{n}\sum_{j=1}^{m}u_{ci}r_{ij}\log w_{cj}, \tag{18}$$

$$\text{s.t.}\quad u_{ci}\in\{0,1\}, w_{cj}\in(0,1],\forall c,i,j, \tag{19}$$

$$\sum_{c=1}^{C}u_{ci}=1,\forall i,\sum_{j=1}^{m}w_{cj}=1,\forall c. \tag{20}$$

In HCCMM, the update rules of $u_{ci}$ and $w_{cj}$ are derived, respectively, as follows:

$$u_{ci}=\begin{cases}1 & \left(c=\arg\max_{1\leq l\leq C}\sum_{j=1}^{m}r_{ij}\log w_{lj}\right),\\ 0 & (\text{otherwize}),\end{cases} \tag{21}$$

$$w_{cj}=\frac{\sum_{i=1}^{n}u_{ci}r_{ij}}{\sum_{l=1}^{m}\sum_{i=1}^{n}u_{ci}r_{il}}. \tag{22}$$

Let $s_{ci}$ be the similarity between cluster $c$ and object $i$:

$$s_{ci}=\sum_{j=1}^{m}r_{ij}\log w_{cj}. \tag{23}$$

Let $s_i^{max}$ be the maximum similarity between object $i$ and all the clusters:

$$s_i^{max}=\max_{1\leq c\leq C}s_{ci}. \tag{24}$$

Then, Eq. (21) can be rewritten by using the inequality of $s_{ci}$ and $s_i^{max}$:

$$u_{ci}=\begin{cases}1 & (s_{ci}\geq s_i^{max}),\\ 0 & (\text{otherwize}).\end{cases} \tag{25}$$

Next, we consider the assignment of object to the upper area of cluster with reference to GRCM to realize the cluster overlap. We relax the condition of object-cluster assignment by decreasing the threshold on the right side of the inequality of Eq. (25) by a linear function of $s_i^{max}$. In RCCMM, $\overline{u}_{ci}$ can be calculated as follows:

$$\overline{u}_{ci}=\begin{cases}1 & (s_{ci}\geq \alpha s_i^{max}+\beta),\\ 0 & (\text{otherwize}),\end{cases} \tag{26}$$

where $\alpha$ ($\alpha\geq 1$), $\beta$ ($\beta\leq 0$) are parameters that adjust the roughness of clustering. We note that the range of $\beta$ differs from that of GRCM because it should have the function of decreasing the threshold. We also note that the larger $\alpha$ decreases the threshold because $s_{ci}$ has a negative value. If $\alpha=1,\beta=0$, an

HCCMM-like assignment is obtained. The larger (smaller) the $\alpha$ ($\beta$) is, the rougher the partition is, and the larger the cluster overlap is. The memberships $\underline{u}_{ci}$ and $\hat{u}_{ci}$ to the lower and boundary areas, respectively, are calculated by using $\overline{u}_{ci}$ in the same manner as GRCM (Eqs. (12) and (13)).

Finally, we consider the calculation of the item membership $w_{cj}$ aggregating three areas of the cluster. We introduce the calculation of the item memberships based on the convex combination of the object memberships of the lower, upper, and boundary areas of the cluster. Let $\tilde{u}_{ci}$ be the convex combination of the lower, upper, and boundary areas:

$$\tilde{u}_{ci} = \underline{t}\,\underline{u}_{ci} + \overline{t}\overline{u}_{ci} + \hat{t}\hat{u}_{ci}, \tag{27}$$

where $\underline{t}, \overline{t}, \hat{t} \geq 0$ s.t. $\underline{t} + \overline{t} + \hat{t} = 1$ are the priority weights of the lower, upper, and boundary areas, respectively. Then, $w_{cj}$ is calculated by:

$$w_{cj} = \frac{\sum_{i=1}^{n} \tilde{u}_{ci} r_{ij}}{\sum_{l=1}^{m} \sum_{i=1}^{n} \tilde{u}_{ci} r_{il}}. \tag{28}$$

A sample algorithm of RCCMM is shown below.

**[Algorithm: RCCMM]**

**Step 1** Set the number of co-clusters, $C$, the roughness parameters, $\alpha \geq 1, \beta \leq 0$, and the priority weights, $\underline{t}, \overline{t}, \hat{t} \geq 0$ s.t. $\underline{t} + \overline{t} + \hat{t} = 1$.
**Step 2** Initialize $w_{cj}$.
**Step 3** Calculate $\overline{u}_{ci}$ by Eqs. (23), (24), and (26).
**Step 4** Calculate $\underline{u}_{ci}$ and $\hat{u}_{ci}$ by Eqs. (12) and (13), respectively.
**Step 5** Calculate $w_{cj}$ by Eqs. (27) and (28).
**Step 6** Repeat **Step 3-5** until $\overline{u}_{ci}$ does not change.

# 4 Numerical Experiments

Through numerical experiments, we confirmed that the proposed RCCMM can appropriately extract the overlapped co-cluster structure from an artificially generated co-occurrence information matrix, in which an overlapped co-class structure is assumed. Furthermore, we observed the change of the object-cluster assignment by setting of the roughness parameter.

## 4.1 Experimental Settings

**Artificial Co-occurrence Information Matrix.** We first created the noiseless co-occurrence information matrix $R_0 = \{r_{ij}^0\}$ composed of $n = 100$ objects and $m = 60$ items (Fig. 2(a)). In Fig. 2(a), black and white cells represent $r_{ij} = 1$ and $r_{ij} = 0$, respectively. Three rectangular co-classes are assumed and some pairs of them have the overlap. Let $Label = \{1, ..., l, ..., L\}$ be the set of class labels. In this case, $|Label| = 3$. Next, a noisy co-occurrence information matrix $R_1 = \{r_{ij}^1\}$ (Fig. 2(b)) was made by adding bit inverting noise to $R_0$. Cells with

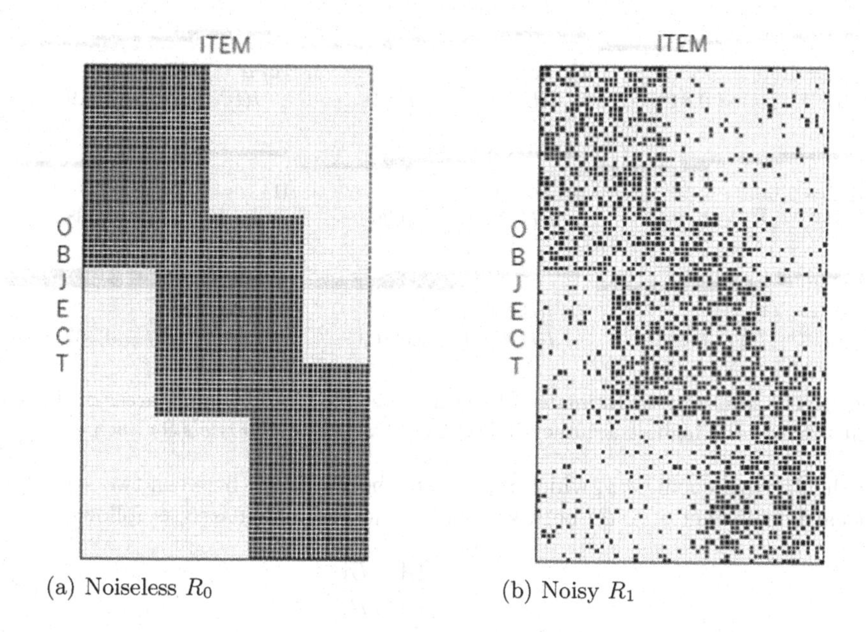


(a) Noiseless $R_0$ (b) Noisy $R_1$

**Fig. 2.** Artificial co-occurrence information matrix $R = \{r_{ij}\}$: $r_{ij}$ represents the co-occurrence relation between object $i$ and item $j$ in grayscale.

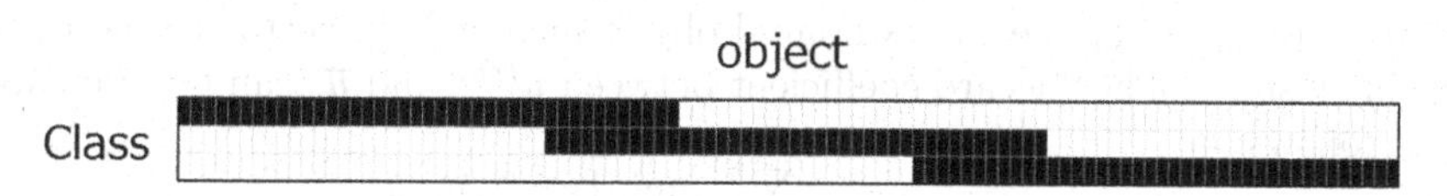


**Fig. 3.** The true membership $U^{true} = \{u_{li}^{true}\}$ of object $i$ to class $l$: $l$-th row represents the true object membership vector $\boldsymbol{u}_l^{true} = (u_{l,1}^{true}, ..., u_{l,100}^{true})$.

$r_{ij}^0 = 1$ were replaced with $r_{ij}^1 = 0$ at a rate of 50%, and cells with $r_{ij}^0 = 0$ were replaced with $r_{ij}^1 = 1$ at a rate of 10%. Figure 3 shows the assumed true membership $U^{true} = \{u_{li}^{true}\}$ of object $i$ to class $l$ in $R_0$. $l$-th row represents the 100-dimensional true object membership vector $\boldsymbol{u}_l^{true} = (u_{l,1}^{true}, ..., u_{l,100}^{true})$. The goal of this experiment is to extract a co-cluster structure that is highly consistent with the true co-class structure that has the overlap.

**Initialization of Item Memberships.** To initialize the item memberships $\boldsymbol{w}_c = (w_{c1}, ..., w_{cm})$, $C$ row vectors $\boldsymbol{r}_c = (r_{c1}^1, ..., r_{cm}^1)$ were chosen from $R_1$ by simple random sampling without replacement and they are normalized so that the sum of the elements is one.

**Accuracy Index.** To evaluate the consistency between the true co-class structure and the estimated co-cluster structure, we consider the accuracy index based

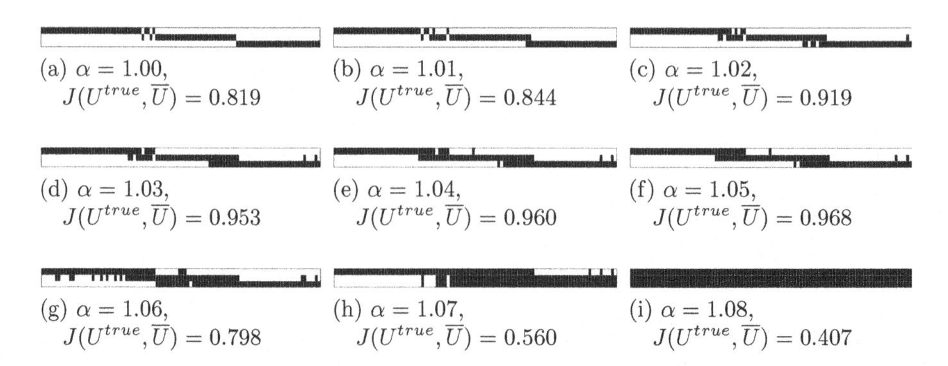

(a) $\alpha = 1.00$, $J(U^{true}, \overline{U}) = 0.819$
(b) $\alpha = 1.01$, $J(U^{true}, \overline{U}) = 0.844$
(c) $\alpha = 1.02$, $J(U^{true}, \overline{U}) = 0.919$
(d) $\alpha = 1.03$, $J(U^{true}, \overline{U}) = 0.953$
(e) $\alpha = 1.04$, $J(U^{true}, \overline{U}) = 0.960$
(f) $\alpha = 1.05$, $J(U^{true}, \overline{U}) = 0.968$
(g) $\alpha = 1.06$, $J(U^{true}, \overline{U}) = 0.798$
(h) $\alpha = 1.07$, $J(U^{true}, \overline{U}) = 0.560$
(i) $\alpha = 1.08$, $J(U^{true}, \overline{U}) = 0.407$

**Fig. 4.** The results of the estimated object membership $\overline{U}$ to the upper areas of clusters that achieves the highest accuracy index $J(U^{true}, \overline{U})$ in $1,000$ trials for each $\alpha$.

on the Jaccard coefficient, which represents the similarity between two sets. The Jaccard coefficient $j(A, B)$ between sets $A$ and $B$ is calculated as follows:

$$j(A, B) = \frac{|A \cap B|}{|A \cup B|}, \tag{29}$$

where $|\cdot|$ represents the cardinality of a set.

Let $\boldsymbol{u}_l^{true} = (u_{l1}^{true}, ..., u_{ln}^{true})$ be the true object membership vector to class $l$. Let $\overline{\boldsymbol{u}}_c = (\overline{u}_{c1}, ..., \overline{u}_{cn})$ be the estimated object membership vector to the upper area of cluster $c$. The Jaccard coefficient between $\boldsymbol{u}_l^{true}$ and $\overline{\boldsymbol{u}}_c$ can be calculated by as follows:

$$j(\boldsymbol{u}_l^{true}, \overline{\boldsymbol{u}}_c) = \frac{\sum_{i=1}^{n} \min\{u_{li}^{true}, \overline{u}_{ci}\}}{\sum_{i=1}^{n} \max\{u_{li}^{true}, \overline{u}_{ci}\}}. \tag{30}$$

For simplicity, we write $j(\boldsymbol{u}_l^{true}, \overline{\boldsymbol{u}}_c)$ as $j(l, c)$.

Let $U^{true} = \{u_{li}^{true}\}$ and $\overline{U} = \{\overline{u}_{ci}\}$ be the true object memberships to classes and the estimated object memberships to the upper areas of clusters, respectively. To evaluate the consistency between the true memberships and the estimated memberships, we introduce the accuracy index $J(U^{true}, \overline{U})$, which is the maximum average Jaccard coefficient in all the class-cluster correspondences:

$$J(U^{true}, \overline{U}) = \max_{(p_1, p_2, ..., p_C) \in P} \frac{\sum_{l \in L} j(l, p_l)}{C}, \tag{31}$$

where $P$ is the set of all permutations $(p_1, p_2, ..., p_C)$ of $C$ cluster indices.

### 4.2 Experimental Results

The proposed RCCMM was applied to $R_1$, where $C = |Label| = 3$. We fixed $\beta = 0$ and changed $\alpha \in \{1.00, 1.01, 1.02, 1.03, 1.04, 1.05, 1.06, 1.07, 1.08\}$. We fixed the priority weights $\underline{t} = \overline{t} = \hat{t} = \frac{1}{3}$.

Figure 4 shows the results of the estimated object membership $\overline{U}$ to the upper areas of clusters that achieves the highest accuracy index $J(U^{true}, \overline{U})$ in $1,000$

trials for each $\alpha$. If $\alpha = 1.00$, a hard object partition was obtained and the cluster overlap cannot be observed (Fig. 4(a)). By observing Figs. 4(a)–(i) in order, we can confirm that the cluster overlap increases as $\alpha$ increases. Finally, each object belongs to all co-clusters (Fig. 4(i)). In Fig. 4(f), the accuracy index reaches the maximum value, $J(U^{true}, \overline{U}) = 0.968$, by $\alpha = 1.05$, and a co-cluster structure with high consistency was extracted by the appropriate roughness setting.

As a result, it is found that the proposed RCCMM can appropriately extract the co-cluster structure considering the cluster overlap by adjusting the roughness parameter.

## 5 Conclusions

In this paper, we proposed a novel rough co-clustering method, rough co-clustering induced by multinomial mixture models (RCCMM), with reference to the conventional rough clustering and fuzzy co-clustering methods, namely, GRCM and FCCMM. RCCMM can extract the overlapped co-cluster structure by assigning objects to the upper areas of their relatively similar co-clusters. Through numerical experiments, we confirmed that the proposed RCCMM can appropriately extract the co-cluster structure considering the cluster overlap by adjusting the roughness parameter.

We plan to consider experiments using real-world datasets and automatic determination of the parameters.

**Acknowledgment.** This work was partly supported by JSPS KAKENHI Grant Numbers JP20K19886.

## References

1. MacQueen, J.B.: Some methods of classification and analysis of multivariate observations. In: Proceedings of 5th Berkeley Symposium on Mathematical Statistics and Probability, pp. 281–297 (1967)
2. Bezdek, J.C.: Pattern Recognition with Fuzzy Objective Function Algorithms. Plenum Press, New York (1981)
3. Lingras, P., West, C.: Interval set clustering of web users with rough K-means. J. Intell. Inf. Syst. **23**(1), 5–16 (2004)
4. Peters, G.: Some refinements of rough K-means clustering. Pattern Recogn. **39**(8), 1481–1491 (2006)
5. Ubukata, S., Notsu, A., Honda, K.: General formulation of rough C-means clustering. Int. J. Comput. Sci. Netw. Secur. **17**(9), 1–10 (2017)
6. Ubukata, S.: A unified approach for cluster-wise and general noise rejection approaches for k-means clustering. PeerJ Comput. Sci. **5**(e238), 1–20 (2019)
7. Oh, C.-H., Honda, K., Ichihashi, H.: Fuzzy clustering for categorical multivariate data. In: Proceedings of Joint 9th IFSA World Congress and 20th NAFIPS International Conference, pp. 2154–2159 (2001)
8. Rigouste, L., Cappé, O., Yvon, F.: Inference and evaluation of the multinomial mixture model for text clustering. Inf. Process. Manage. **43**(5), 1260–1280 (2007)
9. Honda, K., Oshio, S., Notsu, A.: Fuzzy co-clustering induced by multinomial mixture models. J. Adv. Comput. Intell. Intell. Inform. **19**(6), 717–726 (2015)

# Rank Estimators for Robust Regression: Approximate Algorithms, Exact Algorithms and Two-Stage Methods

Michal Černý, Miroslav Rada, and Ondřej Sokol(✉)

Faculty of Informatics and Statistics, Department of Econometrics, University of Economics, Winston Churchill Square 4, 13067 Prague, Czech Republic
{cernym,miroslav.rada,ondrej.sokol}@vse.cz

**Abstract.** Rank estimators for linear regression models have been designed as robust estimators insensitive to outliers. The estimator is defined as a minimizer of Jaeckel's dispersion function. We study algorithms for minimization of the function. Based on P-completeness arguments, we show that the minimization is computationally as hard as general linear programming. We also show that approximate algorithms with controlled error cannot be conceptually simpler since they can be converted into exact algorithms solving the same P-complete problem. Thus, approximate algorithms from literature, which do not use linear programming, cannot be guaranteed to approach the minimizer with a controlled error. Finally, we design two-stage methods combining advantages of both approaches: approximate algorithms with a simple and fast iteration step allow us to get close to the minimizer and exact algorithms, requiring LP techniques, then guarantee convergence and an exact result. We also present computational experiments illustrating the practical behavior of two-stage methods.

**Keywords:** Rank estimator · Robust regression · Computational statistics · Arrangement of hyperplanes

## 1 Introduction and Problem Formulation

Rank estimators for linear regression models have been designed by Jaeckel [11], Jurečková [12], Hettmansperger & McKean [10] and others as estimators robust to outliers. As a reference book see [9] and further references therein. The estimators are also implemented in the R-package `Rfit`, see [13] for details.

Rank estimators find a wide use in finance, economics, biostatistics, medical science, and epidemiology, particularly in estimation of single-index models parameters [3] and in regression of clustered data [6].

This text is devoted to the computational side of the problem: exact algorithms, approximate algorithms and their combination in a form of two-stage methods.

Supported by the Czech Science Foundation under project 19-02773S. Also the discussions with Jaromír Antoch are acknowledged.


V.-N. Huynh et al. (Eds.): IUKM 2020, LNAI 12482, pp. 162–173, 2020.
https://doi.org/10.1007/978-3-030-62509-2_14

**Linear Regression.** Consider the linear regression relationship $y = X\beta^* + \varepsilon$, where $y \in \mathbb{R}^n$ stands for the vector of $n$ observations of the dependent variable, $X \in \mathbb{R}^{n\times p}$ is the matrix of regressors, $\beta^* \in \mathbb{R}^p$ is the vector of (true) regression parameters to be estimated and $\varepsilon \in \mathbb{R}^n$ is the vector of random errors. The $i$th row of $X$ is denoted by $x_i^{\mathrm{T}}$.

**Residuals and Consistent Permutations of Residuals.** If $\beta \in \mathbb{R}^p$ is a candidate estimate of $\beta^*$, the numbers $e_i^\beta = y_i - x_i^{\mathrm{T}}\beta$, $i = 1, \dots, n$ are called *residuals* (w.r.t. $\beta$). If $S_n$ is the group of permutations of $\{1, \dots, n\}$ and $\pi \in S_n$ satisfies $e^\beta_{\pi(1)} \le e^\beta_{\pi(2)} \le \cdots \le e^\beta_{\pi(n)}$, we say that $\pi$ is a $\beta$-*consistent* permutation. The set of all $\beta$-consistent permutations is denoted by $P(\beta)$.

Given $\beta$, let $\pi_\beta$ stand for an arbitrary representative (say, the lexicographically smallest one) of $P(\beta)$.

**Score Function, Jaeckel's Dispersion Function and the *R*-Estimator.** Now fix a *score function* $\varphi : (0,1) \to \mathbb{R}$. Given data $(X, y)$, the *rank estimator* (or *R-estimator* for short) is defined as any minimizer of *Jaeckel's dispersion function*

$$F_{X,y,\varphi}(\beta) \equiv F(\beta) = \sum_{i=1}^{n} \alpha_i e^\beta_{\pi_\beta(i)}, \tag{1}$$

where $\alpha_i := \varphi\left(\frac{i}{n+1}\right)$, $i = 1, \dots, n$.

## 2 Some Properties of $F(\beta)$

We shortly summarize the properties of $F(\beta)$ along the lines of [4,5].

**Basic Properties of $F(\beta)$.** Function $F(\beta)$ is continuous. If the score function $\varphi$ is nondecreasing, then $F(\beta)$ is convex.

If, for a given $\beta$, there exist permutations $\pi, \pi' \in P(\beta)$, $\pi \ne \pi'$, then $\sum_{i=1}^n \alpha_i e^\beta_{\pi(i)} = \sum_{i=1}^n \alpha_i e^\beta_{\pi'(i)}$. It follows that $F(\beta)$ is independent of the choice of the representative of $\pi_\beta \in P(\beta)$.

**Arrangement $\mathfrak{A}$ and its Cells.** The function $F(\beta)$ is piecewise linear. Consider the system ("arrangement", henceforth referred to as $\mathfrak{A}$) of hyperplanes $H_{ij} := \{\beta \mid (x_i - x_j)^{\mathrm{T}}\beta = y_i - y_j\}$, $1 \le i < j \le n$. The hyperplanes divide $\mathbb{R}^p$ into a finite number of polyhedral regions, called *cells*. A cell $C$ has a representation

$$C \equiv C^\pi = \{\beta \in \mathbb{R}^p \mid y_{\pi(1)} - x^{\mathrm{T}}_{\pi(1)}\beta \le y_{\pi(2)} - x^{\mathrm{T}}_{\pi(2)}\beta \le \cdots \le y_{\pi(n)} - x^{\mathrm{T}}_{\pi(n)}\beta\} \tag{2}$$

for some $\pi \in S_n$. It is easily seen that if $C^\pi$ is a cell, then $F(\beta)$ restricted to $C^\pi$ is a linear function.

From now on we will restrict our attention only to the convex case when the score function $\varphi$ is nondecreasing, i.e., $\alpha = (\alpha_1 \le \alpha_2 \le \cdots \le \alpha_n)$.

**Theorem 1 (algorithmic properties of $F(\beta)$).** *Let the data $(X, y, \alpha)$ be rational.*

(a) *It is possible to determine whether* $\inf_{\beta \in \mathbb{R}^p} F(\beta) = -\infty$, *or find a minimizer if it exists, in polynomial time w.r.t. the bit size of* $(X, y, \alpha)$.
(b) *The following problem is P-complete: decide whether* $\inf_{\beta \in \mathbb{R}^p} F(\beta) = -\infty$.
(c) *The following problem is P-complete: given* $\beta$, *decide whether* $\beta$ *is a minimizer of* $F(\beta)$.

See [7] as a reference book on P-completeness theory.

## 3 Exact Algorithms for Minimization of $F(\beta)$ and Their Drawbacks

**CCC-Method.** Currently we are aware of a single algorithm working in polynomial time *and* being able to find a minimizer exactly (or determine that $F(\beta)$ is unbounded): it is the CCC-method from [5]. Although polynomial and exact in theory, it is a kind of algorithm unsuitable for a practical implementation. It is based on the linear programming (LP) formulation

$$\min_{\substack{t \in \mathbb{R} \\ \beta \in \mathbb{R}^p}} t \text{ subject to } t \geq \sum_{i=1}^{n} \alpha_i (y_{\pi(i)} - x_{\pi(i)}^{\mathrm{T}} \beta) \quad \forall \pi \in S_n \tag{3}$$

with $n!$ constraints. (Indeed, if $(t^*, \beta^*)$ is an optimal solution of the LP, then $\beta^*$ is a minimizer of $F(\beta)$.) This LP cannot be solved directly (just listing the inequalities would require superpolynomial time), but the constraints can be handled implicitly by means of the Ellipsoid Method with separation oracle [8]. Unfortunately, the CCC-method suffers from all of the well-known disadvantages concerning the ellipsoid method generally: (i) it is extremely numerically sensitive: the computation involves huge rational numbers, of bitsize $q(L)$, where $q$ is a polynomial and $L$ is the bit-size of the input instance $(X, y, \alpha)$; (ii) the method often requires computation time close to the worst-case bound $q^*(L)$, where $q^*$ is a polynomial of a high degree. (We are not aware of an interior-point method for minimization of $F(\beta)$; this is a challenging open problem.)

**Why We Need Other Algorithms.** Properties (i, ii) justify that in practice, other algorithms than CCC are preferable, possibly at the cost that they do not have a theoretically guaranteed polynomial time (but are often fast in practice) or that they do not produce an exact minimizer, but only an approximate one.

**The Problem is to Determine that a Minimizer has Been Found.** For simplicity assume that $F(\beta)$ has a minimizer. Theorem 1(c) can be loosely interpreted as follows: *determine whether a candidate point* $\beta_0$ *is a minimizer of* $F(\beta)$ *is as difficult as linear programming*, or *we cannot expect that verifying whether or not* $\beta_0$ *is a minimizer could be done with significantly less computational effort than needed for solving an LP*. This is bad news for practical implementations.

Indeed, virtually all algorithms are iterative, constructing a sequence $\beta^1, \beta^2, \ldots$ of improving points, and they would have to solve an LP in each iteration. If the goal is to minimize computation time inside an iteration, it is better to avoid linear programming. Currently we are aware only of the following method for deciding that $\beta_0$ is a minimizer [4,15]:

**Lemma 1.** *Let* $\beta \in \mathbb{R}^p$, *let* $A_\beta = \{(i,j) \in \{1,\ldots,n\}^2 \mid \exists \pi \in P(\beta)\ \pi(i) = j\}$ *and* $e = (1,\ldots,1)^{\mathrm{T}}$. *Then* $\beta$ *is a minimizer of* $F$ *if and only if the linear system*

$$e^{\mathrm{T}} s + e^{\mathrm{T}} r = -1, \quad \alpha_i x_j^{\mathrm{T}} \ell + s_i + r_j \geq 0 \quad \forall (i,j) \in A_\beta \tag{4}$$

*with variables* $\ell \in \mathbb{R}^p, r \in \mathbb{R}^n, s \in \mathbb{R}^n$ *is infeasible.*

**Osborne's Method and WoA-Method.** The recent *Walk-on-Arrangement* (WoA) method [4] can be seen as an extension of Osborne's method [15]. The main step is based on (4): if the system is feasible, then $\ell$ is an improving direction, meaning that $F(\beta + \delta\ell) < F(\beta)$ for some $\delta > 0$. This direction is used for the next step; such steps are iterated.

Both algorithms have serious drawbacks. Osborne's method need not terminate at all (it can end up in an infinite cycle) and WoA, although it provably terminates, has iteration bound $O(n^{2p-2})$. Its practical behavior is much better, but no proof of polynomiality is known.

## 4 Approximate Algorithms: Theory

As argued in the previous section, an exact algorithm must be able to determine whether the current iterate $\beta$ is a minimizer. This test is as difficult as general linear programming (in the sense of P-completeness). Thus, if one wants to avoid linear programming, which is a computationally expensive procedure from the practical viewpoint, it seems unavoidable to admit that the output need not be an exact minimizer, but only an approximate minimizer.

**Methods with Controlled Error.** We say that an approximate algorithm has *controlled error*, if it gets an input instance $(\alpha, X, y, \delta > 0)$ and outputs either a point $\beta$ satisfying $\|\beta - \beta_0\|_\infty < \delta$ for some minimizer $\beta_0$ ("Type-2 controlled error"), or a point satisfying $F(\beta) - F(\beta_0) < \delta$ ("Type-1 controlled error").

**Why it is Difficult to Control the Error of an Approximate Algorithm.** Currently we are not aware of any approximate algorithm with controlled error (except for the exact algorithms described in Sect. 3, which are "approximate" with $\delta = 0$). Virtually all algorithms, namely those studied in statistical literature (which will be discussed in Sect. 5), are *heuristic*, meaning that the error of their output is not controlled at all.

We give an argument why it is difficult to design algorithms with controlled error. We show that with a suitable choice of $\delta > 0$, such an approximate algorithm can be converted into an exact algorithm by a "simple" post-processing step. It follows that such algorithm would be able to solve the P-complete problem from Theorem 1.

**Type-1 Controlled Error.** Let us show the argument for Type-1 controlled error. Let $L$ be the bit-size of rational input data $(X, y, \alpha)$ and let $t_0 = \min_\beta F(\beta)$. From Theorem 1(a) it follows that there exists a polynomial $q$ such that the bitsize of $t_0$ is at most $q(L)$. Thus, $t_0$ can be written down as a rational number $\varrho/\vartheta$ with $\varrho \in \mathbb{Z}$, $\vartheta \in \mathbb{N}$ such that $|\varrho| \leq 2^{q(L)}$ and $\vartheta \leq 2^{q(L)}$. The post-processing step is based on the following lemma.

**Lemma 2 (Diophantine approximation).** *Given $\gamma \in \mathbb{Q}$ and $M \in \mathbb{N}$, there exists at most one rational number $\rho/\theta$ such that $\rho \in \mathbb{Z}$, $\theta \in \mathbb{N}$, $|\gamma - \rho/\theta| < 1/(2M^2)$ and $1 \leq \theta \leq M$. Moreover, the number $\rho/\theta$ (if exists) can be found in polynomial time w.r.t. the bit-size of $\gamma$ and $M$.*

We can run the approximate algorithm with $\delta = 2^{-2q(L)-1}$ (this is a trick similar to [5], see also Chapter 6 of [17]). It finds a number $t_1$ such that $t_1 - t_0 \leq 2^{-2q(L)-1}$. We already know that $t_0$ has a bounded denominator $\vartheta$ and Lemma 2 implies that near $t_1$, there exists at most one rational number with such bounded denominator. So this number must be $t_0$. And the polynomial algorithm from Lemma 2 with $\gamma = t_1$ and $M = 2^{q(L)}$ allows us to find this unique number $\rho/\theta = t_0$. (Indeed, $t_1 - t_0 \leq \delta = 2^{-2q(L)-1} = \frac{1}{2} \cdot \frac{1}{2M^2} < \frac{1}{2M^2}$ and $\vartheta \leq 2^{q(L)} = M$.)

**Type-2 Controlled Error.** For this type of error, a similar argument holds true. Assume for simplicity that a minimizer is unique. Then, the minimizer is a rational vector $\beta_0$ where all elements have a bounded denominator at most $q(L)$ for a suitable polynomial $q$. Then, running an approximate algorithm with Type-2 controlled error with $\delta = 2^{-2q(L)-1}$ allows us to use Diophantine approximation $p$ times to retrieve the exact minimizer per element of $\beta_0$.

**Conclusion of This Section.** We have shown, at least in theory, that an approximate algorithm with a controlled error can be converted into an exact algorithm by specifying a suitable small precision $\delta > 0$ (but still with polynomially bounded size w.r.t. $L$), which allows us to retrieve the exact value by Diophantine approximation. It follows, in a sense, that approximate algorithms with controlled error can effectively solve as difficult problem as finding the minimizer exactly. And we argued before that this problem probably cannot be done without linear programming (in the sense of P-completeness). This is a theoretical reason why we cannot expect approximate algorithms with controlled error which would be significantly "simpler" than the exact algorithms. Or, put otherwise, we can expect that algorithms not relying on LP techniques will be heuristic, with no controlled error. This is what we discuss in the next section.

## 5 Approximate Algorithms: Practice

As far as we are aware, all approximate algorithms studied in statistical literature produce a sequence of points $\beta^1, \beta^2, \ldots$ and stop when the progress is small, either measured by $\|\beta^{i-1} - \beta^i\|$ or $|F(\beta^{i-1}) - F(\beta^i)|$, but the stopping rule is heuristic and the methods do not have a controlled error (or there is no proof thereof).

**A Representative of Approximate Algorithms: IRLS.** There is a rich literature on approximate algorithms, namely in journals on statistics and numerical mathematics. Here we choose Iteratively Reweighted Least Squares (IRLS, see [1,2,14,18,19]) as a representative widely studied in literature. The idea is simple: when $\beta^k$ is the current iterate, we take $\pi \in P(\beta^k)$ (arbitrarily if there are multiple ones) and write

$$F(\beta^k) = \sum_{i=1}^{n} \alpha_i e_{\pi(i)}^{\beta^k} = \sum_{i=1}^{n} \alpha_{\pi^{-1}(i)} e_i^{\beta^k} = \sum_{i=1}^{n} \frac{\alpha_{\pi^{-1}(i)}}{e_i^{\beta^k}} (e_i^{\beta^k})^2 =: \sum_{i=1}^{n} w_i (e_i^{\beta^k})^2, \quad (5)$$

assuming that the residuals $e_i^{\beta^k}$ are nonzero. The minimization of (5) resembles weighted least squares (i.e., minimization of the sum of squared residuals with weights $w_i$). This can be done by solving the least-squares system $X^{\mathrm{T}} W X \beta^{k+1} = X^{\mathrm{T}} W y$ with $W = \mathrm{diag}(w_1, \ldots, w_n)$. Then, $\beta^{k+1}$ is the next iterate. (If $e_i^{\beta^k} = 0$ for some $i$, one can set e.g. $w_i = 0$ or use another choice to make $W$ regular.) When some ending condition is met (e.g. $\|\beta^k - \beta^{k+1}\|$ is sufficiently small in two consecutive iterations), the method terminates.

**The Main Advantage: Little Work Inside an Iteration.** IRLS illustrates the main advantage of many approximate algorithms: the work inside an iteration is simple and fast, reduces essentially to solving a single linear system, with no need of linear programming. But we saw previously that this is at the cost that we have no guarantee of precision of the output. Indeed, [4] shows that there are instances where IRLS *does not terminate at all*: its iterates $\beta^1, \beta^2, \ldots$ end up in an infinite cycle with $\beta^k = \beta^{k+2} = \beta^{k+4} = \cdots$, $\beta^{k+1} = \beta^{k+3} = \beta^{k+5} = \cdots$ and $\beta^k \neq \beta^{k+1}$. See Fig. 1 for an example. This implies that it is necessary to not only rely on changes in $\beta^k$, but it shall also track $F(\beta^k)$ and end when there is not sufficient progress in the objective value.

## 6 Two-Stage Methods

The considerations from the previous sections lead us to two-step methods, combining advantages of both approximate methods (easy and fast iteration steps) and exact methods (guaranteed convergence and exact output, but with a need to solve an LP in an iteration). As far as we are aware, such approach results in the most practically efficient algorithms.

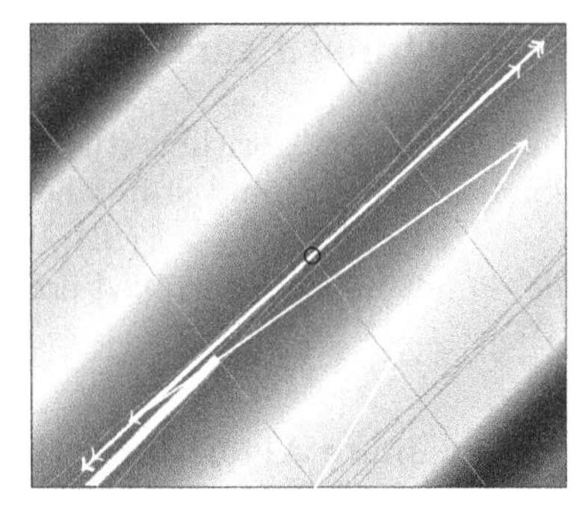

**Fig. 1.** Iterations (white arrows) of IRLS with $p = 2$ and $n = 12$. View zoomed to neighborhood of the optimum (black circle). Note the multiple overlapping arrows that are cummulating in the corners of the figure, demonstrating the cycling behavior of IRLS.

First we use an approximate method, such as IRLS, to find an estimate $\beta$ as close to the minimizer as possible. The progress is measured mostly by $F(\beta^k)$, where $k$ is a number of the current iteration. When the progress (improvement in $F(\beta^k)$) is small in several consecutive iterations, we switch to the exact method which finishes the job. The next section is devoted to computational experiments showing how much computational effort this strategy can save, both in terms of the number of iterations as well as in the number of LPs solved. Note, however, that the ending condition for the first stage is a bit more complex, taking also the total number of iterations and $\|\beta^{k+1} - \beta^k\|$ into account.

**An Analogy with LP Solvers.** The two-stage approach resembles practical implementations of LP solvers. As far as we are aware, the most successful implementations use long-step interior point methods (IPMs) to get close to the optimal solution. Although polynomial in theory, IPMs have a practically fast progress when being far from the optimum; while when being close, the progress is slow and IPMs require a numerically sensitive step, known as *rounding step* [16], to jump from the interior of the feasible space onto the boundary into the exact minimizer. Such solvers implement a *crossover step*: they turn off the IPM and switch to the simplex method, which does not have a guaranteed polynomial bound, but can walk on the boundary and make a step into the optimal point naturally. The implementation is based on the observation that in practice, the IPM approaches the optimum so closely that the number of steps required by the simplex method is usually low (although without a theoretical guarantee).

## 7 Computational Experience with Two-Stage Methods

**WoA as the Exact Method.** We shortly recall the idea of the WoA algorithm [4] which is used as the exact method in the second stage. Given a current iterate $\beta^k$, it determines a permutation $\pi \in P(\beta^k)$ (arbitrarily if multiple exist).

Then it solves the LP problem $\min_{\beta \in C^\pi} F(\beta)$; recall that $C^\pi$ is the cell (2) of arrangement $\mathfrak{A}$. Since $F(\beta)$ is a linear function on $C^\pi$, this is indeed an LP. If the LP is unbounded, then $F(\beta)$ is unbounded. Otherwise, let $\beta^*$ be an optimal solution. Then, system (4) is solved with $\beta := \beta^*$ (this is again an LP problem). If it is infeasible, a minimizer has been found. If it has an optimal solution $(\ell^*, r^*, s^*)$, then $\ell^*$ can be proven to be nonzero and that it can be taken as an improving direction. The algorithm performs line search in the direction $\ell^*$, i.e. it finds $\delta_0$ minimizing the function $\delta \mapsto F(\beta^* + \delta\ell^*)$ (or detects that $F$ is unbounded). Then, the method continues with $\beta^{k+1} = \beta^* + \delta_0\ell^*$ in the next iteration.

**Design of Our Experiments.** The work inside an iteration requires sorting (to get $\pi \in P(\beta^k)$), solving two LPs with $p$ and $2n + p$ variables, respectively, and the line search procedure. The following experiments show how much time is required for these sub-steps per iteration. This is compared to the simple IRLS step (here, IRLS is used as the heuristic stage-one method). We also measure the error of IRLS, i.e. how $\beta_{\text{IRLS}}$ found by IRLS differs from the optimal $\beta_{\text{WoA}}$ found by WoA and also how $F(\beta_{\text{IRLS}})$ differs from $F(\beta_{\text{WoA}})$ to illustrate the tradeoff between computational simplicity of IRLS and guaranteed convergence of WoA. In the experiments, we intentionally choose the initial iterate $\beta^0$ randomly; this is a model for the situation that in practice, the initial guess is sometimes "good" and sometimes "bad". There are two variants for the random choice of $\beta^0$. Firstly, $\beta^0$ is selected in an unbounded cell of $\mathfrak{A}$ (this is a kind of the "worst" possible starting point; this variant is called *start at a borderline point* in Table 2). Secondly, a random point is selected. (However, taking least-squares or least-absolute-deviations fit as the initial guess for $\beta^0$ might be a practically useful idea.)

### 7.1 Computational Time of WoA and IRLS per One Iteration

**Instances.** We measured the computational time per iteration on randomly generated instances. For an instance with $p$ parameters and $n$ observations, we generated an integer matrix $X \in \{-50, -49, \ldots, 49, 50\}^{n \times p}$ and an integer vector $y \in \{-50, -49, \ldots, 49, 50\}^n$. We used 40 combinations of $(n, p)$ in total, namely for each $p \in \{3, 5, 7, 10\}$ and each $n \in \{12, 15, 20, 30, 50, 80, 150, 250, 400, 600\}$. For each $(n, p)$, three different instances were examined.

**Results.** The results are summarized in Table 1. The right part of the table shows how many times is one iteration of IRLS faster than one iteration of WoA. Since IRLS typically stops improving the best found solution after tens or hundreds of iterations, it means that the running time of IRLS is comparable to one iteration of WoA. Note, however, that WoA iteration time scales better than IRLS iteration time with increasing $p$. The reason for this is that iteration time of WoA depends rather on $n$ than on $p$, as shows the left part of the table.

**Table 1.** Left part: WoA times in seconds per iteration for various $n$ and $p$. Right part: ratio of WoA time per iteration and IRLS time per iteration. Note that for large $n$, ratio decreases with $p$. The numbers are average over 3 instances.

| | WoA per iteration [in s] | | | | WoA per it./IRLS per it. | | | |
|---|---|---|---|---|---|---|---|---|
| $n$ | $p = 3$ | $p = 5$ | $p = 7$ | $p = 10$ | $p = 3$ | $p = 5$ | $p = 7$ | $p = 10$ |
| 12 | 0.03 | 0.04 | 0.06 | 0.15 | 355.2 | 486.7 | 669.9 | 1354.7 |
| 15 | 0.04 | 0.05 | 0.06 | 0.12 | 396.5 | 532.2 | 535.4 | 958.7 |
| 20 | 0.05 | 0.05 | 0.06 | 0.10 | 421.7 | 481.2 | 523.4 | 712.9 |
| 30 | 0.06 | 0.08 | 0.08 | 0.10 | 458.3 | 541.3 | 503.9 | 594.3 |
| 50 | 0.11 | 0.11 | 0.12 | 0.14 | 519.6 | 538.8 | 506.7 | 541.0 |
| 80 | 0.18 | 0.19 | 0.19 | 0.21 | 555.4 | 567.5 | 509.4 | 468.1 |
| 150 | 0.36 | 0.36 | 0.37 | 0.38 | 604.3 | 587.4 | 514.4 | 490.9 |
| 250 | 0.68 | 0.68 | 0.67 | 0.71 | 658.7 | 582.7 | 505.6 | 433.8 |
| 400 | 1.25 | 1.26 | 1.24 | 1.34 | 677.9 | 584.2 | 489.2 | 328.3 |
| 600 | 2.38 | 2.39 | 2.15 | 2.29 | 711.5 | 611.1 | 464.8 | 330.3 |

### 7.2 Two-Stage Algorithm: Performance

**Ending Condition for IRLS.** One of the most important implementation details of IRLS is *when to end*. We implemented three ending conditions to the algorithm. IRLS ends if:

- the difference of estimates of $\beta$ in two consecutive iterations is sufficiently small (we implemented this as $\|\beta^{k+1} - \beta^k\| < 10^{-6}$),
- a given maximal number $I_{\max}$ of iterations reached,
- a given number $I_{\text{impr}}$ of iterations is reached without improving $F(\beta)$.

Ending due to the first condition shall indicate that IRLS converges. In our experiments, this occurred only for the instances with $n = 1000$ and $p = 2$, which conforms the good expected asymptotical behavior of IRLS.

For the numbers $I_{\max}$ and $I_{\text{impr}}$, we tested three scenarios:

- IRLS 1 – $I_{\max} = n$, $I_{\text{impr}} = p$,
- IRLS 2 – $I_{\max} = pn$, $I_{\text{impr}} = n$,
- IRLS 3 – $I_{\max} = pn$, $I_{\text{impr}} = p \log n$.

**Instances and Results.** The instances were generated similarly as in Sect. 7.1, however, sizes of the instances were different. For each of numbers of parameters $p = 2, 3, 5, 7, 10$, we use three numbers of observations. The most interesting combinations are tabelated in Table 2. Similarly as in Sect. 7.1, the results, numbers in the table are averages over three instances.

The first row of Table 2 indicates how many iterations are needed by WoA to find optimum from a borderline point. Other rows labeled "# it" give numbers of iterations of WoA in the second stage, i.e. it is the number of iterations WoA needs after IRLS provides $\beta_{\mathrm{IRLS}}$.

Rows labeled "$\beta$ err" contain an absolute error of IRLS, namely $\|\beta_{\mathrm{IRLS}} - \beta_{\mathrm{WoA}}\|$. Relative error $\frac{F(\beta_{\mathrm{IRLS}})-F(\beta_{\mathrm{WoA}})}{F(\beta_{\mathrm{WoA}})}$ of objective value is in rows labeled "$F$ err".

For instances with large $n$ (compared to $p$), IRLS is able to find a very good estimate of $\beta$. Note that in instances with $n = 1000$ and $p = 2$, only one iteration of WoA is needed to find the optimal point, compared to 174 iterations needed to find optimum from a borderline point. For these problem sizes, average relative error in $F(\beta)$ is somewhere between $10^{-6}$ and $10^{-10}$ (numbers less than $10^{-3}$ are indicated by ~0 in Table 2).

For instances with small $n$ (compared to $p$), IRLS provides not very good estimate on $\beta$, see e.g. results for $n = 10$ and $p = 5$. From a random point, IRLS saves only 4 iterations (out of 14). Also, the relative error of $F(\beta_{\mathrm{IRLS}})$ is huge in general.

**Table 2.** Performance of the two-stage method. The first row: number of iterations needed by WoA when run from a borderline point. "$\beta$ err": absolute error of $\beta_{\mathrm{IRLS}}$ returned after the first stage compared to optimal $\beta_{\mathrm{WoA}}$, i.e. $\|\beta_{\mathrm{IRLS}} - \beta_{\mathrm{WoA}}\|$. "$F$ err": relative error of $F(\beta_{\mathrm{IRLS}})$ compared to $F(\beta_{\mathrm{WoA}})$, i.e. $(F(\beta_{\mathrm{IRLS}}) - F(\beta_{\mathrm{WoA}}))/F(\beta_{\mathrm{WoA}})$. The entries "~0" stand for numbers less than $10^{-3}$.

| | | $p$ | 2 | 2 | 3 | 3 | 5 | 5 | 7 | 7 | 10 | 10 |
|---|---|---|---|---|---|---|---|---|---|---|---|---|
| | | $n$ | 10 | 1000 | 10 | 100 | 10 | 200 | 100 | 200 | 30 | 200 |
| WoA | | # it | 11 | 174 | 12 | 53 | 14 | 105 | 155 | 205 | 73 | 173 |
| Start at a random point | IRLS 1 | # it | 6 | 1 | 9 | 27 | 13 | 58 | 71 | 127 | 51 | 121 |
| | | $F$ err | 0.041 | ~0 | 0.211 | 0.001 | 1.373 | ~0 | 0.004 | 0.004 | 0.221 | 0.001 |
| | | $\beta$ err | 0.133 | ~0 | 0.532 | 0.036 | 1.457 | 0.02 | 0.082 | 0.083 | 0.579 | 0.053 |
| | IRLS 2 | # it | 6 | 1 | 7 | 13 | 10 | 46 | 58 | 95 | 53 | 109 |
| | | $F$ err | 0.028 | ~0 | 0.159 | ~0 | 0.956 | ~0 | 0.002 | 0.001 | 0.186 | 0.001 |
| | | $\beta$ err | 0.137 | ~0 | 0.464 | 0.009 | 1.53 | 0.012 | 0.065 | 0.029 | 0.547 | 0.036 |
| | IRLS 3 | # it | 6 | 1 | 7 | 20 | 14 | 57 | 58 | 91 | 48 | 117 |
| | | $F$ err | 0.033 | ~0 | 0.192 | ~0 | 0.758 | ~0 | 0.002 | 0.001 | 0.154 | 0.001 |
| | | $\beta$ err | 0.167 | ~0 | 0.508 | 0.022 | 1.457 | 0.017 | 0.065 | 0.049 | 0.525 | 0.043 |
| Start at a borderline point | IRLS 1 | # it | 6 | 1 | 9 | 32 | 13 | 59 | 63 | 125 | 60 | 109 |
| | | $F$ err | 0.042 | ~0 | 0.197 | ~0 | 0.809 | ~0 | 0.012 | 0.003 | 0.179 | 0.002 |
| | | $\beta$ err | 0.171 | ~0 | 0.508 | 0.018 | 1.094 | 0.018 | 0.114 | 0.081 | 0.606 | 0.06 |
| | IRLS 2 | # it | 6 | 1 | 9 | 18 | 12 | 45 | 69 | 92 | 47 | 123 |
| | | $F$ err | 0.041 | ~0 | 0.197 | ~0 | 0.765 | ~0 | 0.002 | 0.001 | 0.11 | 0.001 |
| | | $\beta$ err | 0.155 | ~0 | 0.508 | 0.009 | 1.697 | 0.012 | 0.052 | 0.04 | 0.53 | 0.045 |
| | IRLS 3 | # it | 6 | 1 | 9 | 31 | 12 | 59 | 55 | 110 | 47 | 123 |
| | | $F$ err | 0.042 | ~0 | 0.197 | ~0 | 0.765 | ~0 | 0.004 | 0.003 | 0.11 | 0.002 |
| | | $\beta$ err | 0.171 | ~0 | 0.508 | 0.017 | 1.697 | 0.018 | 0.076 | 0.062 | 0.53 | 0.057 |

**Visualization for $n = 10$, $p = 2$.** Here, we present an example to illustrate the behavior of WoA and IRLS on a small instance. Figure 2 depicts $\mathfrak{A}$ in the neighborhood of the optimum. Both pure WoA and two-stage method start in a random point outside the figure. WoA (brown arrows) requires 8 iterations

(14 arrows in total as there are two steps inside one iteration), while two-stage method firstly performs 16 iterations of IRLS (white arrows) and then 2 iterations of WoA (black arrows).

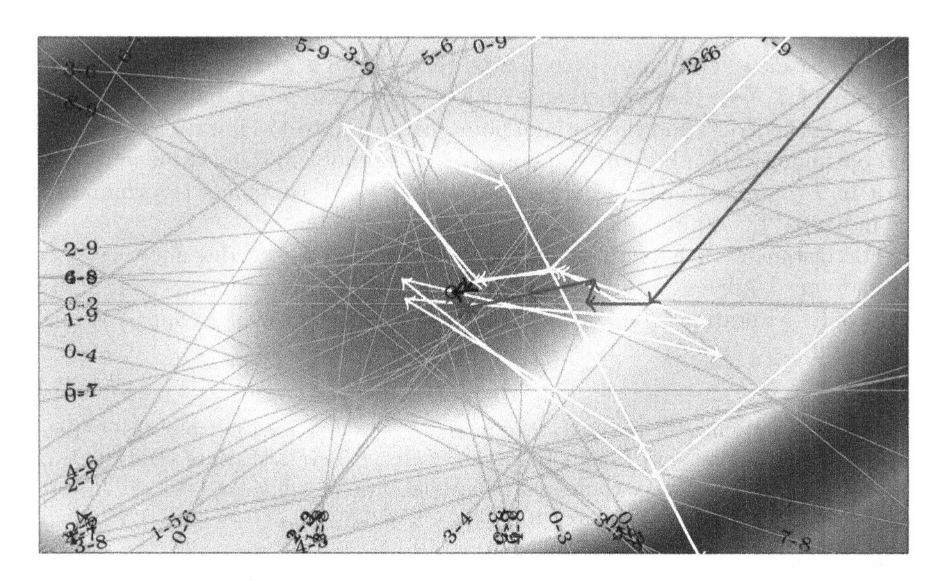


**Fig. 2.** Illustration of two-stage method on an instance with $n = 10, p = 2$. Only neigborhood of the optimum (marked by the black circle) is depicted. IRLS 2 iterations from a random point (far away out of the croped region) are drawn using white arrows (16 iterations). Then, two iterations (black arrows) of WoA are sufficient from the best $\beta$ found by IRLS. Compare this with brown arrows (8 iterations) of WoA when run directly from the same initial random point. (Color figure online)

## 8 Conclusions

IRLS algorithm provides a tight approximation of rank estimator especially for larger instances, where its good asymptotical properties begin to work well. Additionally, our experiments have shown that even for small instances, IRLS is able to find a quite good $\beta$ in a small number of iterations. Since iterations of IRLS are several hundred times cheaper than iterations of WoA, too, our experiments indicate that employing a first heuristic stage (IRLS or any other) can boost up sizes of instances that can be solved to optimality by WoA.

## References

1. Antoch, J., Ekblom, H.: Selected algorithms for robust M- and L-regression estimators. In: Dutter, R., Filzmoser, P., Gather, U., Rousseeuw, P.J. (eds.) Developments in Robust Statistics, pp. 32–49. Physica-Verlag, Heidelberg (2003). https://doi.org/10.1007/978-3-642-57338-5_3
2. Antoch, J., Ekblom, H.: Recursive robust regression computational aspects and comparison. Comput. Stat. Data Anal. **19**(2), 115–128 (1995). https://doi.org/10.1016/0167-9473(93)E0050-E
3. Bindele, H.F., Abebe, A., Meyer, K.N.: General rank-based estimation for regression single index models. Ann. Inst. Stat. Math. **70**(5), 1115–1146 (2017). https://doi.org/10.1007/s10463-017-0618-9
4. Černý, M., Hladík, M., Rada, M.: Walks on hyperplane arrangements and optimization of piecewise linear functions. arXiv:1912.12750 [math], December 2019
5. Černý, M., Rada, M., Antoch, J., Hladík, M.: A class of optimization problems motivated by rank estimators in robust regression. arXiv:1910.05826 [math], October 2019
6. Dutta, S., Datta, S.: Rank-based inference for covariate and group effects in clustered data in presence of informative intra-cluster group size. Stat. Med. **37**(30), 4807–4822 (2018)
7. Greenlaw, R., Hoover, J.H., Ruzzo, W.L.: Limits to Parallel Computation: P-Completeness Theory. Oxford University Press, New York (1995)
8. Grötschel, M., Lovász, L., Schrijver, A.: Geometric Algorithms and Combinatorial Optimization. AC, vol. 2. Springer, Heidelberg (1993). https://doi.org/10.1007/978-3-642-78240-4
9. Hettmansperger, T., McKean, J.: Robust Nonparametric Statistical Methods, 2nd edn. CRC Press (2010). https://doi.org/10.1201/b10451
10. Hettmansperger, T.P., McKean, J.W.: A robust alternative based on ranks to least squares in analysing linear models. Technometrics **19**(3), 275–284 (1977)
11. Jaeckel, L.A.: Estimating regression coefficients by minimizing the dispersion of the residuals. Ann. Math. Stat. **43**, 1449–1458 (1972)
12. Jurečková, J.: Nonparametric estimate of regression coefficients. Ann. Math. Stat. **42**, 1328–1338 (1971)
13. Kloke, J.D., McKean, J.W.: Rfit: rank-based estimation for linear models. R J. **4**(2), 57 (2012). https://doi.org/10.32614/RJ-2012-014
14. O'Leary, D.P.: Robust regression computation using iteratively reweighted least squares. SIAM J. Matrix Anal. Appl. **11**(3), 466–480 (1990). https://doi.org/10.1137/0611032
15. Osborne, M.R.: A finite algorithm for the rank regression problem. J. Appl. Probab. **19**(A), 241–252 (1982). https://doi.org/10.2307/3213564
16. Roos, C., Terlaky, T., Vial, J.P.: Interior Point Methods for Linear Optimization. Springer, Boston (2005). https://doi.org/10.1007/b100325
17. Schrijver, A.: Theory of Linear and Integer Programming. Wiley, New York (2000)
18. Wolke, R., Schwetlick, H.: Iteratively reweighted least squares: algorithms, convergence analysis, and numerical comparisons. SIAM J. Sci. Stat. Comput. **9**(5), 907–921 (1988). https://doi.org/10.1137/0909062
19. Wolke, R.: Iteratively reweighted least squares: a comparison of several single step algorithms for linear models. BIT Numer. Math. **32**(3), 506–524 (1992). https://doi.org/10.1007/BF02074884

# A New Classification Technique Based on the Combination of Inner Evidence

Thanh-Phu Nguyen(✉) and Van-Nam Huynh

School of Advanced Science and Technology, Japan Advanced Institute of Science and Technology, 1-1 Asahidai, Nomi, Ishikawa 923-1292, Japan
{ntphu,huynh}@jaist.ac.jp

**Abstract.** Data with high uncertainty and ambiguity is challenging for the classification task. EKNN is a popular evidence theory-based classification method developed for handling uncertainty data. However, as a distance-based technique, it also suffers from the problem of high dimensionality as well as performs ineffectively with mixed distribution data where closed data points originated from different classes. In this paper, we propose a new classification method that can softly classify new data upon each separate cluster which can remedy the overlapping data problem. Moreover, pieces of evidence induced from the trained clusters are combined using Dempster's combination rule to yield the final predicted class. By defining the mass function of evidence with the weight factor based on the distance between new data points and clusters' centers, it helps reduce the computational complexity which is also a problem in distance-based $k$-NN alike methods. The classification experiment conducted on various real data and popular classifiers has shown that the proposed technique has the results comparable to state-of-the-art methods.

**Keywords:** Classification · Evidence combination · Dempster-Shafer theory

## 1 Introduction

Uncertainty in machine learning is gaining more attention from the research community along with the popularity of artificial intelligence (AI) applications. The need for quantifying and managing the uncertainty has been emphasized in several fields such as healthcare [1], bio-informatics [2] or even pure machine learning research [3,4]. As the problem of uncertainty that cannot be resolved solely by traditional probabilistic framework [5], several uncertainty management theories have been developed. Among them, evidence theory [6] is one of the most popular solutions that can be utilized for handling the uncertainty in data as well as the decision-making process.

In the field of supervised learning, the uncertainty can arise from the labeling data process where some data objects cannot be labeled due to conflicts between


V.-N. Huynh et al. (Eds.): IUKM 2020, LNAI 12482, pp. 174–186, 2020.
https://doi.org/10.1007/978-3-030-62509-2_15

experts. Several approaches that adopt evidence theory for solving the mentioned problem have been proposed. Specifically, some methods have been developed to allow partially supervised learning (labels of data are represented as partial membership of available classes) such as evidence-based induction trees [7] which incorporates evidence-theoretic uncertainty measure to assess the impurity of nodes in decision trees. Recently, a new method named CD-EKNN [8] has been proposed to allow processing partially known labels in training data by adopting the contextual discounting technique as well as using a conditional evidential likelihood for optimizing model's parameters.

Another kind of uncertainty commonly existing inside observed real data is the lack of information (which can cause the sparse data problem) or overlapping (mixed-distributed) data. In order to deal with this kind of uncertainty several methods have been proposed. Among them, EKNN [9] is a notable distance-based classification method which classifies new data points based on the evidence of classes of their neighbors. Despite its robustness in dealing with uncertainty information, EKNN suffers from several problems such as computational complexity like any $k$-NN-based methods which come from the induction of distances of data in the sample space. Several methods have been proposed to remedy this limitation by applying feature selection or dimensional reduction techniques in order to shrink the feature spaces such as ConvNet-BF [10] or REK-NN [11].

Other inherent limitation of EKNN lays in its working mechanism of selecting a fixed number of nearby neighbors to induce evidence for classification process which can lead to misclassifying due to the lack of information or closed data points originating from different classes. Also, the assignment of hard-to-classified data only to an ignorance group is also debatable. In order to reduce those limitations, instead of referring to nearby neighbors, several methods compare new data points with prototypes that are produced from the training process such as ProDS [12], CCR [13]. Other methods consider assigning hard-to-classified data to a various classes-combined group named meta-class in addition to the original ignorance class such as BK-NN [14]. However, there are several criticisms that evidence based on classes of prototypes is unreasonable due to their non-semantic representations. Moreover, the classes of new data points also have to be specified with a degree of certainty rather than merely assigned to some common groups of classes.

In this paper, we make an effort to remedy the above-mentioned problems by proposing a new classification method that can induce the evidence about the classes of new data objects based on groups of data that belong to various distributions. Specifically, by assuming that a data set contains instances that are generated by several different distributions, data generated by each distribution can be represented in the form of heterogeneous clusters. Each distribution has a different mechanism for characterizing the classes of its generated data. For each cluster, we gauge the data distribution by using decision trees on the whole set of data belonging to that cluster. Results from those decision trees could be considered as evidence for determining the classes of new data points. For making a final decision about the predicted class, Dempster's combination rule [6] is used to fuse the evidence collected from previous steps.

The remainder of this paper is presented as follows. In Sect. 2, we remind some backgrounds of evidence theory. In Sect. 3, details of our proposed classification system named *IEBC* (Inner Evidence-Based Classifier) are described. The experimental evaluation is conducted in Sect. 4 to prove the merits of our proposed system. Finally, in Sect. 5, we summarize our work and discuss the limitations and future works.

## 2 Evidence Theory

Evidence theory also known as Dempster-Shafer theory [6,15] encompasses several models for dealing with uncertainty including transferable belief model [16]. In this paper, we will use the term Dempster-Shafer theory when mentioning evidence theory due to its popularity. Considered as a generalization of probability theory and set-membership approaches, Dempster-Shafer theory has proved to be an effective theoretical framework for reasoning with uncertain and imprecise information. In this section, we introduce basic definitions of Dempster-Shafer theory that facilitate as technical and theoretical background for our research.

Let $x$ be a variable taking values in a finite and unordered set $\Omega$ named the frame of discernment which is defined as $\Omega = \{\omega_1, \ldots, \omega_c\}$. Partial knowledge about the actual value of $x$ can be represented by a mass function (or basic probability assignment - *bpa*) $m$, which is defined as a mapping from $2^{\Omega}$ to $[0, 1]$, satisfying that

$$m(\emptyset) = 0$$

$$\sum_{A \subseteq \Omega} m(A) = 1$$

For any subset $A \subseteq \Omega$, $m(A)$ can be interpreted as the belief that one is willing to commit to $A$. For all $A$ that satisfies $m(A) > 0$, they are called the focal elements of mass function $m$. There are two non-additive measures so-called belief (or credibility) function and plausibility function that can equivalently represent $m$ as defined below.

For any subset $A \subseteq \Omega$, the belief function denoted by $Bel(A)$ is defined by

$$Bel(A) = \sum_{B \subseteq A} m(B)$$

and $Bel(A)$ can be interpreted as the degree of belief that the actual value of $x$ is in $A$. The plausibility function $Pl(A)$ is then defined as

$$Pl(A) = \sum_{B \cap A \neq \emptyset} m(B)$$

$Pl(A)$ can be seen as the degree to which the evidence is not contradictory to the proposition that the actual value of $x$ is in $A$.

Furthermore, given two mass functions $m_1$ and $m_2$ on the same frame of discernment that are derived from independent pieces of evidence, they can be fused by using Dempster's combination rule [6] to obtain a new mass function $m_1 \oplus m_2$ which is defined as

$$(m_1 \oplus m_2)(A) = \frac{1}{1-K} \sum_{B \cap C = A} m_1(B) m_2(C)$$

for all $A \subseteq \Omega$, where $K = \sum_{B \cap C = \emptyset} m_1(B) m_2(C)$ is defined as the degree of conflict between $m_1$ and $m_2$. When the degree of conflict $K$ between $m_1$ and $m_2$ is large, the combined evidence obtained by Dempster's combination rule may become unreliable and unintuitive which was pointed out in [17].

# 3 IEBC (Inner Evidence-Based Classifier)

In this section, we will describe the newly proposed classification system based on the combination of evidence about the classes of new data that are induced from the extracted clusters of training data. Specifically, the proposed system contains three main components: fuzzy clustering, decision trees induction and evidence combination.

For the first component of fuzzy clustering, in order to find available clusters from original training data while still be able to model the uncertainty inside data, a fuzzy clustering method is utilized to provide results as a degree of membership of each data point to relevant clusters. Instead of extracting evidence from a discrete number of neighboring data points, new data objects are considered as parts of data distributions (represented in the form of heterogeneous clusters) which determine their characteristics.

In the second component, decision trees are used as a tool to gauge the behaviors of those distributions and provide evidence for assigning testing data into their *"true"* classes. Finally, the collected evidence are combined using Dempster's combination rules to generate final predictions. Illustration of the general process of the proposed system is shown in Fig. 1.

## 3.1 Notations

Before describing the details of our proposed classification model, we define several notations for facilitating the formulation process. Specifically, let $D$ be a dataset of $n$ instances and characterized by $m$ attributes. Denote $D = \{x_i | i = 1, 2, \ldots, n\}$ and $x_i = \{x_{i1}, \ldots, x_{im}\}$. Instances in $D$ are labeled with $\Omega = \{1, \ldots, c\}$. A family of $k$ fuzzy sets (fuzzy clusters) $\{A_j, j = 1, 2, \ldots, k\}$ is defined where each data instance will be assigned a degree of membership to each fuzzy cluster.

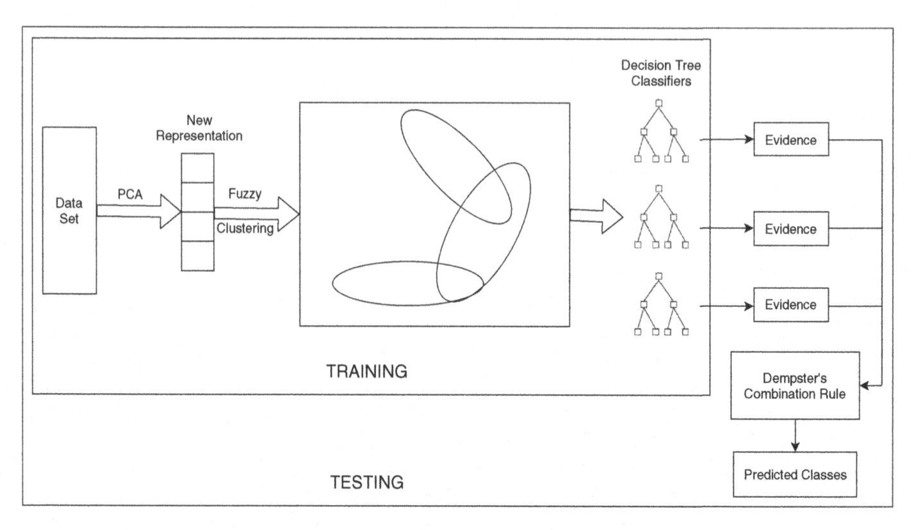


**Fig. 1.** Structure of the proposed classification system *IEBC*

### 3.2 The Proposed Model

The proposed classification system is comprised of three main components: the first part is fuzzy clustering which assigns data into clusters with a degree of membership. While in the second part, decision trees are derived from the trained clusters which aim to mimic the behaviors within each cluster that determine the characteristics of their belonging data instances. In the last component, collected evidence from induced decision trees are combined together for making final decisions.

Specifically, for the first component, clustering is one of the processes that contain high uncertainty [4]. In order to manage the uncertainty in the clustering task, several approaches have been proposed. Among them, fuzzy clustering is a popular approach [18]. Specifically, fuzzy clustering - in contrast with crisp clustering - allows for a degree of membership of data to each cluster. A single data point can have partial membership in more than one cluster. The membership value of a data instance $x_i$ to cluster $A_j$ can be denoted as the following.

$$\mu_{ij} = \mu_{A_j}(x_i) \in [0, 1] \tag{1}$$

The membership values $\mu_{ij}$ in fuzzy clustering have to satisfy the following conditions:

- Sum of all membership values of a single data instance to all clusters has to equal to 1.

$$\sum_{j=1}^{k} \mu_{ij} = 1, \text{ for all } i = 1, 2, \ldots, n$$

– There is no cluster that is empty or contains all of the data instances in $D$.

$$0 < \sum_{i=1}^{n} \mu_{ij} < n$$

A family of fuzzy partition matrices, $M_f$, that contains membership values between $k$ clusters and $n$ data instances can be denoted as:

$$M_f = \left\{ U|\mu_{ij} \in [0,1]; \sum_{j=1}^{k} \mu_{ij} = 1; 0 < \sum_{i=1}^{n} \mu_{ij} < n \right\} \quad (2)$$

Any $U \in M_f$ is a fuzzy $k$-partition, and it follows from the overlapping character of the clusters and the infinite number of membership values possible for describing cluster membership that the cardinality of $M_f$ is also infinity, that is, $\eta_{M_f} = \infty$.

To determine the fuzzy $k$-partition matrix $U$ for grouping a collection of $n$ data sets into $k$ clusters, an objective function $J$ for a fuzzy $k$-partition can be defined as:

$$J(U, v) = \sum_{i=1}^{n} \sum_{j=1}^{k} (\mu_{ij})^{\alpha} d_{ij}^{2} \quad (3)$$

where $d_{ij}$ is the Euclidean distance between the $A_j$ cluster's center and the $x_i$ data instance.

$$d_{ij} = d(x_i - v_j) = \left[ \sum_{l=1}^{m} (x_{il} - v_{jl})^2 \right]^{1/2} \quad (4)$$

Also, $\mu_{ij}$ is the membership of the $x_i$ data instance in the $A_j$ cluster. The parameter $\alpha$ that was introduced by [19] is a weighting parameter. This parameter has a value range of $[1, \infty)$ and controls the amount of fuzziness in the clustering process. While $v_j$ is the center of cluster $A_j$, which is described by $m$ features and can be denoted as $v_j = \{v_{j1}, v_{j2}, \ldots, v_{jm}\}$. Values of clusters' centers $v_j$ are calculated using the following formula:

$$v_j = \frac{\sum_{i=1}^{n} (\mu_{ij})^{\alpha} \times x_i}{\sum_{i=1}^{n} (\mu_{ij})^{\alpha}} \quad (5)$$

In this paper, we use the fuzzy $c$-means clustering algorithm [20] to cluster the original data. Details of the fuzzy $c$-means algorithm are described in [21]. The generated fuzzy clusters then have to be defuzzyfied in order to assign data points to each cluster. After the assignment of data to form hard clusters, decision trees can be induced from each cluster in the second component of our proposed system.

There are two popular methods for hardening the fuzzy partition matrix: the maximum membership method and the nearest center classifier [21]. In the maximum membership method, the largest element in each column of the $U$

matrix is assigned membership of unity and all other elements in each column are assigned a membership value of zero. However, in order to retain fuzzy partitioning characteristics that allow to handle the uncertainty, we revised the maximum membership method so that not just the largest membership but also its nearly equal memberships will be assigned to the same cluster by defining a so-called fuzzy threshold $\beta$ as the following.

$$\mu_{ij} \mapsto \overline{\mu_{ij}} = \begin{cases} 1 & \text{, if } \mu_{ij} = \max_{j'}\{\mu_{ij'}\} \\ & \text{or } |\mu_{ij} - \max_{j'}\{\mu_{ij'}\}| \leq \beta \\ 0 & \text{, otherwise} \end{cases} \tag{6}$$

for $j, j' = 1, \ldots, k$ and $i = 1, 2, \ldots, n$.

By defuzzying partition matrix $U$ into a hard partition matrix $U'$ using Eq. (6), we can obtain $k$ clusters $\{C_1, \ldots, C_k\}$ where

$$C_j = \{x_i | \overline{\mu_{ij}} = 1, i \in \{1, \ldots, n\}\} \text{ with } j \in \{1, \ldots, k\}$$

Along with each cluster $C_j$, a decision tree classifier $dt_j$ is generated using $CART$ algorithm [22] for classifying new data instances. Specifically, for a data instance $x_i$, the results of each classifier is noted as $P_i^j = [p_{i1}^j, \ldots, p_{ic}^j]$ where $p_{il}^j$ with $l \in \Omega$ and $i \in \{1, \ldots, n\}$ is the probability of the data instance $x_i$ which is classified as class $l$ by the classifier $dt_j$.

In the last component of our proposed system, each classifier is considered as an expert that makes independent judgment on the class of a data instances $x_i$. Therefore, the probability matrix $P_i^j$ generated by a classifier $dt_j$ along with the weighting factor $u_{ij} \in U$ can be utilized to define the basic probability assignment ($bpa$) $m_j$ as the following.

$$m_j : 2^{\Omega} \to [0, 1]$$

$$m_j(\{l\}) = p_{il}^j \times u_{ij} \tag{7}$$

$$m_j(A) = 0 \text{ if } |A| \geq 2,\ A \neq \Omega \tag{8}$$

$$m_j(\Omega) = 1 - \sum_{l=1}^{c} p_{il}^j \times u_{ij} \tag{9}$$

Here we only consider mass functions ($bpa$) of focal elements which are singletons. All other imprecision will be assigned to the whole frame of discernment $\Omega = \{1, \ldots, c\}$. Specifically, given the evidence $E_j = \{m_j(\{l\}), m_j(\{\Omega\})\}$ with $j \in \{1, \ldots, k\}$ and $l \in \{1, \ldots, c\}$ that is provided by the classifier $dt_j$, we can combine the set of evidence $E = \{E_1, \ldots, E_j\}$ by using Dempster's combination rule (denoted as $\bigoplus$).

$$m = \bigoplus_{j=1}^{k} m_j$$

Specifically, two mass functions $m_j$ and $m_{j+1}$ derived from two evidence sources $E_j, E_{j+1}$ can be combined by Dempster's rule to obtain a new mass function $m_j \oplus m_{j+1}$, defined as

$$m_j(A) \oplus m_{j+1}(A) = \frac{1}{1-K} \sum_{B \cap C = A} m_j(B) \times m_{j+1}(C)$$

for all nonempty $A \subseteq \Omega$, where $K = \sum_{B \cap C = \emptyset} m_j(B) \times m_{j+1}(C)$ is the degree of conflict between $m_j$ and $m_{j+1}$.

The following table to describe the detailed combination of two evidence sources $E_j, E_{j+1}$ (Table 1).

**Table 1.** Intersection of pieces of evidence from two evidence sources

| $E_{j+1}$ | $E_j$ | | | | |
|---|---|---|---|---|---|
| | $m_j(\{1\})$ | $m_j(\{2\})$ | ... | $m_j(\{c\})$ | $m_j(\Omega)$ |
| $m_{j+1}(\{1\})$ | $m(\{1\})$ | $\emptyset$ | ... | $\emptyset$ | $m(\{1\})$ |
| $m_{j+1}(\{2\})$ | $\emptyset$ | $m(\{2\})$ | ... | $\emptyset$ | $m(\{2\})$ |
| $\vdots$ | $\vdots$ | $\vdots$ | | $\vdots$ | $\vdots$ |
| $m_{j+1}(\{c\})$ | $\emptyset$ | $\emptyset$ | ... | $m(\{c\})$ | $m(\{c\})$ |
| $m_{j+1}(\Omega)$ | $m(\{1\})$ | $m(\{2\})$ | ... | $m(\{c\})$ | $m(\Omega)$ |

Finally, the combined mass function $m$ can be transformed into a probability function for making the last decision using the pignistic probability transformation [16]. However, in this case, because we consider only focal sets which are singletons and the ignorant set $\Omega$, then we have the probability function $m'$ of a mass function $m$ can be represented as below using the pignistic probability transformation.

$$m'(\{l\}) = m(\{l\}) + \frac{m(\Omega)}{|\Omega|} \tag{10}$$

As can be seen from the Eq. (10) that $\frac{m(\Omega)}{|\Omega|}$ is a constant value for any $l \in \Omega$. Therefore, the combined mass $m(\{l\})$ can also be used directly to determine the class of an instance $x_i$ by selecting *bpa* of the class that has the highest mass value without the need to transform into probability function as in the following formulas. Summary of the proposed classification model IEBC is also described in Algorithm 1.

$$predicted\_class = c^{\star} \tag{11}$$

$$c^{\star} = \underset{l}{argmax}(m(\{l\})) \tag{12}$$

Where $c^{\star}, l \in \{1, \ldots, c\}$.

**Algorithm 1. IEBC algorithm**

**Input**: Data set $D = \{D_{train}, D_{test}\}$

**Output**: Predicted classes of $D_{test}$

**Training:**

**Step 1: Transforming data**

1. $D'_{train} = PCA(D_{train})$

**Step 2: Clustering**

2. Set $k, \alpha$ values
   Initialize partition matrix $U^0$
   Set step counter $r = 1$

3: **While** $||U^{r+1} - U^r|| > \epsilon$ :

4: Calculate cluster centers $v_j$ using Eq.(5) on $D'_{train}$

5: Updating partition matrix $U^r$:

6: $$\begin{cases} \mu_{ij}^{(r+1)} = \left[\sum_{j'=1}^{k} \left(\frac{d_{ij}^r}{d_{ij'}^r}\right)^{2/(\alpha-1)}\right]^{-1} & \text{, if } I = \emptyset \\ \mu_{ij}^{(r+1)} = 0 & \text{, if } j \in \neg I \end{cases}$$

where $I = \{j | 2 \leq k \leq n; d_{ij}^r = 0\}$
$\neg I = \{1, 2, ..., k\} - I$

7: Defuzzy clusters using Eq.(6)

8: **Return:** $U = \{u_1, ..., u_k\}$, $C = \{C_1, ..., C_k\}$

**Step 3: Build classifiers**

9: **For each** $C_j$ **in** $C$:

10: $dt_j = Decisiontree(C_j)$

11: **Return:** $DT = \{dt_1, ..., dt_k\}$

**Testing:**

**Step 4: Cluster referencing**

12: **For each** $x$ **in** $D_{test}$:

13: **For each** $u$ **in** $U$:

14: Compute distances $d = Euclid(x, u)$

15: Update $U$ as in Step 2 without recalculating $v_j$

16: Assign $W_x = U_x$

17: **Return:** $W_x = [w_1, ..., w_k]$

**Step 5: Assigning mass of evidence**

18: **For each** $x$ **in** $D_{test}$:

19: **For each** $dt_j$ **in** $DT$:

20: $l = dt_j(x)$

21: Assign mass $m(\{l\}) = 1$

22: Apply weighting factor $m_w(\{l\}) = w_j \times m(\{l\})$

23: **Return:** $M_w^x = [m_{w\_1}, ..., m_{w\_k}]$

**Step 6: Combining evidence**

24: **For each** $x$ **in** $D_{test}$:

25: $m_{w\_1}(\Omega) = 1 - \sum_{l=1}^{c} m_{w\_1}(\{l\})$

26: Initialize combined mass $m_{comb} = m_{w\_1}$

27: **For** $i$ **in range** $(2, k)$:

28: $m_{w\_i}(\Omega) = 1 - \sum_{l=1}^{c} m_{w\_i}(\{l\})$

29: $m_{comb} = Dempster(m_{comb}, m_{w\_i})$

30: $y_x = \underset{l}{argmax}(m_{comb}(\{l\}))$

31: **Return:**
   Predicted classes $Y = [y_1, ..., y_{|D_{test}|}]$

# 4 Experimental Evaluation

## 4.1 Testing Datasets

In this work, we use a subset of MIMIC III dataset that contains admissions suspected of sepsis for the in-hospital mortality prediction task [23]. MIMIC III (Medical Information Mart for Intensive Care III) [24] is one of the most popular and biggest healthcare data sources that was developed by MIT Lab for Computational Physiology. MIMIC contains de-identified health data of nearly 60000 ICU admissions in the Beth Israel Deaconess Medical Center. Besides, we selected 9 other real datasets collected from UCI [25] to test with our proposed system to prove its effectiveness with a variety of data. Details of the testing datasets are described in Table 2.

**Table 2.** Detailed characteristics of 10 datasets collected from UCI and MIMIC III

| No. | Name | Inst. | Attr. | Classes | Data types | Missing values |
|---|---|---|---|---|---|---|
| 1 | Chess | 3196 | 36 | 2 | Categorical | No |
| 2 | Credit-approval | 690 | 15 | 2 | Categorical, Integer, Real | Yes |
| 3 | Diabetes | 769 | 9 | 2 | Real | No |
| 4 | Haberman | 306 | 3 | 2 | Integer | No |
| 5 | Heart | 303 | 75 | 2 | Categorical, Integer, Real | Yes |
| 6 | Heart-statlog | 270 | 13 | 2 | Categorical, Real | No |
| 7 | Ionosphere | 351 | 34 | 2 | Integer, Real | No |
| 8 | Liver-disorder | 345 | 7 | 2 | Categorical, Real, Integer | No |
| 9 | Magic | 19020 | 11 | 2 | Real | No |
| 10 | Sepsis | 11791 | 29 | 2 | Categorical, Real | No |

### 4.2 Experimental Setups and Final Results

In order to evaluate the performance of our proposed system IEBC, we conducted a binary classification task on a wide range of real datasets. In which, our system is compared with popular classification methods including CART, evidence-based classifiers (EKNN [9], ProDS [12]) and strong supervised learners (Neural network, Random Forest, AdaBoost and RBF SVM). Generally, strong learners group is expected to give better performance than others due to their capability of capturing complex and diverse patterns inside the data. Hyper-parameters of baseline classifiers are optimized using the GridSearch method. For IEBC, there are three main hyper-parameters including the number of fuzzy clusters $k$, the depth of decision trees defined for each cluster and the fuzzy threshold $\beta$. Classifiers are run with 5 folds cross-validation for each dataset and final results are the average of results of 5 folds cross-validation run-times as in Table 3 and 4. A part of the results is reused from our previous work [26].

**Table 3.** AUC of classification results of 10 testing datasets

| Method | Sepsis | Haberman | Magic | Chess | Diabetes |
|---|---|---|---|---|---|
| Neural Net | $\mathbf{0.764 \pm 0.021}$ | $0.579 \pm 0.063$ | $0.848 \pm 0.007$ | $0.994 \pm 0.004$ | $0.703 \pm 0.024$ |
| Random Forest | $0.713 \pm 0.014$ | $0.574 \pm 0.047$ | $\mathbf{0.856 \pm 0.008}$ | $0.989 \pm 0.004$ | $0.733 \pm 0.031$ |
| AdaBoost | $0.711 \pm 0.012$ | $0.603 \pm 0.059$ | $0.821 \pm 0.003$ | $0.967 \pm 0.008$ | $0.708 \pm 0.031$ |
| RBF SVM | $0.664 \pm 0.011$ | $0.529 \pm 0.03$ | $0.838 \pm 0.005$ | $\mathbf{0.995 \pm 0.002}$ | $0.722 \pm 0.046$ |
| CART | $0.592 \pm 0.066$ | $0.606 \pm 0.051$ | $0.711 \pm 0.024$ | $0.891 \pm 0.033$ | $0.698 \pm 0.039$ |
| ProDS | $0.526 \pm 0.013$ | $0.571 \pm 0.044$ | $0.789 \pm 0.007$ | $0.940 \pm 0.009$ | $0.723 \pm 0.050$ |
| EKNN | $0.567 \pm 0.013$ | $0.546 \pm 0.053$ | $0.821 \pm 0.004$ | $0.957 \pm 0.005$ | $0.714 \pm 0.016$ |
| IEBC | $0.761 \pm 0.013$ | $\mathbf{0.665 \pm 0.039}$ | $0.833 \pm 0.007$ | $0.978 \pm 0.016$ | $\mathbf{0.748 \pm 0.026}$ |

**Table 4.** AUC of classification results of 10 testing datasets

| Method | Liver-disorder | Credit-approval | Heart-statlog | Ionosphere | Heart |
|---|---|---|---|---|---|
| Neural Net | $0.678 \pm 0.052$ | $0.86 \pm 0.024$ | $\mathbf{0.835 \pm 0.031}$ | $0.913 \pm 0.024$ | $0.83 \pm 0.049$ |
| Random Forest | $0.662 \pm 0.056$ | $\mathbf{0.874 \pm 0.019}$ | $0.823 \pm 0.065$ | $\mathbf{0.93 \pm 0.037}$ | $0.808 \pm 0.032$ |
| AdaBoost | $0.633 \pm 0.026$ | $0.865 \pm 0.008$ | $0.782 \pm 0.049$ | $0.905 \pm 0.042$ | $0.817 \pm 0.046$ |
| RBF SVM | $0.629 \pm 0.033$ | $0.863 \pm 0.025$ | $0.827 \pm 0.042$ | $0.929 \pm 0.049$ | $0.812 \pm 0.083$ |
| CART | $0.635 \pm 0.066$ | $0.749 \pm 0.037$ | $0.776 \pm 0.057$ | $0.861 \pm 0.046$ | $0.741 \pm 0.05$ |
| ProDS | $0.669 \pm 0.023$ | $0.867 \pm 0.019$ | $0.823 \pm 0.041$ | $0.937 \pm 0.015$ | $\mathbf{0.845 \pm 0.022}$ |
| EKNN | $0.637 \pm 0.053$ | $0.864 \pm 0.037$ | $0.813 \pm 0.031$ | $0.920 \pm 0.042$ | $0.796 \pm 0.051$ |
| IEBC | $\mathbf{0.689 \pm 0.031}$ | $0.842 \pm 0.098$ | $0.822 \pm 0.042$ | $0.909 \pm 0.047$ | $0.83 \pm 0.043$ |

For the binary classification task, AUC (Area Under The Curve) score is usually used as a major metric to evaluate the performance. As can be seen in the final results, strong classifiers such as neural network or random forest mostly achieve higher performance compared with CART and other evidence-based methods in terms of AUC metric. For our proposed method IEBC, its performance is competitive with the others while still keeps a reasonable degree of intelligibility. Specifically, IEBC performed equally or even out-performed other state-of-the-art methods in medical-related datasets such as Sepsis, Haberman, Diabetes, Liver-disorder and Heart. The final results proved our assumption about the effectiveness of the proposed method with healthcare data (which is usually generated by various distributions). Moreover, values of IEBC's hyper-parameters are restricted to short ranges that support the interpretability of the proposed model - which is an important characteristic of machine learning models for healthcare applications.

## 5 Conclusion

In this paper, we proposed a new evidence-based classification system named IEBC that has comparative results with popular classification models. The proposed method is developed to tackle the problem of classifying mixed and uncertain data which are generated by various distributions. To that end, we applied a fuzzy clustering technique to detect available heterogeneous clusters inside testing data that can represent their generated distributions. After that, decision trees are used to capture the behaviors inside each cluster that determine the characteristics of belonging objects. Finally, collected pieces of evidence about the class of testing data are combined to generate the last result. The binary classification experiment on various types of real data has proved the effectiveness of our proposed method, especially with healthcare datasets. Moreover, due to the structure of IEBC which combines interpretable machine learning methods such as decision trees and intuitive evidence combination, the proposed system can also provide users with knowledge on underlying training data. The extracted knowledge can support the decision-making process in fields which require to make high-stakes decisions such as healthcare, medicine or banking. For our future work, a more extensive experiment will be conducted to evaluate the proposed system. Specifically, a multi-classes prediction experiment on various datasets as well as a detailed analysis of the impact of the model's parameters will be performed. Moreover, an investigation in various use cases and applications in healthcare or business fields will be done which further explores the potential of our proposed models.

**Acknowledgment.** This research was supported in part by the US Office of Naval Research Global under Grant no. N62909-19-1-2031.

## References

1. Begoli, E., Bhattacharya, T., Kusnezov, D.: The need for uncertainty quantification in machine-assisted medical decision making. Nat. Mach. Intell. **1**(1), 20–23 (2019). Nature Publishing Group
2. Vluymans, S., Cornelis, C., Saeys, Y.: Machine learning for bioinformatics: uncertainty management with fuzzy rough sets, p. 2 (2016)
3. Huynh, V.N.: Uncertainty management in machine learning applications. Int. J. Approx. Reason. **107**, 79–80 (2019)
4. Hüllermeier, E.: Uncertainty in clustering and classification. In: Deshpande, A., Hunter, A. (eds.) SUM 2010. LNCS (LNAI), vol. 6379, pp. 16–19. Springer, Heidelberg (2010). https://doi.org/10.1007/978-3-642-15951-0_6
5. Jousselme, A.L., Maupin, P., Bosse, E.: Uncertainty in a situation analysis perspective. In: 2003 Proceedings of the 6th International Conference of Information Fusion, vol. 2, pp. 1207–1214 (July 2003)
6. Shafer, G.: A Mathematical Theory of Evidence. Princeton University Press, Princeton (1976)
7. Denoeux, T., Bjanger, M.: Induction of decision trees from partially classified data using belief functions. In: SMC 2000 Conference Proceedings of the 2000 IEEE International Conference on Systems, Man and Cybernetics. 'Cybernetics Evolving to Systems, Humans, Organizations, and their Complex Interactions' (cat. no. 0), vol. 4, pp. 2923–2928 (October 2000). ISSN: 1062–922X
8. Denœux, T., Kanjanatarakul, O., Sriboonchitta, S.: A new evidential K-nearest neighbor rule based on contextual discounting with partially supervised learning. Int. J. Approx. Reason. **113**, 287–302 (2019)
9. Denoeux, T.: A k-nearest neighbor classification rule based on Dempster-Shafer theory. IEEE Trans. Syst. Man Cybern. **25**(5), 804–813 (1995). Conference Name: IEEE Transactions on Systems, Man, and Cybernetics
10. Tong, Z., Xu, P., Denœux, T.: ConvNet and Dempster-shafer theory for object recognition. In: Ben Amor, N., Quost, B., Theobald, M. (eds.) SUM 2019. LNCS (LNAI), vol. 11940, pp. 368–381. Springer, Cham (2019). https://doi.org/10.1007/978-3-030-35514-2_27
11. Su, Z., Hu, Q., Denaeux, T.: A distributed rough evidential K-NN classifier: integrating feature reduction and classification. IEEE Trans. Fuzzy Sys., 1 (2020). Conference Name: IEEE Transactions on Fuzzy Systems
12. Denoeux, T.: A neural network classifier based on Dempster-Shafer theory. IEEE Trans. Syst. Man Cybern. Part A Syst. Hum. **30**(2), 131–150 (2000). Conference Name: IEEE Transactions on Systems, Man, and Cybernetics - Part A: Systems and Humans

13. Liu, Z., Pan, Q., Dezert, J., Mercier, G.: Credal classification rule for uncertain data based on belief functions. Pattern Recogn. **47**(7), 2532–2541 (2014)
14. Liu, Z., Pan, Q., Dezert, J.: A new belief-based K-nearest neighbor classification method. Pattern Recogn. **46**(3), 834–844 (2013)
15. Dempster, A.P.: Upper and lower probabilities induced by a multivalued mapping. In: Yager, R.R., Liu, L. (eds.) Classic Works of the Dempster-Shafer Theory of Belief Functions. Studies in Fuzziness and Soft Computing, vol. 219, pp. 57–72. Springer, Heidelberg (2008). https://doi.org/10.1007/978-3-540-44792-4_3
16. Smets, P., Kennes, R.: The transferable belief model. Artif. Intell. **66**(2), 191–234 (1994)
17. Zadeh, L.A.: Review of books: a mathematical theory of evidence. AI Mag. **5**(3), 81–83 (1984)
18. Dunn, J.C.: A fuzzy relative of the ISODATA process and its use in detecting compact well-separated clusters. J. Cybern. **3**(3), 32–57 (1973). https://doi.org/10.1080/01969727308546046. Taylor & Francis
19. Bezdek, J.C.: Pattern Recognition with Fuzzy Objective Function Algorithms. Advanced Applications in Pattern Recognition. Springer, New York (1981). https://doi.org/10.1007/978-1-4757-0450-1
20. Bezdek, J.C., Ehrlich, R., Full, W.: FCM: the fuzzy c-means clustering algorithm. Comput. Geosci. **10**(2), 191–203 (1984)
21. Ross, T.J.: Fuzzy Logic with Engineering Applications, 3rd edn. Wiley, Chichester (2010). oCLC: ocn430736639
22. Breiman, L., Friedman, J.H., Olshen, R.A., Stone, C.J.: Classification and Regression Trees. Wadsworth and Brooks, Monterey (1984)
23. Johnson, A.E.W., et al.: A comparative analysis of sepsis identification methods in an electronic database. Crit. Care Med. **46**(4), 494–499 (2018)
24. Johnson, A.E.W., et al.: MIMIC-III, a freely accessible critical care database. Sci. Data **3**, 160035 (2016)
25. Dua, D., Graff, C.: UCI machine learning repository (2017). http://archive.ics.uci.edu/ml
26. Nguyen, T.P., Nguyen, S., Alahakoon, D., Huynh, V.N.: GSIC: a new interpretable system for knowledge exploration and classification. IEEE Access **8**, 108544–108554 (2020)

# SOM-Based Visualization of Potential Technical Solutions with Fuzzy Bag-of-Words Utilizing Multi-view Information

Yasushi Nishida and Katsuhiro Honda(✉)

Graduate School of Engineering,
Osaka Prefecture University, Sakai, Osaka 599-8531, Japan
nishida@iao.osakafu-u.ac.jp, honda@cs.osakafu-u.ac.jp

**Abstract.** SOM-based visualization is a promising approach for revealing potential technical solutions varied in patent documents. In this paper, we try to improve the quality of visual representation by constructing Fuzzy Bag-of-Words (FBoW) matrices through utilization of multi-view information. *F-term* is the special theme code given by the examiners of Japan Patent Office (JPO) for Japanese patent documents and is expected to be a potential candidate of second-view information. Document × *F-term* co-occurrence information is utilized for improving the quality of FBoW representation such that semantical similarities among words are measured by considering *F-term* semantics. The advantage of utilizing *F-term* semantics in constructing FBoW is demonstrated through analysis of patent document data.

**Keywords:** Patent documents · Technical solution · Fuzzy Bag-of-Words · Multi-view information

## 1 Introduction

How to utilize public wisdom is an important issue in supporting innovation acceleration, and patent documents are regarded as huge success cases accumulated by ancestors from past to present. With the goal of finding mutual connection among various technical solutions varied in patent documents, we have developed some schemes of SOM-based visualization of potential technical solutions in Japanese patent documents [1,2], in order to intuitively represent the mutual connection among technical terms.

Self-Organizing Maps (SOM) [3] is a powerful tool in visualizing mutual connection among various objects such that a low-dimensional representation reveals intrinsic cluster tendencies. In our schemes, non-structured patent documents are summarized into Bag-of-Words (BoW) vector representation [4] considering frequencies of basis terms (keywords) in each document, and SOMs were constructed visualizing word features of two different types of multi-dimensional


V.-N. Huynh et al. (Eds.): IUKM 2020, LNAI 12482, pp. 187–198, 2020.
https://doi.org/10.1007/978-3-030-62509-2_16

features: co-occurrence probability vectors or correlation coefficient vectors. Our previous results demonstrated that co-occurrence probability vectors were utilized for revealing connections, which were directly utilized in existing patents, while correlation coefficient vectors were expected to be useful for revealing potential connections, which can be utilized in future developments.

In general document analysis tasks, each document often contains only a very small portion of all basis terms and some intrinsic connections among basis terms may be concealed by the sparse nature of BoW vectors. Fuzzy Bag-of-Words (FBoW) [5] is a technique for handling sparse characteristics of word feature values, which extends the conventional BoW based on a fuzzy mapping function such that it introduces vagueness in the matching between words and the basis terms. Besides the hard (complete) word matching, fuzzy mapping enables a word semantically similar to a basis term to be activated in the BoW model and was shown to be useful in document classification. Usually, semantical similarities among words are implemented by word2vec representation [6] utilizing some common dictionaries or original document datasets. In our previous work, we proposed to utilize FBoW representation in SOM-based visualization and demonstrated that the enriched information of FBoW representation is useful in revealing intrinsic potential techniques [7].

In this paper, we propose to enhance our SOM-based visualization model by utilizing multi-view information in constructing FBoW representations. Multi-view data analysis has been shown to be useful in such fields as multi-view clustering [8–10] for extracting intrinsic knowledge varied in several different data sources stored under different viewpoints. In patent document analysis, we have not only *document* × *keyword* co-occurrences representing the mutual connection among patents and technical words but also *document* × *F-term* co-occurrences. F-term classification system [11] was developed by the Japan Patent Office (JPO), in which entire technical area is divided into small areas called *themes* and patent documents are analyzed in each *theme*. Then, *F-term* is expected to be useful for utilizing the professional knowledge analyzed by the examiners of the JPO.

The main contribution of this paper is summarized as follows:

- We try to utilize *document* × *F-term* co-occurrence information for improving the quality of FBoW representation such that semantical similarities among words are measured by considering *F-term* semantics.
- Through experimental comparisons, we confirm that we can construct a slightly different Fuzzy BoW matrix by utilizing expert knowledge of JPO examiners.

The remaining parts of this paper are organized as follows: Section 2 briefly reviews our SOM-based visualization model with patent documents and Sect. 3 presents the basis of FBoW representation and the proposed approach of utilizing *document* × *F-term* co-occurrence information in FBoW construction. Experimental results are shown in Sect. 4 and the summary of this research is given in Sect. 5.

# 2 SOM-Based Visualization of Potential Technical Solutions in Patent Documents

## 2.1 Data Preprocessing

We have generated a patent document dataset composed of 6213 publications of patent applications in Japan [1,2], where 6213 publications were stored by a patent search tool on May 9, 2017. These patents were selected such that they satisfy the following two conditions:

**Condition 1:** Num. of *information provision* is more than 1.
**Condition 2:** Application date is January 1, 2005 or later.

*Information provision* is a systematic process for providing evidence information, which is often awakened from competitors, who want to prevent the right of the corresponding invention. Then, the patent having *information provision* is expected to be "an invention considered to be a threat for competitors".

In our study, the sentences of *means for solving the problem* section in each publication of patent applications were extracted to be analyzed with the goal of finding potential technical solution means, and were divided into word units by performing morphological analysis utilizing MeCab software (mecab-ipadic-NEologd version [12]). Here, each sentence was regarded as a datum to be characterized by a BoW vector considering the frequencies of occurrences of nouns, and then, summarized into a dataset having 55358 sentences (data instances) in total.

In the preprocessing phase, words having the occurrence frequency of 30% or more are regarded as general words and are filtered out. Additionally, words having the frequency of 0.3% or less are also filtered out in order to eliminate meaningless (rare) terms. Finally, the numerical data matrix to be analyzed is composed of $55358 \times 100$ elements with 55358 sentences and 100 terms. Data dimensionality was reduced by selecting only 100 representative words having the highest TF-IDF scores [13], where the score is a numerical statistic based on term frequency-inverse document frequency.

## 2.2 Construction of *Term* × *Term* Feature Matrices

Before applying SOM-based visualization, we constructed two types of *term* × *term* feature matrices for summarizing the characteristic features among technical words [1,2].

**Co-occurrence Probability Vectors.** Because meaningful pairs of elements (words) such as an important constituent and its behavior are expected to co-occur in the *means for solving the problem* in the same patent documents, we first constructed word co-occurrence probability vectors for revealing mutual connection among technical words varied in conventional patent documents. Word feature vectors in the original BoW matrix are processed into co-occurrence probability vector among technical words in order to focus on co-occurrence frequency among technical words.

Each element of a co-occurrence probability vector is calculated as follows: Let the number of sentences be $L$, the number of co-occurrences of word $w_i$ and word $w_j$ in sentence $l$ be $N_{ij}(l)$, and the total appearances of word $w_i$ in sentence $l$ be $M_i(l)$. Then, the co-occurrence probability $P_{ij}$ of words $w_i$ and $w_j$ is given as:

$$P_{ij} = \frac{\sum_{l=1}^{L} N_{ij}(l)}{\sum_{l=1}^{L} M_i(l)} \tag{1}$$

For each word $i$, co-occurrence probability degrees $P_{ij}$ are summarized into an $n$-dimensional vector $\boldsymbol{P_{wi}} = (P_{i1}, \cdots, P_{in})$ and are merged into a matrix $P_{word}$ considering all $n$ representative words as:

$$P_{word} = \begin{pmatrix} P_{11} & P_{12} & \cdots & P_{1n} \\ P_{21} & P_{22} & \cdots & P_{2n} \\ \vdots & & \ddots & \vdots \\ P_{n1} & P_{n2} & \cdots & P_{nn} \end{pmatrix} = \begin{pmatrix} \boldsymbol{P_{w1}} \\ \boldsymbol{P_{w2}} \\ \vdots \\ \boldsymbol{P_{wn}} \end{pmatrix} \tag{2}$$

Since technical words having similar co-occurrence characteristics often have similar vectors, we can expect that meaningful pairs of technical words are located mutually near in SOMs.

**Correlation Coefficient Vectors.** Second, we constructed correlation coefficient vectors for revealing intrinsic connection among technical words. While each element of word co-occurrence probability vectors represent the frequency of the combinatorial use of representative words in conventional patents, the correlation coefficient of co-occurrence probability vectors are expected to be utilized for finding the potential combination of technical word pairs.

Correlation coefficient $C_{ij}$ of two co-occurrence probability vectors $\boldsymbol{P_{wi}}$ and $\boldsymbol{P_{wj}}$ of representative words $w_i$ and $w_j$ can be expressed as follows:

$$C_{ij} = \frac{\sum_{x=1}^{n} (P_{ix} - \overline{P_{wi}})(P_{jx} - \overline{P_{wj}})}{\sqrt{\sum_{x=1}^{n} (P_{ix} - \overline{P_{wi}})^2} \sqrt{\sum_{x=1}^{n} (P_{jx} - \overline{P_{wj}})^2}} \tag{3}$$

where $\overline{P_{wi}}$ is the arithmetic mean of co-occurrence probabilities of word $w_i$.

The correlation coefficient for each word is summarized into an $n$-dimensional vector $\boldsymbol{C_{wi}} = (C_{i1}, \cdots, C_{in})$ and are merged into a symmetric matrix $C_{word}$ as:

$$C_{word} = \begin{pmatrix} C_{11} & C_{12} & \cdots & C_{1n} \\ C_{21} & C_{22} & \cdots & C_{2n} \\ \vdots & & \ddots & \vdots \\ C_{n1} & C_{n2} & \cdots & C_{nn} \end{pmatrix} = \begin{pmatrix} \boldsymbol{C_{w1}} \\ \boldsymbol{C_{w2}} \\ \vdots \\ \boldsymbol{C_{wn}} \end{pmatrix} \tag{4}$$

Even when the co-occurrence probability of the two words is not so high, the potential combinations of technical word pairs are expected to be revealed in correlation analysis because they may be related to common technical issues.

### 2.3 SOM-Based Visualization

The word feature vectors generated in the previous step are processed for the input to SOM, respectively. A SOM of co-occurrence probability vectors is constructed with $n$ vectors $\boldsymbol{P_{w1}}, \ldots, \boldsymbol{P_{wn}}$, each of which consists of $n$ co-occurrence probabilities $P_{ij}$. In a similar manner, a SOM of correlation coefficient vectors is constructed with $n$ vectors $\boldsymbol{C_{w1}}, \ldots, \boldsymbol{C_{wn}}$, each of which consists of $n$ correlation coefficients $C_{ij}$. Through the SOM-based unsupervised learning, a low-dimensional (typically two-dimensional) feature map visualizes the intrinsic mutual connections among words, which are varied in the multi-dimensional feature vectors of $\boldsymbol{P_{wi}}$ or $\boldsymbol{C_{wi}}$.

In this study, SOM was implemented in Python Programming Language with Somoclu software [14].

## 3 Fuzzy Bag-of-Words and Multi-view Information Utilization

### 3.1 Fuzzy Bag-of-Words

In general document analysis tasks, we can have very sparse BoW vectors because each document often contains only a very small portion of all basis terms. Indeed, the BoW vectors, which we constructed in the previous section, have 96.9% zero elements. Then, intrinsic connections among basis terms may be concealed by the spares nature of BoW vectors.

In order to tackle the sparse feature of the conventional BoW model, Fuzzy BoW was proposed considering the fuzzy mapping function [5]. Assume that $A_{t_i}(w)$ is a mapping function and the term frequency of a basis term $t_i$ is calculated by summing up $A_{t_i}(w)$ for all words $w$ in a sentence. Usually, the BoW model adopts the frequency of each representative term by counting the number of exact word matching by employing the following membership function:

$$A_{t_i}(w) = \begin{cases} 1, & \text{if } w \text{ is } t_i \\ 0, & \text{otherwise} \end{cases} \tag{5}$$

Then, each term occurrence can activate only a single term frequency and causes a very sparse nature in the BoW matrix.

On the other hand, the FBoW model adopts semantic matching or fuzzy mapping to project the words occurred in documents to the basis terms. To implement semantic matching, a fuzzy membership function is considered as follows:

$$A_{t_i}(w) = \begin{cases} \cos(W[t_i], W[w]), & \text{if } \cos(W[t_i], W[w]) > 0 \\ 0, & \text{otherwise} \end{cases} \tag{6}$$

where $W[w]$ denote word embeddings for word $w$ such that they represent mutual semantic similarities among words. In [5], *word2vec* [6] was adopted in word embeddings, and the fuzzy membership degree that measures the similarity between attribute (words in documents) and set (basis terms in BoW space) was approximated by their cosine similarity score.

Then, the numerical vector representation $\boldsymbol{z}$ of a document under fuzzy BoW model is given by

$$\boldsymbol{z} = \boldsymbol{x}H, \tag{7}$$

where $\boldsymbol{x}$ is a vector composed of the number of occurrence of words and $H$ is a square matrix composed of fuzzy memberships of Eq. (6) as $H = \{A_{t_i}(w_j)\}$, whose dimensions are equivalent to the number of the basis terms.

Supported by the fuzzy representation, each term occurrence can activate multiple term frequencies with various similarity weights, and the FBoW representation is expected to have richer information than the original (sparse) BoW representation.

### 3.2 Multi-view Information Utilization in FBoW Construction

In FBoW construction, how to evaluate mutual semantic similarities among words is an important issue. In [5], the pre-trained word2vec vectors published by Google was utilized, where word embeddings were trained on a Google News corpus (over 100 billion words) and have a dimensionality of 300. For Japanese document analysis, we also have some general word embeddings such as the Asahi Shimbun wordvectors[1] trained on Japanese newspapers and Shiroyagi wordvectors[2] trained on Wikipedia data. Although these common word embeddings are useful for evaluating general semantic similarities, they may not suit some special purpose analysis such as patent document analysis where some terms are used in a specific manner in the field. Indeed, in our previous work [7], general purpose corpuses are reported not to be suitable in patent document analysis because general terms are overemphasized while other potential connections are concealed.

In order to reveal intrinsic semantic similarities under a specific purposes, in [7], we constructed a specific word embedding by training on the patent document gathered to be analyzed. However, some potential connections might be still concealed by the influences of non-technical words.

[1] https://cl.asahi.com/api_data/wordembedding.html.
[2] https://aial.shiroyagi.co.jp/2017/02/japanese-word2vec-model-builder/.

These days, multi-view data analysis has been shown to be useful in such fields as multi-view clustering [8–10] for extracting intrinsic knowledge varied in several different data sources stored under different viewpoints. In this paper, we propose a novel approach of constructing FBoW matrices by utilizing multi-view information. In patent document analysis, we have not only *document* $\times$ *keyword* BoW representation but also *document* $\times$ *F-term* co-occurrences. F-term classification system [11] developed by the Japan Patent Office (JPO) divided entire technical area into small areas called *themes* and each patent document is related to some *themes* by the examiners of the JPO. Then, *F-term* is expected to be useful for utilizing the professional knowledge.

In the followings, we try to utilize *document* $\times$ *F-term* co-occurrence information for improving the quality of FBoW representation such that semantical similarities among words are measured by considering *F-term* semantics. Assume that $L \times n$ matrix $X$ is a BoW representation among documents and basis words, whose element $X_{ij}$ depicts the frequency (TF-IDF value) of basis term $j$ in document $i$, and $L \times q$ matrix $F$ is the second view information of *document* $\times$ *F-term* co-occurrence data, whose element $F_{ij} \in \{1, 0\}$ indicates the presence/absence of *F-term* $j$ in patent document $i$. In the proposed approach, a novel supplemental *word* $\times$ *F-term* co-occurrence information matrix $S = \{S_{ij}\}$ is constructed as:

$$S = X^\top F, \tag{8}$$

where element $S_{ij}$ represents the total appearance of word $i$ in conjunction with *F-term* $j$ as:

$$S_{ij} = \sum_{k=1}^{L} X_{ki} F_{kj}. \tag{9}$$

Then, the sparse BoW matrix $X$ is enhanced into a Fuzzy BoW matrix by adopting fuzzy mapping $A_{t_i}(w)$ of word $w$ for a basis term $t_i$ with the following membership function under the similar concept with [5]:

$$A_{t_i}(w) = \begin{cases} \mathrm{cor}(S[t_i], S[w]), & \text{if } \mathrm{cor}(S[t_i], S[w]) > 0 \\ 0, & \text{otherwise} \end{cases} \tag{10}$$

where $S[w] = (S_{w1}, \cdots, S_{wq})$ denote feature vectors for word $w$ such that they represent mutual *F-term* semantic similarities among words. That is, the fuzzy membership degree for measuring the similarity between attribute (words in documents) and set (basis terms in BoW space) is approximated by their Pearson correlation coefficient $\mathrm{cor}(\cdot)$ under the second view of *F-term* semantics.

In our study, we selected 302 *F-terms* that are most frequently related to the patent documents (55358 sentences) where the last 12 *F-terms* had tie scores at the 291st rank. Then, $55358 \times 302$ matrix $F$ was transformed into $100 \times 302$ matrix $S = X^\top F$, where $S$ represents co-occurrence information among 100 basis terms and 302 *F-terms*.

## 4 Experimental Results

In this experimental section, we demonstrate the effects of utilizing the second view of *F-term* semantics in Fuzzy BoW construction under the task of SOM-based visualization of potential technical solutions in patent documents.

### 4.1 Comparison of Characteristics of *Document* × *term* matrix and *term* × *F-term* matrix

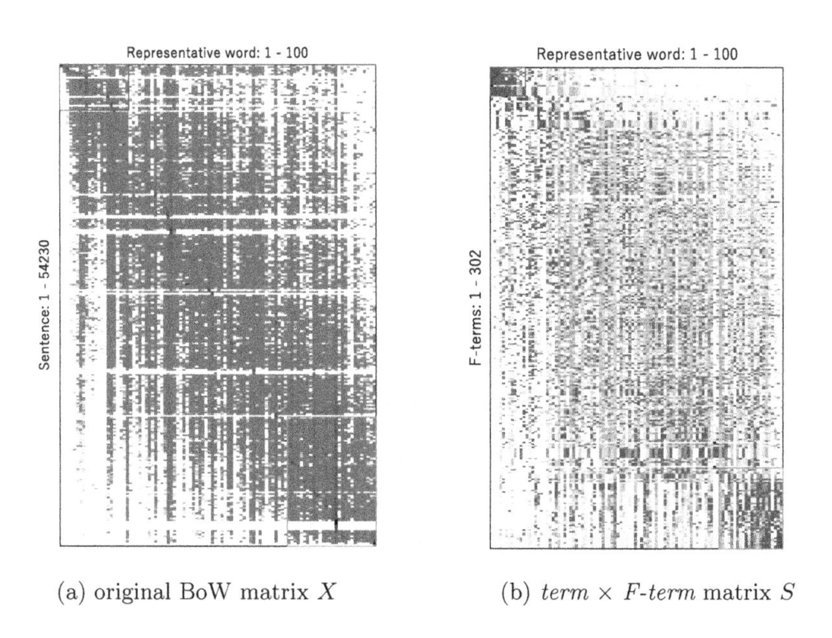


(a) original BoW matrix $X$ (b) *term* × *F-term* matrix $S$

**Fig. 1.** Comparative co-cluster visualization of original BoW matrix $X$ and *term* × *F-term* matrix $S$.

First, we compare the characteristics of *term* × *F-term* matrix $S$, which is introduced in this study, with those of the original *document* × *term* matrix $X$. In order to reveal the intrinsic co-cluster structures varied in the two matrices, spectral ordering-based co-cluster visualization [15] was applied to $X$ and $S$, where co-occurrence-sensitive ordering of both row and column elements was performed so that co-cluster structures are emphasized in the diagonal blocks. Figure 1 compares the visual images after ordering, where the grayscale cells depict the values of matrix elements such that black and white indicate the maximum and zero values. Figure 1-(a) was drawn from [2], where only 54230 sentences were considered by rejecting some very short sentences and the typical co-cluster areas (left-top and right bottom) are indicated by red and blue rectangles, respectively. Figure 1-(b) was also constructed in a similar way with matrix $S$. In both figures, we can weakly find typical co-cluster structures in the left-top and right-bottom areas, and then, the first and last several words seems to be typical words related mainly to the representative topics.

**Table 1.** Comparison of cluster-wise typical technical words

(a) original BoW matrix $X$

| cluster | typical technical words |
|---|---|
| red | 説明 (description), 後述 (see below), 空気 (air), 規定 (regulation), 意味 (meaning), 次 (next), 解決 (solution), 的 (like), 手段 (procedure), 製品 (product), 適用 (apply), 採用 (adopt), 有利 (advantage), 情報 (information), テーブル (table) |
| blue | 飲料 (beverage), 界面活性剤 (surfactant), 官能基 (functional group), 炭素 (carbon), 紫外線 (ultraviolet), 透明 (transparency), 塩 (salt), バイオマス (biomass), フィラー (filler), 類 (kind), 基 (group), 殺菌 (sterilization), 0 (zero), 質量 (mass), 物質 (material), 系 (system), 配合 (mix), 接着 (glueing), 提供 (offer), 樹脂 (rosin), 一般 (general), 由来 (origin), 下記 (as follows), 溶液 (liquor), 添加 (addition), 数 (number), 層 (lamina), 度 (degree) |

(b) *term* × *F-term* matrix $S$

| cluster | typical technical words |
|---|---|
| red | 外周 (periphery), 本体 (main body), 画像 (image), 領域 (area), 複数 (multiple), 空間 (space), 有利 (advantage), 手段 (procedure), データ (data), 情報 (information), テーブル (table) |
| blue | 飲料 (beverage), 殺菌 (sterilization), 素材 (ingredient), 交換 (exchange), 類 (kind), 添加 (addition), 配合 (mix), 由来 (origin), 知見 (knowledge), 水 (water), 製造 (manufacturing), 塩 (salt), 調整 (adjustment), 型 (model), 製品 (product), 提供 (offer), バイオマス (biomass), 工程 (process), 程度 (degree), 界面活性剤 (surfactant), 質量 (mass) |

Table 1 compares the typical words related to the red and blue co-clusters, and implies that the clusters have some different typical words in the document-induced semantics of the first view information and the second view of *F-term* semantics. Additionally, in visualization of *term* × *F-term* matrix $S$, the blue cluster was related to such *F-terms* as '2C*** (manufactural aspect)' while the red cluster was related to '4B*** (foods)' and '4C*** (medicals)'. Here, we can confirm that the second view information of *term* × *F-term* matrix $S$ has slightly different characteristics from the original *document* × *term* matrix $X$ under the consideration of expert knowledge of the JPO examiners. In this way, it is expected that the original sparse BoW matrix $X$ can be enriched into a Fuzzy BoW matrix by introducing the expert knowledge if we consider the *F-term* semantics.

### 4.2 Comparison of SOMs Derived by SOM-Based Visualization

Second, the SOMs given by our SOM-based visualization using the FBoW method with/without *F-term* semantics are compared. In this paper, due to

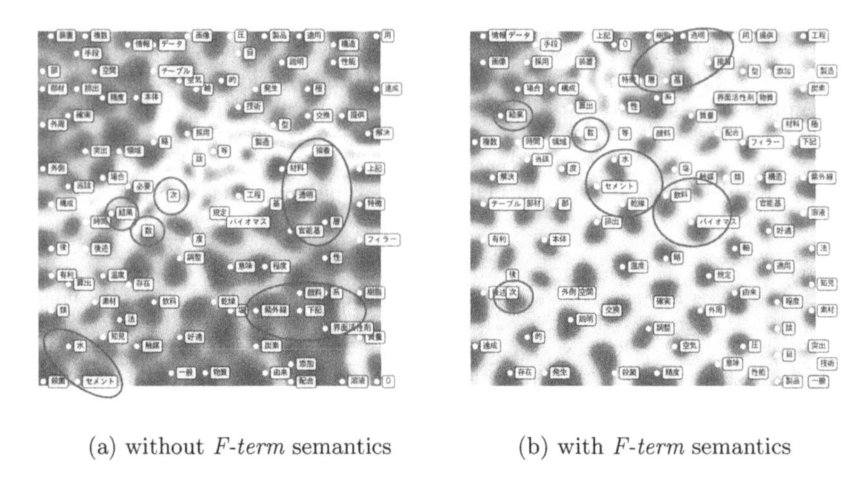

(a) without *F-term* semantics (b) with *F-term* semantics

**Fig. 2.** SOM from the Fuzzy BoW with/without *F-term* semantics consideration

the page limit, we only compare the SOMs derived from correlation coefficient vectors, i.e., the second model of Sect. 2.2.

Figure 2 compares the SOMs from the Fuzzy BoW with/without *F-term* semantics consideration, where mutually related technical words are expected to be located nearby. Figure 2-(a) was constructed in the same manner with the previous work [7] such that SOM was implemented to the FBoW representation constructed under consideration of semantic similarity in *word2vec* features of the patent documents. As implied in [7], the FBoW-induced approach was useful for finding such possible intrinsic connections as "“接着, 透明 (glueing, transparency)’, "“紫外線, 界面活性剤 (ultraviolet, surfactant)” and "“水, セメント (water, cement)”, some of which were concealed in the SOM derived from the original sparse BoW representation. However, such general terms as “数(number)”, “次 (next)” and “結果 (result)” were also located in the center area such that they are mutually relevant. It may be because the Fuzzy BoW representation was constructed under consideration of the *general features* of whole the patent documents.

On the other hand, Figure 2-(b) was derived from the FBoW matrix, which was constructed by the proposed multi-view consideration supported by the expert knowledge varied in *F-term* semantics. Then, we could locate such typical connections as “水, セメント (water, cement)” in the center area such that they are also relevant to such operation as “乾燥 (drying)” by kicking away the general terms, which were located in the center area in Fig. 2-(a). Additionally, we can newly find the possible connection of “飲料, バイオマス (beverage, biomass)”, which seem to be irrelevance at a glance but can be a promising field as implied in several articles[3] [4].

In this way, utilizing the second view of expert knowledge seems to be a promising approach in enriching the sparse BoW matrix into a Fuzzy BoW matrix.

## 5 Conclusions

In this paper, we proposed a novel approach to SOM-based visualization of potential technical solutions in patent documents, where the original sparse BoW matrix is enriched into Fuzzy BoW matrix by utilizing the second view information of *term* $\times$ *F-term* semantic co-occurrence matrix. Through experimental comparisons, we confirmed that we can construct a slightly different Fuzzy BoW matrix by utilizing expert knowledge of JPO examiners, and other potential pairs of technical terms can be found rather than the conventional single-view analysis.

A potential future work is to improve the readability of SOMs considering rejection of meaningless words [16] or constructing multi-dimensional, spherical or hierarchical SOMs. Another direction is to combination with multi-view co-clustering [17] so that we can find field-related knowledge by adopting SOM with cluster-wise BoW matrices.

**Acknowledgment.** This work was achieved through the use of large-scale computer systems at the Cybermedia Center, Osaka University, and was supported in part by JSPS KAKENHI Grant Number JP18K11474.

## References

1. Nishida, Y., Honda, K.: Visualization of potential technical solutions by self-organizing maps and co-cluster extraction. In: Joint 10th International Conference on Soft Computing and Intelligent Systems and 19th International Symposium on Advanced Intelligent Systems, pp. 820–825 (2018)
2. Nishida, Y., Honda, K.: Visualization of potential technical solutions by self-organizing maps and co-cluster extraction. J. Adv. Comput. Intell. Intell. Inf. **24**(1), 65–72 (2020)
3. Kohonen, T.: Self-organizing Maps, 3rd edn. Springer, Heidelberg (2000). https://doi.org/10.1007/978-3-642-56927-2
4. Lan, M., Tan, C.L., Su, J., Lu, Y.: Supervised and traditional term weighting methods for automatic text categorization. IEEE Trans. Pattern Anal. Mach. Intell. **31**(4), 721–735 (2009)
5. Zhao, L., Mao, K.: Fuzzy bag-of-words model for document representation. IEEE Trans. Fuzzy Syst. **26**(2), 794–804 (2018)

[3] https://sustainablebrands.com/read/cleantech/trending-beverage-makers-turning-production-waste-into-biomass-carbonation.

[4] https://iopscience.iop.org/article/10.1088/1755-1315/395/1/012090/meta.

6. Mikolov, T., Chen, K., Corrado, G., Dean, J.: Efficient estimation of word representations in vector space. In: International Conference on Learning Representations (2013). https://arxiv.org/pdf/1301.3781.pdf
7. Nishida, Y., Honda, K.: A comparative study on SOM-based visualization of potential technical solutions using fuzzy bag-of-words and co-occurrence probability of technical words. In: Seki, H., Nguyen, C.H., Huynh, V.-N., Inuiguchi, M. (eds.) IUKM 2019. LNCS (LNAI), vol. 11471, pp. 360–369. Springer, Cham (2019). https://doi.org/10.1007/978-3-030-14815-7_30
8. Bickel, S., Scheffer, T.: Multi-view clustering. In: 4th IEEE International Conference on Data Mining, pp. 19–26 (2004)
9. Xu, Y.M., Wang, C.D., Lai, J.H.: Weighted multi-view clustering with feature reduction. Pattern Recogn. **53**, 25–35 (2016)
10. Yang, Y., Wang, H.: Multi-view clustering: a survey. Big Data Min. Anal. **1**, 83–107 (2018)
11. Schellner, I.: Japanese file index classification and F-terms. World Patent Inf. **24**, 197–201 (2002)
12. Sato, T.: Neologism dictionary based on the language resources on the Web for mecab (2015). https://github.com/neologd/mecab-ipadic-neologd
13. Salton, G., Buckley, C.: Term-weighting approaches in automatic text retrieval. Inf. Process. Manage. **24**(5), 513–523 (1988)
14. Wittek, P., Gao, S.-C., Lim, I.-S., Zhao, L.: Somoclu: an efficient parallel library for self-organizing maps. J. Stat. Softw. **78**(9), 1–21 (2017)
15. Honda, K., Sako, T., Ubukata, S., Notsu, A.: Visual co-cluster assessment with intuitive cluster validation through co-occurrence-sensitive ordering. J. Adv. Comput. Intell. Intell. Inf. **22**(5), 585–592 (2018)
16. Honda, K., Yamamoto, N., Ubukata, S., Notsu, A.: Noise rejection in MMMs-induced fuzzy co-clustering. J. Adv. Comput. Intell. Intell. Inf. **21**(7), 1144–1151 (2017)
17. Chen, T.-C.T., Honda, K.: Fuzzy Collaborative Forecasting and Clustering. SpringerBriefs in Applied Sciences and Technology. Springer, Heidelberg (2019). https://doi.org/10.1007/978-3-030-22574-2

# Machine Learning Applications

# Relative Humidity Estimation Based on Two Nested Kalman Filters with Bicubic Interpolation for Commercial Cultivation of Tropical Orchids

Nutchanon Siripool[1,2(✉)], Kraithep Sirisanwannakul[1,2], Waree Kongprawechnon[1], Prachumpong Dangsakul[2], Anuchit Leelayuttho[2], Sommai Chokrung[2], Jakkaphob Intha[2], Suthum Keerativittayanun[2], and Jessada Karnjana[2]

[1] Sirindhorn International Institute of Technology, Thammasat University, 131 Moo 5, Tiwanon Rd., Bangkadi, Muang 12000, Pathum Thani, Thailand
{6022800245,6022792111}@g.siit.tu.ac.th

[2] NECTEC, National Science and Technology Development Agency, 112 Thailand Science Park, Khlong Luang 12120, Pathum Thani, Thailand
prachumpong.dan@nectec.or.th

**Abstract.** This paper aims to reduce the number of polymer dielectric-based humidity sensors used in the orchid greenhouse monitoring system by replacing some of them with a mathematical model. Therefore, this paper proposes an interpolation technique based on two nested Kalman filters with bicubic interpolation. Our objective is not only to accurately estimate the value at the location of interest but also to make it possible for practical usage. Thus, the computational time complexity is one of the criteria. The humidity values are estimated by an outer Kalman filter, of which its prediction is made based on another inner Kalman filter that fuses information obtained from surrounded sensors. The bicubic interpolation technique generates data used in the measurement update stage of the outer Kalman filter. Experimental results show that the proposed method can improve the interpolated data accuracy by 16.88%, compared with using interpolation techniques alone. This paper also discusses the possibility of applying the proposed scheme in improving the accuracy of the sensor used for a long time.

**Keywords:** Kalman filtering · Bicubic interpolation · Sensor fusion · Orchid greenhouse

## 1 Introduction

The orchid is one of the highly commercial agricultural products in the world [1]. In 2019, the world market value of orchids was around 400 million US dollars, and more than 69 nations are sharing in the market [1]. The major contributors in the market are from Thailand, Taiwan, Singapore, Malaysia, Vietnam,


V.-N. Huynh et al. (Eds.): IUKM 2020, LNAI 12482, pp. 201–212, 2020.
https://doi.org/10.1007/978-3-030-62509-2_17

etc. Among those countries, Thailand is the world's biggest exporter of the cut orchids with valued around 72 million US dollars [3]. There are several species of orchid worldwide. In Thailand, the notable species are Cattleya, Phalaenopsis, Vandas, etc. [2]. Each species has characteristics that require different environmental factors to be controlled, such as temperature, humidity, nutrients, air quality, and light intensity [15]. Nevertheless, most orchid farms in Thailand are traditional farming. They operate in an open and uncontrolled environment. Therefore, productivity depends on the expertise and experience of the gardener. To increase the quality and productivity, ICT and embedded system technologies play a crucial role and push forward intensive farming. The core idea of these technologies is to monitor and control the environmental factors automatically to match orchids' needs [5,17]. In general, the monitoring system's performance depends not only on the quality but also on the quantity of the sensors. Because it is necessary to monitor the variation of interested parameters spatially to precisely control the environment. Thus, in smart greenhouse, many sensors are deployed, and as a consequence, it causes the system more expensive.

In order to reduce the system cost while keeping its performance, one of the straightforward solutions is to replace some sensors by mathematical models, such as linear interpolation [12], bicubic interpolation [7], kriging [4], and Kalman filter [16] to estimate unknown values. For example, Moon Taewon *et al.* uses machine learning to interpolate data [10]. Their results are stable and accurate, but one drawback is that it requires a large number of data. This research aims to replace some sensors with a mathematical model based on Kalman filtering with a bicubic interpolation technique so that the system's total cost is reduced without losing necessary information.

This paper is organized as follows. Section 2 describes an overview of the monitoring system deployed in an experiment orchid-greenhouse, the concept of the one-dimensional Kalman filtering algorithm, and bicubic interpolation. Section 3 provides detail of the proposed method. Experimental results and performance evaluation are made in Sect. 4. Discussion and conclusion are provided in Sects. 5 and 6, respectively.

## 2 Background

This section is divided into three subsections. First, an overview of the monitoring system that was deployed to collect data used in this work is introduced. Second, for the completeness in itself, the one-dimensional Kalman filter, on which this work is based, is shortly reviewed. Third, the bicubic interpolation technique used to generate data for Kalman's measurement update is briefly explained.

### 2.1 Overview of an Orchid Greenhouse Monitoring System

The architecture of the orchid greenhouse monitoring system for collecting data is shown in Fig. 1. The system is installed at a $6 \times 20$ m$^2$ commercial greenhouse located in Ratchaburi province, Thailand. This system consists of 21 humidity and temperature sensors, three dataloggers, a gateway, a number of fan and foggy controllers, and a weather station. Each humidity-temperature sensor is spatially separate to cover $2 \times 2.5$ m$^2$, as shown in Fig. 2. Data from the sensors are read via the I$^2$C bus protocol by a datalogger, of which its function is to re-format the collected data and forward them to the gateway via an XBee-based network. The collected data are scheduled to upload to the database by the gateway through a GPRS network. By connecting to the Internet, the user can access uploaded data via the web browser. The humidity-temperature sensor used in the project is a digital sensor (SHT31) from the Sensirion company. It operates based on the principle of capacitance measurement.

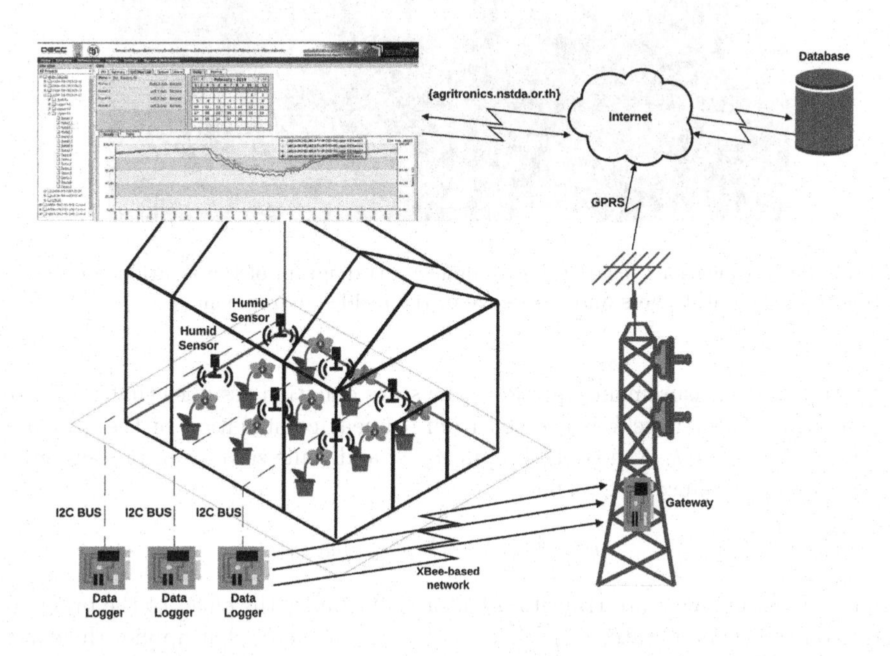


**Fig. 1.** Architecture of the orchid monitoring system.

## 2.2 One-Dimensional Kalman Filtering

The Kalman filtering is an optimal estimation algorithm that estimates some non-measurable states of a linear system with random behavior by combining observed measurements and estimates from a mathematical model of the system [6]. Kalman filtering algorithm has been applied in various applications, such as navigation, control system, signal processing, and time-series analysis [11], and one of those is constructing new data points within a range of known data points [8,9,14]. The Kalman filtering consists of two stages, which are prediction and measurement update, as shown in Fig. 3. According to the Kalman filtering, we can describe any linear system in the steady-state by two equations: state

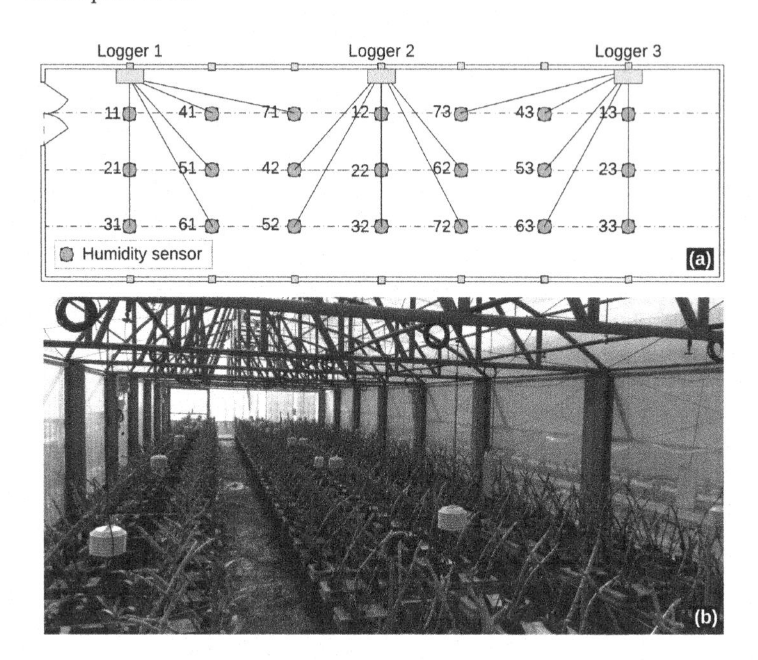


**Fig. 2.** Sensor locations inside the greenhouse: (a) diagram of the locations in association with the dataloggers and (b) image of the inside environment.

equation and measurement equation. The state equation assumes that the state $x_k$ of a system at time $k$ is evolved from a linear combination of the previous state $x_{k-1}$ at time $k-1$, a control input $u_{k-1}$, and some zero-mean process noise $w_{k-1}$ with a variance of $Q$. Thus,

$$x_k = Ax_{k-1} + Bu_{k-1} + w_{k-1}, \tag{1}$$

where $A$ is a known state-transition factor that applies the effect of the previous state on the current state, and $B$ is a control input factor that applies the effect of the input on the state.

The measured output $y_k$ of the system is assumed to be a linear combination of the state $x_k$ that we want to estimate and some measurement white noise $v_k$ with a variance of $R$. Thus,

$$y_k = Cx_k + v_k, \tag{2}$$

where $C$ is a factor that relates the state $x_k$ to the measurement $y_k$.

The Kalman filter estimates $x_k$, which is denoted by $\hat{x}_k$ and called *a posteriori estimate*, by combining a predicted state estimate $\hat{x}_k^-$ (*a priori estimate*) and the difference between the measurement $y_k$ and a predicted measurement $\hat{y}_k$. That difference is sometimes called a *residual* or a *measurement innovation.* Thus, the state estimate can be expressed mathematically as

$$\hat{x}_k = \hat{x}_k^- + K(y_k - \hat{y}_k) = \hat{x}_k^- + K(y_k - C\hat{x}_k^-), \tag{3}$$

where $K$ is *Kalman gain.*

The Kalman filtering algorithm calculates the Kalman gain $K$ such that the expectation of the square of the difference between $x_k$ and $\hat{x}_k$ is minimized. The difference (i.e., $x_k - \hat{x}_k$) is called *a posteriori estimate error*, and the expectation is *a posteriori error variance*, denoted by $P_k$.

As shown in Fig. 3, the Kalman filtering algorithm consists of five steps as follows. First, given the previous state estimate $\hat{x}_{k-1}$, it projects the state ahead from the equation

$$\hat{x}_k^- = A\hat{x}_{k-1} + Bu_{k-1}. \tag{4}$$

Note that, for $k=1$, $\hat{x}_{k-1}$ is $\hat{x}_0$, which is called the *initial state.*

Second, *a priori error variance* $P_k^-$, which is defined by the expectation of the square of the difference between the state $x_k$ and the predicted state estimate $\hat{x}_k^-$, is projected ahead from the equation

$$P_k^- = AP_{k-1}A + Q, \tag{5}$$

where $P_{k-1}$ is the previous (*a posteriori*) error variance. For $k=1$, $P_{k-1}$ is $P_0$, which is called the *initial error variance.* Note that $P_k^-$ is a measure of the uncertainty in the state estimate $\hat{x}_k$ due to a process noise and the propagation of the uncertainty of the previous predicted state estimate $\hat{x}_{k-1}^-$. The first two steps form the prediction stage of the algorithm.

Third, the Kalman gain $K$ is computed by

$$K = \frac{P_k^- C}{CP_k^- C + R}. \tag{6}$$

Fourth, the Kalman gain $K$ is used to scale the measurement innovation (i.e., $y_k - C\hat{x}_k^-$), and the state estimate $\hat{x}_k$ is updated by adding the scaled measurement innovation to the predicted state estimate $\hat{x}_k^-$, i.e.,

$$\hat{x}_k = \hat{x}_k^- + K(y_k - C\hat{x}_k^-). \tag{7}$$

Last, the state error variance is updated by the equation

$$P_k = (1 - KC)P_k^-. \tag{8}$$

The last three steps form the measurement update stage of the Kalman filtering algorithm, and it can be seen that given some initial state estimate $\hat{x}_0$ and some initial error variance $P_0$, the Kalman filter can estimate the state $x_k$ for any $k$.

### 2.3 Bicubic Interpolation

The bicubic interpolation is an interpolation method for a two-dimensional regular grid. In this study, relative humidity values are considered as function values of a surface of which its domain is a regular grid. In the bicubic interpolation,

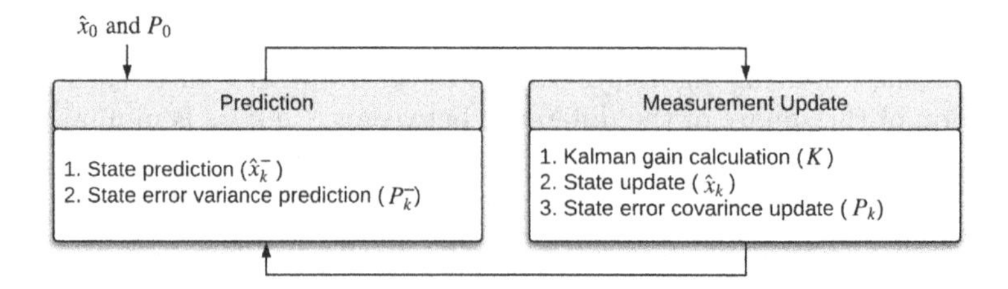


**Fig. 3.** Kalman filtering algorithm.

the surface value $p(x, y)$ assigned to a point $(x, y)$ is obtained by the following equation.

$$p(x, y) = \sum_{i=0}^{3} \sum_{j=0}^{3} a_{ij} x^i y^j, \tag{9}$$

where $a_{ij}$, for integers $i$ and $j$ in $\{0, 1, 2, 3\}$, are 16 coefficients to be determined by solving 16 equations, given that four corner values of the unit square (i.e., $p(0,0)$, $p(0,1)$, $p(1,0)$, and $p(1,1)$) and their derivatives (i.e., $\frac{\partial p(x,y)}{\partial x}$, $\frac{\partial p(x,y)}{\partial y}$, and $\frac{\partial^2 p(x,y)}{\partial x \partial y}$) are known or can be approximated.

The bicubic interpolation method is used in this work because it can produce the smoother surface compared to those obtained from bilinear interpolation or nearest-neighbor interpolation [7,13].

## 3 Proposed Method

As mentioned in the previous section, a regular grid of 21 sensors is deployed in the greenhouse. In this work, we aim to reduce the number of sensors by removing some of them and applying an estimator based on nested Kalman filters with bicubic interpolation, as shown in Fig. 4(a), to interpolate data at the locations where sensors are removed. An example of such a location is illustrated in Fig. 4(b). The Kalman filter used to interpolate the missing data points requires a model for the prediction stage and data for the measurement update stage, and it is referred to as the outer Kalman filter in Fig. 4(a). In our proposed method, another Kalman filter, hereafter referred to as the inner Kalman filter, is used as the model for the prediction stage. It estimates the humidity value by assuming that the humidity data read from neighbor sensors (i.e., those that surround the location of interest) are a sequence of its measurements, and its prediction is the average value of the humidity data read from the neighbor sensors. In the measurement update stage of the outer Kalman filter, the interpolated data points obtained from the bicubic interpolation applied to data read from the neighbor sensors are used as the measured $y_k$ in Eq. (3). An example of an interpolated data point in the measurement update stage of the outer Kalman filter is shown in Fig. 4(c).

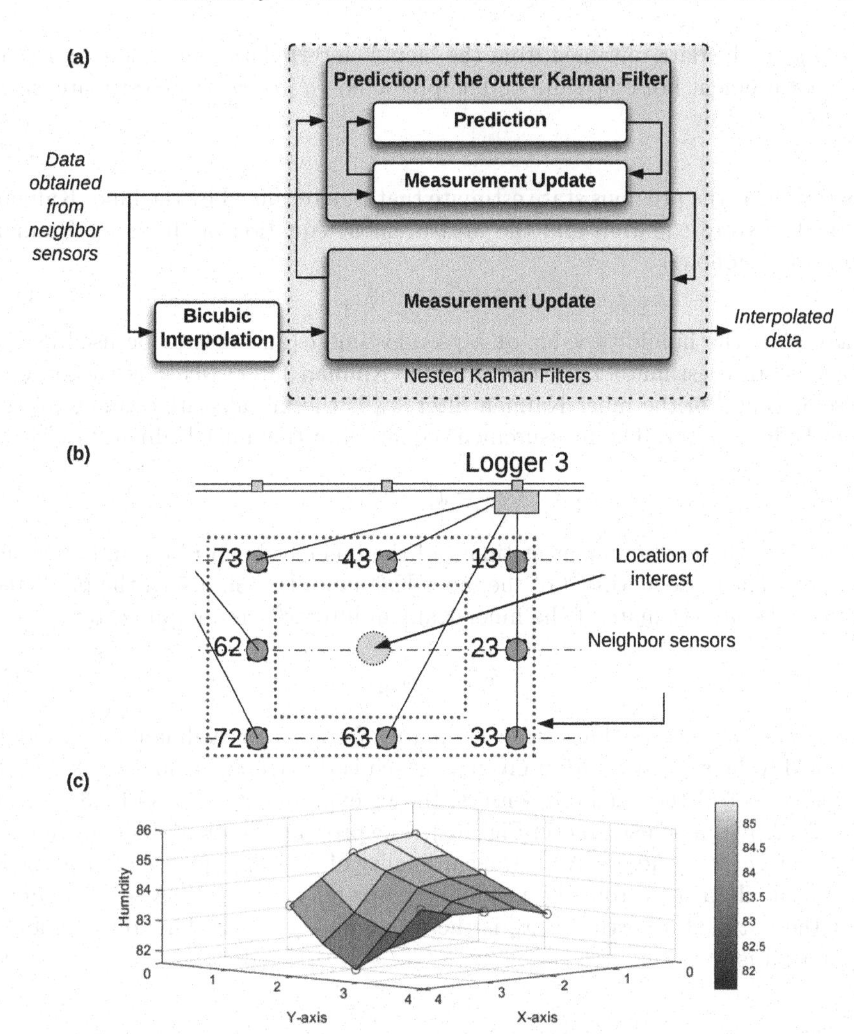


**Fig. 4.** Proposed method: (a) the diagram of two nested Kalman filters, (b) positions of sensors used in this paper, (c) interpolated value by bicubic interpolation, which is used in the measurement update.

In this work, the outer and inner Kalman filters are designed as follows. The state equation and the measurement equation of the outer Kalman filter are as follows.

$$x_k = x_{k-1} + w_{k-1}, \tag{10}$$

where $x_k$ is the humidity value to be interpolated at time $k$, and $w_{k-1}$ is the process noise at time $k-1$. The measurement equation is

$$y_k = x_k + v_k, \tag{11}$$

where $y_k$ is the data obtained from the bicubic interpolation at time $k$ and $v_k$ is the measurement noise at time $k$. The predicted *(a priori)* state estimate $\hat{x}_k^-$ is the computed by

$$\hat{x}_k^- = \hat{x}_{k-1}, \tag{12}$$

where $\hat{x}_{k-1}$ is the previous state estimate that is determined by the inner Kalman filter. The state equation and the measurement equation of the inner Kalman filter are as follows.

$$x_l^p = x_{l-1}^p + w_{l-1}^p, \tag{13}$$

where $x_l^p$ is the humidity value at a pseudo time $l$ (and it is to be used as the previous state estimate $\hat{x}_{k-1}$ of the outer Kalman filter), $w_{l-1}^p$ is the process noise at time $l$ of the inner Kalman filter, $x_0^p$ is the initial state estimate of the inner Kalman filter. The measurement equation of the inner Kalman filter is

$$y_l^p = x_l^p + v_l^p, \tag{14}$$

where $y_l^p$ is the measurement at time $l$ of the inner Kalman filter, and $v_l^p$ is the measurement noise at time $l$ of the inner Kalman filter. Similarly, the predicted *(a priori)* state estimate of the inner Kalman filter $\hat{x}_l^{p-}$ is computed by

$$\hat{x}_l^{p-} = \hat{x}_{l-1}^p, \tag{15}$$

where $\hat{x}_{l-1}^p$ is the previous state estimate of the inner Kalman filter and is assumed to be a constant for a given $k$, which is the average value of $y_l^p$s. Let $N$ be indices set of the neighbor sensors. In our experiment, $N = \{13, 23, 33, 43, 62, 63, 72, 73\}$, as illustrated in Fig. 4(a). The proposed method assumes that $y_l^p$ for $l=1$ to $n(N)$, where $n(N)$ is the cardinality of $N$, is a sequence of $y_{k,i}$, where $y_{k,i}$, is the data read from the neighbor sensor no. $i$ for $i \in N$. Also, it assumes that the initial state estimate $x_0^p$ is the average value of the humidity values of $y_{k,i}$ for all $i$s.

# 4 Experiment and Results

We conducted experiments to verify the concept of nested Kalman filters with bicubic interpolation to interpolate humidity values at the location of interest. The details of the experimental setup and result are as follows.

## 4.1 Experimental Setup

In this study, humidity data are measured and send to the database every 5 min. An example of data in one day is shown in Fig. 5. Therefore, in total there are 288 data points in one day. This experiment's duration is about 102 days, i.e., from 21 September 2018 to 31 December 2018. This study aims to apply the proposed method to interpolate the humidity value at the middle location marked in Fig. 4(a). To evaluate the performance of the proposed method, we compared the interpolated value with the one that is actually read from a sensor installed at that location.

Also, we compared the proposed method with the same prediction model but in different sources of information for the measurement update stage: kriging and linear interpolation. All simulations were done on ASUS ROG STRIX GL503 (Laptop), INTEL CORE I7-8750H, 12 GB DDR4 RAM, 1 TB HDD, 5400 RPM + 128 GB SSD, GEFORCE GTX1050TI 4 GB GDDR5.

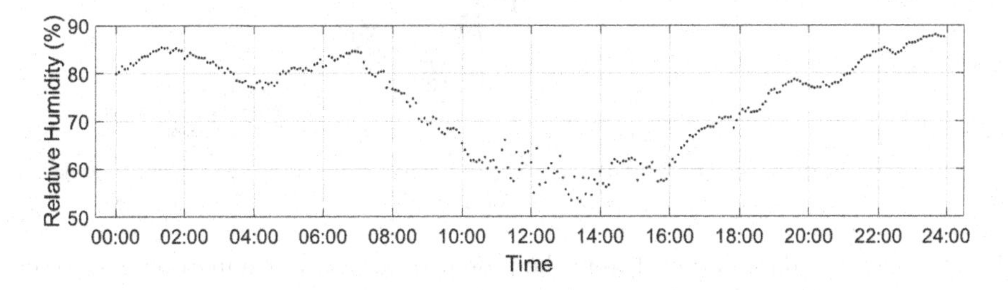


**Fig. 5.** Example of humidity data in one day.

### 4.2 Experimental Result

Figure 6 shows an example of interpolated values obtained from the proposed method, compared to data read from a sensor at that location, those from kriging, and those linearly interpolated. In this study, we use the root-mean-square error (RMSE) to evaluate the error between the actual data and those from other methods. The comparison of the RMSEs among different methods is shown in Table 1.

When the nested Kalman filtering was deployed with the bicubic interpolation, the average RMSE dropped from 1.60 to 1.33, i.e., the average RMSE dropped approximately by 16.88%. It can also be seen that the average RMSE of using kriging interpolation together with two nested Kalman filters was 1.32, which is 7.69% less than the average RMSE of Kriging interpolation alone. Similarly, the average RMSE of using linear interpolation with Kalman filtering was 1.34, which is 16.25% lesser. Even though the average RMSE of the proposed method was slightly greater than that of the two nested Kalman filters with kriging interpolation (i.e., approximately by 0.75%), it outperformed those without Kalman filtering. It should be noted that, the 0.75% improvement when combining our two nested Kalman filters with kriging comes with computational cost, which is to be discussed in the next section.

## 5 Discussion

There are two points to be discussed in this section. First, the computational time of the proposed method was 138.68 s, where as that of our two nested

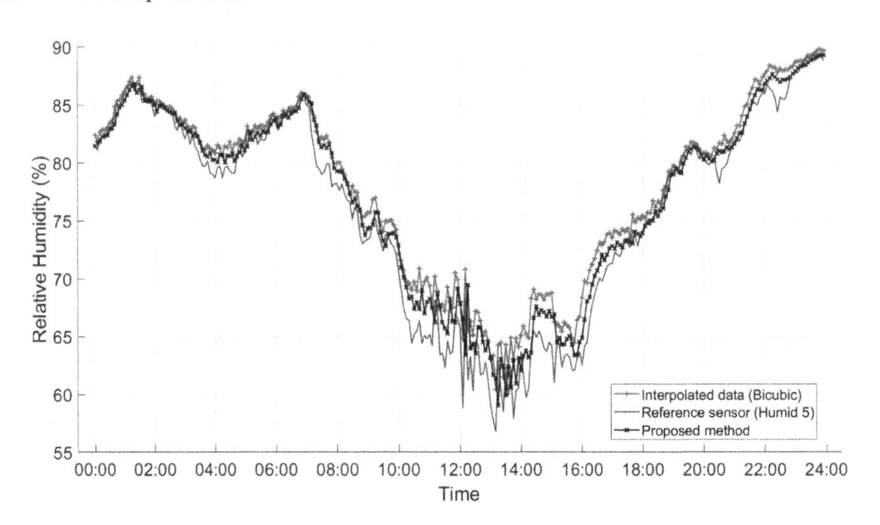


**Fig. 6.** Humidity data obtained from the proposed method in comparison with those obtained from other methods.

**Table 1.** Average RMSE comparison among different method.

| Date | With Kalman filter | | | Without Kalman filter | | |
|---|---|---|---|---|---|---|
| | Bicubic | Kriging | Linear | Bicubic | Kriging | Linear |
| 21–30 September 2018 | 1.14 | 1.14 | 1.15 | 1.33 | 1.26 | 1.34 |
| 1–31 October 2018 | 1.16 | 1.14 | 1.17 | 1.48 | 1.30 | 1.52 |
| 1–30 November 2018 | 1.25 | 1.26 | 1.26 | 1.43 | 1.51 | 1.45 |
| 1–31 December 2018 | 1.78 | 1.71 | 1.77 | 2.15 | 1.66 | 2.07 |
| Average RMSE | **1.33** | **1.32** | **1.34** | **1.60** | **1.43** | **1.60** |

Kalman filters with kriging was 471.36 s. That is, the proposed method is 3.4 times faster. For the practical usage, we think the 3.4 times less computational cost has a higher weight than 0.75% improvement in RMSE.

Second, the proposed method can estimate the actual humidity value of the location of interest and has the potential for improving the accuracy of the sensor when the sensor has been used for a long period of time, as shown in Fig. 7. Note that this experiment was done based on back analysis of the collected data by hand, not by an automatic procedure, which is to be investigated further in the future.

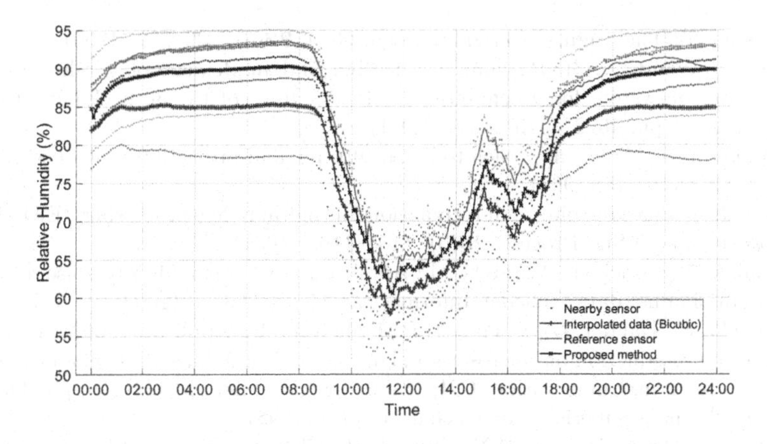


**Fig. 7.** Applying the proposed method to sensor when it is used for a long time.

## 6 Conclusion

This paper proposed a method for estimating the actual humidity values at the interested location by using nested Kalman filters with bicubic interpolation. The outer Kalman filter is used to interpolate the value at a specific location, and the inner Kalman filter is used as the prediction model of the outer one. The experimental results showed that this technique could precisely interpolate the values compared to actual one. Also, it is effective in terms of the computational time complexity.

**Acknowledgement.** This work is the output of an ASEAN IVO (http://www.nict.go.jp/en/asean_ivo/index.html) project, titled 'A Mesh-topological, Low-power Wireless Network Platform for a Smart Watering System,' and partially financially supported by NICT (http://www.nict.go.jp/en/index.html). The authors of this paper would like to express their sincere gratitude to Thai Orchids Co., Ltd., for the experiment greenhouse.

## References

1. Department of international trade promotion: The orchid world situation. https://ditp.go.th/contents_attach/539560/539360.pdf/. Accessed 24 July 2020
2. Orchid types - the top 10 most-popular types of orchids. https://www.nationthailand.com/breakingnews/30360623?fbclid=IwAR1_RA9UZ620plpMlWS020I556XC2RXXVnHxOGG02MNsjNAQyxZh-Fdodk/. Accessed 24 July 2020
3. Thai orchid exports to get boost under memo with ali auctions (2020). https://www.nationthailand.com/breakingnews/30360623?fbclid=IwAR1_RA9UZ620plpMlWS020I556XC2RXXVnHxOGG02MNsjNAQyxZh-Fdodk/. Accessed 24 July 2020

4. Cressie, N.: The origins of kriging. Math. Geol. **22**(3), 239–252 (1990)
5. Du, X., Wang, J., Ji, P., Gan, K.: Design and implement of wireless measure and control system for greenhouse. In: Proceedings of the 30th Chinese Control Conference, pp. 4572–4575. IEEE (2011)
6. Kalman, R.E.: A new approach to linear filtering and prediction problems. J. Basic Eng. **82**(1), 35–45 (1960)
7. Keys, R.: Cubic convolution interpolation for digital image processing. IEEE Trans. Acoust. Speech Sig. Process. **29**(6), 1153–1160 (1981)
8. Lesniak, A., Danek, T., Wojdyla, M.: Application of kalman filter to noise reduction in multichannel data. Schedae Informaticae **17**(18), 63–73 (2009)
9. Marselli, C., Daudet, D., Amann, H.P., Pellandini, F.: Application of kalman filtering to noisereduction on microsensor signals. In: Proceedings du Colloque interdisciplinaire en instrumentation, C2I, 18–19 Novembre 98, pp. 443–450, pp. 443–450. Ecole Normale Supérieure de Cachan, France (1998)
10. Moon, T., Hong, S., Choi, H.Y., Jung, D.H., Chang, S.H., Son, J.E.: Interpolation of greenhouse environment data using multilayer perceptron. Comput. Electron. Agric. **166**, 105023 (2019)
11. Musoff, H., Zarchan, P.: Fundamentals of Kalman Filtering: A Practical Approach. American Institute of Aeronautics and Astronautics, Reston (2009)
12. Needham, J.: Science and Civilisation in China, vol. 3, Mathematics and the Sciences of the Heavens and the Earth with Wang Ling (1959)
13. Rajarapollu, P.R., Mankar, V.R.: Bicubic interpolation algorithm implementation for image appearance enhancement. Int. J. **8**(2), 23–26 (2017)
14. Rhudy, M.B., Salguero, R.A., Holappa, K.: A kalman filtering tutorial for undergraduate students. Int. J. Comput. Sci. Eng. Surv. **8**(1), 1–9 (2017)
15. Stuckey, I.H.: Environmental factors and the growth of native orchids. Am. J. Bot. **54**(2), 232–241 (1967)
16. Welch, G., Bishop, G., et al.: An Introduction to the Kalman Filter (1995)
17. Xing, X., Song, J., Lin, L., Tian, M., Lei, Z.: Development of intelligent information monitoring system in greenhouse based on wireless sensor network. In: 2017 4th International Conference on Information Science and Control Engineering (ICISCE), pp. 970–974. IEEE (2017)

# Customized Dynamic Pricing for Air Cargo Network via Reinforcement Learning

Kannapha Amaruchkul(✉)

Graduate School of Applied Statistics, National Institute of Development Administration (NIDA), Bangkok 10240, Thailand
kamaruchkul@as.nida.ac.th

**Abstract.** A customized dynamic pricing problem for air cargo network management is proposed and formulated as a Markov decision process. Our model combines a customized pricing model for a B2B setting and a booking control problem for an air cargo network. Our solution approach employs reinforcement learning. In the numerical example based on historical records in 2016–2019 at one of the largest European carriers, the performance of reinforcement learning is evaluated. The policy from the reinforcement learning clearly outperforms the myopic policy and is within 10% of optimal actions in most cases.

**Keywords:** Air cargo · Revenue management · Reinforcement learning

## 1 Introduction

Despite its sharp drop due to the COVID-19 pandemic, air cargo is crucial to keep global supply chains going, especially for time-sensitive essentials including medical supplies, food products and ICT goods [13]. Other demand drivers include e-commerce [4] and the emerging implementation of supply chain management strategies, which emphasize on short lead times, e.g., lean management and just-in-time (JIT).

The objective of this paper is to propose a customized pricing model for an air cargo network. The problem is formulated as a Markov decision problem (MDP). A customer is clustered/segmented into groups with different willingness to pay. When a booking request arrives, the carrier decides a price to quote for each customer. Our solution approach employs an off-the-shelf reinforcement learning (RL) library in R. The applicability of reinforcement learning (RL) is illustrated via numerical experiments. The literature review and our contribution are given below.

In contrast to the business to consumer (B2C) list pricing in the passenger airline industry, air cargo pricing is business to business (B2B). Most of the carrier's customers are freight forwarders and integrators such as FedEx and DHL. In the passenger airline industry, the airline sells to lots of customers,


V.-N. Huynh et al. (Eds.): IUKM 2020, LNAI 12482, pp. 213–224, 2020.
https://doi.org/10.1007/978-3-030-62509-2_18

each of which represents a small proportion of total revenue. On the other hand, in the air cargo industry, the carrier sells to a few customers, each of which represents a large proportion of total revenue. Optimizing a quote price in the B2B setting is referred to as bid price/quote optimization or customized pricing. Overview of customized pricing can be found in [8,16] and [3]. They consider a single-period (static) problem, in which a price is quoted for a single individual customer. We extend it to a multi-period (dynamic) problem, in which requests for quotation (RFQ) arrive according to a stochastic process.

Since air cargo capacity can be sold at different prices to heterogeneous customers but cannot be sold after the flight departure, it is a prime candidate for revenue management (RM) strategies. Overview of RM theory and practice can be found in [15], [19], [1] and [5]. Despite an extensive literature on passenger RM, literature on air cargo RM is fairly limited, and a survey paper can be found in [20] and [6].

**Table 1.** Selected articles on dynamic booking control and scheduling problems for air cargo

| Articles | Spatial network | Temporal network | Cust.'s choice | Bid-price control |
|---|---|---|---|---|
| [14] | 0 | 1 | 0 | 1 |
| [2] | 0 | 0 | 0 | 0 |
| [7] | 0 | 0 | 0 | 1 |
| [10] | 1 | 1 | 0 | 0 |
| [9] | 0 | 1 | 1 | 1 |
| [12] | 0 | 0 | 0 | 0 |
| Ours | 1 | 0 | 1 | 1 |

Key air cargo operational decisions are a booking control problem and a network routing/scheduling problem. Table 1 is provided for comparison among some articles on the air cargo operational problems. A spatial network problem is considered in [10] and ours, whereas the rest consider a single-leg problem. The temporal network scheduling decision is presented in [10,14] and [9]. In [9], a customer exhibits a choice behavior among different flights, whereas in ours each customer group exhibits different willingness to pay. For the bid-price control, the bid price can be calculated by solving a deterministic multi-dimensional knapsack problem in [14], a stochastic programming problem in [7], a dynamic program value function approximation in [9]. Our solution approach is to use RL. A detailed introduction to RL can be found in [18].

The main contribution of this paper is as follows: First, we combine the dynamic booking control problem with the customized pricing problem to capture the customer's different willingness to pay. In contrast to the B2C's list pricing in the passenger airline industry, the B2B's customized pricing in the air cargo industry is explicitly incorporated in our model. Second, as shown in Table 2, our model includes other important air cargo characteristics, which are distinct from passenger. Third, we identify an air cargo booking control problem as a potential use case of RL. The rest of the paper is organized as follows: The

**Table 2.** Key features of our air cargo booking control model

| Passenger booking control | Air cargo booking control |
|---|---|
| B2C (list pricing) | B2B (customized pricing) |
| One-dimensional | Multi-dimensional (e.g., volume and weight) |
| Showup rate $\leq 100\%$ | Showup rate $\neq 100\%$ |
| Deterministic capacity | Stochastic cargo capacity |

model formulation and the solution approach are presented in Sect. 2. In Sect. 3, we illustrate the applicability of RL through numerical examples based on real world data set. Section 4 gives a summary and some extensions.

## 2 Formulation and Solution Methodology

Below, the customized air cargo pricing problem is formulated as a discrete-time, finite-horizon Markov decision problem (MDP) with expected total reward criteria. The booking horizon is divided into $\tau$ periods such that, in each period, the number of booking requests is at most one. The time indices run forward, so $t = 0$ is the beginning of the booking horizon, and $t = \tau$ is the time of service (flight departure time). In each period, if an RFQ arrives, the carrier either gives a quotation or rejects the request. In practice, the quotation may depend on the volume and weight already sold in all legs of a network, the customer's shipment specification such as its due date, its volume and weight, the type of goods (e.g., general or dangerous goods), the current market condition and the competitors' pricing strategies as well as the customer's relationship with the carrier (e.g., booking history and other data for segmentation). In our model, we assume that the carrier dynamically determines the quote price, based on the current volume and weight sold in the network, the shipment's volume and weight and the customer's group.

### 2.1 Markov Decision Problem

Consider an air cargo carrier that operates in a network with $n_L$ legs and $n_P$ origin-destination (OD) pairs. The network structure is specified by an $n_P \times n_L$ *incident matrix* $\mathbf{M}$, whose element $m_{h\ell} = 1$ if leg $\ell$ is used by the OD pair $h$ and $m_{h\ell} = 0$ otherwise. Let $\mathbf{M}_h$ denote the $h$-th row vector. The volume and weight capacities on all legs are denoted by the row vectors $\mathbf{K}_v = (K_{v\ell};\ \ell = 1, \ldots, n_L)$ and $\mathbf{K}_w = (K_{w\ell}; \ell = 1, \ldots, n_L)$, respectively. These capacities are random at the time of the booking, and they are materialized at the flight departure time. Uncertainty in cargo capacity arises from various factors such as the weather condition at the flight departure time, the length of the runway, and the total volume and weight of passenger bags. The carrier sells volume up to $\kappa^e_{v\ell}$ and weight up to $\kappa^e_{w\ell}$. Denote $\boldsymbol{\kappa}^e = (\boldsymbol{\kappa}^e_v, \boldsymbol{\kappa}^e_w)$, where $\boldsymbol{\kappa}^e_v = (\kappa^e_{v\ell}; \ell = 1, \ldots, n_L)$ and $\boldsymbol{\kappa}^e_w = (\kappa^e_{w\ell}; \ell = 1, \ldots, n_L)$. Borrowing the term from the passenger airline industry, we refer to $(\boldsymbol{\kappa}^e_v, \boldsymbol{\kappa}^e_w)$ as the volume and weight *authorization* (AU) levels. If the amount of cargo actually shows up is smaller than the booked amount, overbooking may be needed, and the AU may be greater than the largest value of the capacity. All combinations of volume and weight sold in the network are referred to as the *resource space*:

$$\mathbb{X} = (\prod_{\ell=1}^{n_L} \{0, 1, \ldots, \kappa^e_{v\ell}\}) \times (\prod_{\ell=1}^{n_L} \{0, 1, \ldots, \kappa^e_{w\ell}\}).$$

Let $n_G$ be the number of customer groups. The volume requirement belongs to one of the $n_v$ types, and the weight requirement belongs to one of the $n_w$ types. All combinations of the attributes are referred to as the *request attribute space*

$$\mathbb{P} = \{1, 2, \ldots, n_P\} \times \{1, 2, \ldots, n_G\} \times \{1, 2, \ldots, n_v\} \times \{1, 2, \ldots, n_w\}.$$

Let $\mathbf{0} = (0, 0, 0, 0)$ denote no arrival. The *request arrival space* is defined as $\mathbb{Q} = \mathbb{P} \cup \{\mathbf{0}\}$. The *state space* is defined as $\mathbb{S} = \mathbb{X} \times \mathbb{Q}$. Let $\mathbf{x}_v = (x_{v\ell}; \ell = 1, 2, \ldots, n_L)$, $\mathbf{x}_w = (x_{w\ell}; \ell = 1, 2, \ldots, n_L)$ and $\mathbf{x} = (\mathbf{x}_v, \mathbf{x}_w) \in \mathbb{X}$. The state $s \in \mathbb{S}$ is a vector

$$(x_{v1}, \ldots, x_{v,n_L}, x_{w1}, \ldots, x_{w,n_L}, h, g, i, j) = (\mathbf{x}_v, \mathbf{x}_v, h, g, i, j) = (\mathbf{x}, h, g, i, j).$$

Define $\chi : \mathbb{S} \to \mathbb{X}$ such that $\chi(s)$ selects the state's resource sold $(\mathbf{x}_v, \mathbf{x}_w)$. Define $\Theta : \mathbb{S} \to \mathbb{Q}$ such that $\Theta(s)$ selects the state's arrival request type $(h, g, i, j)$. The carrier is said to be in state $s \in \mathbb{S}$, if is current volume and weight sold is $\chi(s) = \mathbf{x}$, and the attribute of the arrival RFQ is $\Theta(s) = (h, g, i, j)$ where $h$ is the OD pair, $g$ is the customer's group, and $(i, j)$ are the shipment's volume and weight indices.

Let $U_{hgij}$ be the willingness to pay of the type-$(h, g, i, j)$ request. The bid response function, the probability that a quote price (chargeable rate) $y$ is accepted by the type-$(h, g)$ customer with the type-$(i, j)$ shipment, is $P(U_{hgij} > y)$. The *action space* is denote by $\mathbb{A} = \{0, 1, 2, \ldots, n_R\}$ where $n_R$ is the number of possible chargeable rates, and action $a = 0$ corresponds to a rejection of a request. Let $\rho_a$ denote the chargeable rate associated with action $a = 1, 2, \ldots, n_R$.

For each leg $\ell$, let $(T_{v\ell}, T_{w\ell})$ be the random show-up rate for volume and weight. Given the state $s$ at the departure time, the terminal reward function is

$$r_\tau = \begin{cases} 0; & \text{if} \quad T_{v\ell} x_{v\ell} \le K_{v\ell}, T_{w\ell} x_{w\ell} \le K_{w\ell} \quad \text{for all } \ell \\ -\sum_{\ell=1}^{n_L} \left[\tilde{r}^v_\ell (T_{v\ell} x_{v\ell} - K_{v\ell})^+ + \tilde{r}^w_\ell (T_{w\ell} x_{w\ell} - K_{w\ell})^+ + \tilde{r}^f_\ell\right]; & \text{otherwise} \end{cases} \tag{1}$$

where $\tilde{r}_\ell^v$ is the penalty cost per unit volume, $\tilde{r}_\ell^w$ is the penalty cost per unit weight, and $\tilde{r}_\ell^f$ is the fixed penalty cost on leg $\ell$. The penalty cost (1) is piecewise linear in the volume and weight excesses; this assumption can be found in [2] and [12].

The type-$(i,j)$ booking request has volume $v_i$ and weight $w_j$. If $v_i$ is given in cubic meter (cbm) and $w_j$ is given in kilogram (kg), then the *chargeable weight* is calculated as $w_{ij}^c = \max\{w_j, v_i/166.67\}$, where 166.67 is the standard density defined by the International Air Transport Association (IATA). For each time period $t < \tau$, let $r(s,a)$ be the reward if the current state is $s$ and action $a$ is taken. If there is no request ($\Theta(s) = \mathbf{0}$) or the request is rejected ($a = 0$), then $r(s,a) = 0$. For $\Theta(s) \neq \mathbf{0}$ and $a \neq 0$, then

$$r(s,a) = \begin{cases} P(U_{hgij} > \rho_a)\rho_a w_{ij}^c & \text{if all } \left(\mathbf{x} + (v_i\mathbf{M}_h, w_j\mathbf{M}_h) \leq \boldsymbol{\kappa}^e\right) \\ -\infty & \text{if any } \left(\mathbf{x} + (v_i\mathbf{M}_h, w_j\mathbf{M}_h) > \boldsymbol{\kappa}^e\right). \end{cases} \quad (2)$$

In (2), if the resource sold does not exceed the AU, the single-period reward is the expected revenue: The customer accepts the quote price $\rho_a$ with probability $P(U_{hgij} > \rho_a)$, and the carrier earns a revenue of $\rho_a w_{ij}^c$.

Let $N_{hg}(t)$ denote the total number of a type-$(h,g)$ customer arrived during $[0,t]$, i.e., from the beginning of the booking horizon until period $t$. Assume that the type-$(h,g)$ customer arrives according to a Poisson process $\{N_{hg}(t) : t \geq 0\}$ with rate $\lambda_{hg}$. Assume that the arrival process $\{N_{hg}(t) : t \geq 0\}$ is independent across all OD pairs and groups. Let $N(t) = \sum_{h=1}^{n_P}\sum_{g=1}^{n_G} N_{hg}(t)$ be the total arrivals from all OD pairs and groups. By superposition of Poisson processes, $\{N(t) : t \geq 0\}$ is a Poisson process with rate $\Lambda = \sum_{h=1}^{n_P}\sum_{g=1}^{n_G}\lambda_{hg}$. Given that an arrival of the $\{N(t) : t \geq 0\}$ process occurs at time $t$, then independent of what occurred prior to $t$, the arrival comes from the $\{N_{hg}(t) : t \geq 0\}$ process with probability $\frac{\lambda_{hg}}{\Lambda}$ by decomposition of Poisson processes. The probability that the type-$(h,g)$ customer with the type-$(i,j)$ shipment arrives is as follows:

$$q(h,g,i,j) = \begin{cases} e^{-\Lambda} & \text{if } (h,g,i,j) = \mathbf{0} \\ \left[(1 - e^{-\Lambda})\frac{\lambda_{hg}}{\Lambda}\right] f(i,j) & \text{otherwise} \end{cases} \quad (3)$$

where $f(i,j)$ is the probability that the volume and weight requirement is type-$(i,j)$, defined below. We could have defined the arrival process as a nonhomogeneous Poisson process with intensity function $\lambda_{hg}(t)$. Nevertheless, in our numerical example, the homogeneous Poisson process sufficiently describes the arrival process, since the booking horizon is only 7 d.

Assume that the volume $V$ and weight $W$ of a shipment follow the marginal distributions $F_v$ and $F_w$ respectively, and they are joined by copula $C$: It follows from Sklar's theorem that their joint distribution is $F(v,w) = C(F_v(v), F_w(w))$. We discretize volume $V$ into $n_v$ bins and weight $W$ into $n_w$ bins. Let $(b_i^v; i = 0, 1, 2, \ldots, n_v, n_{v+1})$ and $(b_j^w; j = 0, 1, 2, \ldots, n_w, n_{w+1})$ be increasing sequences of volume and weight breakpoints where $b_0^v = b_0^w = -\infty$ and $b_{n_v+1}^v = b_{n_w+1}^w = \infty$. The discretized volume $v_i$ is chosen such that $v_i \in (b_{i-1}^v, b_i^v]$ for $i < n_v$ and

$v_{n_v} \in (b^v_{n_v-1}, b^v_{n_v+1})$. Similarly, the discretized weight $w_j$ is chosen such that $w_j \in (b^w_{j-1}, b^w_j]$ for $j < n_w$ and $w_{n_w} \in (b^w_{n_w-1}, b^w_{n_w+1})$. Then, $f(i,j)$ is the probability of a two-dimensional vector $(V, W)$, distributed according to a given copula $C$, to fall in a hypercube with the following lower and upper bounds:

$$f(i,j) = \begin{cases} P\Big((b^v_{i-1}, b^w_{j-1}) \le (V,W) \le (b^v_i, b^w_j)\Big) & \text{if } i < n_v,\ j < n_w \\ P\Big((b^v_{i-1}, b^w_{j-1}) \le (V,W) \le (b^v_{i+1}, b^w_{j+1})\Big) & \text{if } i = n_v,\ j = n_w \\ P\Big((b^v_{i-1}, b^w_{j-1}) \le (V,W) \le (b^v_i, b^w_{j+1})\Big) & \text{if } i < n_v,\ j = n_w \\ P\Big((b^v_{i-1}, b^w_{j-1}) \le (V,W) \le (b^v_{i+1}, b^w_j)\Big) & \text{if } i = n_v,\ j < n_w. \end{cases} \tag{4}$$

Note that in (4), we assume that $(V, W)$ is independent of the arrival process $\{N_{hg}(t)\}$.

Let $p(s'|s,a)$ denote the transition probability given that the current state is $s \in \mathbb{S}$ and the action $a \in \mathbb{A}$ is taken. Let $\mathbf{x} = \chi(s)$ and $\mathbf{x}' = \chi(s')$. Let $1\{E\}$ be an indicator function: $1\{\mathbf{x}' = \mathbf{x}\} = 1$ if $x'_{v\ell} = x'_{v\ell}$ and $x'_{w\ell} = x_{w\ell}$ for all $\ell$.

If there is no arrival ($\Theta(s) = \mathbf{0}$) or the request is rejected ($a = 0$), then

$$p(s'|s,a) = 1\{\mathbf{x}' = \mathbf{x}\} q(\Theta(s')). \tag{5}$$

Suppose that $\Theta(s) \neq \mathbf{0}$ and $a \neq 0$. If the resource sold exceeds the overbooking level ($\mathbf{x}_v + v_i \mathbf{M}_h > \boldsymbol{\kappa}^e_v$ or $\mathbf{x}_w + w_j \mathbf{M}_h > \boldsymbol{\kappa}^e_w$), then

$$p(s'|s,a) = 1\{\mathbf{x}' = \mathbf{x}\} q(\Theta(s')). \tag{6}$$

For $\Theta(s) \neq \mathbf{0}$ and $a \neq 0$, if all $\big(\mathbf{x} + (v_i \mathbf{M}_h, w_j \mathbf{M}_h) \le \boldsymbol{\kappa}^e\big)$, then

$$p(s'|s,a) = \begin{cases} P(U_{hgij} > \rho_a) q(\Theta(s')) & \text{if } \mathbf{x}' = \mathbf{x} + (v_i \mathbf{M}_h, w_j \mathbf{M}_h) \\ P(U_{hgij} \le \rho_a) q(\Theta(s')) & \text{if } \mathbf{x}' = \mathbf{x}. \end{cases} \tag{7}$$

From (5)–(7), the volume and weight sold would remain the same if (i) there is no request, or (ii) the carrier rejects the request, or (iii) the resource sold exceeds the AU level, or (iv) the customer's willingness to pay does not exceed the quote price.

The 4-tuple $\{\mathbb{S}, \mathbb{A}, p(\cdot|s,a), r(s,a)\}$ is a Markov decision process. For each realization of the history

$$S_0, A_0,\ S_1, A_1,\ S_2, A_2,\ \ldots, S_{\tau-1}, A_{\tau-1},\ S_\tau$$

there corresponds a sequence of rewards

$$r(S_0, A_0),\ r(S_1, A_1),\ r(S_2, A_2),\ \ldots, r(S_{\tau-1}, A_{\tau-1}),\ r_\tau(S_\tau).$$

Let $R_t = r(S_{t-1}, A_{t-1})$ denote the random reward at time $t$ due, stochastically to, $S_{t-1}$ and $A_{t-1}$, and $R_{\tau+1} = r_\tau(S_\tau)$ denote the terminal reward. Let $\pi$ be a randomized policy, i.e., $\sum_{a \in \mathbb{A}} \pi(a|s) = 1$ and $\pi(a|s) \ge 0$. A deterministic policy is a special case of a randomized policy, which assigns the probability of one to exactly one action for all $s \in \mathbb{S}$. Let $\gamma \in [0,1]$ be the discount factor. In our problem, the discount factor is near one, because most air cargo booking requests arrives within one month from the departure date. For a given policy $\pi$ and the initial state $S_0 = s$, the expected total discounted reward is

$$v_\pi^\tau(s) = E[\sum_{t=0}^{\tau} \gamma^{t-1} r(S_t, A_t) + \gamma^\tau r_\tau(S_\tau)] = E[\sum_{t=1}^{\tau+1} \gamma^{t-1} R_t].$$

We have formulated our problem as a finite-horizon MDP with expected total reward criteria.

### 2.2 Reinforcement Learning

Solving the MDP is computationally intensive because of the *curse of dimensionality*. The size of the request arrival space is $\|\mathbb{Q}\| = n_P n_G n_v n_w + 1$. The size of the resource space is $\|\mathbb{X}\| = \prod_{\ell=1}^{n_L} (\kappa_{v\ell} + 1)(\kappa_{w\ell} + 1)$. The size of the state space is $\|\mathbb{S}\| = \|\mathbb{X}\| \|\mathbb{Q}\|$. Even for a small network, the size of the state space can be very large.

With a slight abuse of notation, we let

$$p(s', r|s, a) = P(S_t = s', R_t = r|S_{t-1} = s, A_{t-1} = a)$$

denote the conditional probability that the state is $s'$ and the reward is $r$ given that the preceding state and action are $s$ and $a$, respectively. Markov decision process $\{\mathbb{S}, \mathbb{A}, p(\cdot|s, a), r(s, a)\}$ can also be specified by $\{\mathbb{S}, \mathbb{A}, p(\cdot, \cdot|s, a)\}$. The *action-value function* for a policy $\pi$ is the revenue to go from time period $t + 1$ until the end of the horizon

$$q_\pi(s, a) = E_\pi[\sum_{k=0}^{\tau-t} \gamma^k R_{t+k+1}|S_t = s, A_t = a]$$

for all $s \in \mathbb{S}$. Let $q_*(s, a)$ denote the action-value function associated with an optimal policy. The Bellman optimality equations are

$$q_*(s, a) = \sum_{s'} \sum_{r} p(s', r|s, a)[r + \gamma \max_{a'} q_*(s', a')] \quad (8)$$

for all $s \in \mathbb{S}, a \in \mathbb{A}$. If $q_*(s, a)$ is known, the optimal policy $\pi^*(a|s)$ is given by the action, $\text{argmax}_{a \in \mathbb{A}}\{q_*(s, a)\}$. Nevertheless, $q_*(s, a)$ is generally not known due to the curse of dimensionality. Thus, the learning problem of the agent (carrier) is to maximize the expected reward by learning a policy function $\pi^*(a|s)$.

Let $Q(s, a)$ be the estimate of action-value function $q_*$. An RL algorithm updates the action-value function $Q(s, a)$ based on the past experience. The RL algorithms can be roughly categorized into *model-free* and *model-based* methods. The model-free RL learns $Q(s, a)$ directly from experiences, whereas the model-based RL also learns the environmental dynamics model. The model-free RL method includes $Q$-learning and Sarsa, whereas the model-based RL method includes Dyna-Q. A literature review of model-based RL can be found in the work of [11].

## 3 Numerical Experiment

We will evaluate the performance of RL. Consider a very small MDP so that an optimal policy can be obtained by solving the Bellman optimality Eq. (8). Let $n_P = 3$ and $n_L = 2$. The network structure is given by the incident matrix $\mathbf{M} = [\mathbf{M}_1\ \mathbf{M}_2]$ where $\mathbf{M}_1 = [1\,1\,0]^\top$ and $\mathbf{M}_2 = [1\,0\,1]^\top$. Our data set consists of booking requests during 2016–2019 from a carrier in Europe. In each year, there are two seasons, referred to as 1 and 2. All computations are done in R.

For each OD, all customers are divided into two groups ($n_G = 2$) by k-means clustering (function `kmeans` in R) based on the total chargeable weight and the log of the total booking revenue. We discretize the volume and weight such that 1 unit volume is 25 cbm and 1 unit weight is 5000 kg. The capacity is deterministic, and we do not overbook: $n_v = \kappa_{v\ell} = \kappa_{v\ell}^e = 2$ and $n_w = \kappa_{w\ell} = \kappa_{w\ell}^e = 2$ for all $\ell$. The booking horizon is divided such that 1 time period is 1 min. Assume that the booking horizon is 7 d, so the number of decision periods is $\tau = (7)(24)(60) = 10080$. In our data set, more than 75% of all bookings occur within seven days. In our control problem, the task is *episodic*, and the episode ends when the flight departs. Let the discount factor be $\gamma = 0.999$. The dependency of volume and weight can be specified using a copula (function `findCopula` in `VineCopula` package). The copula and the marginal distributions of the volume and weight as well as the expected number of arrivals $E[N(\tau)]$ are given in Table 3. (The marginal distributions are fitted using `fitdistr` package).

**Table 3.** Copula and marginal distributions of volume and weight. The mean volume $E[V]$ is given in cbm, and the mean weight $E[W]$ is given in kg.

| Sea. | Volume Distr. | Weight Distr. | Copula | $E[V]$ | $E[W]$ | $E[N(\tau)]$ |
|---|---|---|---|---|---|---|
| 1 | Lognorm(0.737, 1.215) | Lognorm(6.120, 1.316) | Frank(20.9) | 25.5 | 5171.2 | 6.1 |
| 2 | Lognorm(−0.001, 1.769) | Lognorm(5.016, 1.939) | Gauss.(0.9) | 25.9 | 5177.5 | 6.7 |

Let the discretized volume be $v_i = i$ for $i = 1, 2$, and the discretized weight be $w_j = j$ for $j = 1, 2$. Then, the physical volume is $\bar{v}_i = 25i$ cbm for $i = 1, 2$, and the physical weight is $\bar{w}_j = 5000j$ kg for $j = 1, 2$. Volume breakpoints are $(b_0^v, b_1^v, b_2^v, b_3^v) = (-\infty, \bar{v}_1, \bar{v}_2, \infty)$ and weight breakpoints are $(b_0^w, b_1^w, b_2^w, b_3^w) = (-\infty, \bar{w}_1, \bar{w}_2, \infty)$. The probability $f(i, j)$ in (4) is calculated using `copula` package.

Let $\bar{w}_{ij}^c = \max\{\bar{w}_j, \bar{v}_i/166.67\}$ denote the chargeable weight of the type-$(i, j)$ shipment. Assume that the willingness to pay $U_{hgij}$ is normally distributed with the mean of

$$\mu_{hgij} = \beta_0 + \beta_1 \log(\bar{w}_{ij}^c) + \beta_2 1\{g = 2\} + \beta_3 1\{h = 2\} + \beta_4 1\{h = 3\} \qquad (9)$$

and the standard deviation of $\sigma$. (Recall a random-utility model, the utility is $U = u + \xi$, where a representative component $u$ is deterministic and a mean-zero random component is $\xi$. For a probit model, $\xi$ is normally distributed with the mean of zero and the standard deviation of $\sigma$.) To estimate the coefficients $\beta$'s, we perform the linear regression of the chargeable rate as in (9) using function `lm`, and obtain $(\beta_0, \beta_1, \beta_2, \beta_3, \beta_4) = (182.8, -16.8, 2.6, -19.9, 22.6)$ for Season 1 and $(385.3, -49.0, -4.4, -31.5, -25.5)$ for Season 2. In both seasons, $\beta_1 < 0$ due to the quantity discount. This implies the negative relationship between the willingness to pay and the chargeable weight. The standard error of the estimate is used to estimate $\sigma \approx 113.2488$. For each $(h, g, i, j) \in \mathbb{P}$, we maximize the expected revenue per kilogram. Let

$$y^*(h, g, i, j) = \text{argmax}_{y=1,2,\ldots} \{y\bar{\Phi}((y - \mu_{hgij})/\sigma)\}.$$

The set of all possible quote prices (in THB/kg) is

$$\bigcup\{y^*(h, g, i, j)\} = \{35, 37, 41, 42, 46, 48, 53, 55, 60, 61, 62, 67, 68, 69, 70, 75, 76\}.$$

The MDP is solved using the `MDPtoolbox` package. Since the booking horizon is divided into a very large number of periods, the problem can be approximated as the infinite-horizon MDP, $v_\pi(s) = E[\sum_{t=1}^{\infty} \gamma^{t-1} R_t]$. Specifically, the optimal action is determined by the R's `mdp_policy_iteration` function, which requires three arguments, namely the transition matrix given in (5)–(7), the reward matrix given in (2), and the discount factor of $\gamma = 0.999$.

As a baseline, define the myopic policy as follows: For a state $s$, let $(h, g, i, j) = \Theta(s)$ and $\mathbf{x} = \chi(s)$,

$$\text{action} = \begin{cases} \text{offer price } y^*(h, g, i, j) & \text{if all } \big(\mathbf{x} + (v_i\mathbf{M}_h, w_j\mathbf{M}_h) \leq \boldsymbol{\kappa}\big) \\ \text{reject} & \text{if any } \big(\mathbf{x} + (v_i\mathbf{M}_h, w_j\mathbf{M}_h) > \boldsymbol{\kappa}\big). \end{cases}$$

If there is enough capacity to accommodate the request, then under the myopic policy, we offer the chargeable rate that maximizes the single-period reward. The myopic policy is nearsighted, since the future reward is not considered.

The RL is carried out using the `ReinforcementLearning` package; see details in [17]. In the current package version, only the $Q$-learning with *experience replay* method is implemented. We use a model-free $Q$-learning, instead of a model-based RL, because it is readily available in the one-line code in R. The R's `ReinforcementLearning` function requires the table consisting of historical states, actions and rewards, and the control parameters, which are the exploration rate $\epsilon$ and the step size $\alpha$.

Figure 1 shows revenues as the number of episodes increases to 365 (one year). The optimal solution from solving the MDP is shown in the dotted line. The lower bound (shown in the dash-dotted line) is the revenue associated with the myopic policy. It can be seen from Fig. 1 that the RL outperforms the myopic policy in both seasons. Figure 2 shows the percent optimal action. In Season 1, the myopic policy attains 26% optimal action, while the RL attains 85% in just the second episode. In Season 2, the RL starts from 46% in the third episode and learns very fast within the first 10 episodes. In Figs. 1 and 2, the step size is $\alpha = 0.1$, and the exploration rate is $\epsilon = 0.1$.

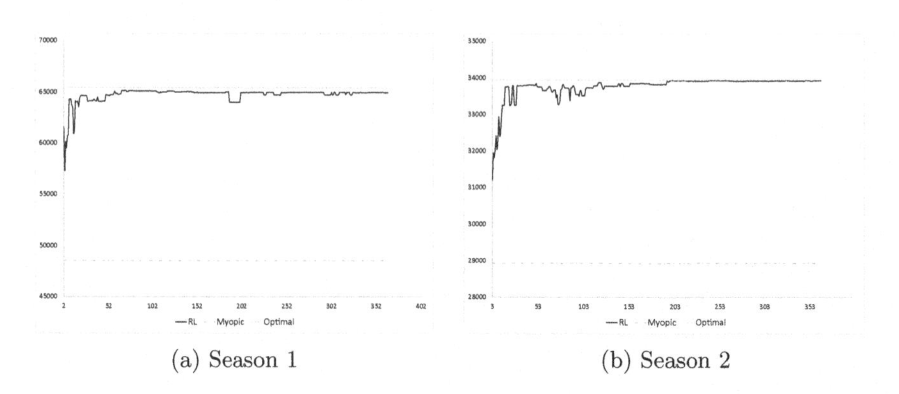


(a) Season 1 (b) Season 2

**Fig. 1.** Revenues as the number of episodes increases

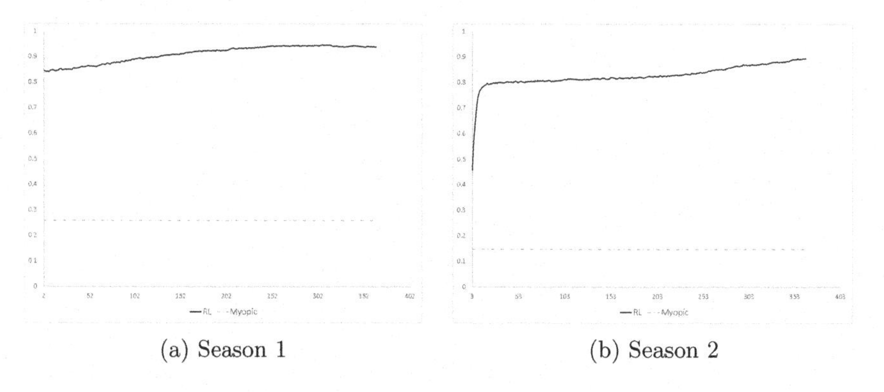


(a) Season 1 (b) Season 2

**Fig. 2.** Percent optimal actions as the number of episodes increases

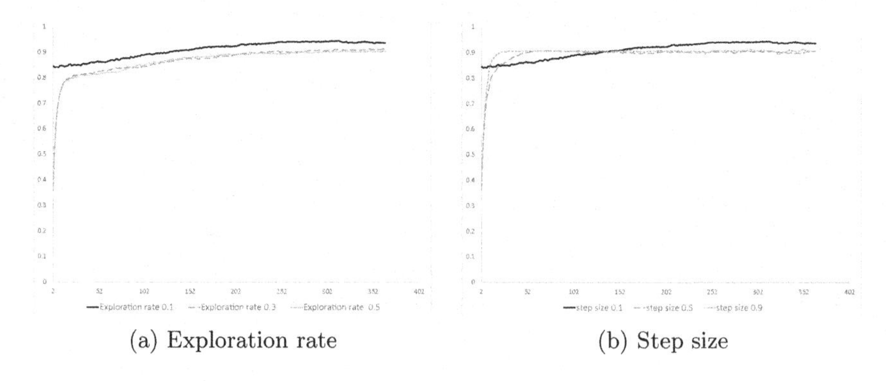


(a) Exploration rate (b) Step size

**Fig. 3.** Effects of exploration rate $\epsilon$ and step size $\alpha$. In Fig. 3a, $\alpha = 0.1$, whereas in Fig. 3b, $\epsilon = 0.1$.

In Fig. 3, we investigate the effect of the exploration rate and the step size in Season 1. (A similar pattern is found in Season 2, so figures are omitted.) It can be seen from Fig. 3a that the exploration rate $\epsilon = 0.1$ performs best. It can be seen from Fig. 3b that although the RL with a large step size learns very fast at the beginning, the RL with the small step size of $\alpha = 0.1$ eventually outperforms those with the larger step sizes.

## 4 Concluding Remarks

In summary, we formulate the customized dynamic pricing problem for an air cargo network as an MDP. RL is employed as our solution approach, and its performance is evaluated in the numerical example. A few extensions are as follows: The concept of charging different prices to different customer groups can be extended to different types of cargo, e.g., flowers, auto part, dry food and pharmaceuticals. In addition to a customer group, a shipment type may specify a type of goods, expedite/premium service, and so on. Furthermore, our MDP can be extended to include competition. A sequential game (a.k.a. stochastic/Markov game) is a multi-person decision process, in which each participant makes a sequence of decisions. Air cargo carriers compete on not only price but service quality. A customer choice model may be needed. Finally, we can explicitly model a terminal reward using a scheduling/routing problem, in which an offloaded shipment is assigned to a future itinerary, given the current flight schedule. We hope to pursue these or related problems.

## References


1. Amaruchkul, K.: Revenue Optimization Models. National Institute of Developement Administration Press, Bangkok (2018)
2. Amaruchkul, K., Cooper, W., Gupta, D.: Single-leg air-cargo revenue management. Transp. Sci. **41**(4), 457–469 (2007)
3. Bodea, T., Ferguson, M.: Segmentation, Revenue Management, and Pricing Analytics. Routledge, New York (2014)
4. Boeing Company: World air cargo forecast 2018–2037 http://www.boeing.com/commercial/market/cargo-forecast/ (2018). Accessed 01 July 2020
5. Chiang, W., Chen, J., Xu, X.: An overview of research on revenue management: Current issues and future research. Int. J. Revenue Manag. **1**(1), 97–128 (2007)
6. Feng, B., Li, Y., Shen, Z.: Air cargo operations: Literature review and comparison with practices. Transp. Res. Part C **56**, 263–280 (2015)
7. Han, D.L., Tang, L.C., Huang, H.C.: A Markov model for single-leg air cargo revenue management under a bid-price policy. Eur. J. Oper. Res. **200**, 800–811 (2010)
8. Higbie, J.A.: B2B price optimization analytics. In: Yeoman, I., McMahon-Beattie, U. (eds.) Revenue Management: A Practical Pricing Perspective. Palgrave Macmillian, New York (2011)
9. Levin, Y., Nediak, M., Topaloglu, H.: Cargo capacity management with allotments and spot market demand. Oper. Res. **60**(2), 351–365 (2012)

10. Levina, T., Levin, T., McGill, J.: Network cargo capacity management. Oper. Res. **59**(4), 1008–1023 (2011)
11. Moerland, T.M., Broekens, J., Jonker, C.: Model-based reinforcement learning: A survey. ArXiv abs/2006.16712 (2020)
12. Moussawi-Haidar, L.: Optimal solution for a cargo revenue management problem with allotment and spot arrivals. Transp. Res. Part E **72**, 173–191 (2014)
13. Organization for Economic Co-operation and Development (OECD): COVID-19 and international trade: Issues and actions http://www.oecd.org/coronavirus/policy-responses/covid-19-and-international-trade-issues-and-actions-494da2fa/ (2020). Accessed 7 July 2020
14. Pak, K., Dekker, R.: Cargo revenue management: Bid-prices for a 0-1 multi knapsack problem. ERIM Report Series Research in Management: School of Management. Erasmus Universiteit Rotterdam, Rotterdam, The Netherlands (2004)
15. Phillips, R.: Pricing and Revenue Optimization. Stanford University Press, Stanford (2005)
16. Phillips, R.: Customized pricing. In: Working Paper Series: No. 2010–1. Columbia University, Center for Pricing and Revenue Management (2010)
17. Prollochs, N., Feuerriegel, S.: Reinforcement learning in R https://arxiv.org/pdf/1810.00240.pdf (2018). Accessed 30 June 2020
18. Sutton, R.S., Barto, A.G.: Reinforcement Learning: An Introduction. The MIT Press, Cambridge, Massachusetts (20202)
19. Talluri, K., van Ryzin, G.: The Theory and Practice of Revenue Management. Kluwer Academic Publishers, Boston (2004)
20. Yeung, J., He, W.: Shipment planning, capacity contracting and revenue management in the air cargo industry: A literature review. In: Proceedings of the 2012 International Conference on Industrial Engineering and Operations Management, Istanbul, July 2012

# Analysis of the Determinants of $CO_2$ Emissions: A Bayesian LASSO Approach

Heng Wang[1,2](✉), Jianxu Liu[3], and Songsak Sriboonchitta[1]

[1] Faculty of Economics, Chiang Mai University, Chiang Mai 50200, Thailand
sysherr@163.com
[2] School of Foreign Languages for Business, Guangxi University of Finance and Economics, Nanning 530003, China
[3] School of Economics, Shandong University of Finance and Economics, Jinan 250000, China
liujianxu1984@163.com

**Abstract.** $CO_2$ emissions are recognized as the main driving factor to climate change. This study applies Bayesian LASSO approach to investigate the main determinants of $CO_2$ emissions in 56 countries from 1995 to 2014. In a multivariate framework, this study examines two hypotheses, including Environmental Kuznets curve (EKC) hypothesis and Pollution haven hypothesis (PHH). The sample is divided into two subperiods to compare the different determinants of $CO_2$ emissions before and after Kyoto Protocol came into effect in 2005. The results show that $CO_2$ emissions are mainly affected by energy consumption while using renewable energy and public transportation can reduce $CO_2$ emissions. Although economic development and urbanization are two factors opposite to the demand of emission reduction, technology and international trade, as well as international political cooperation, can mitigate $CO_2$ emissions. Education has a positive impact before 2005 and become negative on $CO_2$ emissions after 2005, which supports the EKC hypothesis, but no strong evidence for the PHH.

**Keywords:** $CO_2$ emissions · Bayesian LASSO · Energy consumption

## 1 Introduction

For the past several decades, there exists a broad consensus among countries worldwide that climate change is caused by environmental degradation. The pollution produced by humans is mainly from the combustion of fossil fuels to improve living standards. Therefore, governments confront with the challenge between the energy consumption and economic growth. New economies emerged since 1990s. Their quick growths have been driven by the soaring demand in energy consumption, causing a rapid increase in $CO_2$ emissions (Apergis and Payne 2010). Although these new economies take the rapid pace of the development, most of them have not yet finished the task of urbanization industrialization. In the future years, they will face the pressure to reduce $CO_2$ emissions.

Numerous hypothetical investigations formally model an immediate connection between energy and development. The principal viewpoint of the pollution haven hypothesis (PHH) is that dirty industries from industrialized countries damage natural ecology


V.-N. Huynh et al. (Eds.): IUKM 2020, LNAI 12482, pp. 225–237, 2020.
https://doi.org/10.1007/978-3-030-62509-2_19

in developing countries (Taylor 2004). But the so-called PHH effect yields contradictory results. Some scholars have provided the evidence that international trade from the developed countries to developing countries has increased global $CO_2$ emissions. On the contrary, other researchers have found that international trade slows down the emissions, as the improved environmental efficiency in resource-intensive countries offset the rise in emissions by trade openness.

Besides, the relationship between GDP and environmental pollution is described through the Environmental Kuznets curve (EKC) hypothesis. At an early stage, technological investments and financial development lead to pollution by the consumption of all kinds of natural resources. With further growth, technology improves with the increase of income; consequently, pollution starts to decline (Carson 2010).

Further, the paper is organized as follow: the next part is "Literature review", data and econometric method are shown in "Data description" and "Methodology", results and discussion are given in "Empirical results", and finally it is "Conclusions".

## 2 Literature Review

The factors influencing $CO_2$ emissions are associated with not only intensive energy use but also economic growth. Renewable energy production is related to environmental and economic goals. Apergis and Payne (2010) explored the roles of nuclear energy and renewable energy consumption on $CO_2$ emissions in 19 countries from 1984 to 2007. They found that nuclear energy use effects are significantly negative, while renewable energy consumption is not effective in reducing $CO_2$ emissions.

Baghdadi et al. (2013) used GDP per capita, population and openness, which was measured as the sum of imports and exports over GDP, to estimate their efforts on $CO_2$ emissions per capita and found that population is significantly positively related to the emissions gap between a pair of countries. Saboori et al. (2014) studied the relationship between GDP, energy and $CO_2$ emissions in OECD countries during 1960–2008. They summarized that there is a positive bi-directional relation between GDP, energy and $CO_2$ emissions in the long run. Wang et al. (2019) studied the role of participation in global value chains on $CO_2$ emissions per capita in 62 countries from 1995 to 2011. They stated that GDP per capita has an N-shaped relationship with $CO_2$ emissions per capita, indicating that $CO_2$ emissions per capita rises when the countries develop, declines as the threshold is reached, and then goes up when GDP per capita continues to climb.

International tourism in the EKC hypothesis gives a new vision of study that promotes sustainable tourism and environmental protection around the world. Gao et al. (2019) found the tourism-induced EKC hypothesis in the Mediterranean countries, that is, northern region confirms the hypothesis between tourism demand and economic growth, while southern region reveals the tourism-led growth hypothesis, which shows sound policy for building tourism infrastructure.

Urbanization has a negative effect in the global panel. Lv and Xu (2018) examined the effects of urbanization and trade openness on $CO_2$ emissions in 55 countries during 1992–2012, and implied that environmental quality is improved by urbanization. Rüstemoğlu and Andrés (2016) analyzed the determinants of $CO_2$ emissions in Russia and Brazil from 1992 to 2011. They proposed that population and urbanization are two key determinants and improved energy intensity decreases $CO_2$ emissions.

Mobile cellular telephone subscriptions per 100 people and patent application are used to measure technological factor, but there are opposite conclusions (Raheem et al. 2019; Shahbaz et al. 2020). Besides, education has also been considered. The variable on the total number of students in higher education at graduate and postgraduate levels in Australia, which proxies environmental education, is clearly significant (Balaguer and Cantavella 2018).

Globalization has a positive impact on economic efficiency, which promotes environmental improvement. KOF globalization index is used to reflect political cooperation. Increasing globalization is more environmental-friendlier, as it is an instrument for improving efficiency and technical progress. Chen et al. (2020) also studied the effect of globalization on $CO_2$ emissions growth in 36 OECD countries during 1970–2016 and found that political globalization decreases $CO_2$ emissions growth.

Many models are applied to decompose factors affecting $CO_2$ emissions, such as FMOLS, ARDL, GMM, VECM and SUR. Omri et al. (2014) studied the relationship between FDI, economic growth, energy use and $CO_2$ emissions in 54 countries from 1990 to 2011. The Cobb-Douglas type production function dynamic simultaneous-equation model was used to find that the main determinants of $CO_2$ emissions are energy, GDP and FDI. Zhu et al. (2016) used the panel quantile regression model to explore the effort of FDI, energy consumption and economic growth on $CO_2$ emissions in 5 ASEAN countries during 1981–2011. The results of this paper revealed that FDI decreases $CO_2$ emissions in the ASEAN countries, and population size and economic growth have a negative impact on $CO_2$ emissions. Hakimi and Hamdi (2016) applied co-integration techniques and VECM to study the relationship among FDI, trade openness, and $CO_2$ emissions in Morocco and Tunisia from 1970 to 2012. They showed that there exists bi-directional causal association between FDI and $CO_2$ emissions, and trade openness has a negative impact on the environmental quality. The similar studies can be found in the energy economics literature.

Concerning other covariates in the relationship between economic growth and $CO_2$ emissions, scholars tested a small set of determinants on the subject. Therefore, the objective of this study is to explore the effects of eight categories (See Table 1) in 56 countries[1] from 1995 to 2014 to investigate the determinants of $CO_2$ emissions by Bayesian LASSO regression model.

This paper contributes to the related literature. Firstly, it compensates for the lack of studies on so many factors of $CO_2$ emissions in many countries. Secondly, it further compares the different factors of $CO_2$ emissions over different periods. It contributes to providing more suggestions for $CO_2$ emissions and helps to find out the changing characteristics after Kyoto Protocol came into effect in 2005. Thirdly, the Bayesian LASSO regression model helps to avoid the problem with multicollinearity. Although Best Subset regression can be used to eliminate multicollinearity, the number of candidate explanatory variables should be less than 26 (Tamura et al. 2017).

[1] Argentina, Australia, Austria, Bangladesh, Belarus, Belgium, Brazil, Bulgaria, Chile, China, Colombia, Croatia, Czech Republic, Denmark, Egypt, Finland, France, Georgia, Germany, Greece, Hungary, India, Indonesia, Iran, Israel, Italy, Jamaica, Japan, Kazakhstan, Kenya, Korea, Rep., Latvia, Malaysia, Mexico, Moldova, Morocco, Netherlands, New Zealand, Norway, Pakistan, Peru, Philippines, Poland, Portugal, Romania, Russian Federation, Slovak Republic, South Africa, Spain, Sri Lanka, Sweden, Switzerland, Thailand, Turkey, Ukraine, United Kingdom

**Table 1.** Explained and explanatory variables

| Categories | Variables | Measurement | Notation |
|---|---|---|---|
| Explained variable | $CO_2$ emission | metric tons per capita | *co2emis* |
| $CO_2$ sources | $CO_2$ from public transportation | % of total fuel combustion | *co2t* |
| | $CO_2$ from electricity and heat production | % of total fuel combustion | *co2eh* |
| | $CO_2$ from other sectors, excluding residential buildings | % of total fuel combustion | *co2o* |
| | $CO_2$ from residential buildings | % of total fuel combustion | *co2bs* |
| Energy consumption | Renewable energy consumption | % of total final energy consumption | *renew* |
| | Energy use | kg of oil equivalent per capita | *enuse* |
| | Energy intensity level of primary energy | MJ/$2011 PPP GDP | *enin* |
| Economic growth | GDP per capita | constant 2010 US$ | *gdp* |
| | Inflation | annual % | *infl* |
| | Gross fixed capital formation | % of GDP | *gfcf* |
| | Total natural resources rents | % of GDP | *nrent* |
| Population | Population | Total | *pop* |
| | Population growth | annual % | *popg* |
| | Urban population | % of total population | *upop* |
| Education and technology | Education expenditure | current US$ | *edu* |
| | Patent applications | | *papp* |
| | Mobile cellular subscriptions | per 100 people | *mob* |
| | Individuals using the Internet | % of population | *int* |
| Import and export | Merchandise exports | current US$ | *merex* |
| | Trade openness | % of GDP | *open* |
| | Foreign direct investment | % of GDP | *fdi* |
| | International tourism, receipts | current US$ | *tour* |

(*continued*)

**Table 1.** (*continued*)

| Categories | Variables | Measurement | Notation |
|---|---|---|---|
| Political cooperation | Political Globalization Index (PGI) | | *pgi* |
| | PGI (de facto) | | *pgidf* |
| | PGI (de jure) | | *pgidj* |
| Financial Development | Financial Development Index | | *fd* |
| | Domestic credit to private sector | % of GDP | *dcre* |
| | Private credit to GDP | % | *pcre* |

## 3 Data Description

This paper uses the $CO_2$ emissions per capita in 56 countries as the explained variable. It refers to metric tons per capita. Based on the related literature, 28 explanatory variables are selected to investigate the determinants of $CO_2$ emissions in 56 countries. The variables are described in Table 1. The data of *co2emis, co2t, co2eh, co2o, co2bs, renew, enuse, enin, gdp, gfcf, nrent, pop, upop, edu, papp, mob, int, merex, open, tour, pgi, pgidf, pgidj, dcre* and *pcre* are transformed into natural logs before analysis. All data are obtained from World Bank, except that *pgi, pgidf* and *pgidj* from KOF Swiss Economic Institute and *fd* from International Monetary Fund.

Table 2 shows that multicollinearity is observed among explanatory variables. The VIF (Variance Inflation Factor) of 12 variables are greater than 10, which indicates high correlation. So it is not suitable to use Ordinary Least Squares.

**Table 2.** VIF for each explanatory variable

| | *co2t* | *co2eh* | *co2o* | *co2bs* | *renew* | *enuse* | *enin* | *gdp* | *Infl* | *gfcf* |
|---|---|---|---|---|---|---|---|---|---|---|
| VIF | 14.20 | 10.05 | 1.57 | 3.54 | 7.25 | 11.77 | 3.01 | 12.65 | 1.20 | 1.78 |
| | *pop* | *popg* | *upop* | *nrent* | *edu* | *papp* | *mob* | *int* | *Merex* | *open* |
| VIF | 4.05 | 1.66 | 3.73 | 1.58 | 10.67 | 3.66 | 4.50 | 6.62 | 10.71 | 2.60 |
| | *fdi* | *tour* | *pgi* | *pgidf* | *pgidj* | *fd* | *dcre* | *pcre* | | |
| VIF | 1.30 | 6.75 | 143.37 | 48.28 | 33.37 | 12.97 | 36.60 | 36.07 | | |

## 4 Methodology

Although LASSO is an efficient method, it cannot provide interval estimates for parameters estimated by exactly zero. To overcome this problem, Park and Casella (2008) considered LASSO from a Bayesian viewpoint, having prior distributions for the coefficient parameter $\beta$ and variance $\sigma^2$.

We assume a conditional Laplace prior on $\beta$ of the form

$$\pi\left(\beta|\sigma^2\right) = \prod_{j=1}^{p} \frac{n\lambda}{2\sqrt{\sigma^2}} \exp\left[-\frac{n\lambda|\beta_j|}{\sqrt{\sigma^2}}\right]$$

where $\lambda$ is a regularization parameter in LASSO, $n$ refer to the dimension of response vector $y$ in a linear regression model $y = X\beta + \varepsilon$, and $p$ refers to the dimension of coefficient vector $\beta$, and the noninformative scale-invariant marginal prior $\pi\left(\sigma^2\right) \propto 1/\sigma^2$ or inverse-gamma prior $\pi\left(\sigma^2\right) = IG\left(\frac{\nu_0}{2}, \frac{\eta_0}{2}\right)$ on $\sigma^2$, where $\frac{\nu_0}{2}$ is a shape parameter, $\frac{\eta_0}{2}$ is a scale parameter, and both of these parameters are positive. Conditioning on $\sigma^2$ for the prior distribution is crucial, because it guarantees a unimodal full posterior distribution (see Appendix A in Park and Casella (2008)). As an alternative specification for the prior distribution, Park and Casella (2008) put forward the hierarchical representation

$$\pi\left(\beta|\sigma^2, \tau_1^2, \ldots, \tau_p^2\right) = \prod_{j=1}^{p} \frac{n}{\sqrt{2\pi\sigma^2\tau_j^2}} \exp\left[-\frac{n^2\beta_j^2}{2\sigma^2\tau_j^2}\right],$$

$$\pi\left(\tau_1^2, \ldots, \tau_p^2\right) = \prod_{j=1}^{p} \frac{\lambda^2}{2} \exp\left[-\frac{\lambda^2\tau_j^2}{2}\right].$$

The prior distributions are represented by the Laplace distribution as a scale mixture of normals:

$$\pi\left(\beta|\sigma^2\right) = \int_0^\infty \ldots \int_0^\infty \pi\left(\beta|\sigma^2, \tau_1^2, \ldots, \tau_p^2\right)\pi\left(\tau_1^2, \ldots, \tau_p^2\right)d\tau_1^2 \ldots d\tau_p^2.$$

We implement the Gibbs sampler for $\beta, \sigma^2 \text{and} \tau_1^2, \ldots, \tau_p^2$. Supposing an inverse-gamma prior distribution $\pi\left(\sigma^2\right) = IG\left(\frac{\nu_0}{2}, \frac{\eta_0}{2}\right)$ on $\sigma^2$, the full conditional posterior distributions of $\beta$, $\sigma^2$ and $1/\tau_j^2$ (j = 1, …, p) are, respectively, given by

$$\beta|y, X, \sigma^2, \tau_1^2, \ldots, \tau_p^2 \sim N_p\left(A^{-1}X^T y, \sigma^2 A^{-1}\right),$$

$$A = X^T X + n^2 D_\tau^{-1}, D_\tau = diag\left(\tau_1^2, \ldots, \tau_p^2\right),$$

$$\sigma^2|y, X, \beta, \tau_1^2, \ldots, \tau_p^2 \sim IG\left(\frac{\nu_1}{2}, \frac{\eta_1}{2}\right),$$

$$\nu_1 = n + p + \nu_0, \eta_1 = ||y - X\beta||^2 + n^2\beta^T D_\tau^{-1}\beta + \eta_0,$$

$$\frac{1}{\tau_j^2}|\beta_j, \sigma^2, \lambda \sim IGauss(\mu', \lambda'),$$

$$\mu' = \sqrt{\frac{\lambda^2\sigma^2}{n^2\beta_j^2}}, \lambda' = \lambda^2, j = 1, \ldots, p,$$

where $IGauss(\mu, \lambda)$ represents the inverse Gaussian distribution with density function

$$f(x|\mu, \lambda) = \sqrt{\frac{\lambda}{2\pi}}x^{-3/2}\exp\left[-\frac{\lambda(x-\mu)^2}{2\mu^2 x}\right], x > 0.$$

According to these posterior distributions, generating MCMC samples, we get the estimates of parameters $\beta$ and $\sigma^2$ numerically. In this study, Bayesian LASSO regression was conducted on the *monomvn* package in R. Normal priors are used for regression coefficients $\beta$ as normal distribution is the closest to Laplace distribution for Bayesian LASSO regression (Kawano et al. 2015), and inverse-gamma prior is used for the variance $\sigma^2$. *monomvn* package provides the posterior probability that is nonzero to select variables. 0.95 is defined as the lower bound of posterior probability.

# 5 Empirical Results

To test whether there is any significant difference in different countries and periods, we add a country dummy and a period dummy in the regression. We can express the categorical variables *country* and *period* as single dummy variables respectively, so D1 = 1 for developed countries, D1 = 0 for developing countries, and D2 = 1 for 2005–2014, D2 = 0 for 1995–2004. We conduct Bayesian LASSO regression and find that these two dummy variables' estimated posterior probabilities that are nonzero are 0.060 and 0.995. So we do not divide the sample by countries but by periods. And the coefficient of *period* is −0.076 with a credible interval of [−0.102, −0.051], which shows that the $CO_2$ emissions slowed down after Kyoto Protocol came into effect in 2005.

This paper firstly investigates the determinants of $CO_2$ emissions for the full sample of the 56 countries. Then we conduct regressions in different periods, before and after 2005. The regression results are shown in Table 3. The posterior mean values are used as the point estimators of Bayesian LASSO estimation.

**Table 3.** Regression results of the determinants of $CO_2$ emissions

| Variables | (1) 1995–2014 | (2) 1995–2004 | (3) 2005–2014 |
|---|---|---|---|
| *co2t* | −0.346<br>[−0.375, −0.314] | −0.353<br>[−0.394, −0.314] | −0.430<br>[−0.470, −0.391] |
| *co2eh* | 0.246<br>[0.218, 0.274] | 0.224<br>[0.186, 0.265] | 0.203<br>[0.163, 0.240] |

*(continued)*

**Table 3.** (*continued*)

| Variables | (1) 1995–2014 | (2) 1995–2004 | (3) 2005–2014 |
|---|---|---|---|
| *co2bs* | | −0.055<br>[−0.076, −0.036] | |
| *renew* | −0.062<br>[−0.074, −0.051] | −0.052<br>[−0.067, −0.039] | −0.054<br>[−0.070, −0.038] |
| *enuse* | 1.239<br>[1.168, 1.310] | 1.256<br>[1.156, 1.360] | 1.148<br>[1.064, 1.239] |
| *enin* | −0.590<br>[−0.657, −0.521] | −0.608<br>[−0.711, −0.508] | −0.607<br>[−0.692, −0.523] |
| *gdp* | −0.203<br>[−0.251, −0.154] | −0.258<br>[−0.326, −0.196] | −0.154<br>[−0.216, −0.095] |
| *gfcf* | 0.166<br>[0.126, 0.204] | 0.273<br>[0.221, 0.323] | |
| *nrent* | 0.015<br>[0.009, 0.020] | | 0.028<br>[0.020, 0.036] |
| *upop* | 0.425<br>[0.384, 0.466] | 0.460<br>[0.408, 0.516] | 0.403<br>[0.347, 0.454] |
| *edu* | −0.071<br>[−0.086, −0.056] | 0.067<br>[0.041, 0.094] | −0.105<br>[−0.121, −0.089] |
| *papp* | −0.020<br>[−0.028, −0.011] | −0.026<br>[−0.037, −0.016] | |
| *int* | −0.020<br>[−0.026, −0.015] | | |
| *merex* | | −0.067<br>[−0.037, −0.016] | |
| *open* | −0.158<br>[−0.181, −0.130] | | −0.128<br>[−0.162, −0.093] |
| *fdi* | | −0.006<br>[−0.009, −0.003] | |
| *tour* | 0.115<br>[0.104, 0.126] | 0.082<br>[0.068, 0.096] | 0.136<br>[0.119, 0.150] |
| *pgi* | −0.325<br>[−0.394, −0.255] | | −0.454<br>[−0.572, −0.338] |
| *pgidj* | | −0.284<br>[−0.380, −0.189] | |
| *pcre* | 0.053<br>[0.040, 0.066] | | 0.041<br>[0.026, 0.057] |

Note: 95% equal-tailed credible intervals in parentheses.

### 5.1 Regression Results for the Full Sample

In the full sample, 16 variables out of the 28 potential variables are selected by posterior probability (see Table 4). From model (1), the coefficient of $CO_2$ emissions from public transportation (*co2t*) is −0.346 with a confidence interval of [−0.375, −0.314]. The assumption of the Bayesian LASSO model that all parameters are random implies that in this study the mean of the posterior distribution for *co2t* is −0.346, and the parameters of *co2t* fall between −0.375 and −0.314, and the estimated posterior probability that is nonzero is greater than 0.998. We can infer that the $CO_2$ emissions are negatively related to the public transportation. Public transportation substantially saves fuels in low-carbon economy. It can reduce the need for many private vehicles in the cities, and emit less carbon pollution on average. Meanwhile, the $CO_2$ emissions has a negative relationship with the $CO_2$ from renewable energy consumption (*renew*), because renewable energy has a lower influence on our environment.

**Table 4.** The estimated posterior probability that the individual components of the regression coefficients beta is nonzero in model (1)

| Variables | *co2t* | *co2eh* | *renew* | *enuse* | *enin* | *gdp* | *Gfcf* | *nrent* |
|---|---|---|---|---|---|---|---|---|
| Probability | 0.998 | 0.999 | 0.999 | 0.999 | 0.998 | 0.998 | 0.998 | 0.999 |
| Variables | *upop* | *edu* | *papp* | *int* | *open* | *tour* | *Pgi* | *pcre* |
| Probability | 0.998 | 0.999 | 0.998 | 0.998 | 0.998 | 0.998 | 0.997 | 0.998 |

The $CO_2$ from electricity and heat production (*co2eh*), use of primary energy before transformation to other end-use fuels (*enuse*), gross fixed capital formation (*gfcf*) and total natural resources rent (*nrent*) have a positive impacts on $CO_2$ emissions. They are the major challenges because the demands for the energy and economic growth and environmental protection are two contradictory goals for the countries. Our finding corresponds to Salim and Rafiq (2012) who presented that the rise in $CO_2$ emissions is due to energy consumption that is considered as the main source. According to the definition of gross fixed capital formation by World Bank, it mainly refers to the infrastructure, which is a driving force for economic growth. Total natural resources rents provide incentives to exploration companies to utilize oil, coal, natural gas, minerals and forest for energy consumption. It is noted that the variable *enuse* has the largest positive coefficient, 1.239. Thus energy consumption is the essential factor for the reduction in $CO_2$ emissions.

Energy intensity has a negative impact on $CO_2$ emissions. It is generally declining due to convergence and economic development (Atalla and Bean 2017). Theoretically decreasing energy intensity is good for the environment and economy, but here negative impact means that primary energy consumption is growing inefficiently. And there exists a significantly negative correlation between $CO_2$ emissions and GDP per capita. Thus our result does not support the EKC hypothesis.

$CO_2$ emissions are also positively related to urban population proportion of total population (*upop*). Its coefficient as the second largest illustrates that urban population is considered to be one of the key sources of carbon emissions. The $CO_2$ emissions tend to rise when the urban population proportion increases, since the pollution is caused by urbanization with excessive consumption of energy.

Education expenditure (*edu*) and technology represented by *papp* and *int*, negatively influenced $CO_2$ emissions, but the small absolute values of their coefficients show these variables are not the main factors. And there is also a negative relationship between $CO_2$ emissions and openness, so high trade openness with low $CO_2$ emissions encourages countries to be more exposed to open markets. The significantly positive value of the coefficient for expenditures by international inbound visitors (*tour*) indicates the positive correlation between $CO_2$ emissions and inbound tourism.

Political Globalization Index (*pgi*) has a significantly negative influence on the $CO_2$ emissions because political globalization promotes international cooperation and builds up standard environmental regimes. (Chen et al. 2020)

Although Foreign Direct Investment and Financial Development Index is insignificant to affect $CO_2$ emissions, the ratio of private credit to GDP (*pcre*) significantly increase $CO_2$ emissions with small coefficient 0.053. Private credit can boost the scale of economic activities and the amount of energy usage in the country.

For the other variables, since their posterior probability that is nonzero is less than 0.95 or their credible intervals contain zero, we cannot draw conclusions that there are significant correlations between these variables and $CO_2$ emissions.

## 5.2 The Determinants of $CO_2$ Emissions in Different Periods

There are several obvious differences in terms of the factors of reducing $CO_2$ emissions in two subperiods through the comparison of model (2) and model (3). 15 variables are selected in the first subperiod while 13 variables in the second.

First of all, the posterior mean of the coefficients for electricity and heat production (*co2eh*) and primary energy use (*enuse*) are significantly positive for two subperiods, and the posterior mean of the coefficients for public transportation (*co2t*), renewable energy consumption (*renew*) and energy intensity (*enin*) are still significantly negative for both subperiods. In contract, individuals using the Internet (*int*) have no relation with $CO_2$ emissions in two subperiods, but patent application (*papp*) still has a negative impact on $CO_2$ emissions in the first subperiod. With the advance of technology, it is no longer an important factor or is transformed into other factor for reducing $CO_2$ emissions in the second subperiod.

Secondly, four new factors, $CO_2$ emissions from residential buildings (*co2bs*), merchandise export (*merex*), foreign direct investment (*fdi*) and Political Globalization Index (de jure) (*pgidj*), become negative and significant during 1995–2004, but are not significant during 2005–2014. In fact, *merex* and *pgidj* are replacements of *open* and *pgi* in the full sample, but the effects of *merex* and *pgidj* become weakened. *co2bs* can show the residents' habits for energy use in households. Furthermore, although the impact of FDI is significantly negative, its coefficient, −0.006, indicates that its effect is small. However, the coefficients of *gfcf* and *upop* become greater, implying that infrastructure and urbanization increase energy use. The expansion of the middle classes will be made in the areas of infrastructure including upgrades and urbanization in the countries. Infrastructure spending will boost economic prospects in as many regions as possible and underpin growth through job creation, which encourages urbanization.

Thirdly, during 2005–2014, total natural resources rents (*nrent*) and private credit to GDP (*pcre*) have positive relationship with $CO_2$ emissions, and openness (*open*) and Political Globalization Index (*pgi*) have negative relationship. But these four variables are not significant during 1995–2004. This implies that after Kyoto Protocol came into effect in 2005, international political cooperation and international trade have contributed to reduction in $CO_2$ emissions. Moreover, the process of extracting natural resources involves the use of heavy machinery which not only means the use of more energy but also implies increased pollution and carbon emissions. It also hints that environmental sustainability depends on natural resources rent. And it is remarkable that the effect of inbound tourism is increasing after 2005.

At last, the effect of education is more complex. In the first subperiod, its effect is positive, while in the second subperiod, the effect becomes negative. The result is similar to the discussion in the research on Australian data during 1950–2014. The first effect takes place when expansion in education intensifies pollution emissions. The second effect is given when education is provided for more people to have environmental awareness that could help improve environmental quality (Balaguer and Cantavella 2018). Comparing the coefficients in two subperiods, the second effect is increasing importantly.

## 6 Conclusions

We have explored the relationship between $CO_2$ emissions and energy consumption, economic growth, population, education and technology, international trade, tourism, political cooperation and financial development for 56 countries during 1995–2014 by Bayesian LASSO regression model. The following major conclusions are obtained:

First of all, $CO_2$ emissions are mainly affected by energy consumption. Properly controlling the scale of electricity and heat production is conducive to reduction in $CO_2$ emissions reduction, along with using renewable energy and encouraging people to choose public transportation. In addition, improving energy efficiency is also an effective way to reduce $CO_2$ emissions.

Secondly, economic development and urbanization are contradictory factors influencing the goal of emission reduction. All countries are facing the pressure of reducing $CO_2$ emission in terms of infrastructure construction and increasing urban population and inbound tourists. The development of science and technology, as well as strengthening international political cooperation and international trade, will have a positive impact on emission reduction, reflecting the purpose of Kyoto Protocol.

Third, after 2005 when Kyoto Protocol came into force in 2005, the impact of primary energy use on $CO_2$ emissions decreases, but is still positive. Overdependence on natural resources rent leads to increase in $CO_2$ emissions. In two subperiods, education and $CO_2$ emissions have an inverted U-shaped relationship, which validates the EKC hypothesis when considering education in the model. But it is difficult to detect the significant relation between FDI and pollution, so PHH cannot be proved.

## References

Apergis, N., Payne, E.: Renewable energy consumption and economic growth: evidence from a panel of OECD countries. Energy Policy **38**(1), 656–660 (2010)

Atalla, T., Bean, P.: Determinants of energy productivity in 39 countries: an empirical investigation. Energy Econ. **62**, 217–229 (2017)

Baghdadi, L., Martinez-Zarzoso, I., Zitouna, H.: Are RTA agreements with environmental provisions reducing emissions? J. Int. Econ. **90**(2), 378–390 (2013)

Balaguer, J., Cantavella, M.: The role of education in the Environmental Kuznets Curve: evidence from Australian data. Energy Econ. **70**(C), 289–296 (2018)

Carson, T.: The Environmental Kuznets Curve: seeking empirical regularity and theoretical structure. Rev. Environ. Econ. Policy **4**(1), 3–23 (2010)

Chen, T., Gozgor, G., Koo, K., Lau, M.: Does international cooperation affect $CO_2$ emissions? evidence from OECD countries. Environ. Sci. Pollut. Res. **27**, 8548–8556 (2020)

Gao, J., Xu, W., Zhang, L.: Tourism, economic growth, and tourism induced EKC hypothesis: evidence from the Mediterranean region. Empirical Economics **4**, 1–23 (2019)

Hakimi, A., Hamdi, H.: Trade liberalization, FDI inflows, environmental quality and economic growth: a comparative analysis between Tunisia and Morocco. Renew. Sustain. Energy Rev. **58**(C), 1445–1456 (2016)

Kawano, S., Hoshina, I., Shimamura, K., Konishi, S.: Predictive model selection criteria for Bayesian LASSO regression. J. Jpn. Soc. Comput. Stat. **28**(1), 67–82 (2015)

Lv, Z., Xu, T.: Trade openness, urbanization and $CO_2$ emissions: dynamic panel data analysis of middle-income countries. J. Int. Trade Econ. Dev. **28**(3), 317–330 (2018)

Omri, A., Nguyen, K., Rault, C.: Causal interactions between $CO_2$ emissions, FDI, and economic growth: evidence from dynamic simultaneous-equation mode ls. Econ. Model. **42**, 382–389 (2014)

Park, T., Casella, G.: The Bayesian lasso. J. Am. Stat. Assoc. **103**, 681–686 (2008)

Raheem, I., Tiwari, A., Balsalobre-Lorente, D.: The role of ICT and financial development in $CO_2$ emissions and economic growth. Environ. Sci. Pollut. Res. **27**(1), 1912–1922 (2019)

Rüstemoğlu, H., Andrés, A.: Determinants of $CO_2$ emissions in Brazil and Russia between 1992 and 2011: a decomposition analysis. Environ. Sci. Policy **58**(C), 95–106 (2016)

Saboori, B., Sapri, M., Baba, M.: Economic growth, energy consumption and $CO_2$ emissions in OECD (Organization for Economic Co-operation and Development)'s transport sector: a fully modified bi-directional relationship approach. Energy **66**(C), 150–161 (2014)

Salim, R.A., Rafiq, S.: Why do some emerging economies proactively accelerate the adoption of renewable energy? Energy Econ. **34**(4), 1051–1057 (2012)

Shahbaz, M., Raghutla, C., Song, M., Zameer, H., Jiao, Z.: Public-private partnerships investment in energy as new determinant of $CO_2$ emissions: the role of technological innovations in China. Energy Econ. **86**(C), 1–45 (2020)

Tamura, R., Kobayashi, K., Takano, Y., Miyashiro, R., Nakata, K., Matsui, T.: Best subset selection for eliminating multicollinearity. J. Oper. Res. Soc. Jpn. **60**(3), 321–336 (2017)

Taylor, M.: Unbundling the pollution haven hypothesis. Adv. Econ. Anal. Policy **3**(2), 1–28 (2004)

Wang, J., Wan, G., Wang, C.: Participation in GVCs and $CO_2$ emissions. Energy Econ. **84**(C), 1–43 (2019)

Zhu, H., Duan, L., Guo, Y., Yu, K.: The effects of FDI, economic growth and energy consumption on carbon emissions in ASEAN-5: evidence from panel quantile regression. Econ. Model. **58**, 237–248 (2016)

# Comparison Between Original Texts and Their Revisions to Improve Foreign Language Writing

RuiLin Liur and Tomoe Entani(✉)

University of Hyogo, Kobe 651-0047, Japan
{ab19s201,entani}@ai.u-hyogo.ac.jp

**Abstract.** A text written by non-native speakers makes readers feel unusual, even if there are no evident errors in the sense of basic grammar. We discuss such an intangible error in this paper. Therefore, the goal is to discover the similarities of writing among foreign languages by Japanese native speakers. We compare the original texts written in a foreign language and their revisions by a native speaker to identify the parts which s/he feels unusual and gives the options. To not rely on the subjective context and the uniqueness of a language, we replace a word into a part-of-speech by the morphological analysis. Some parts-of-speech are deleted from the original text by revising a text, and some parts-of-speech are added instead. We consider a set of different parts-of-speech, which is added or deleted by the revision, and find common trends with multiple languages from the numbers and ratios of sentences. These findings are not always familiar to the experts, and on the other hand, they may give new viewpoints to language learners and teachers to improve their activities.

**Keywords:** Comparison analysis · Morphological analysis · Association rule

## 1 Introduction

In school educations, most of us learn foreign languages for several years. At that time, we write texts in foreign languages as class assignments and exams. In an increasingly globalized world, not only at schools but also in business and private, we face more often to communicate with others in a foreign language. Accordingly, some of us benefit from various useful translating tools. Such tools as Google Translate and DeepL Translator are beneficial. They have recently been remarkably improved for quality, speed, and cost and will undoubtedly be enhanced by the contribution of the research in natural language processing.

On the other hand, there are many language learners whose motivations are varied so that some are willing, and the others are forced to learn. Regardless of the reasons, it is essential for them to express themselves in the language for communication. They would like to write their texts in a foreign language as


V.-N. Huynh et al. (Eds.): IUKM 2020, LNAI 12482, pp. 238–248, 2020.
https://doi.org/10.1007/978-3-030-62509-2_20

native speakers do, even with helpful translating tools. It is not easy to exclude the influence of mother-tongue and master a foreign language since it profoundly relates to our thinking. The experience and interpersonal relationships are transformed into meaning, and the meaning is further transformed into texts. The relationship between grammar and discourse semantic presents a model of language in context in Systemic Functional Linguistics [5].

In our mother-tongues, we sometimes figure out whether the text is written by a native or non-native speaker, even if there are no evident errors in the sense of basic grammar. A native speaker can understand what the non-native writer mentions, still s/he may feel unusual to the whole text. It is difficult for the native speaker to explain exactly where and why. Instead, s/he can show some alternative expressions if s/he asked to revise it. Our research question is to find out the difference between texts by non-native and native speakers. The goal is to derive some common factors in Japanese native speakers' texts in which native speakers feel unusual and corresponding options given by them. Therefore, we compare the original text and its revisions by sentence by sentence.

In linguistic researches in each language, many aspects to characterize the language are pointed out and reflected in its language education. Conventionally, they often focus on a few targets, such as specific words or grammatical issues, and analyze sample data from the viewpoints. Recently, statistical research is becoming popular, owing to big data and powerful computation. Various kinds of corpora to reveal structures of a language have been built focusing on spoken language [12] and clinical documents [6]. A large web corpus is built from the texts collected by crawling web sites. TenTen Corpus Family is a set of web corpora in various languages, including Japanese, German, and Chinese [2]. Multiple languages are compared from grammar rules and culture background [4,7]. Although the volume of data is larger and larger, the same kinds of texts are usually compared and analyzed. For instance, the standard and correct texts, such as in novels and newspapers, are analyzed to make proper texts reflecting the ordinary rules of a language. The texts by non-native speakers are analyzed to detect some trends of the errors from the educational viewpoint [8,9].

Our approach is data-oriented, so we first analyze the data and then find the viewpoints that may be worth discussing. The data we use has to be a pair of the original and revised texts to detect where and how the original texts is revised from the viewpoint of intangible errors. Furthermore, our analysis is not from a particular foreign language and consider multiple languages for Japanese native speakers. To compare different languages, we apply morphological analysis, one of the underlying technology in the field of natural language processing.

This paper outlines as follows. In the next section, we show the data and approach for comparing two kinds of texts. In Sect. 3, we show the example of two foreign languages to Japanese native speakers to illustrate the proposed approach and some findings. Then, we draw the conclusion in Sect. 4.

# 2 Approach

## 2.1 Original and Revised Texts

Instead of grammatical mistakes, we consider no evident errors in a text written in a foreign language. For that purpose, this study's original text is an essay written by a non-native speaker based on his/her opinion or thinking. A writer explains a fact in her/his preferred expression, so the different writers use different sentences for the same event. Such a writer's writing style appears in an essay rather than in a translation from a mother-tongue text. The translation tends to be a word to word replacement, so the sentences for the same fact may be different when the given sentences in his/her mother-tongue are different. A native speaker feels unusual in the original text and can revise the sentences in it into other structures with other words. Specifically, we compare the original sentence written in a foreign language to its revised sentence by a native speaker.

It should be noted that the number of sentences of the original and revised ones is not always equal since a long sentence may be divided into two, while two short sentences may be merged into one. Therefore, we determine the sentences' correspondence in the original and revised texts based on a similarity of sentences. The similarity of a pair of sentences is defined, for instance, based on Word2vec technology [10]. We consider merging the original sentence to its next one or the revised one to its next one when the original sentences and the revised sentences are not similar.

The merge algorithm is based on the idea that the two sentences are merged if it makes the original and revised sentences more similar. It is illustrated in Fig. 1, where $I = \{0, 1, \ldots, i^*\}$ and $J = \{0, 1, \ldots, j^*\}$ represent the numbers of sentences of the original and revised texts, respectively, and $\alpha$ and $\beta$ are the thresholds of similarities. For the $i$th sentence of original text $Y_{oi}$ and the $j$th sentence of the revised one $Y_{rj}$, whose similarity is $S(Y_{oi}, Y_{rj})$, we assume three possibilities for a comparison pair: no merging, merging original sentences as $Y_{oi} \leftarrow Y_{oi} + Y_{o(i+1)}$, or merging revised ones as $Y_{rj} \leftarrow Y_{rj} + Y_{y(j+1)}$. In other words, the original and revised sentences become a comparison sentence $(Y_{oi}, Y_{rj})$ when they or their next ones are similar enough, or merging does not improve their similarities. The former and latter thresholds are $\alpha$ and $\beta$, respectively, such as $\alpha = 0.95$ and $\alpha = 0.93$. Otherwise, original or revised sentences are merged such as $(Y_{oi} + Y_{o(i+1)}, Y_{rj})$ or $(Y_{oi}, Y_{rj} + Y_{r(j+1)})$. Starting merging from the first sentences of both texts does not certify ending with their last sentences. However, since the number of sentences in the revised text is not very different from the original one, the algorithm is practically acceptable.

## 2.2 Morphological Analysis

The original texts, which are essays used in this study, are very personal, and their contexts depend on the writers who are language learners. Moreover, our research is from the viewpoint of foreign languages rather than a particular language, so we consider multiple foreign languages. Therefore, we replace a

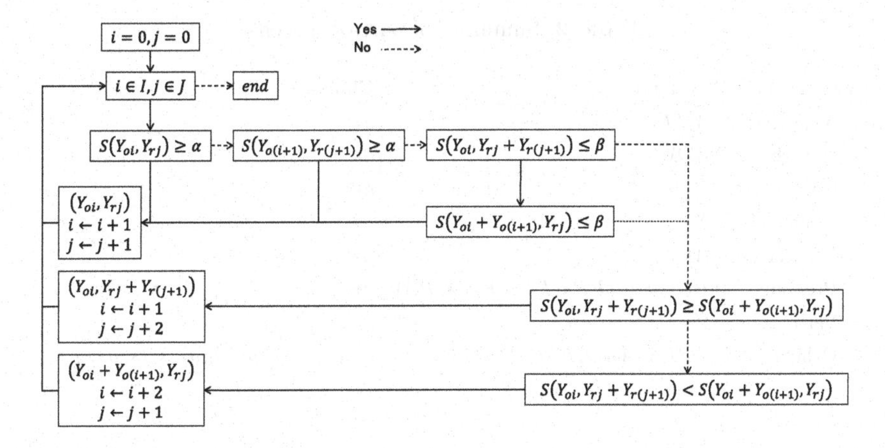


**Fig. 1.** Merge algorithm

word into a part-of-speech by the morphological analysis not to rely on the context and language.

The writers are Japanese, and the foreign languages are Chinese and German. In some tools for Morphological analysis, we use TreeTagger for German [3,11] and NLPIR-ICTCLAS for Chinese [1]. Table 1 shows the example sentences in German and Chinese, and they are replaced into parts-of-speech by TreeTagger and NLPIR-ICTCLAS, respectively. The numbers of parts-of-speech by these tools are 50 and 84 in German and Chinese, respectively. In Table 2, we divided into 7 categories of adjective, adverb, conjunction, noun, preposition, pronoun, and verb, and others. Note that the category, others, includes the frequently used article, $ART$, in German and time word, $t*$, and particle, $u$, in Chinese.

We denote each sentence of the original and revised texts with parts-of-speech. We then count each part-of-speech as $C(Y,x)$, where $Y$ and $x$ denote a sentence and part-of-speech, respectively. For instance, sentence $G_1$ has two $NN$, which is denoted as $C(G_1, NN) = 2$, and similarly, it has two words in verb category denoted as $C(G_1, V*) = 2$.

**Table 1.** Examples of morphological analysis

| language | sentence | part-pf-speech |
|---|---|---|
| German $G_1$ | In dem Text wird folgende Inhalt geschrieben. | $APPR, ART, NN, VAFIN, ADJA, NN, VVPP$ |
| Chinese $C_1$ | 人类智慧的结晶 | $n, n, ude1, n$ |

**Table 2.** Summary of part-of-speech

| Category | German | Chinese |
|---|---|---|
| Adjective | $ADJ*$ | $a*$ |
| Adverb | $ADV$ | $d*$ |
| Conjunction | $KO*$ | $c*$ |
| Noun | $N*$ | $n*$ |
| Preposition | $AP*$ | $p*$ |
| Pronoun | $PD*, PI*, PP*, PR*, PW*, PAV$ | $r*$ |
| Verb | $V*$ | $v*$ |
| Others | $ART, CARD, FM, PT*, T*$ | $b*, e, f, h, k, m*, o, q*, s, t*, u*, y, z$ |

## 2.3 Comparison of Original and Revised Sentences

Some parts-of-speech in the original sentence are deleted in a text revision, and the others are added to it. The sentence does not mention one sentence but two sentences if they are merged for comparison. Table 3 shows the example of the original Chinese sentence and the corresponding revised sentence, which consists of two sentences. We find that $r$ is deleted while $ude1, p, t$, and $ns$ are added by the revision. Hence, this pair of $Co_1$ and $Cr_1$ is summarized as their difference, $D(C_1) = \{r_d, ude1_a, p_a, t_a, ns_a\}$, where $C_1$ is a pair of $Co_1$ and $Cr_1$, and $\cdot_d$ and $\cdot_a$ represent delete and addition, respectively. In the case of $D(C_i) = \emptyset$, the original and revised sentences in the $i$th comparison sentence are the same in the sense of part-of-speech: $D(Co_i) = D(Cr_i)$, though they are not always the same in the sense of the words.

**Table 3.** Comparison of original and revised Chinese sentences

| | Part-of-speech |
|---|---|
| $Co_1$ | $t, vshi, ns, n, n, wd, r, vshi, n, v, n, wd, n, vi, ude1, n$ |
| Deleted | $r$ |
| $Cr_1$ | $t, vshi, ns, ude1, n, n, p, t, n, vi, wd, v, n, vshi, ns, n, ude1, n$ |
| Added | $ude1, p, t, ns$ |

The number of comparison sentences where a group of elements $X = \{x_i, ...\}$ is counted as $CD(Y, X) = \sum_i CD(Y_i, X)$, where $Y$ is a whole text and $Y_i$ is a sentence in $Y$. For instance, $CD(C_1, \{p_a\}) = 1$ and $CD(C_1, \{p_a, t_a\}) = 1$, where $C_1$ mentions a comparison sentence which is a pair of the original and revised sentences, $Co_1$ and $Cr_1$. Moreover, we define a ratio of their coincide, which is similar to the well-known index, confidence, in the association rule, as $c(Y, X, x_i) = CD(Y, X)/CD(Y, \{x_i\})$, where $x_i \in X$ is deleted or added.

**Table 4.** Original and revised texts in Chinese and German

| | Chinese | German |
|---|---|---|
| original | 中秋节是中国传统节日,那天是人们吃月饼,全家人聚会的日子.我是日本人,不过中国人的节日.我不回家见,我一年不在家.我和中国人过中秋节,很开心.中秋节那天,我给父亲打电话.我中午和朋友一起去了一家百货公司,我买衣服了.我主要去买月饼,我的朋友也买了很多月饼.中秋节放假三天.我辛苦劳累也要完成作业.我不想总是起床.我知道这是一个坏习惯,所以我想改变这个习惯.我每天早上起床学习.中秋节非常快.我以前没有吃过月饼,这是我第一次吃月饼,味道很香,很美味.它有一种陌生的味道,让我很想回家. | In diesem Text geht es um die Ausdehnung von dem digitalen Du in dem Internet. Früher waren.Außengrenzen des Privaten gut geschützt, doch heute dehnt die in dem Internet mit der Globalisierung aus. Mit dem Netz kam die digitale Nähe. Das besteht aus vielen winzigen Gruppen. Darin war man schnell beim Du, denn die Mitglieder die Fremde im Geiste waren. Damit greift das Du im Internet immer mehr um sich. Zum Beispiel duzen junge Firmen den Autor in E-Mails, als sie seinen Freund sein wollen. Sie lesen viel auf Englisch.Durch die Globalisierung schleift Gebrauch des you eine weitere Grenze. Noch ist es aber nicht so weit, dass das digitale Duzen überall gern gesehen ist. Ich denke dass in Japan die Internetwelt selbst stärker als die Globalisierung ist. Nämlich erweiterte die Grenze des Privaten sich in die ohne einen Siegeszug des you. Ich möchte über die zwei Punkte sagen.Erstens duzen Japaner in des Netz, als das die Anonymität und die Leichtigkeit hat. Man gebraucht in Japan die einige 2. Personen. Die stimmt nicht ganz mit dem deutschen Du oder dem deutchen Sie überein. Hier möchte aber ich die japanische private 2. Personen das Du für der Bequemlichkeit nennen.Wie sagt man diese japanische wörter kimi, omae, und anta auf Deutsch Die alle Antworten sind du. Für japanische Leute ist wohl unnatürlich einen Fremde zu duzen.In dem Internet duzen doch viele Japaner den Gesprächspartner, wenn auch er der Premierminister um die Wahrheit zu sagen war.Zwar siezen ihn andere viele Menschen, aber besteht in Twitter oderinAnschlagbrett vom Internet weit die Kultur.Die Anonymität und Leichtigkeit des Netz müsssen einen großen Einfluss ausüben. Zweitens benutzen in Japan viele Leute kaum Englisch. Meines Erachtens lesen sie keinen englischen Satz. Internatonale Menschen, schüler oder der intelektuelle Kreise sind die Ausnahmen. Als die folgende Gründen die Leute beeinflussen, brauchen sie extra nicht Englisch. Japanishe2.Personen unterscheiden ziemlich sich von dem englischen Person. Ich kann die Gemeinsamkeit finden. Und sind wahrsheinlich noch Chancen, dass sie Englisch sprechen, gering. In solchem Ort verbreitet das you nicht sich. Also ist nach meiner Ansicht die japanischen Internetwelt selbst stärker als Globalisierung. |
| revised | 中秋节是中国的传统节日.在中秋节全家团聚,吃月饼是中国人的习俗.我是日本人,不过中国的节日,不回家团聚.但是这个节日氛围让我想家.我离家已有一年之久.虽然无法回家与家人团聚,但是和中国人一起过节,我也很开心.中秋节当日,我与父亲通了电话,随后中午和朋友去了百货公司购物.主要是为了买月饼,还顺带买了衣服.朋友也买了很多月饼.中秋假期三天.我一如既往的睡了懒觉,我不愿意早起.我不管多累也要完成作业,以后每天早上早起学习.假期过得非常快,马上就要结束了.期间我第一次吃了月饼,月饼的味道很香,很美味.这种陌生的味道让我有些想家. | In dem Text wird uber die Medien in der Zeit des Nationalsozialismus geschrieben. Vor der Machtubernahme Adolf Hitlers durften Berichte noch relativ frei erstattet werden. Einige Zeitungen hatten eine bestimmte Richtung, aber andere hatten keine politische Richtung. In der Region Rostock gab es funf Zeitungen. Mecklenburgische Volkszeitung und Niederdeutsche Beobachter wurden bereits in einer politischen Richtung veroffentlicht. Dagegen war Rostocker Anzeiger eine Zeitung ohne politische Richtung. Nach der Machtubernahme veranderte sich die Presselandschaft. Aus der freien Berichterstattung wurde die Presse zur Propaganda. Die Verordnung vom 4.Februar 1933 griff in die Medien ein. Die Presse, die keine politische Richtung hatte, geriet in die Krise. Das Verlagsgebaude wurde geschlossen. Die Zeitungen wurden eingestellt, und die Maschienen sowie Materialien wurden konfiziert. Nach dem Reichstagsbrand gab es kein Presse- und Meinungsfreiheit mehr. Beispielsweise wurde die Presse in ihrer Freiheit beschrankt, was und wie in den Zeitungen zu berichten war. Nachdem der SA-Strumbannfuhrer Klaus Gundlach installiert wurde, standen die Printmedien, Radio und Kino vollig unter der Kontrolle. Nach dem Text halte ich die Medien fur unglaubhaft. Die Informationen aus den Medien konnten uns kontrollieren. Wenn wir eine Auswahl unparteiisch treffen, sollen wir keine Informationen aus den Medien bekommen. Wir mussen leben wie in fruheren Zeit. Wir brechen alle Mediumapparate. Wir werfen Handys weg, schneide PCs und bohren Fernsehen durch. Dann konnen wir primitiv leben. Ein primitives Leben bringt uns freiere Meinung. Es macht vielleicht auch unsere Gesellschaft friedlich, padagogisch und umweltlich besser. |

It is noted that when we consider a category instead of each part-of-speech, the sum of its elements does not always equal to the number of sentences. Assuming that $x_{1d}, x_{21d}$, and $x_{22d}$ are deleted in comparison sentence $Y_i$ in text $Y$, and $x_{21d}$ and $x_{22d}$ belong to the same category of part-of-speech, such as $vshi_d$ and $v_d$, respectively. When we consider a category $x_{2 \cdot d}$ as a sum, it happens $\sum_{i,j} CD(Y_i, \{x_{1d}, x_{2jd}\}) \geq 2$ since sentence $Y_i$ is counted twice, $CD(Y_i, \{x_{1d}, x_{21d}\}) = CD(Y_i, \{x_{1d}, x_{22d}\}) = 1$.

Assume there are $m$ parts-of-speech in a language, the number of deleted or added single element in $X$ is $2m$, and that with two elements, which are both added, both deleted, or deleted and added is $3m(m-1)/2$. The number of elements in $X$ can increase until corresponding to each sentence and $C(Y, X) = 1$. The small number of sentences satisfying $X$ makes the finding too particular. Hence, we consider one or two elements in this study. Furthermore, we analyze the comparison among categories and then focus on the parts-of-speech in a specific category to reduce possible combinations. The set of parts-of-speech $X$, especially from the viewpoint of $x_i \in X$, is worth analyzing, when numbers $CD(Y^k, X), \forall k$ and ratios $c(Y^k, X, x_i), \forall k$, where $k$ denotes a language, are large and high.

# 3 Example

## 3.1 Text Summary

Ten Japanese students wrote the Chinese texts in our experiment, and another ten Japanese students wrote German texts. Each group of students has studied Chinese or German for two to three years. In Chinese class, the students watched the short video about Mid-Autumn Festival, which is a harvest festival and holiday, and they had the assignment "Write how you spent the Mid-Autumn Festival holiday,". In German class, the students read an article on the media's influence, and they had the assignment "Write your opinion about the media dependence of Japanese youth". Then, the tangible errors in these texts are corrected beforehand, and they are used as the original texts in the following.

Each of the original Chinese texts consists of about 11 sentences, and a sentence consists of approximately 15 words. The original German text has an average of 25 sentences, each of which has approximately 15 words. The Chinese or German native speakers revised the given original texts to sound natural for him/her. Table 4 shows examples of the original and revised texts in Chinese and German. The numbers of sentences in the original and revised Chinese texts are 15 and 14, respectively. Following the merge algorithm in Sect. 2.1, the original and revised sentences are merged 3 and 2 times, respectively. The number of comparison sentences becomes 12. Similarly, in the German texts, the number of comparison sentences is 24 by merging the 21st and 22nd sentences in the original one. We have nine more pairs of texts in each language. The original and revised Chinese texts consist of 113 sentences with 1764 words and 106 sentences with 1501 words, respectively, and 96 comparison sentences. The original and revised texts in German consist of 213 sentences with 2849 words and 210 sentences with 2863 words and 200 comparison sentences.

**Table 5.** Summary of texts

| | | Number | | Ratio | | Number | | Ratio | |
|---|---|---|---|---|---|---|---|---|---|
| | | Revised | Original | Used | Used | Added | Deleted | Added | Deleted |
| German | Adjective | 237 | 248 | 0.08 | 0.09 | 29 | 40 | 0.12 | 0.16 |
| | Adverb | 99 | 95 | 0.03 | 0.03 | 27 | 23 | 0.27 | 0.24 |
| | Conjunction | 175 | 182 | 0.06 | 0.06 | 28 | 35 | 0.16 | 0.19 |
| | Noun | 825 | 829 | 0.29 | 0.29 | 55 | 59 | 0.07 | 0.07 |
| | Preposition | 288 | 270 | 0.10 | 0.09 | 59 | 41 | 0.20 | 0.15 |
| | Pronoun | 265 | 249 | 0.11 | 0.11 | 73 | 57 | 0.28 | 0.23 |
| | Verb | 446 | 451 | 0.16 | 0.16 | 83 | 88 | 0.19 | 0.20 |
| | others | 470 | 466 | – | – | 77 | 73 | – | – |
| Chinese | Adjective | 76 | 66 | 0.03 | 0.02 | 22 | 12 | 0.29 | 0.18 |
| | Adverb | 154 | 145 | 0.05 | 0.05 | 45 | 36 | 0.29 | 0.25 |
| | Conjunction | 63 | 48 | 0.02 | 0.02 | 31 | 16 | 0.49 | 0.33 |
| | Noun | 409 | 390 | 0.14 | 0.14 | 68 | 48 | 0.17 | 0.12 |
| | Preposition | 49 | 34 | 0.02 | 0.01 | 24 | 9 | 0.49 | 0.26 |
| | Pronoun | 166 | 162 | 0.06 | 0.06 | 52 | 48 | 0.31 | 0.30 |
| | Verb | 407 | 360 | 0.14 | 0.13 | 120 | 73 | 0.29 | 0.20 |
| | Others | 440 | 296 | – | – | 230 | 82 | – | – |

The Chinese and German original and revised texts are summarized in Table 5 on seven common categories. The left half shows the numbers of parts-of-speech used in the original and revised texts and their ratios. The right half shows the added and deleted ones in the native speaker's revision and their ratios to the used ones. It is no wonder that nouns and verbs are the two most frequently used. The used ratios of adjectives and prepositions in German are more than those in Chinese, which may characterize each language.

### 3.2 Discussion

We know that a fact or event is expressed in various ways, and a verb plays a vital role in deciding the structure of the sentence. Therefore, first, we focus on verbs. In Table 5, verbs account for approximately 15% of used words, and the revised ratios to the used ones are 20% and more. The details are shown in the left half of Table 6, where the most to least used verbs are lined from top to bottom, and the numbers of the deleted and added verbs are shown. All $VVIZU$ and $vl$, which are scarcely used in the original and revised texts, are all deleted and added. The students avoided using these verbs and corresponding sentence structures, though the native speakers found a few sentences that are preferably suitable. In the right half of Table 6, the numbers of significant pairs of delete and addition of verbs are shown. The upper subscript of a number, $^a$ or $^d$, represents the pair's ratio to a verb at the row. Similarly, the lower subscript of a number, $_a$ or $_d$, represents the pair's ratio to a verb at the column. For instance, $CD_d$ in the case of $c(Y, \{x_{id}, x_{ja}\}, x_{id}) \geq 0.25$, and $CD^a$ in the case of $c(Y, \{x_{id}, x_j^a\}, x_{ia}) \geq 0.25$. For instance, in Table 6, $VMFIN$ is added only in 7 sentences. So, all the ratios to it are more than 0.25, and the numbers in its column have the upper subscripts $^a$. The most used verbs, $VVFIN$ and $v$, which are simple and basic verbs, are added more than deleted. In German, adding $VVFIN$ coincides with the deleting $VAFIN$ and $VVPP$, respectively, in 7 sentences. Furthermore, $VVPP$ and $VAFIN$ are simultaneously deleted in 5 sentences so that $VVFIN$ seems to replace these two kinds of verbs. Besides, they both are added in 5 sentences. They are used together for the past perfect, so this pair is reasonable in grammar. From the viewpoint of $VVPP$, it is used in 114 sentences and 30% less frequent than $VVINF$, while it is added and deleted in 29 sentences and more than $VVINF$. The ratio of the pairs of simultaneous addition or delete of $VVPP$ and $VAFIN$ to $VVPP$ is $c(G, \{VAFIN_d, VVPP_d\}, VVPP_d) = 0.33$ or $c(G, \{VAFIN_a, VVPP_a\}, VVPP_a) = 0.29$, i.e., 30% of the revisions of $VVPP$ may be caused by past perfect. The students have learned the usage of past perfect, though they have difficulty to manage it properly. In Chinese, the additions of $v$ coincide with the delete of the sophisticated verbs $vi$ and $vshi$ in 10 and 5 sentences, respectively. In adding $v$, more than two $v$ are simultaneously added in 12 sentences so that two $v$ are the replacement of $vi$ or $vshi$ by modifying the sentence structures. Students chose the verbs other than basic ones, $VVFIN$ and $v$, for a rich express and variation, though the native speaker felt that basic verbs were more suitable and covered a more comprehensive range. It might be because, for the native speaker, such a sentence was too redundant or complicated for the context. It is difficult but essential for a student to extend the expression range of the basic verbs, which s/he is already familiar with to some extent.

**Table 6.** Numbers of sentences regarding to verbs

| | Add | Delete | Sum | Used |
|---|---|---|---|---|
| VVFIN | 21 | 17 | 38 | 272 |
| VAFIN | 12 | 21 | 33 | 228 |
| VVINF | 14 | 13 | 27 | 175 |
| VVPP | 14 | 15 | 29 | 114 |
| VMFIN | 7 | 11 | 18 | 76 |
| VAINF | 3 | 3 | 6 | 22 |
| VVIZU | 3 | 1 | 4 | 4 |
| VAPP | 0 | 0 | 0 | 2 |
| VMINF | 0 | 0 | 0 | 2 |
| VVIMP | 1 | 1 | 2 | 2 |

| | Add | | | | | Delete |
|---|---|---|---|---|---|---|
| Delete | VVFIN | VAFIN | VVINF | VVPP | VMFIN | VAFIN |
| VVFIN | 0 | $4^a$ | $6^a_d$ | $4^a$ | $2^a$ | 1 |
| VAFIN | $7^a_d$ | 0 | 1 | 3 | $5^a$ | 0 |
| VVINF | 3 | 1 | 0 | 1 | 0 | 3 |
| VVPP | $7^a_d$ | 0 | 1 | 0 | $2^a$ | $5_d$ |
| VMFIN | 2 | 2 | 1 | 0 | 0 | 2 |
| Add | | | | | | |
| VAFIN | $4_a$ | 0 | 0 | $5^a$ | – | |

| | Add | Delete | Sum | Used |
|---|---|---|---|---|
| v | 40 | 16 | 56 | 449 |
| vi | 15 | 18 | 33 | 105 |
| vshi | 16 | 11 | 27 | 87 |
| vf | 11 | 8 | 19 | 59 |
| vyou | 6 | 5 | 11 | 47 |
| vn | 2 | 2 | 4 | 14 |
| vl | 5 | 1 | 6 | 6 |

| | Add | | | |
|---|---|---|---|---|
| Delete | v | vi | vshi | vf |
| v | 0 | $4^a$ | 4 | 1 |
| vi | $10_d$ | 0 | 3 | 1 |
| vshi | $5_d$ | $3_d$ | 0 | 0 |
| vf | 2 | 2 | 1 | 0 |
| Add | | | | |
| v | $12^a$ | $6^a$ | $4^a$ | $7^a$ |

Next, we choose pronouns since from Table 5, around 30% of them used in the original and revised texts are deleted and added, respectively. Since nouns and verbs are the most frequently used, Table 7 shows the pairs regarding pronoun except them. In both languages, more than 30% of the deleted and added pronouns seem to be the replacements to and from the other pronouns: $c(C, \{x^p_d, x^p_a\}, x^p_d) = 0.36, c(C, \{x^p_d, x^p_a\}, x^p_a) = 0.32, c(G, \{x^p_d, x^p_a\}, x^p_d) = 0.45, c(G, \{x^p_d, x^p_a\}, x^p_a) = 0.36$, where $x^p$ represents a pronoun category. The replacement may be because of a wrong choice or a result of the modification of sentence structure. Above all, pronouns in Japanese thinking are not similar to those in both languages. Hence, the students need to be careful with choosing pronouns, and the revisers need to understand that the exchange of pronouns is not easy and straightforward for the students to understand.

In Table 7, subscript * is added when the ratio of pronoun to the delete and addition of the others is more than 0.6: $c(Y, \{x^p, x^i\}, x^i) \geq 0.6$, where $x^i$ is part-of-speech other than pronoun category. In both languages, the ratios of the pairs of pronoun delete and adverb delete and pronoun delete and conjunction delete are high. In the following, we consider details from the viewpoints of adverb and conjunction.

In Table 5, the deleted and added adverb ratios are around 25% high, and an adverb seems to be preferably added than deleted. An adverb is not always necessary for a context, though it enriches the expression and helps communication. Understandably, the students cannot afford to use such words in adverbs, and the native speaker feels such texts too flat. It is a good option for the students to add adverbs to make a sentence natural, on the one hand. On the other hand, the delete of adverb coincides with the delete of pronoun; it needs to be careful to use an adverb if it is likely to be used with a pronoun. In addition to both pronoun and adverb delete, in both languages, they are often added simultaneously. In German, their simultaneous addition is in 17 sentences with $c(G, \{pron._a, adv._a\}, adv._a) = 0.71$, and in Chinese, it happens in 18 sentences with $c(C, \{pron._a, adv._a\}, adv._a) = 0.55$. Hence, there could be a relation between pronoun and adverb, so that in general, pronoun and adverb in a sentence give a chance to reconsider the other ways of expression.

The conjunction's role is to link two items, so the sentence with it tends to be complicated in logic, and it needs to be careful how they connect. In Table 5, the ratios of the revisions regarding conjunctions in Chinese are almost twice as high as in German. Moreover, it seems that in Chinese, conjunction tends to be added, while in German, it tends to be deleted. The original Chinese texts seem to be divided into too small pieces, and, on the contrary, the original German texts are overly descriptive and redundant. In this way, the evaluation of conjunctions in the texts in both languages is different. It is natural to explain something logically by sentences with proper lengths in a mother-tongue due to a match of thinking and language. However, it becomes difficult in a foreign language, which seems to appear in conjunction. The pronoun and conjunction tend to be added and deleted simultaneously from both the numbers of sentences and the conjunction ratios. It is noted that in German, the ration is less than 0.6, which means that the addition of conjunction is accompanied by the addition of pronouns and the addition or delete of the others. In terms of reducing linking items with conjunctions, pronouns play an essential role so that they need to be carefully considered. When the students find a sentence with a pronoun and conjunction in their drafts, it is worth reconsidering the structure to be a more natural expression for a native speaker.

**Table 7.** Numbers of sentences regarding to pronouns

| Add | Add | | | | | Delete | | | | |
|---|---|---|---|---|---|---|---|---|---|---|
| Pronoun | Adj. | Adv. | Prep. | Conj. | Pron. | Adj. | Adv. | Prep. | Conj. | Pron. |
| Ge. 69 | 9 | 17* | 29 | 14 | 13 | 18 | 7 | 14 | 17 | 25 |
| Ch. 47 | 12* | 18 | 11 | 15* | 8 | 8* | 15* | 6* | 9* | 15 |
| Del | Add | | | | | Delete | | | | |
| Pronoun | Adj. | Adv. | Prep. | Conj. | Pron. | Adj. | Adv. | Prep. | Conj. | Pron. |
| Ge. 55 | 16 | 10 | 17 | 18* | 25 | 13 | 14* | 20 | 19* | 13 |
| Ch. 22 | 9 | 12 | 9 | 9 | 15 | 9* | 15* | 8* | 10* | 7 |

# 4 Conclusion

This paper proposed the approach to find out the difference between texts by non-native and native speakers. A group of Japanese students wrote the texts in German or Chinese. The native speakers revised some sentences when they feel something unusual, and they gave better possible options to be more natural expressions. We used two foreign languages to reduces the influence of uniqueness to a language and replaced words to parts-of-speech to compare different languages. Then, we summarized the difference between the original texts and their revisions by comparing sentence by sentence. A comparison sentence consists of one or two sentences so that two sentences are merged if merging improves the similarity of the sentences in the original and revised texts. The findings are about replacing verbs, pronouns, and the relations of pronouns to adverbs and conjunction.

# References


1. NLPIR Chinese lexical analysis system. http://ictclas.nlpir.org/index_e.html
2. TenTen corpus family. https://www.sketchengine.eu/documentation/tenten-corpora/
3. TreeTagger - a part-of-speech tagger for many languages. https://www.cis.uni-muenchen.de/~schmid/tools/TreeTagger/
4. Graf, T., Philipp, M., Xu, X., Kretzschmar, F., Primus, B.: The interaction between telicity and agentivity: experimental evidence from intransitive verbs in German and Chinese. Lingua **200**, 84–106 (2017)
5. Halliday, M.A.K., Matthiessen, C.M.I.M.: An Introduction to Functional Grammar. Hodder Arnold, London (2004)
6. He, B., et al.: Building a comprehensive syntactic and semantic corpus of Chinese clinical texts. J. Biomed. Inform. **69**, 203–217 (2017)
7. Hsieh, S.C.: A corpus-based study on animal expressions in Mandarin Chinese and German. J. Pragmat. **38**(12), 2206–2222 (2006)
8. Isobe, M.: Wie tolerant muss das Korrekturlesen sein? Zur systematischen Korrektur im deutschen Sprachunterricht. NU-Ideas **6**, 45–51 (2017)
9. Manuel, K.: Quantitative und qualitative Fehleranalyse Japanischer Deutschlerner (JDL) bei Aufsatzübungen in deutscher Sprache mit Schwerpunkt auf dem Artikelgebrauch. Cult. Rev. Waseda Commercial Stud. Assoc. **56**, 55–81 (2019)
10. Mikolov, T., Corrado, G., Chen, K., Dean, J.: arXiv:1301.3781v3 [cs.CL], pp. 1–12 (2013)
11. Schmid, H.: Improvements in part-of-speech tagging with an application to German. In: Proceedings of the ACL SIGDAT-Workshop, pp. 1–9 (1995)
12. Schuppler, B., Hagmüller, M., Zahrer, A.: A corpus of read and conversational Austrian German. Speech Commun. **94**, 62–74 (2017)

# Motion Time Study with Convolutional Neural Network

Jirasak Ji[1], Warut Pannakkong[1(✉)], Pham Duc Tai[2], Chawalit Jeenanunta[1], and Jirachai Buddhakulsomsiri[1]

[1] Sirindhorn International Institute of Technology, Thammasat University, Pathum Thani, Thailand
warut@siit.tu.ac.th
[2] King Mongkut's Institute of Technology Ladkrabang, Bangkok, Thailand

**Abstract.** Manufacturing and service industries use motion time study to determine work element time and standard time for production planning and process improvement. Traditional time study is performed by human analysts with stopwatches, and therefore, is subject to uncertainties from human errors. This study proposes an automated time study model featuring a convolutional neural network. The trained model can analyze a video footage of an operation to time the work elements. The timing data are compared with reference values using a statistical analysis. The result shows effectiveness of the model in accurately and consistently estimating the work element times and standard time of the operation.

**Keywords:** Motion time study · Convolutional neural network

## 1 Introduction

There are three major types of uncertainty that arise in manufacturing contexts and that can affect a production plan [1]: uncertainty in demand forecast, external supply process, and internal supply process. Internal supply process is further comprised of internal manufacturing, transportation and supply processes. With more accurate timing of these processes within one's organization, the uncertainty in internal supply process can be reduced, thus allowing for a more accurate production planning. Time study is a technique that can accurately time the process and reduce the uncertainty of the internal supply process.

Time study is an important step in production planning of every industries. Although many work processes are being automated in modern times, there are still many manual work processes, especially in developing countries. The time of each processes in a production should be known for operational planning and resource management. The process time can be found by finding the difference in time between events. These events are defined either from the operator's motion (i.e., grabbing, holding, or releasing an object) or the shape of an object that undergoes change. The traditional method of time study requires a person to measure and record the time of each process manually, which causes uncertainty. This work is redundant and prone to errors. The errors can be caused


V.-N. Huynh et al. (Eds.): IUKM 2020, LNAI 12482, pp. 249–258, 2020.
https://doi.org/10.1007/978-3-030-62509-2_21

by boredom of repetitive timing, disturbance from other factors, and the need to sample a small number of processes due to the limited working time of the worker timing the processes. As it is economically impossible to time every process, workers are advised to sample a small number of processes for timing, which can cause a large amount of bias that does not represent the true behavior of the data. For example, an operator can work faster when they are being timed, then works slower after the timing is done. Uncertain process time data lead to uncertainty in production planning, as management cannot accurately predict the expected output.

However, this mundane process can be automated with computers. Wang et al. [2] reviewed various papers that studied sensors in activity recognition. They explained that traditional activity recognition methods heavily relied on heuristic hand-crafted feature extraction, which could hinder their generalization performance. Many papers have succeeded in activity recognition with motion sensors. However, sensors are considered to be invasive to operators and can hinder their performance. We have decided to use cameras with computer vision algorithms instead as they are non-invasive, can detect objects from longer range and with finer granularity. In this research, we try to answer the question how effective computer vision is in identifying and timing the working processes.

Computer vision is a field of computer science that enables computers to identify and process objects in images and videos. As the number of available data and computational power of machines exponentially increase in the recent years, computer vision algorithms are able to detect objects with promising accuracy and performance. Many companies in both the service and industrial sectors are utilizing computer vision to support and improve their processes. To the best of our knowledge, no studies have used computer vision in motion time study and, therefore, will be discussed in this paper. The computer vision algorithm that will be used in this study is Convolutional Neural Networks (CNN) [3]. In recent years, CNN has made significant breakthroughs in the field of computer vision and has proven to be effective in object detection. Many real-world applications such as self-driving cars and surveillance use CNN to identify objects. However, CNN also possess a risk in falsely identifying objects and motions. The CNN model should be sufficiently trained with variant objects and motions to ensure an acceptable accuracy. In this study, we will use CNN to identify the motion and time the work processes in the industrial sector.

## 2 Related Works

There are many papers that study computer vision and its application in industries. The most popular applications are found to be quality control and measurement. An example of computer vision's application for quality control can be found in the study of Kazemian et al. [4]. The authors develop a real-time computer-based vision system to monitor and control the quality of an extrusion process in construction industry. By using image blurring and binarization for contour extraction, the vision system produces a reliable feedback, based on which, the extrusion rate is automatically adjusted to maintain the quality of the extrusion process. In another research, Frustaci et al. [5] presents an embedded computer vision system for an in-line quality check of assembly processes. Their system can detect possible geometrical defects, which are planar and/or rotational shifts by using a combinations of image processing techniques such as Region of Interest and binarization. An application of computer vision for measurement can be found in the study of Gadelmawla [6], in which the author develops an algorithm to classify many types of screws and threads according to extracted features from their images.

In recent years, the research of computer vision has been rapidly expanded with the advancement of Artificial Neural Networks (ANN). By harnessing a vast amount of data, an ANN model can generalize and learn how to classify images more quickly and accurately. The application of ANN in the field of computer vision is demonstrated in the study of Arakeri and Lakshmana [7]. A computer vision-based fruit grading system is developed to evaluate tomatoes. Features from images of tomatoes are extracted and input into an ANN model to determine if a tomato is defective, ripe, or unripe. The system is capable of processing 300 tomatoes per hours. The throughput of this system was further improved by Costa et al. [8]. In specific, the processing speed is 172,800 tomatoes per hour, which is equivalent to 48 tomatoes per second. The significant improvement is attributed to the implementation of a Residual Neural Network (ResNet), a variant of CNN. Comparatively, CNN is similar to traditional ANN in an aspect that they contain neurons that self-optimize through learning [3]. The only notable difference between CNN and the traditional ANN is that CNN is primarily used for pattern recognition within images. A CNN model allows one to capture the spatial features from images and encode image-specific features into the architecture. This make the model more suitable for image-focused tasks, while further reducing the parameters required to set it up. In a different context, Raymond et al. [9] design a model to manage all assets on electrical grid such as supports, insulators, transformers, and switches by using Faster Region-based Convolutional Neural Networks (Faster R-CNN), another variant of CNN. The authors deploy drones to take pictures, from which the numbers of assets are counted by their Faster R-CNN model to generate inventory report. Shi et al. [10] also use Faster R-CNN to detect hydrocarbon leakage in a Ethane cracker plant. Kumar et al. [11] use CNN to identify defects in centrifugal pumps. Watanabe et al. [12] use CNN to monitor and analyze failure of power devices. Yang et al. [13] use CNN to detect the laser welding defects of safety vent on power battery. These studies show that CNN is gaining popularity in industrial application. However, most of the studies use CNN to detect object-based events, such as defects. No study has proposed using CNN to detect action-based events in the industries. As a result, it is reasonable to use CNN in this study and contribute our knowledge to the industry.

## 3 Methodology

A production/manufacturing process contains several working elements, each of which consists a series of motions. Therefore, to determine the duration of working element, a typical time-study analyst/operator must observe and identify which motions indicate the starting and ending of an element. To enable a computer to replicate this timing process through a video feed, a computer vision-based model, is developed. This model is made up of two models, which are CNN and time study model. The purpose of CNN model is to determine the element, to which a motion in a frame of the video feed belongs. The CNN model is similar to that of Goodfellow et al. [14], while the time study model is developed to analyze the element detected by the CNN model. Since our CNN model considers each frame of a video as independent from the others, this results in a difficulty to detect if there is a change from an ending motion of an element to a starting motion of the next element. As a result, time duration of a working element cannot be accurately recorded. To overcome this difficulty, rolling-prediction averaging [15] is integrated with the time study model to enhance the detection of the starting and ending motions in a working element such that its time duration is precisely captured.

The development of our vision-based model consisted of two steps. The first step is to train the CNN model to recognize motion of a working element from a video feed. The second step is to construct the time study model with rolling-prediction averaging to determine the time duration of each working element based on the output of the CNN model. The procedure for training the model is shown in Fig. 1. A video of a process, which contains several working elements, is collected by setting up a camera to capture all of the motions in that process. Once the video is obtained, it is converted into individual frames, each of which is equivalent to an image. These images are later labelled according to the working elements of the process. There are 6 label classes in our model: 5 working elements and 1 idle element. After the labelling is finished, the images are split into training and testing sets. 80% of the images are randomly assigned to the training set, while the remaining 20% of them are assigned to the testing set for evaluation purpose. The model is then trained until classification accuracy stagnates or overfitting starts to occur.

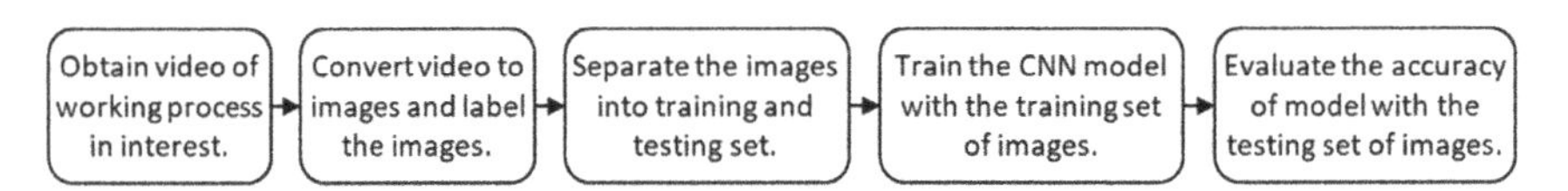


**Fig. 1.** Procedure of training our CNN model.

After the CNN model is trained, time study model is constructed and integrated with the trained CNN model. The integration is illustrated in Fig. 2. Firstly, each frame of a video is fed to the trained CNN model to obtain a prediction of the element based on the motion observed in the frame. Secondly, the prediction is added to a list of the last K predictions. The average of the last K predictions is then computed and the label with the largest corresponding probability is selected as the classification for the current frame. Finally, the prediction of the current frame obtained from the rolling-prediction model is evaluated against that of the previous frame. If the predictions are the same, we consider that the element is still on progress and do nothing with the timing. Otherwise, it is regarded that the previous element has ended, and the next element has started. Based on the comparison between the current and previous predictions, the timer is activated accordingly. Particularly, timing for an element is started at its beginning and is stopped at the end of that element. The time duration of each element is recorded into a data file for further analysis. All the frames from the video are processed individually until the end of the video has been reached.

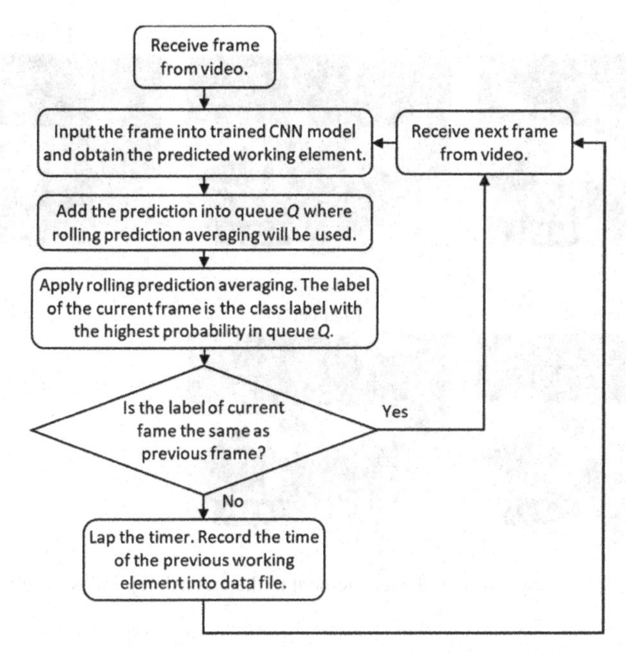


**Fig. 2.** Procedure in detecting and timing each working element in a working process.

## 4 Experiment

In this section, the effectiveness of our model is evaluated through an extension joint assembly process (see Fig. 3). A video clip of this process is used for analysis. The video was originally made to teach time study to students in our institution. The process contains 5 elements and the description of each element is as follows:

1. Getting the joint from the bin and placing it in the center of workspace
2. Getting the rubber ring from the bin and assembling it with the joint
3. Getting the plastic core from the bin and assembling it with the joint
4. Getting the joint cover from the bin and assembling it with the joint by running it down through the thread
5. Disposing the extension joint into the bin.

The video contains 20 working cycles. After converting this video into images (or individual frames), 1,014 images are obtained and sequentially analyzed by our model. The result from the algorithm is compared with the reference timing done by us, where we timed the processes in slow motion to obtain the time as accurately as possible. Ideally, the timing from the algorithm should be the same as the reference timing. A statistical test is then performed to check if there are any significant difference between the time from our proposed algorithm and the reference time.

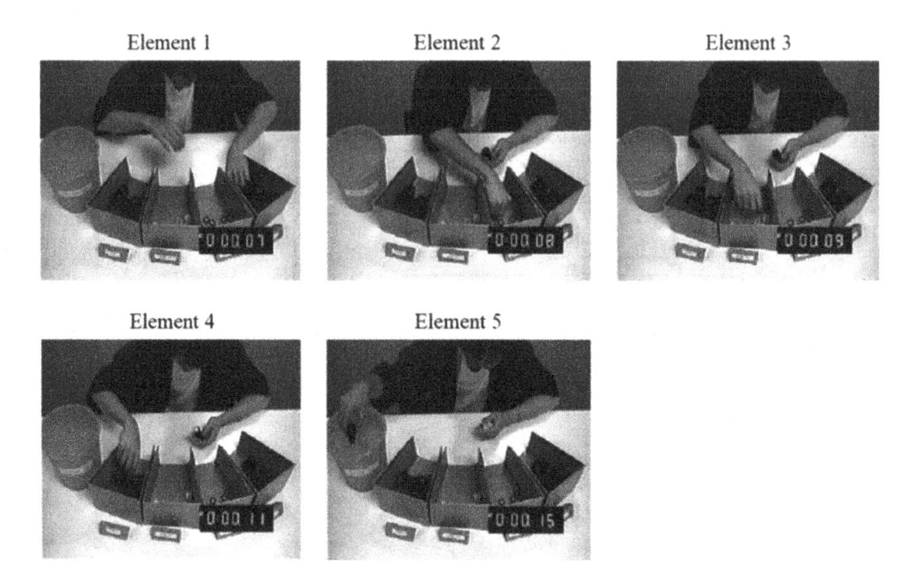


**Fig. 3.** Montage of a sample frame of each element in the extension joint assembly process video clip.

## 5 Results

We timed the process from the video clip with our proposed algorithm. The video clip of the working process is input into our algorithm. The outputs of the algorithm are the same video clip with additional annotations for evaluation to check if the algorithm misclassifies any element (see Fig. 4) and a .csv file that contains the element time of each working cycle (Table 1). The annotations in the output video file are:

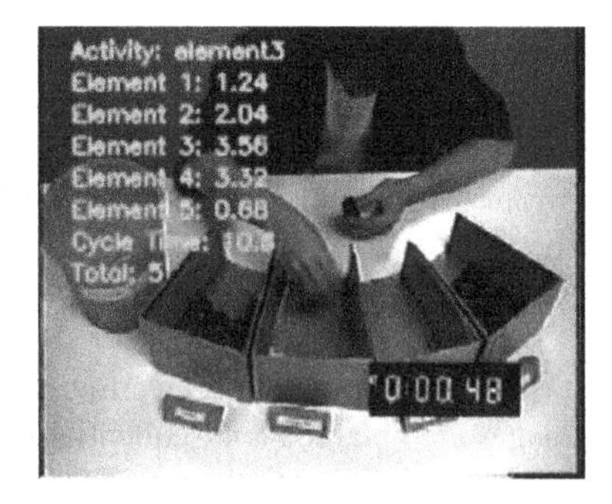


**Fig. 4.** A sample frame of the video clip output from our algorithm.

1. The active element in the current frame
2. Previous cycle's element 1 time
3. Previous cycle's element 2 time

4. Previous cycle's element 3 time
5. Previous cycle's element 4 time
6. Previous cycle's element 5 time
7. Previous cycle's cycle time
8. Total working cycles

**Table 1.** Working element time (s) of each cycle obtained from our algorithm.

| Cycle | Element 1 | Element 2 | Element 3 | Element 4 | Element 5 | Cycle time |
|---|---|---|---|---|---|---|
| 1 | 0.56 | 2.24 | 1.84 | 3.48 | 0.76 | 8.88 |
| 2 | 0.48 | 1.96 | 1.2 | 4.08 | 0.84 | 8.56 |
| 3 | 1.04 | 0.4 | 1.8 | 3.76 | 0.88 | 7.88 |
| 4 | 0.76 | 2.2 | 1.96 | 4 | 0.76 | 9.68 |
| 5 | 1.2 | 2.04 | 3.56 | 3.32 | 0.68 | 10.8 |
| 6 | 1.24 | 2.04 | 1.6 | 3.6 | 0.76 | 9.24 |
| 7 | 1.08 | 2.2 | 1.52 | 4 | 0.8 | 9.6 |
| 8 | 0.84 | 2.96 | 1.76 | 4.44 | 0.64 | 10.64 |
| 9 | 1.04 | 2.44 | 1.6 | 3.92 | 0.6 | 9.6 |
| 10 | 0.92 | 2.44 | 1.88 | 4.6 | 0.6 | 10.44 |
| 11 | 0.88 | 2.16 | 1.64 | 3.56 | 0.68 | 8.92 |
| 12 | 0.88 | 2.2 | 1.48 | 3.72 | 0.72 | 9 |
| 13 | 1.32 | 2.8 | 1.44 | 3.92 | 0.76 | 10.24 |
| 14 | 0.84 | 2.24 | 1.92 | 4.44 | 0.64 | 10.08 |
| 15 | 1 | 2.36 | 1.8 | 5.76 | 0.8 | 11.72 |
| 16 | 1.12 | 2.32 | 1.6 | 4.28 | 0.56 | 9.88 |
| 17 | 1.08 | 2.16 | 1.84 | 4.44 | 0.68 | 10.2 |
| 18 | 1.12 | 2.08 | 1.92 | 4.32 | 0.64 | 10.08 |
| 19 | 0.96 | 2.28 | 1.68 | 4.04 | 0.68 | 9.64 |
| 20 | 1.12 | 2.72 | 2 | 5.44 | 0.84 | 12.12 |
| Avg. | 0.97 | 2.21 | 1.80 | 4.16 | 0.72 | 9.86 |
| S.D. | 0.21 | 0.49 | 0.45 | 0.59 | 0.09 | 1.01 |

From Table 1, we can see that there is an abnormality in Element 2 of Cycle 3. The time is too short compared to other cycles, so we checked the annotation in the output video file. It is realized that the algorithm misclassifies Element 2, invaliding the timing at that instance. Therefore, Cycle 3 should be excluded from the analysis as other elements in Cycle 3 that Element 2 had misclassified as is also invalid.

The results from our proposed model are compared with that from the reference element time dataset. The reference element and cycle times are shown in Table 2. A statistical test, namely Analysis of Variance (ANOVA), is performed to determine whether the timing results from our proposed algorithm is significantly different from the reference time. The result of the test is shown in Table 3.

**Table 2.** Actual element and cycle times.

| Cycle | Element 1 | Element 2 | Element 3 | Element 4 | Element 5 | Cycle time |
|---|---|---|---|---|---|---|
| 1 | 0.73 | 1.87 | 2.03 | 3.37 | 0.77 | 8.77 |
| 2 | 0.76 | 1.7 | 1.74 | 4.1 | 0.86 | 9.16 |
| 3 | 1.27 | 2.1 | 1.9 | 3.77 | 0.83 | 9.87 |
| 4 | 0.73 | 2.14 | 2.13 | 3.93 | 0.67 | 9.6 |
| 5 | 1.63 | 1.64 | 1.7 | 3.4 | 0.8 | 9.17 |
| 6 | 1.23 | 1.97 | 1.8 | 3.63 | 0.83 | 9.46 |
| 7 | 1.14 | 2.06 | 1.6 | 4.04 | 0.73 | 9.57 |
| 8 | 1.1 | 2.93 | 1.7 | 4.64 | 0.56 | 10.93 |
| 9 | 1.1 | 2.5 | 1.64 | 4 | 0.5 | 9.74 |
| 10 | 1.2 | 2.13 | 2.23 | 4.54 | 0.5 | 10.6 |
| 11 | 1.06 | 2.07 | 1.77 | 3.6 | 0.6 | 9.1 |
| 12 | 1 | 2.23 | 1.63 | 3.74 | 0.7 | 9.3 |
| 13 | 1.43 | 2.73 | 1.57 | 3.9 | 0.7 | 10.33 |
| 14 | 1.03 | 2.2 | 2.04 | 4.53 | 0.6 | 10.4 |
| 15 | 1.07 | 2.3 | 2.06 | 5.84 | 0.56 | 11.83 |
| 16 | 1.27 | 2.3 | 1.73 | 4.27 | 0.6 | 10.17 |
| 17 | 1.13 | 2.17 | 1.93 | 4.5 | 0.74 | 10.47 |
| 18 | 1.13 | 2 | 2.1 | 4.33 | 0.54 | 10.1 |
| 19 | 1.13 | 2.23 | 1.8 | 4.1 | 0.6 | 9.86 |
| 20 | 1.27 | 2.67 | 2.1 | 5.5 | 0.53 | 12.07 |
| Avg | 1.12 | 2.20 | 1.86 | 4.19 | 0.66 | 10.03 |
| SD | 0.21 | 0.31 | 0.20 | 0.62 | 0.11 | 0.85 |

From Table 3, there is no significant difference between our proposed algorithm and the reference time (p-value = 0.519), which indicates the algorithm's effectiveness in measuring the element times. In addition, there is no interaction between the timing method and work elements, which means that the algorithm's effectiveness is the same for all work elements.

**Table 3.** ANOVA test results.

| Source | DF | Adj SS | Adj MS | F-value | P-value |
|---|---|---|---|---|---|
| Method | 1 | 0.054 | 0.0544 | 0.42 | 0.519 |
| Element | 4 | 296.472 | 74.1181 | 569.76 | 0.000 |
| Cycle | 19 | 6.066 | 0.3192 | 2.45 | 0.001 |
| Method*Element | 4 | 0.236 | 0.0589 | 0.45 | 0.770 |
| Error | 171 | 22.245 | 0.1301 | | |
| Total | 199 | 325.073 | | | |

## 6 Conclusion and Future Work

In this study, we have proposed a computer vision-based model using CNN to detect motion and time the working elements from a work process video. The study has proved that using computer vision can result in an accurate and consistent timing of work processes. Based on our model, if a new operator is to perform the operations, the model might not be able to detect the elements correctly. Therefore, if the algorithm is to be applied, the model should be further trained with more variant dataset. Another limitation is when the operator performs an unordinary motion, the current algorithm might not be able to identify the motion and incorrectly label it as an "idle" motion. As a future work, we will design methods in dealing with potential errors caused by computer vision. We will increase complexity in the working process, such as increasing the number of operators and motions. Incorrect or unrelated motions will also be purposely introduced and tested if our algorithm can accurately capture and label these motions. The potential of using computer vision in industries are countless, and motion time study is just one of them.

## References

1. Graves, S.: Uncertainty and production planning. In: Planning Production and Inventories in the Extended Enterprise, pp. 83–101 (2011)
2. Wang, J., Chen, Y., Hao, S., et al.: Deep learning for sensor-based activity recognition: a survey. Pattern Recogn. Lett. **119**, 3–11 (2019). https://doi.org/10.1016/j.patrec.2018.02.010
3. O'Shea, K., Nash, R.: An introduction to convolutional neural networks (2015)
4. Kazemian, A., Yuan, X., Davtalab, O., Khoshnevis, B.: Automation in construction computer vision for real-time extrusion quality monitoring and control in robotic construction. Autom. Constr. **101**, 92–98 (2019). https://doi.org/10.1016/j.autcon.2019.01.022
5. Frustaci, F., Perri, S., Cocorullo, G., Corsonello, P.: An embedded machine vision system for an in-line quality check of assembly processes. Procedia Manuf. **42**, 211–218 (2020). https://doi.org/10.1016/j.promfg.2020.02.072
6. Gadelmawla, E.S.: Computer vision algorithms for measurement and inspection of external screw threads. Measurement **100**, 36–49 (2017). https://doi.org/10.1016/j.measurement.2016.12.034

7. Arakeri, M.P., Lakshmana: Computer vision based fruit grading system for quality evaluation of tomato in agriculture industry. Procedia – Comput. Sci. **79**, 426–433 (2016). https://doi.org/10.1016/j.procs.2016.03.055
8. Costa, A., Figueroa, H.E.H., Fracarolli, J.A.: Computer vision based detection of external defects on tomatoes using deep learning. Biosyst. Eng. **190**, 131–144 (2019). https://doi.org/10.1016/j.biosystemseng.2019.12.003
9. Raymond, J., Didier, K., Kre, M., et al.: Assets management on electrical grid using Faster-RCNN. Ann. Oper. Res. (2020). https://doi.org/10.1007/s10479-020-03650-4
10. Shi, J., Chang, Y., Xu, C., et al.: Real-time leak detection using an infrared camera and Faster R-CNN technique. Comput. Chem. Eng. **135** (2020). https://doi.org/10.1016/j.compchemeng.2020.106780
11. Kumar, A., Gandhi, C.P., Zhou, Y., et al.: Improved deep convolution neural network (CNN) for the identification of defects in the centrifugal pump using acoustic images. Appl. Acoust. **167**, 107399 (2020). https://doi.org/10.1016/j.apacoust.2020.107399
12. Watanabe, A., Hirose, N., Kim, H., Omura, I.: Convolutional neural network (CNNs) based image diagnosis for failure analysis of power devices. Microelectron. Reliab. **100–101**, 113399 (2019). https://doi.org/10.1016/j.microrel.2019.113399
13. Yang, Y., Yang, R., Pan, L., et al.: A lightweight deep learning algorithm for inspection of laser welding defects on safety vent of power battery. Comput. Ind. **123**, 103306 (2020). https://doi.org/10.1016/j.compind.2020.103306
14. Goodfellow, I., Bengio, Y., Courville, A.: Deep Learning. MIT Press, Cambridge (2016)
15. Rosebrock, A.: Video classification with Keras and Deep Learning. PyImageSearch (2019)

# Unsupervised Change Detection in Multi-temporal Satellite Images Based on Structural Patch Decomposition and $k$-means Clustering for Landslide Monitoring

Asadang Tanatipuknon[1,2](✉), Pakinee Aimmanee[1], Suthum Keerativittayanun[2], and Jessada Karnjana[2]

[1] Sirindhorn International Institute of Technology, Thammasat University, Tiwanon Rd, 131 Moo 5, Bangkadi, Muang 12000, Pathum Thani, Thailand
5822770525@g.siit.tu.ac.th, pakinee@siit.tu.ac.th

[2] NECTEC, National Science and Technology Development Agency, 112 Thailand Science Park, Khlong Luang 12120, Pathum Thani, Thailand
suthum.kee@ncr.nstda.or.th, jessada.karnjana@nectec.or.th

**Abstract.** This paper proposes a new, simple, and effective unsupervised change detection method for multi-temporal satellite imagery in the landslide monitoring system. The method combines a modified version of $k$-means clustering, so called the adaptive $k$-means clustering algorithm and a structural consistency map to binarize input multi-temporal satellite images. The absolute differences between pairs of contiguous binary images are used to construct a change map. The structural consistency map is generated by a method based on the structural patch decomposition. The evaluation results show that the proposed method is considerably better than a conventional method based on principal component analysis (PCA) and $k$-means clustering in terms of pixel error rate, correctness, and robustness against noise addition. For the pixel error rate and the correctness, the improvements are approximately 42.61% and 13.92%, respectively. Also, when noises are added into the input images, the pixel error rate of the proposed method is improved by 98% from that of the PCA-based method, and the correctness is improved by 77.27%.

**Keywords:** Change detection · Landslide monitoring · Structural patch decomposition · $k$-means clustering · Multi-temporal satellite imagery

## 1 Introduction

One of severe natural disasters that is difficult to prevent is the landslide. It causes damages widely especially in mountainous areas. For example, according to the World Health Organization, landslides affected approximately 4.8 million


V.-N. Huynh et al. (Eds.): IUKM 2020, LNAI 12482, pp. 259–269, 2020.
https://doi.org/10.1007/978-3-030-62509-2_22

people and caused more than 18,000 deaths from between 1998 and 2017 [1]. Therefore, to predict, mitigate, prevent, and manage the situation effectively, Information and Communication Technology and remote sensing technologies have been deployed. There are many approaches applied to tackle landslide-related problems. For example, Intrieri *et al.* described the design and implementation of a landslide early warning system with many components, such as extensometers, thermometers, rain gauges, and cameras [9], which aimed to reduce the system's false alarm. The advantage of this local-sensor-based monitoring system is that it can acquire data accurately in real-time. However, such a system must be installed in a specific and limited area. Hence, to cover a larger area, a more extensive system is required. Another approach is based on remote sensing [12]. For example, Niethammer *et al.* equipped digital cameras to an unmanned aerial vehicle (UAV) to map landslides at a high ground resolution [13]. They showed that the UAV with cameras can generate landslide data. However, the covered area of this approach is bounded by the UAV's maximum flight distance. In order to detect changes in a large area, techniques for analyzing satellite imagery can be adopted [6].

This work aims to analyze multi-temporal satellite images to detect changes caused by landslides. The fundamental concept of change detection is analyzing change data. In general, the change data can be constructed from the following methods: image differencing, normalized difference vegetation index, change vector analysis, principal component analysis (PCA), and image rationing [5]. Machine learning techniques, such as the support vector machine and neural networks, can be applied but may suffer from a high computational cost due to its high algorithm complexity [3,7]. To avoid unnecessary cost and complexity, more simple and effective methods based on image differencing have been proposed. In literature, there are three conventional change detection approaches with an inexpensive cost: expectation-maximization-based (or EM-based) thresholding, Markov-random-field-based (or MRF-based) change detection, and the method based on PCA [4,5]. It has been experimentally demonstrated that the PCA-based method outperformed EM-based method and MRF-based method [5]. This work is motivated by the simplicity and effectiveness of the PCA-based method. It aims not only to keep the spirit of simplicity but also to further reduce computational time complexity since PCA's computational time complexity increases with the dimension. Also, according to our preliminary experimental results, we found that the PCA-based method tends to false-positively detect changes between input images. Thus, another objective of this work is to reduce such false positives.

The rest of this paper is organized as follows. Section 2 describes the detail of the proposed method. Section 3 provides simulation conditions, experiments, results, and evaluation. Remarks and discussions are made in Sect. 4. Section 5 concludes this work.

## 2 Proposed Method

Given two satellite images from different times as the input, the proposed method produces a binary map that locates changes between them, which is called a *change map*. The change map is generated based on $k$-means clustering [2] and structural patch decomposition [11], which are detailed in this section. The proposed method consists of four steps: adaptive $k$-means clustering, structural consistency map generation, bitwise AND operation, and subtraction, as shown in Fig. 1 (left).

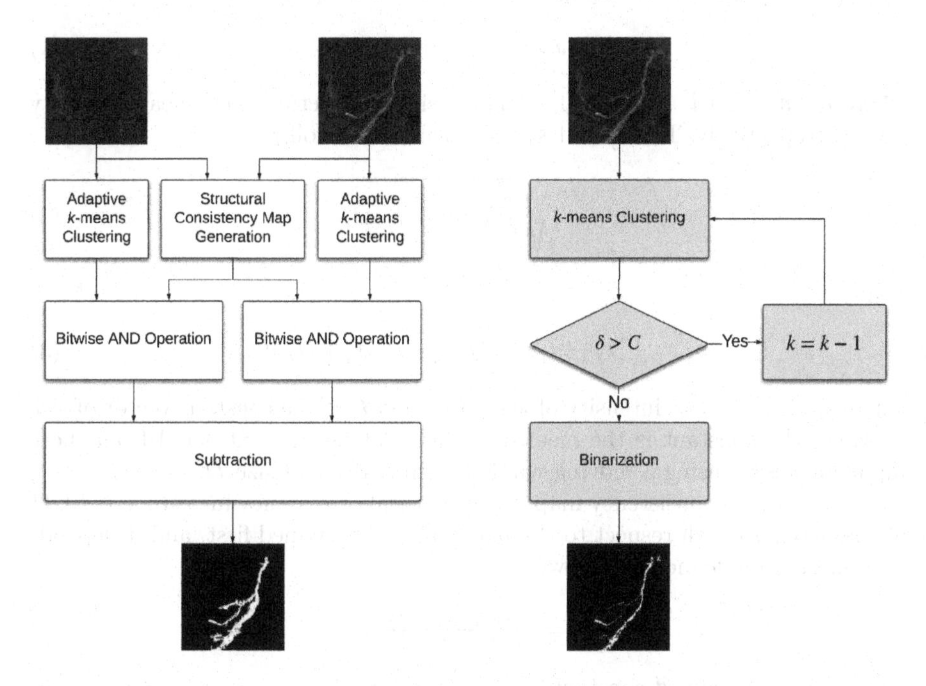


**Fig. 1.** Overall framework of the proposed method (left) and procedure in the adaptive $k$-means clustering algorithm (right).

First, two input images are separately clustered by using an algorithm based on $k$-means clustering, which is called adaptive $k$-means clustering, as shown in Fig. 1 (right). In this step, the value of $k$ is firstly set to a predefined value. Then, the intensities of the image are classified into $k$ groups. Let $G_1$, $G_2$, $G_3$, ..., and $G_k$ be groups sorted in ascending order of their intensity centers. Let $I_c(G_i)$ denotes the intensity center of the $i$th group, for $i \in \{1, 2, 3, \ldots, k\}$, and $\delta$ denotes the difference between the last two groups $(I_c(G_k) - I_c(G_{k-1}))$. In our adaptive $k$-means clustering algorithm, when $\delta > C$, where $C$ is a predefined constant, the image is re-clustered into $k-1$ groups; otherwise, the $k$-clustered image is binarized. To binarize the $k$-clustered image, values of all pixels in $G_k$

are set to 1, and values of pixels in other groups are set to 0. The outputs of this step are two binary maps $\boldsymbol{\mathcal{B}}_1$ and $\boldsymbol{\mathcal{B}}_2$.

Second, a structural consistency map is generated by applying the structural patch decomposition (SPD) to both input images [11]. According to the SPD approach, each RGB input image is used to construct a set of image patch vectors, which is a set of column vectors in $\mathbb{R}^d$, where $d = 3M^2$ and $M$ is the spatial size of a square patch.

Let $\boldsymbol{x}_1$ and $\boldsymbol{x}_2$ denote patch vectors of the first and the second input images ordered by time, respectively. That is, $\boldsymbol{x}_j = [x_{j,1} \ \ x_{j,2} \ \ldots \ x_{j,d}]^{\mathrm{T}}$ for $j \in \{1, 2\}$. The patch $\boldsymbol{x}_j$ can be decomposed by the following equations.

$$\boldsymbol{x}_j = c_j \cdot \boldsymbol{s}_j + \boldsymbol{l}_j, \tag{1}$$

where $c_j$, $\boldsymbol{s}_j$, and $\boldsymbol{l}_j$ are signal strength, signal structure, and mean intensity vector, respectively. The $c_j$ and $\boldsymbol{s}_j$ are calculated as follows.

$$c_j = \sqrt{\sum_{i=1}^{d} (x_{j,i} - l_{j,i})^2}, \tag{2}$$

and

$$\boldsymbol{s}_j = \tfrac{1}{c_j}[x_{j,1} - l_{j,1} \ \ x_{j,2} - l_{j,2} \ \ldots \ x_{j,d} - l_{j,d}]^{\mathrm{T}}, \tag{3}$$

where $l_{j,i}$ is the mean intensity of the patch and $\boldsymbol{l}_j$ is the constant vector of size $d$, where the constant is the mean intensity. Patches are extracted from those input images by using a moving window with a stride of size $D$.

To generate a consistency map, the structural consistency measure $\zeta \in \{0, 1\}$ of the patch $\boldsymbol{x}_2$ with respect to the patch $\boldsymbol{x}_1$ is determined first, and it depends on a measure $\rho$ defined as follows.

$$\rho = \tfrac{(c_1 \cdot \boldsymbol{s}_1)^{\mathrm{T}}(c_2 \cdot \boldsymbol{s}_2) + \epsilon}{\|c_1 \cdot \boldsymbol{s}_1\| \|c_2 \cdot \boldsymbol{s}_2\| + \epsilon}, \tag{4}$$

where $\epsilon$ is a predefined constant.

The structural consistency measure is determined by the following relation: $\zeta = 1$ when $\rho \geq t_1$ and $|l_2 - l_2'| < t_2$, where $t_1$ and $t_2$ are two predefined constants, $\rho$ is a measure that can be used to detect inconsistent changes, and $l_2'$ is the mean value of all entries of a patch vector constructed from a latent image of the second input image, where the latent image is created by applying the intensity mapping function between intensity values of those two input images [8]; otherwise, $\zeta = 0$.

Remark that it can be seen that, when $\epsilon = 0$, $\rho$ is the cosine similarity of vectors of signal structure vectors $\boldsymbol{s}_1$ and $\boldsymbol{s}_2$. The constant $\epsilon$ is added to ensure that the structural consistency is robust against noise [11]. The structural consistency map ($\boldsymbol{\mathcal{M}}$) is then generated by the structural consistency measures of all patches.

Third, each binary image $\boldsymbol{\mathcal{B}}_j$ obtained from the adaptive $k$-means clustering is updated by applying bitwise logical AND operator ($*$) to $\boldsymbol{\mathcal{B}}_j$ and the structural

**Table 1.** Details of datasets used in the experiment.

| Set No. | Covered area size ($km^2$) | Location | Image size (pixel×pixel) | Actual/Artificial landslide event |
|---|---|---|---|---|
| #1 | 3.6 × 2.1 | Hiroshima, Japan | 1208 × 764 | Actual |
| #2 | 25.6 × 14.5 | Na-Khao, Thailand | 864 × 486 | |
| #3 | 25.6 × 14.5 | Tainan, Taiwan | 1116 × 632 | |
| #4 | ∼13.75 × 13.75 | Lake Tahoe, USA | 200 × 200 | |
| #5 | 25.6 × 14.5 | Kathun, Thailand | 1116 × 632 | |
| #6 | 10 × 10 | Tainan, Taiwan | 434 × 434 | Artificial |
| #7 | 25.6 × 14.5 | Tainan, Taiwan | 1116 × 632 | |
| #8 | 3.6 × 2.1 | Hiroshima, Japan | 1208 × 764 | |
| #9 | 10 × 10 | Tainan, Taiwan | 434 × 434 | |
| #10 | 25.6 × 14.5 | Na-Khao, Thailand | 864 × 486 | |

consistency map $\boldsymbol{\mathcal{M}}$. In other words, the updated binary images $\boldsymbol{\mathcal{B}}_1^*$ and $\boldsymbol{\mathcal{B}}_2^*$ are $\boldsymbol{\mathcal{B}}_1 * \boldsymbol{\mathcal{M}}$ and $\boldsymbol{\mathcal{B}}_2 * \boldsymbol{\mathcal{M}}$, respectively.

Finally, the change map is constructed by calculating the absolute difference between $\boldsymbol{\mathcal{B}}_1^*$ and $\boldsymbol{\mathcal{B}}_2^*$.

## 3 Evaluation

Ten sets of satellite images used in our experiments are from Google Earth and the U.S. Geological Survey [6]. Each set consists of two images, which are before- and after-landslide events. Five sets are of actual landslide events, and others are of artificial landslide events edited by hand in order to evaluate change detection methods more precisely. Because the real landslide image has many details, it is not easy to identify the affected areas pixel-by-pixel. Details of these datasets are provided in Table. 1.

In this work, the parameters $M$, $D$, $t_1$, $t_2$, $\epsilon$, $C$, and the initial value of $k$ are set to 3, 2, 0.8, 0.1, 0.00045, 95, and 4, respectively. An example of input images in the set no. 6 are shown in Fig. 2(a) and Fig. 2(b). The change maps obtained from the conventional PCA-based method [5] and from the proposed method are shown in Fig. 2(c) and Fig. 2(d), respectively.

Four measures are used to evaluate the performance of the proposed method: pixel error rate (PER), correctness ($\Omega$), completeness ($\Theta$), and $F_\beta$-measure. The PER is defined as the total number of incorrectly detected pixels divided by the total number of the image pixels. The lower is the PER, the better is performance. The correctness and the completeness are defined by the following equations [14].

$$\Omega = \frac{TP}{TP+FN}, \tag{5}$$

and

$$\Theta = \frac{TP}{TP+FP}, \tag{6}$$

where $TP$ denotes the number of true positives, $FN$ denotes the number of false negatives, and $FP$ denotes the number of false positives.

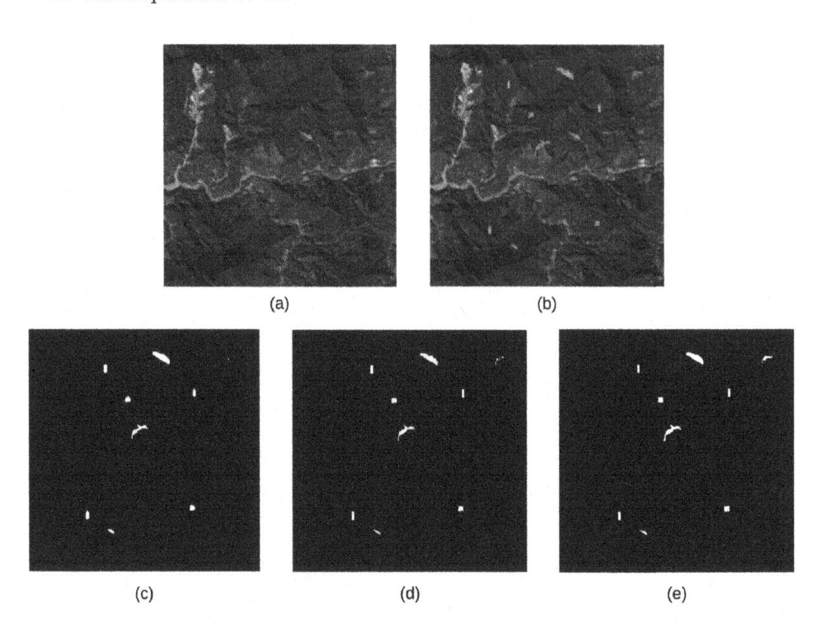


**Fig. 2.** Example of the dataset #6: (a) before-landslide events and (b) after-landslide events. The change maps obtained from the PCA-based method (c) and from the proposed method (d). (e) is a picture of the ground truth.

The $F_\beta$-measure is defined by

$$F_\beta = (1 + \beta^2)\frac{\Omega \cdot \Theta}{\beta^2 \cdot \Omega + \Theta}, \tag{7}$$

where the constant $\beta$ is used to weight between completeness and correctness [10]. In our experiments, we used $F_{0.5}$, $F_1$, and $F_2$, which represent lower, equal, and greater weight of completeness, respectively [10]. For the completeness, the correctness, and the $F_\beta$-measure, the higher the score, the better.

We compared the performance of the proposed method with a conventional change detection method based on principle component analysis (PCA) and $k$-means clustering [5]. Also, the robustness of the scheme is evaluated by adding two kinds of noise, which are the white Gaussian noise and the speckle noise, into the input images. The added noises are controlled so that the signal-to-noise ratios (SNR) vary from 0 dB to 15 dB.

The evaluation results are shown in Table 2. It can be seen that our proposed method outperform the standard $k$-means-based algorithm in all measures except completeness, of which the scores are comparable. It also outperform the conventional PCA-based method in terms of PER, completeness, and $F_{0.5}$, whereas the correctness, $F_1$, and $F_2$ scores are comparable. That is, the PCA-based method weights completeness more than correctness, whereas the proposed method weights them more equally.

Table 3 and Table 4 show the comparisons of averages of evaluation scores when the Gaussian noises with various SNRs are added into the input images and

**Table 2.** Evaluation results comparison among the proposed method, the standard $k$-means based method, and the conventional PCA-based method [5]. Note that PER is shown in percent.

| Dataset | PCA-based method | | | | | | $k$-means-based method | | | | | | Proposed method | | | | | |
|---|---|---|---|---|---|---|---|---|---|---|---|---|---|---|---|---|---|---|
| | PER | $\Theta$ | $\Omega$ | $F_{0.5}$ | $F_1$ | $F_2$ | PER | $\Theta$ | $\Omega$ | $F_{0.5}$ | $F_1$ | $F_2$ | PER | $\Theta$ | $\Omega$ | $F_{0.5}$ | $F_1$ | $F_2$ |
| #1 | 33.56 | 0.93 | 0.04 | 0.05 | 0.04 | 0.13 | 1.60 | 0.81 | 0.45 | 0.50 | 0.29 | 0.56 | 2.67 | 0.81 | 0.32 | 0.36 | 0.23 | 0.50 |
| #2 | 8.09 | 0.18 | 0.41 | 0.33 | 0.12 | 0.16 | 63.59 | 0.09 | 0.01 | 0.01 | 0.01 | 0.03 | 8.05 | 0.41 | 0.46 | 0.45 | 0.22 | 0.33 |
| #3 | 7.58 | 0.59 | 0.65 | 0.64 | 0.31 | 0.48 | 14.92 | 0.95 | 0.41 | 0.46 | 0.28 | 0.60 | 7.28 | 0.43 | 0.77 | 0.66 | 0.28 | 0.38 |
| #4 | 3.80 | 0.45 | 0.72 | 0.64 | 0.28 | 0.39 | 3.81 | 0.37 | 0.80 | 0.65 | 0.25 | 0.33 | 3.72 | 0.40 | 0.79 | 0.66 | 0.26 | 0.35 |
| #5 | 19.60 | 0.41 | 0.53 | 0.50 | 0.23 | 0.34 | 20.90 | 0.14 | 0.47 | 0.32 | 0.11 | 0.13 | 19.13 | 0.20 | 0.60 | 0.43 | 0.15 | 0.19 |
| Average | 14.53 | 0.51 | 0.47 | 0.43 | 0.20 | 0.30 | 20.96 | 0.47 | 0.47 | 0.39 | 0.19 | 0.33 | 8.17 | 0.45 | 0.59 | 0.51 | 0.23 | 0.35 |
| #6 | 0.16 | 0.85 | 0.88 | 0.87 | 0.43 | 0.69 | 0.18 | 0.80 | 0.89 | 0.87 | 0.42 | 0.65 | 0.07 | 0.89 | 1.00 | 0.97 | 0.47 | 0.73 |
| #7 | 0.04 | 0.87 | 0.84 | 0.85 | 0.43 | 0.69 | 0.13 | 0.36 | 0.49 | 0.46 | 0.21 | 0.31 | 0.03 | 0.78 | 0.99 | 0.94 | 0.44 | 0.65 |
| #8 | 0.23 | 0.74 | 0.96 | 0.91 | 0.42 | 0.62 | 0.20 | 0.78 | 0.97 | 0.93 | 0.43 | 0.65 | 0.19 | 0.78 | 0.98 | 0.93 | 0.43 | 0.65 |
| #9 | 0.07 | 0.93 | 0.86 | 0.87 | 0.45 | 0.73 | 0.03 | 0.95 | 0.95 | 0.95 | 0.48 | 0.76 | 0.02 | 0.94 | 1.00 | 0.99 | 0.49 | 0.76 |
| #10 | 0.56 | 0.75 | 0.92 | 0.88 | 0.41 | 0.63 | 1.67 | 0.36 | 0.55 | 0.50 | 0.22 | 0.31 | 1.16 | 0.36 | 0.99 | 0.73 | 0.27 | 0.33 |
| Average | 0.21 | 0.83 | 0.89 | 0.88 | 0.43 | 0.67 | 0.44 | 0.65 | 0.77 | 0.74 | 0.35 | 0.54 | 0.29 | 0.75 | 0.99 | 0.91 | 0.42 | 0.62 |

when the speckle noises with various SNRs are added into them, respectively. In our simulations, there are three cases of the noise addition: (1) noises are added into only the before-landslide event image, (2) noises are added into only the after-landslide event image, and (3) noises are added into both. Also, in the simulations regrading noise addition, only images from datasets no. 6 to no. 10 are used.

Examples of Gaussian-noise-added images and speckle-noise-added images are shown in Figs. 3(a)–(b) and Figs. 3(c)–(d), respectively. Examples of the change maps obtained from the PCA-based method and from the proposed method when the Gaussian noise with the SNR of 5 dB is added are shown in Fig. 3(e) and Fig. 3(f), respectively. Examples of the change maps obtained from the PCA-based method and from the proposed method when the speckle noise with the SNR of 5 dB is added are shown in Fig. 3(g) and Fig. 3(h), respectively.

It can be seen from the tables that our proposed method outperform the conventional PCA-based method in terms of PER and correctness when noises were added. For the pixel error rate and the correctness on average, the improvements are approximately 42.61% and 13.92%, respectively. Also, when noises are added into the input images, the pixel error rate of the proposed method is improved by 98% from that of the PCA-based method, and the correctness is improved by 77.27%.

Even though the averaged completeness scores of the proposed method are lower, it does not imply that the proposed scheme cannot completely detect the changes, compared to the PCA-based method. The explanation for the ineffectiveness of using only the completeness score in the noise addition scenario is be discussed in the next section.

**Table 3.** Evaluation results comparison when the Gaussian noise with different SNR is added into the input images.

| | SNR = 0 dB | | | SNR = 5 dB | | | SNR = 10 dB | | | SNR = 15 dB | | |
|---|---|---|---|---|---|---|---|---|---|---|---|---|
| | PCA | $k$-means | Proposed | PCA | $k$-means | Proposed | PCA | $k$-means | Proposed | PCA | $k$-means | Proposed |
| PER | 48.57 | 31.75 | 1.89 | 50.45 | 18.76 | 0.36 | 33.73 | 25.27 | 0.26 | 6.08 | 12.08 | 0.23 |
| $\Theta$ | 0.97 | 0.53 | 0.40 | 0.95 | 0.64 | 0.67 | 0.95 | 0.55 | 0.79 | 0.80 | 0.64 | 0.80 |
| $\Omega$ | 0.01 | 0.04 | 0.68 | 0.01 | 0.04 | 0.90 | 0.20 | 0.04 | 0.93 | 0.80 | 0.08 | 0.94 |
| $F_{0.5}$ | 0.02 | 0.04 | 0.57 | 0.02 | 0.05 | 0.82 | 0.19 | 0.05 | 0.90 | 0.76 | 0.10 | 0.91 |
| $F_1$ | 0.01 | 0.03 | 0.23 | 0.01 | 0.03 | 0.37 | 0.10 | 0.04 | 0.43 | 0.36 | 0.07 | 0.43 |
| $F_2$ | 0.05 | 0.08 | 0.33 | 0.05 | 0.10 | 0.56 | 0.18 | 0.10 | 0.65 | 0.55 | 0.19 | 0.66 |

**Table 4.** Evaluation results comparison when the speckle noise with different SNR is added into the input images.

| | SNR = 0 dB | | | SNR = 5 dB | | | SNR = 10 dB | | | SNR = 15 dB | | |
|---|---|---|---|---|---|---|---|---|---|---|---|---|
| | PCA | $k$-means | Proposed | PCA | $k$-means | Proposed | PCA | $k$-means | Proposed | PCA | $k$-means | Proposed |
| PER | 38.34 | 24.03 | 0.51 | 24.48 | 18.89 | 0.39 | 16.03 | 18.21 | 0.36 | 8.89 | 15.82 | 0.25 |
| $\Theta$ | 0.95 | 0.53 | 0.49 | 0.95 | 0.61 | 0.59 | 0.94 | 0.54 | 0.67 | 0.91 | 0.64 | 0.78 |
| $\Omega$ | 0.02 | 0.04 | 0.82 | 0.03 | 0.05 | 0.92 | 0.08 | 0.05 | 0.94 | 0.26 | 0.08 | 0.94 |
| $F_{0.5}$ | 0.02 | 0.05 | 0.70 | 0.04 | 0.06 | 0.82 | 0.10 | 0.05 | 0.86 | 0.27 | 0.10 | 0.90 |
| $F_1$ | 0.02 | 0.03 | 0.30 | 0.03 | 0.04 | 0.36 | 0.06 | 0.04 | 0.38 | 0.15 | 0.07 | 0.42 |
| $F_2$ | 0.06 | 0.09 | 0.42 | 0.11 | 0.12 | 0.51 | 0.18 | 0.12 | 0.57 | 0.30 | 0.19 | 0.64 |

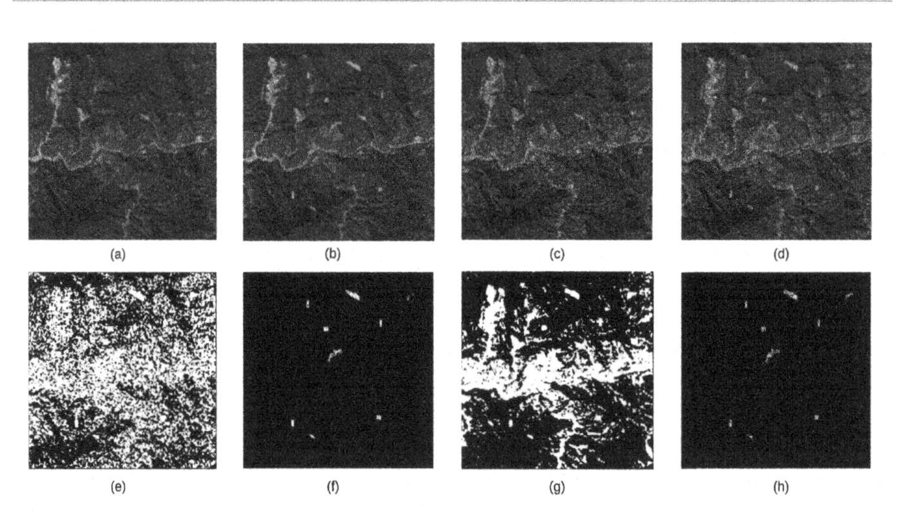


**Fig. 3.** Change map comparison when noise is added into input images: (a) and (b) are the input images with the Gaussian noise (5 dB SNR); (c) and (d) are the input images with the speckle noise (5 dB SNR); (e) and (g) are the change maps from the PCA-based method; and (f) and (h) are the change maps from the proposed method. The ground truth of this figure is the same as the Fig. 2

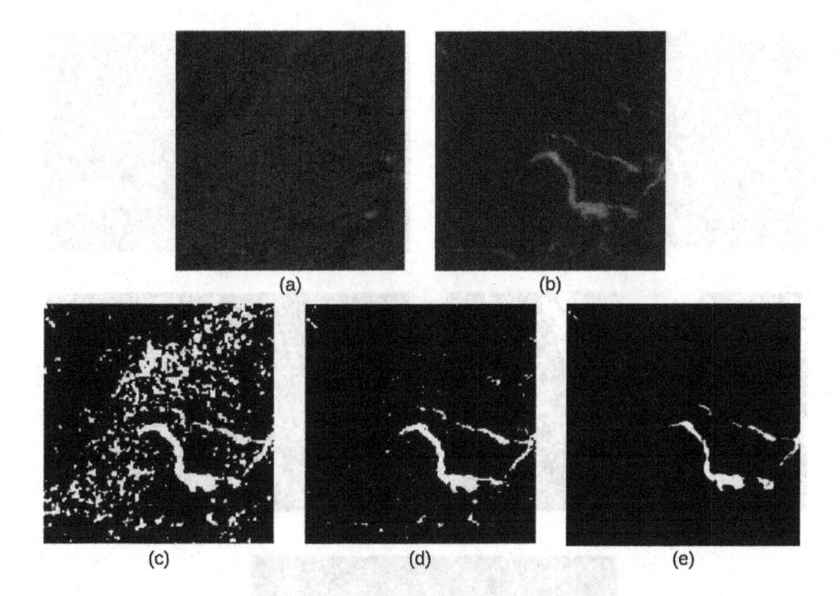


**Fig. 4.** False positive detection of the PCA-based method: (a) and (b) are the input images from the dataset #1, (c) is the change map of the PCA-based method. (d) is the proposed method's change map. (e) is a picture of the ground truth.eps

## 4 Discussion

Three points concerning the overall performance of the proposed method are discussed in this section. First, although the proposed method and the PCA-based method are comparably equal in many aspects except the correctness and robustness against noise addition, the computational time of the proposed method is less than that of the PCA-based method. The average value of computational times of the proposed method and that of the PCA-baed method are approximately 6.01 and 11.07 seconds, respectively, per collection. Hence, the proposed method is 1.84 times faster.

Second, when comparing the PCA-based method and the proposed method in terms of completeness, PCA-based method is better than ours as its solution is often over-segment. This is proven by a high value of the false-positive detection rate of the PCA-based method (shown in Fig. 4).

Third, in some actual events where landslides occurred, our proposed method could detect the changes while the PCA-based method could not, as shown in Fig. 5. In this case, we notice that some details in the after-landslide event image lost when the resolutions of both input images are different. Due to our experiments, we found that in-completeness of the proposed method occurs in the process of segmentation. Some pixels of landslide areas do not show a significant difference in intensity, so the algorithm fails to segment it.

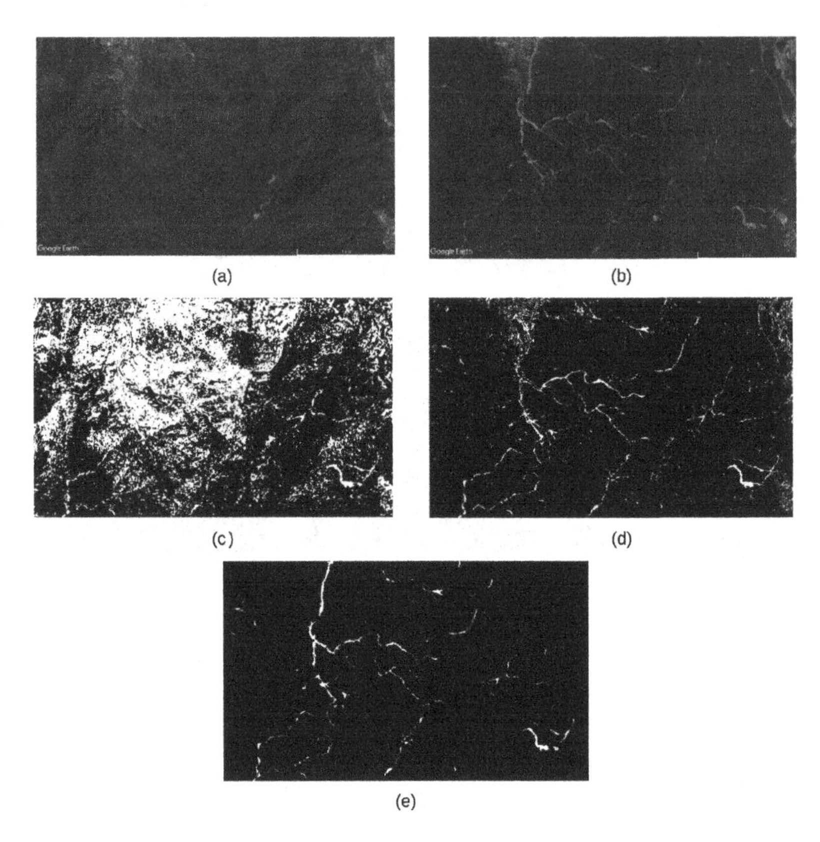


**Fig. 5.** Dataset #1: (a) and (b) are input images, (c) is the change map of the PCA-based method. (d) is the proposed method's change map. Note that the PCA-based method could not detect change properly. (e) is a picture of the ground truth.

## 5 Conclusion

This paper propose a new, simple, and effective method for change detection for the landslide monitoring system. The proposed method is based on an adaptive $k$-means algorithm with a structural consistency map generated by a method based on the structural patch decomposition. The evaluation results show that the proposed method is better than the conventional PCA-based method in terms of pixel error rate (PER), correctness, and robustness against noise addition. It is comparable to the PCA-based method in terms of completeness. Moreover, the computational time is 1.84 times faster.

**Acknowledgment.** This research is supported by Thailand Advanced Institute of Science and Technology (TAIST), National Science and Technology Development Agency (NSTDA), and Tokyo Institute of Technology under the TAIST-Tokyo Tech program.

Also, this work has been partially supported by the ASEAN Committee on Science, Technology and Innovation (COSTI) under the ASEAN Plan of Action on Science, Technology and Innovation (APASTI) funding scheme and by e-Asia JRP funding scheme.

## References

1. Landslides (who hompage). https://www.who.int/health-topics/landslides. Accessed 11 Sept 2020
2. Arthur, D., Vassilvitskii, S.: k-means++: the advantages of careful seeding. Tech. rep., Stanford (2006)
3. Bovolo, F., Bruzzone, L., Marconcini, M.: A novel approach to unsupervised change detection based on a semisupervised SVM and a similarity measure. IEEE Trans. Geosci. Remote Sens. **46**(7), 2070–2082 (2008)
4. Bruzzone, L., Prieto, D.F.: Automatic analysis of the difference image for unsupervised change detection. IEEE Trans. Geosci. Remote Sens. **38**(3), 1171–1182 (2000)
5. Celik, T.: Unsupervised change detection in satellite images using principal component analysis and $k$-means clustering. IEEE Geosci. Remote Sens. Lett. **6**(4), 772–776 (2009)
6. Djuric, D., Mladenovic, A., Pevsic-Georgiadis, M., Marjanovic, M., Abolmasov, B.: Using multiresolution and multitemporal satellite data for post-disaster landslide inventory in the Republic of Serbia. Landslides **14**(4), 1467–1482 (2017)
7. Ghosh, S., Bruzzone, L., Patra, S., Bovolo, F., Ghosh, A.: A context-sensitive technique for unsupervised change detection based on hopfield-type neural networks. IEEE Trans. Geosci. Remote Sens. **45**(3), 778–789 (2007)
8. Grossberg, M.D., Nayar, S.K.: Determining the camera response from images: what is knowable? IEEE Trans. Pattern Anal. Mach. Intell. **25**(11), 1455–1467 (2003)
9. Intrieri, E., Gigli, G., Mugnai, F., Fanti, R., Casagli, N.: Design and implementation of a landslide early warning system. Eng. Geol. **147**, 124–136 (2012)
10. Lu, P., Qin, Y., Li, Z., Mondini, A.C., Casagli, N.: Landslide mapping from multi-sensor data through improved change detection-based Markov random field. Remote Sens. Environ. **231**, 111235 (2019)
11. Ma, K., Li, H., Yong, H., Wang, Z., Meng, D., Zhang, L.: Robust multi-exposure image fusion: a structural patch decomposition approach. IEEE Trans. Image Process. **26**(5), 2519–2532 (2017)
12. Mantovani, F., Soeters, R., Van Westen, C.: Remote sensing techniques for landslide studies and hazard zonation in Europe. Geomorphology **15**(3–4), 213–225 (1996)
13. Niethammer, U., James, M., Rothmund, S., Travelletti, J., Joswig, M.: UAV-based remote sensing of the Super-Sauze landslide: evaluation and results. Eng. Geol. **128**, 2–11 (2012)
14. Trinder, J., Salah, M.: Airborne lidar as a tool for disaster monitoring and management. In: Proceedings of the GeoInformation for Disaster Management, Antalya, Turkey, pp. 3–8 (2011)

# Nowcasting and Forecasting for Thailand's Macroeconomic Cycles Using Machine Learning Algorithms

Chukiat Chaiboonsri(✉) and Satawat Wannapan

Modern Quantitative Economics Research Center (MQERC), Faculty of Economics, Chiang Mai University, Ching Mai, Thailand
chukiat1973@gmail.com, lionz1988@gmail.com

**Abstract.** With the complexity of social-economic distributional variables, the introduction of artificial intelligent learning approaches was the major consideration of this paper. Machine learning algorithms were fully applied to the multi-analytical time-series processes. Annual macroeconomic variables and behavioral indexes from the search engine database (Google Trends) were observed and they were limited at 2019. The exploration of up-to-date data by the nowcasting calculation based on the Bayesian structural time-series (BSTS) analysis was the solution. To understand Thailand economic cycles, yearly observed GDP was categorized as cyclical movements by the unsupervised learning algorithm called "k-Mean clustering". To predict three years beforehand, categorized cyclical GDP was estimated with the updated data by using supervised algorithms. Linear Discriminant Analysis (LDA) and k-Nearest Neighbors (kNN) are the two predominant learning predictors contain the highest Kappa's coefficients and accuracies. The two findings from the two learning approaches are different. The linear-form learning model (LDA) hints expansion periods are still the predictive sign for Thai economy. Conversely, the non-form algorithm (kNN) gives recession signs for Thai economic cycles during the next three years.

**Keywords:** Nowcasting · Machine learning · Macroeconomics · Google Trends · Business cycle · Bayesian structural time-series analysis

## 1 Introduction

Predictive macroeconomics is the way to visualize economic scenarios and decide suitable frameworks for policy implementing. Historically walked through the story of traditional statistics, macroeconomic estimations and predictions chronically depended on restricted statistical tools with conditional assumptions. The ability of data exploration has been conditionally supposed to be normally distributional inferences. A fixed parameter is suspiciously used to be a proxy for random samples. The consequence is critical. Predictive economics nearly closes to the edge of conceptual reconsideration.

It is undeniable that the lack of observations in time-series analyzing has been the major concern for economists since *big data* is becoming more mentioned. The question


V.-N. Huynh et al. (Eds.): IUKM 2020, LNAI 12482, pp. 270–282, 2020.
https://doi.org/10.1007/978-3-030-62509-2_23

is not only the number of sets of samples, but it is how updated information and relative variables can be abruptly obtained. Nowcasting calculations are potentially the solution for this difficult task. The introduction of Google trends stated by Scott and Varian (2014) can be retrieved almost real-time, much earlier than the time that current period's economic indicators actually come out, incorporating Google Trends information into the nowcasting model can improve the predictions (Nakavachara and Lekfuangfu 2017). For Fig. 1, nowcasting calculations are graphically reported by seasonal trends, predictively being as economic cycles. Thus, nowcasting for big data starts the idea of this paper to challenge the novel way for the macroeconomic cyclical investigation.

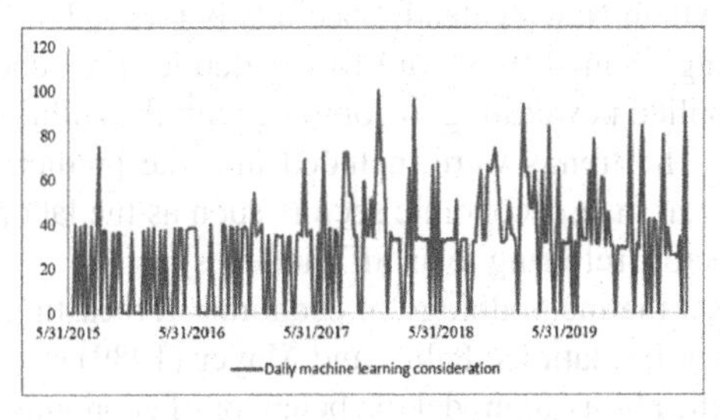


**Fig. 1.** The daily frequencies regarding nowcasting calculations by Google Trend (Source: www.google.trend.com)

To modernize econometric estimations and forecasting, the ML approaches have been being focused comparing with the counterpart during five years. With the flexibility for dealing with missing information, containing the capacity for calculating millions of data sets, and having the ability for eliminating multicollinearity problems in linear models, the ML methods which are unsupervised and supervised algorithms are employed to find the solution for clustering an economic cyclical stage, classifying the most suitable distribution, validating observed data for a better prediction. This paper is therefore divided into two main objects:

- To explore up-to-date sets of macroeconomic variables by the nowcasting calculation.
- To predictively estimate economic cylical movements by machine learning approaches during 2020 to 2022.

## 2 Literature Review

### 2.1 Nowcasting and Artificial Intelligence for Macroeconomics

Back to the inspiration of up-to-date calculations, nowcasting was fundamentally described in the online handbook written by BańBura et al. (2012). It is the major role plays for macroeconomic variables to obtain "early estimates" of such key observed indicators, nowcasters use information from data that are the relationship with the target variable but collected at higher frequencies, typically monthly details, and released recent manner. Historically, BańBura et al. (2013) applied nowcasting to real-time macroeconomic data flows. This working paper was supported by European Central Bank. Bragolia and Modugno (2016) presented the empirical findings that including U.S. data in a nowcasting model for Canada dramatically improves its predictive accuracy for GDP forecasting, mainly because of the absence of timely production data for Canada. In 2017, Bok et al. (2017) conceptually reviewed the way for tracking U.S. economic conditions using big data have evolved over time. Also, econometric nowcasting techniques had advanced to mimic and automate the best practices of forecasters on market-monitoring roles underlying the New York Fed Staff Nowcast. In Asia, Richardson et al. (2018) contributed the idea that applying machine learning algorithms could improve nowcasts of real GDP growth in New Zealand. Specifically focused on Thailand's economic forecasting, nowcasting seemed to be rare for academic researches. Nakavachara and Lekfuangfu (2017) applied nowcasting to combine with the additive variables observed from Google Trends. The trends were included into the predictive regression model for estimating the crucial macroeconomic sectors such as the labor market, automobile sales, and financial sector (referring to as SET indexes).

It is undeniable that the methodology of economic forecasting relies on econometric models with evident limitations. Paliés and Mayer (1989) strongly stated Artificial Intelligence (AI) has the ability to model the behavior of economists. It deals with making models work as a knowledge base for the phenomenon's observer, produces the observed state: the economic forecast (Bourgine 1984). The AI computation is interestingly employed to many research branches. Zheng et al. (2017) did the panel discussion for literally describing the situation that econometrics met big data. In the same expression, Charpentier et al. (2018) stated an interesting phrase that learning models were indicated to be more effective than traditional econometric methods. Moreover, Athey and Imbens (2019) concluded that machine learning was important to include in the core graduate econometrics sequences. To apply for Thai economy, Chaiboonsri and Wannapan (2019) employed supervised machine learning algorithms with macroeconomic variables and Google Trends to predict an economic cycle in a short moment. Hence, in this paper, ML approaches including both unsupervised and supervised algorithms are used in nowcasting calculations, clustering cycle stages, and forecasting Thai economic cycles. More expressly, Fig. 2 shows the conceptual framework of the research.

## 3 Data Review

To practically use the high dimensions of observed data in learning models, macroeconomic variables were observed from official databases such as World Bank Data, Bank of Thailand, and accessible websites. The range of annual data was between 2005 and 2019. The variables were transformed to be as a percent change, and these were all stationary. The collected variables are GDP, foreign direct investment (FDI), exchange rates, agricultural value-added volumes, consumer price index (CPI), gross domestic investments (GDI), exports, imports, industrial value-added volumes, international reserves, military expenditures, public debts, population growth rates, services (import and export sides), annual temperature, international tourist arrivals, and unemployment rates.

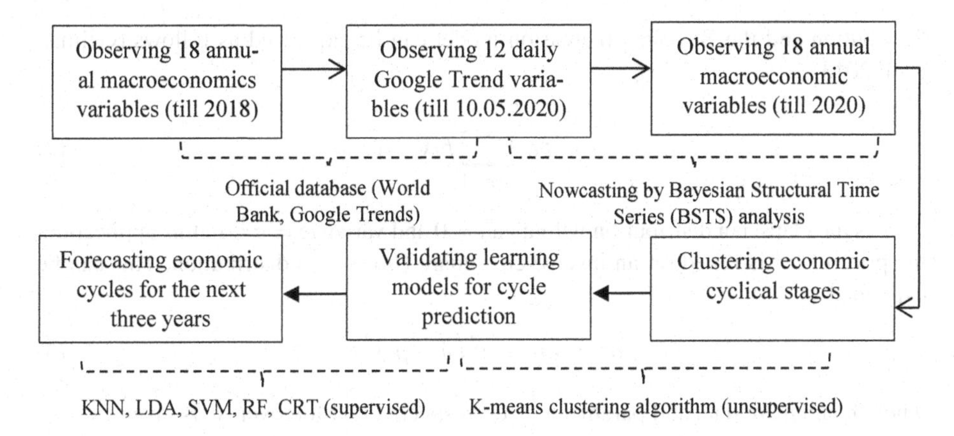


**Fig. 2.** Multi-process calculating for Thai economic cycles

In modern economics, to predictively estimate the economic cycle needs more than using macroeconomic factors restrictly. Behavior indicators are becoming more crucial. With the advanced technology for data databases, Google Trends information is practically used to be the main component for data nowcasting. Variable details are investment mention frequencies, stock market issues, business circumstances, employment issues, international trades and investments, agriculture affairs, industrial affairs, tourism issues, social network trends, environmental pinpoints, banking situations, and political atmospheres.

# 4 Methodology

## 4.1 The Bayesian Structural Time Series (BSTS) Analysis for Nowcasting

Various Bayesian inference approaches have been used in economics. For the BSTE model, Scott and Varian (2014) started the non-seasonally adjusted initial claims for predicting US unemployment benefits by using behavioral indexes for Google trends as an additive-effect factor in the BSTS model. Additionally, Jun (2019) proposed a sustainable technical analysis applying a Bayesian structural time series Forecasting (BSTSF) model depended on time series data to provide the set of observations in the regression model. As a result, the BSTS model is based on the Bayes theorem as follows (Koduvely 2015; Jun 2019),

$$P(\theta|x) = \frac{P(x|\theta)P(\theta)}{P(x)} \tag{1}$$

$x$ is observed information and $\theta$ is the model parameter. $P(x|\theta)$ presents the prior and likelihood functions. $P(x|\theta)$ is the posterior function. This is updated by learning the observed data $x$ given $\theta$ which is a likelihood. For setting the prior, the Gaussian distribution and the Bayesian regression model can be expressed as follows (Gelman et al. 2013)

$$y = \beta_0 + \sum_{p=1}^{P} \beta_i x_i + e. \tag{2}$$

$e$ is the Gaussian distribution with mean = 0 and variance = $\sigma^2$. In this application, the prior of $\sigma^2$ is relied on an inverse-chi square ($Inv - \chi^2$) distribution. This can be explained as

$$\sigma^2 \sim Inv - \chi^2\left(n - p, s^2\right), \tag{3}$$

where $n$ and $p$ are data and parameter sizes, respectively. Moreover, $s^2$ is given by

$$s^2 = \frac{1}{n-p}\left(y - X\hat{\beta}\right)^T\left(y - X\hat{\beta}\right). \tag{4}$$

To practically use the BSTS model for nowcasting, *Local linear trend*, the first component of the model without behavioral information from Google Trends is defined by the pair of two formulas,

$$\mu_{t+1} = \mu_t + \delta_t + \eta_{\mu,t}, \tag{5.1}$$

$$\delta_{t+1} = \delta_t + \eta_{\delta,t}, \tag{5.2}$$

$\eta_{\mu,t}$ and $\eta_{\delta,t}$ are supposed to be normal distributions, $N\left(0, \sigma^2_{\mu,t}\right)$ and $N\left(0, \sigma^2_{\delta,t}\right)$, respectively. The component $\mu_t$ is the value of the trend at time $t$. The component $\delta_t$ is the expected increase in $\mu$ between times $t$ and $t + 1$. Thus, it can be implied as a slope at time $t$. On the other hand, the model including the data from Google Trends is the modification of the formula (2), which is given by

$$y_t = \mu_t + \beta^T x_t + \varepsilon_t, \tag{6.1}$$

$$\mu_{t+1} = \mu_t + \delta_t + \eta_{0,t}, \tag{6.2}$$

$$\delta_{t+1} = \delta_t + \eta_{1,t}. \tag{6.3}$$

$\beta^T x_t$ stands for the dimension of behavioral factors included into the structural time-series model.

### 4.2 Unsupervised Classification by the k-Means Clustering Algorithm

To specifying this algorithm to be acceptable is a function of the type of available data and the selective purpose of analysis (Oyelade et al. 2010). Technically, Euclidean distance as the criterion is adapted since the calculation of Euclidean distance in both hyperspace and two-dimensional space are almost identical (Li and Wu 2012). According to the methodology of traditional k-Means algorithms, the development of its performance is expressed as five following steps:

- Step 1: to accept the amount of clusters to categorize data into and the set of data to cluster as input values and minimize variance clustering of the data into $k$ groups.
- Step 2: to initialize the first $k$ groups by taking the first $k$ instance or random sampling of $k$ elements.
- Step 3: to estimate the arithmetical mean of each cluster formed in the set of data, which can be designed as a gradient descent procedure.
- Step 4: the k-Means algorithm allocates each point in the dataset to specific class of the initial clusters. Each recorded point is assigned to the nearest cluster using a measurement of Euclidean distance.
- Step 5: the algorithm repeatedly assigns each recorded point of the dataset.

## 4.3 Supervised Machine Learning Algorithms for Big Data Prediction

### 4.3.1 Linear Discriminant Analysis (LDA)

The linear discriminant analysis (LDA) is a basic learning algorithm to identify the linear features that maximize the between-class separation of data, while minimizing the within-class scatter (Bishop 1995; Ioffe 2006). In terms of mathematical expressions, let $n$ be the number of clusters (4 characteristic levels such as "downfall", "expansion", "boom", and "recession" are used), $\mu$ be the mean of all samples, $N_i$ be the number of observations in the $i^{th}$ class, $\mu_i$ the mean of observations in the $i^{th}$ class, and $\sum_i$ be the scatter matrix of observations in the $i^{th}$ class. The solution of the within-class scatter matrix, $S_W$, can be defined as the following formula:

$$S_W = \sum_{i=1}^{n} \Sigma_i, \tag{7}$$

and the solution of the between-class scatter matrix, $S_B$, can be also defined as

$$S_B = \sum_{i=1}^{n} N_i(\mu_i - \mu)(\mu_i - \mu)^T. \tag{8}$$

With the condition of diagonal matrix, $S_W^{-1} S_B$, eigenvalues and eigenvectors are solved.

### 4.3.2 Classification and Regression Trees (CART)

Not similar to logistic and linear regression, CART does not develop a prediction equation (Krzywinksi and Altman 2017). Conceptually, the process is applied for clustering on the data in each child node. The splitting process stops if the relative reduction in impurity is below a pre-specified threshold. Algorithm 1 provides with the pseudo-code for the basic steps (Loh 2011).

**Algorithm 1.** *Pseudocode for tree constructions by exhaustive search*

I. Start at the root node.

II. For each $X$ variables, search for the set $S$ that minimizes the sum of the node impurities in the two child nodes and select the split $\{X^* \in S^*\}$ which yields to the minimum overall $X$ and $S$.
III. If a stopping criterion is reached, exit. Differently, apply the second step to each child node in turn.

### 4.3.3 The Concept of k-Nearest Neighbors (kNN)

$k$-Nearest Neighbors (kNN) is a non-form artificial algorithm for both classification and linear regression problems (Chakraborty and Joseph 2017). Data was performed as its clusters to others in the feature space. To deal with the problem, observations are allocated. The majority class of its nearest samples is selected. In other words, the average value of its nearest neighbors in the linear regression model is the key to group $K$ neighbors. The performance of $k$-NN algorithm can analyze the error rate of miss-classified samples for classification or squared errors similar in regression problems. The first crucial process is the computation searching the distance $x_i$, a single sample, related to all other points in the feature space. This stage controls its $k$ closet neighbors $\{x_j\}_i^k$. Euclidean distance is generally used to measure distance. The second step is to assign the result of $y_i$ to the class membership (for instance, $y_i \in \{C_1, C_2, \ldots, C_c\}$ $\forall i = 1, 2, \ldots, n$, where $c$ is a number of class levels) by the majority vote of its $k$ nearest neighbors. The $k$-NN regression, $y_i$, represents the average value of its single nearest neighbor, $y_i = 1/k \sum_{j=1; x_i \in \{x_j\}_i^k}^{k} x_j$. One can be also defined as a distance-weighted average.

### 4.3.4 Support Vector Machine (SVM)

The support vector machine (SVM) is a promising application for both classification and regression estimations (Chakraborty and Joseph 2017). Basically, the two-class classification problems are modeled as the interval based on logistic regression (Logit model). In other words, the position to hyperplane in the feature space is projected to a binary (0 to 1) interval, which can be described as probabilities of class memberships. One stands for the supporting vector and 0 refers to the error function. The SVM algorithm would computationally search a decision boundary to separate these two classes by the maximal margin. The first idea is the group to solve the spaces in which the data are linearly separable. The second is to identify the points in the input space which defines, or supports, the maximum margin in the vectors.

### 4.3.5 Tree Models and Random Forest (RF) Algorithm

The Random Forest (RF) algorithm (Tree modeling) is one of favorite supervised machine learning approaches for yield with regression estimations and classification in big-data issues. The goal is to maintain the minimum of the entropy; $H(Y|X)$, which is the objective function, within areas of the baskets constrained by the features of $X$ (Galton 1907). At the initial stage, the completed set $X$ of $m$ samples is recognized for controlling the forms $x$, which lead to the highest information gain ($I$). The three random algorithms can be expressed as follows;

$I(Y|x) = H(Y|x) - \sum_{v \in \{x\}} \frac{|X_v|}{|m|} H(Y\}X_v)$: Informative gain (classification),
$H(Y|X) = -\sum_{c=1}^{C} p(Y = c|X) \log(p(Y = c|X))$: Entropy (classification),
$H(Y|X) = \frac{1}{m} \sum_{j=1}^{k} \sum_{i=1}^{m_j} \left(y_i - \mu_j\Big|_{x_i \in X_j}\right)^2$: MSE (regression).

$p(Y = c|X)$ is the joint existences of class $c$ samples in $X$. $|X_v|$ is the set of samples which takes on each value. In the regression model, the entropy can be alternatively substituted by the measurement of mean squared errors (MSE), and the separated branches are performed along the dimensions which are the most minimum residual. Thus, decision trees and random forests can be useful for addressing sensible forecasts and generating insights.

# 5 Empirical Results

## 5.1 Nowcasting Information by the Bayesian Structural Time Series Approach

To improve the accuracy for time-series forecasting is the main task for the first section. Thai GDP is plotted as the line graph in Fig. 3. The predicted line is moved along scatter points by univariate estimating. In other words, the predicted rate of economic expansions in this scenario is estimated by its lags, no additively multiple variables (without Google Trends data). As seen the black shade along the line, the efficiency of prediction is low that implies the predicted line seems to not be unable to accurately explain the observed data distribution since an enormous amount of information is ignored. The nowcasting calculation by using the BSTS model with Google Trends data is graphically shown in Fig. 4. With the power of behavioral factors added into the structural time-series model, the black shade is evidently visualized wider than the previous case. Expressly, the ability of data distributional estimations is improved. As a result, the frequency of Google Trends indexes and Bayesian inference are becoming more necessary for the econometric prediction in this current movement.

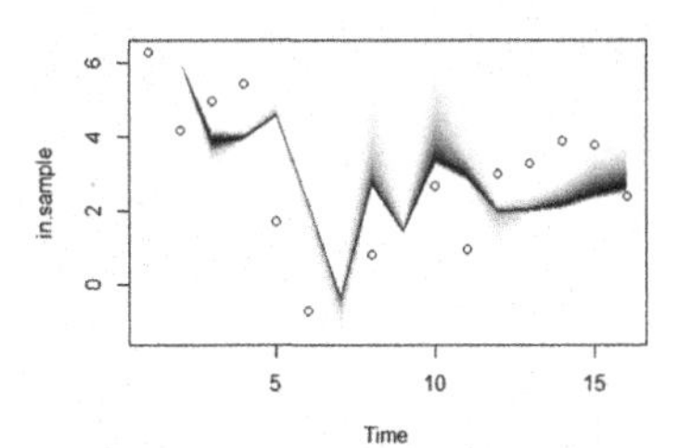


**Fig. 3.** The nowcasting calculation without Google Trends indexes (Source: authors' calculation)

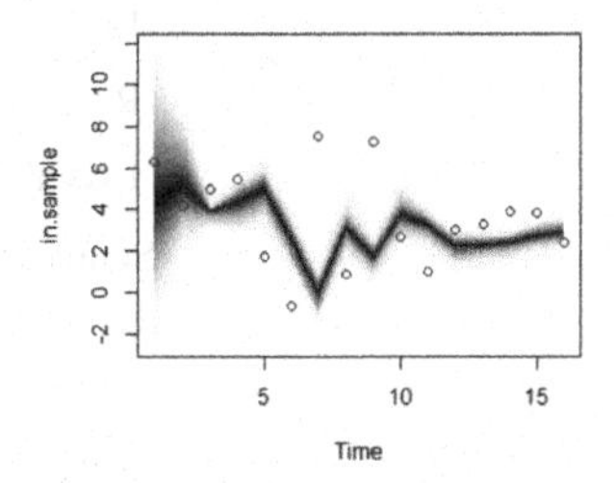


**Fig. 4.** The nowcasting calculation with Google Trends indexes (Source: authors' calculation)

## 5.2 The Clustering Result by the *k*-Means Algorithm

The target of this section is the cluster stages of economic cycles for Thai historical economy during 2005 to 2019. Two endogenous factors including the growth rate of exports and imports of commodities are added to be the effecting indicator for cycle regimes. As stated by Apaitan et al. (2019), international trades are a crucial activity of an economy and are inseparable from economic development. Thailand's balance of payment has chronically depends on these sectors since the country got recovery from the Asian financial crisis in 1998. In this paper, the clustering regimes (expansion, boom, recession, and depression) are the multi-color points displayed in Fig. 5. The clustering result is clear that exports and imports can be the efficient cycling predictor for Thailand GDP. Empirically, there are two years standing for expansion periods, six years are predictively denoted boom stages, four times indicate the years of recession, and two years are referred as depression. Moreover, the cycling regimes are also detailed in Table 1.

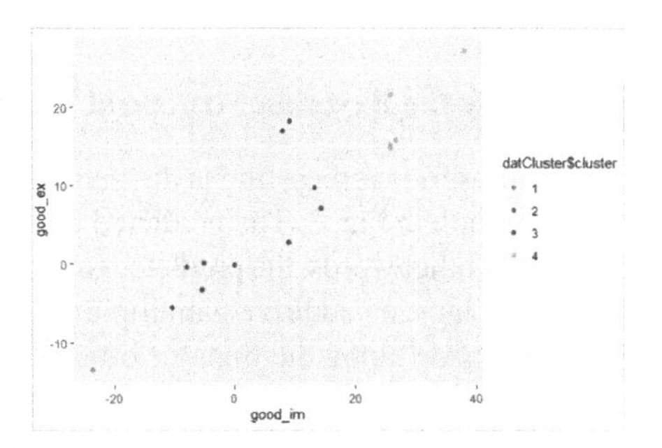


**Fig. 5.** Clustering plots by *k*-Means algorithm (Source: authors' calculation)

**Table 1.** Cycling regimes clustered by the *k*-Means algorithm

| Description | Frequencies (years) |
|---|---|
| Expansion | 6 |
| Peak | 2 |
| Recession | 4 |
| Downfall | 2 |

Source: authors

## 5.3 Thailand Predictive Economic Cycles by Using Machine Learning Algorithms with Nowcasting Information

As the observations are up to date and dependent variable (GDP) is already classified as the economic cycling stages, the goal in this section is to employ five learning algorithms such as linear discriminant modelling (LDA), Classification and Regression Trees (CART), k-Nearest Neighbors (kNN), Support Vector Machine (SVM), and Random Forrest (RF) to predictively indicate an economic picture in the upcoming three years ahead (2020 to 2022). Before the prediction is confirmed, the five algorithms are validated by the Kappa's coefficients for assuring the potential learning predictor. As displayed in Fig. 6, LDA and kNN are chosen by the first and second highest coefficient validation. The former stands for the linear-form learning algorithm and the second refers to as non-form learning algorithm. Coefficients and accuracies are additionally detailed in Table 2.

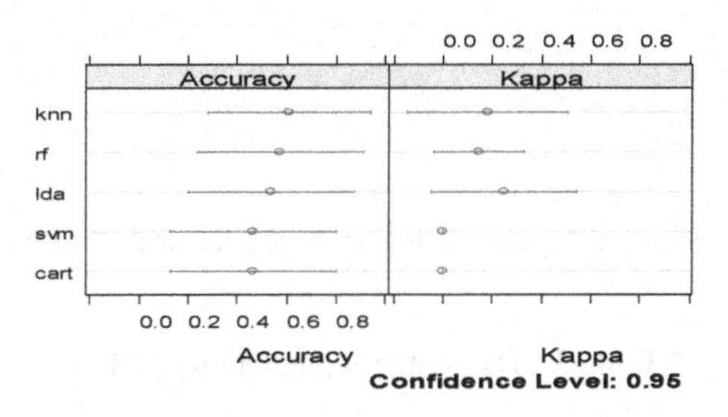


**Fig. 6.** Kappa's coefficient validation (Source: authors' calculation)

**Table 2.** The comparison result of machine learning approaches for forecasting Thailand economic cycles by Kappa's coefficients

| Algorithm | Cross-validation method | Accuracy | Kappa coefficients |
|---|---|---|---|
| LDA** | Cohen's kappa | 0.6666667 | 0.2460317 |
| CART | Cohen's kappa | 0.3333333 | 0.0000000 |
| *k*-NN** | Cohen's kappa | 0.6666667 | 0.1809524 |
| SVM | Cohen's kappa | 0.3333333 | 0.0000000 |
| Random forest | Cohen's kappa | 0.5000000 | 0.1452381 |

Noted: ** indicates the chosen algorithm
Source: authors

Considering into the economic cycling prediction by LDA, Fig. 7 graphically results the linear learning category of cycling stages. The interesting point is the grouping data which is intercept for each other. With observed macroeconomic variables and some of sets of behavioral indexes, it is evident that the linear learning algorithm has a limitation for classifying indistinguishable movements between expansion and recession periods. The algorithm can only group the predominant occurrences, which are the boom and depressing periods. As a result, LDA finds Thailand economic cycles still stay on the expansion periods during the next three years (see details in Table 3).

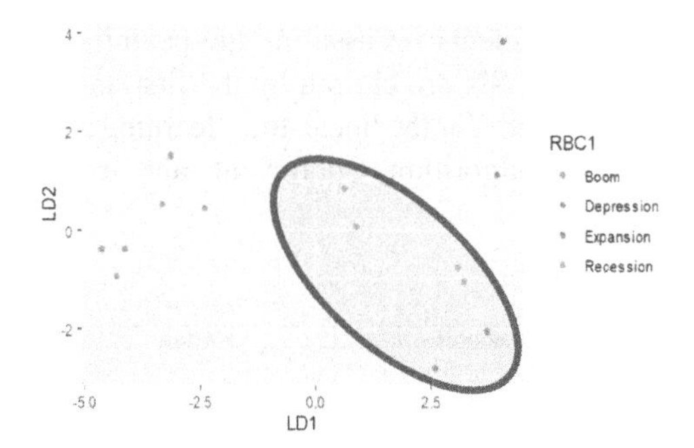


**Fig. 7.** The economic cycling category by the linear discriminant analysis

**Table 3.** The posterior prediction by LDA

| Year/Stages | Expansion | Boom | Recession | Depression |
|---|---|---|---|---|
| 2020 | 1 | 0 | 0 | 0 |
| 2021 | 1 | 0 | 0 | 0 |
| 2022 | 1 | 0 | 0 | 0 |

Source: authors

Interestingly, the predictive result by the kNN algorithm is different. The optimal neighbor calculation repeats the highest predictive accuracy at seven neighbors. In other words, the graphical line shown in Fig. 8 confirms this non-form learning algorithm can efficiently test and train observed data, and this potentially provides the conclusion for sensible economic predictions. Empirically, Table 4 represents three years Thai economy stays on recession periods.

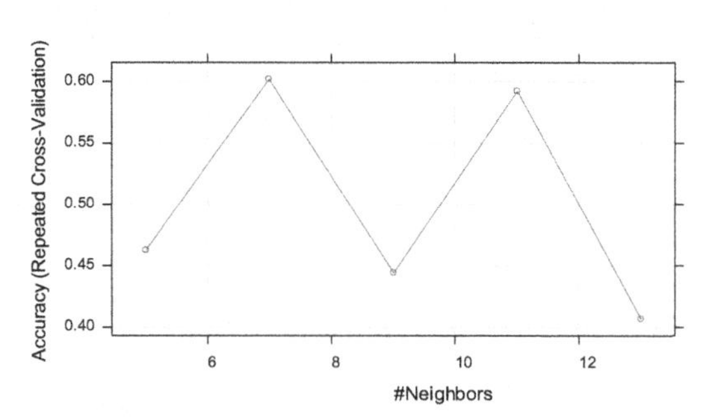


**Fig. 8.** The optimal neighborhood calculation by the kNN algorithm (Source: authors' calculation)

**Table 4.** The cycle prediction by kNN neighbor distancing

| Year/Stages | Expansion | Boom | Recession | Depression |
|---|---|---|---|---|
| 2020 | 0 | 0 | 1 | 0 |
| 2021 | 0 | 0 | 1 | 0 |
| 2022 | 0 | 0 | 1 | 0 |

Source: authors

# 6 Discussion and Conclusion

Although machine learning computations still have many critical issues on their sensibility and verification, findings conclude that the learning models are becoming more skillful for various types of data distributions in reality. Interestingly, this data procedure is the heart of machine learning approaches. Hence, the substantiality of the findings in this paper regards some critically predictable trends for Thai economic fluctuations by fully using artificial intelligent computations for nowcasting, clustering, and forecasting.

The range of official time-series information of Thailand macroeconomics was limited to the period during 2019. However, Thai economy is currently facing the great economic depression caused by the CORONAVIRUS-19 epidemic appearing in 2020. The nowcasting calculation with updating behavioral information estimated by the Bayesian structural time-series learning model is the way to approximately catch reals. This tool is becoming one of crucial parts for observing misreading up-to-date data that is ignored or assumed to be insignificant by traditional econometric methods.

The performance of machine learning additionally covers the major roles in econometrics such as a classificatory tool, cluster calculator, parametric or non-parametric estimator, and predictor. To computationally cluster Thai economic cycles, the unsupervised *k*-Mean algorithm can address the complexity of data distributions and group regimes of the cycles. Historical trends are all categorized as literal definitions for allowing machines to predictively tell the upcoming cyclical situations in Thai economy. On the other process, the capability of supervised machine learning approaches to test and train an enormous amount of high dimensional samples is the advantage that traditional econometric methods cannot be installed. This data procedure raises the machines are meticulous and powerful to learn every single detail of information that can be possibly observed. Moreover, the flexibility of machine learning to allow non-form calculations inside the processes of the k-nearest neighbor algorithm is the strong point to challenge antiquarian parametric estimations. The findings are evident that the k-neighbor learning algorithm answers the unfavorable recession periods for Thai economy in the next three years (2020–2022).

In conclusion, the reliable computation by machine learning is becoming more indispensible for giving a smart solution and sensible policy implementation. The suggestion is clear that Thai economy does not need pro-cyclical policies, but it does require some intelligent applications to counter this economic downsizing situation. Budget allocations should be carefully provided to every local sector without any corruption, especially healthcare systems and small-to-medium enterprises (SME).

## References

Scott, S.L., Varian, H.R.: Predicting the present with Bayesian structural time series. Int. J. Math. Mod. Num. Optim. **5**(1/2), 4–23 (2014)

Nakavacharaa, V., Lekfuangfu, N.W.: Predicting the present revisited: the case of Thailand. Discussion Paper No. 70. Puey Ungphakorn Institute for Economic Research (2017)

BańBura, M., Giannone, D., Reichlin, L.: Nowcasting. Oxford University Press, Oxford (2012). Edited by Clements, M.P., Hendry, D.F

Bańbura, M., Giannone, D., Modugno, M., Reichlin, L.: Now-casting and the real-time data flow. Working Paper Series: No. 1564. European Central Bank (ECB) (2013)

Bragolia, D., Modugno, M.: A nowcasting model for Canada: do U.S. variables matter? Finance and Economics Discussion Series 2016-036. Board of Governors of the Federal Reserve System, Washington (2016)

Bok, B., Caratelli, D., Giannone, D., Sbordone, A., Tambalotti, A.: Macroeconomic nowcasting and forecasting with big data. Staff Report No. 830. Federal Reserve Bank of New York (2017)

Richardson, A., Florenstein, T.V., Vehbi, M.T.: Nowcasting New Zealand GDP using machine learning algorithms. In: IFC – Bank Indonesia International Workshop and Seminar on "Big Data for Central Bank Policies/Building Pathways for Policy Making with Big Data", Bali, Indonesia. Reserve Bank of New Zealand (2018)

Paliés, O., Mayer, J.: Economics reasoning by econometrics or artificial intelligence. Theory Decis. **27**(1–2), 135–146 (1989)

Bourgine, P.: La M.A.O.: mod61isation assist6e par ordinateur. Interfaces AFCET **21**, 3–9 (1984)

Zheng, E., et al.: When econometrics meets machine learning. Data Inf. Manag. **1**(2), 75–83 (2018)

Charpentier, A., Flachaire, E., Ly, A.: Econometrics and machine learning. Economie et Statistique **505–506**, 147–169 (2018)

Athey, S., Imbens, G.W.: Machine learning methods that economists should know about. Ann. Rev. Econ. **11**, 685–725 (2019)

Chaiboonsri, C., Wannapan, S.: Big data and machine learning for economic cycle prediction: application of Thailand's economy. In: Seki, H., Nguyen, C.H., Huynh, V.-N., Inuiguchi, M. (eds.) IUKM 2019. LNCS (LNAI), vol. 11471, pp. 347–359. Springer, Cham (2019). https://doi.org/10.1007/978-3-030-14815-7_29

Koduvely, H.M.: Learning Bayesian Models with R. Packt, Birmingham (2015)

Jun, S.H.: Bayesian structural time series and regression modeling for sustainable technology management. Sustainability **11**, 4945 (2019)

Oyelade, O.J., Oladipupo, O.O., Obagbuwa, I.C.: Application of k-Means clustering algorithm for prediction of students' academic performance. Int. J. Comput. Sci. Inf. Secur. **7**(1), 292–295 (2010)

Li, Y., Wu, H.: A clustering method based on K-Means algorithm. Phys. Procedia **25**, 1104–1109 (2012)

Bishop, C.: Neural Networks for Pattern Recognition. Oxford University Press, Oxford (1995)

Ioffe, S.: Probabilistic linear discriminant analysis. In: Leonardis, A., Bischof, H., Pinz, A. (eds.) ECCV 2006. LNCS, vol. 3954, pp. 531–542. Springer, Heidelberg (2006). https://doi.org/10.1007/11744085_41

Krzywinksi, M., Altman, N.: Classification and regression trees. Nat. Methods **14**(8), 757–758 (2017)

Chakraborty, C., Joseph, A.: Machine learning at central banks. Staff Working Paper No. 647. Bank of England (2017)

Galton, F.: Vox populi. Nature **75**, 450–451 (1907)

Apaitan, T., Disyatat, P., Samphantharak, K.: Dissecting Thailand's international trade: evidence from 88 million export and import entries. Asian Dev. Rev. **36**(1), 20–53 (2019)

# Econometric Applications

# Dependence of Financial Institutions in China: An Analysis Based on FDG Copula Model

Yangnan Cheng[1], Jianxu Liu[2,3](✉), Mengjiao Wang[1], and Songsak Sriboonchitta[1,3]

[1] Faculty of Economics, Chiang Mai University, Chiang Mai 50200, Thailand
[2] School of Economics, Shandong University of Finance and Economics, Jinan 250000, China
liujianxu1984@163.com
[3] Puey Ungphakorn Center of Excellence in Econometrics, Chiang Mai University, Chiang Mai 50200, Thailand

**Abstract.** This paper presents a novel methodology for the dependence measurement of financial industry in China. We apply the ARMA-GJR-GARCH model and one-Factor with Durante Generators (FDG) copula to a dataset of stock prices of 42 financial institutions. Considering the impact of financial crisis to the financial market, we conducted our analysis in three periods—pre-crisis, crisis and post-crisis, respectively. The data ranges from September 2003 to May 2020 and the crisis period is from January 2007 to September 2008. Our results tell that dependence coefficients during the crisis period were higher than that in the other two periods. Dependence coefficients between securities companies were the highest in all periods. Moreover, some securities companies were found to be highly correlated with all the other companies in the financial industry.

**Keywords:** Dependence · FDG copula · Financial industry · Financial crisis · China

## 1 Introduction

The modern portfolio theory tells that when a group of securities are not perfectly positive correlated, the yield may remain the same and the variance/risk can be reduced by decentralized investment. Theoretically, the non-systematic risk can be reduced to zero by fully diversified investment [1]. Inspired by this theory, many studies on dependence among stocks or stock groups have been conducted, mainly on correlations between stock markets of different regions or different sectors. The dependence among the stocks not only affects the investment risk and return for individual investors, but also helps policy makers and market participants better understand and evaluate stock market. A variety of methods have been applied to measure the correlations or dependence of the stocks but some of them have been proved ineffective in real life. The advent of the 2007 financial crisis showed that some methods of risk measurement did not perform as well as expected. Poor performances of inappropriate methods may bring serious consequences to the stability of financial markets and even cause adverse effects on the national economy. As an inevitable part of financial risk measurement, dependence


V.-N. Huynh et al. (Eds.): IUKM 2020, LNAI 12482, pp. 285–296, 2020.
https://doi.org/10.1007/978-3-030-62509-2_24

analysis of financial market has been paid much attention. Various methods are available for dependence analysis but only some are applicable to financial data. Thereinto, copula is one of the most popular methods in studies on risk measurement. Copulas that are suitable for high-dimensional data, such as Vine copulas, Archimedean copulas and elliptical copulas have been widely used in the literature. Examples can be found in Liu et al. (2020) [2], Kenourgios et al. (2011) [3], Aloui et al. (2011) [4], Reboredo and Ugolini (2015) [5], and Shahzad et al. (2018) [6]. Yet, weaknesses of these models must not be overlooked. For instance, elliptical copulas only allow for tail symmetry and the number of parameters exponentially increase with the increasing of variables in Vine copulas. These models are either flexible or tractable but rarely both. Hence, the one-Factor copula with Durante Generators (FDG) model appears to be a good alternative. Proposed by Mazo et al. [7], this model suited well for high-dimensional data. Besides, this class of one-factor copula is non-parametric, and, therefore, encompasses many distributions with different features. Unlike elliptical copulas, the members of this class allow for tail asymmetry. Furthermore, the associated extreme-value copulas can be derived to carry out analysis of extreme values. Finally, this model is theoretically well-grounded, and practically fast and accurate, thanks to its ability of calculating explicitly the dependence coefficients.

To the best of our knowledge, this paper is the first to apply the FDG copula model to financial data and to measure the dependence between stocks in financial industry of China. China's stock market has experienced a big bull market in 2007 and a bear market in 2008. We examine the correlations between the stock market in three stages. One stage represents the drastic shock period during the financial crisis in 2007 and 2008, and others are defined as the pre- and post-crisis period. Accurate assessment of the dependence will help investors optimize their investment portfolios and may provide policy makers and market participants with valuable information about the financial industry in China.

The rest of this paper is organized as follows. Section 2 describes the ARMA-GJR-GARCH (Auto-regressive moving-average Glosten-Jagannathan-Runkle generalized autoregressive conditional heteroskedasticity) model and FDG copula model. Section 3 gives the explanation of data and presents the empirical results. Section 4 details the conclusions.

## 2 Methodology

As mentioned above, ARMA-GJR-GARCH model was adopted to obtain the cumulative distribution function of each stock in three periods. The best distribution was selected from skew-norm, skew-t and skew-GED by BIC. Then, the selected cumulative distribution functions of all stocks were used in the FDG copula model to obtain the dependence coefficients. Details of ARMA-GJR-GARCH model and FDG copula are introduced below.

### 2.1 ARMA-GJR-GARCH Model

The GJR-GARCH model was proposed by Glosten et al. [8] in 1993. This model performs better in the analysis of financial market due to its ability of capturing an empirical phenomenon that the simplest ARMA-GARCH model cannot capture. That is the negative impact from time *t-1* to the variance of time *t* is greater than positive impact. It is believed that the negative impact increases the leverage, and hence causes the increase in the risk. The ARMA-GJR-GARCH model consists of ARMA (p,q) and GJR-GARCH (m,n) processes. The ARMA (p,q) is defined as

$$y_t = c + \sum_{i=1}^{p} \varphi_i y_{t-i} + \sum_{j=1}^{q} \rho_j \varepsilon_{t-j} + \varepsilon_t \tag{1}$$

where $y_t$ is the conditional mean and $\varepsilon_t$ denote the error terms (return residuals, with respect to mean process). Here, $\varepsilon_t$ are split into two parts, $z_t$, a random variable, and $\sigma_t$, the standard deviation

$$\varepsilon_t = z_t \sigma_t \tag{2}$$

where $z_t$ are i.i.d variables of standard innovation.

The GJR-GARCH (m,n) model is specified as

$$\sigma_t^2 = \alpha_0 + \sum_{i=1}^{n} \left[\alpha_i \varepsilon_{t-i}^2 + \gamma_i \mathbb{I}_{t-i} \varepsilon_{t-i}^2\right] + \sum_{j=1}^{m} \beta_j \sigma_{t-j}^2 \tag{3}$$

where $\gamma$ is the lever effect and

$$\mathbb{I}_{t-i} = \begin{cases} 0 \text{ } if \text{ } \varepsilon_{t-i} \geq 0, \\ 1 \text{ } if \text{ } \varepsilon_{t-i} < 0. \end{cases}$$

## 2.2 FDG Copula Models

Compared with other copula models, FDG copula is a flexible and tractable class of one-factor copulas [9]. Detailed introduction is following the review of one-factor copula model.

Factor copula model is useful when the dependence in observed variables is based on a few unobserved variables, and there exists tail asymmetry or tail dependence in the data, so that the multivariate normality assumption is not valid. In one-factor copula model, we assume there is one latent variable. More specifically, let $U = (U_1, \ldots, U_d)$ be a random vector with $U_i \sim U(0, 1)$. And $U_1, \ldots, U_d$ are assumed to be conditionally independent given the latent variable $U_0$. Let $C_{0i}$ be the joint distribution of $(U_0, U_i)$ and $C_{i|0}(\cdot|u_0)$ be the conditional distribution of $U_i$ given $U_0 = u_0$ for $i = 1, \ldots, d$. Then the one factor copula is given by

$$C(u_1, \ldots, u_d) = \int_0^1 C_{1|0}(u_1|u_0) \ldots C_{d|0}(u_d|u_0) du_0. \tag{4}$$

Note that

$$C_{i|0}(\cdot|u_0) = \frac{\partial C_{0i}}{\partial u_0}.$$

The copulas $C_{0i}$ are called the linking copulas because they link the factor $U_0$ to the variables of interest $U_i$. The class of FDG copulas is constructed by choosing appropriate linking copulas for the one-factor copula model. The class of linking copulas which served to build the FDG copulas is referred to as the Durante class [10] of bivariate copulas. The Durante class takes the form of

$$C(u, v) = \min(u, v)f(\max(u, v)) \quad (5)$$

where $f : [0, 1] \to [0, 1]$ called the generator of C, is a differentiable and increasing function such that $f(1) = 1$ and $t \mapsto f(t)/t$ is decreasing. The FDG acronym thus stands for "one-Factor copula with Durante Generators". This model possess the advantages of one-factor copula, such as nonexchangeability, parsimony and easy data generation from the copulas, and of Durante linking copulas: the integral in Eq. (4) can be calculated and the resulting multivariate copula is nonparametric. Four examples of families indexed by a real parameter for the generators $f_1, \ldots, f_d$ are given by Mazo, et al. [6]. Among those, two parametric families, FDG-CA and FDG-F, along with their extreme value copula—EV-FDG-CAF, were employed in this paper.

(1) FDG copula with Cuadras-Augé generators

In Eq. (5), let

$$f_i(t) = t^{1-\theta_i}, \quad \theta_i \in [0, 1]. \quad (6)$$

A copula belonging to the Durante class with generator (6) gives rise to the well-known Cuadras-Augé copula with parameter $\theta_i$ [11].

The Spearman's rho is given by

$$\rho_{ij} = \frac{3\theta_i\theta_j}{5 - \theta_i - \theta_j}.$$

The lower and upper tail dependence coefficients are given by

$$\lambda_{ij}^{(L)} = 0 \; and \; \lambda_{ij}^{(U)} = \theta_i\theta_j.$$

The Kendall's tau is given by

$$\tau_{ij} = \begin{cases} \frac{\theta_i\theta_j(\theta_i\theta_j+6-2(\theta_i+\theta_j))}{(\theta_i+\theta_j)^2-8(\theta_i+\theta_j)+15} & if \; \theta_i + \theta_j \neq 1 \\ \frac{\theta(\theta-1)(\theta^2-\theta-4)}{8} & if \; \theta = \theta_i = 1 - \theta_j. \end{cases}$$

(2) FDG copula with Fréchet generators

In Eq. (5), let

$$f_i(t) = (1 - \theta_i)t + \theta_i, \theta_i \in [0, 1]. \quad (7)$$

A copula belonging to the Durante class with generator (7) gives rise to the well-known Fréchet copula with parameter $\theta_i$ [12]. The Spearman's rho, the lower and upper tail dependence coefficients are respectively given by

$$\rho_{ij} = \lambda_{ij}^{(L)} = \lambda_{ij}^{(U)} = \theta_i\theta_j.$$

The Kendall's tau is given by

$$\tau_{ij} = \frac{\theta_i\theta_j\left(\theta_i\theta_j + 2\right)}{3}.$$

Since the FDG copulas with Cuadras-Augé and Fréchet generators lead to the same EV-FDG copula, totally three types of FDG copulas are used in this paper. The Cuadras-Augé and Durante-sinus families allow for upper but no lower tail dependence, the Durante-exponential family allows for lower but no upper tail dependence, and the Fréchet family allows for both. In the Fréchet case, furthermore, the lower and upper tail dependence coefficients are equal: this is called tail symmetry, a property of elliptical copulas.

We first applied this model to our data with three possible distributions and calculated their maximum likelihood values, respectively. Then BIC was used to select the best distribution for each stock. Thus, we obtained the most suitable cumulative distribution function and all the parameters in the model. Before fitting the data to copula functions, we must ensure that all variables used in the copula function are in the range of [0,1]. Therefore, the error terms are filtered by

$$\hat{z}_t = \frac{\hat{\varepsilon}_t}{\hat{\sigma}_t} \tag{8}$$

to form a new series of error terms that are to be used in the FDG copula functions. Three types of FDG copula were applied to the filtered series with the help of estimation method for Inference functions of margins (IMF) [13]. To select the best from these three, mean absolute percentage error (MAPE) are calculated by

$$MAPE_r = \frac{1}{p}\sum_{i<j}\left|\frac{r_{i,j} - r(\theta_i,\theta_j)}{r(\theta_i,\theta_j)}\right| \tag{9}$$

# 3 The Data and Empirical Results

## 3.1 The Data

Following Erkens et al. [14], we define the period from January 1, 2007 to September 30, 2008 as the financial crisis period. To cover as much companies as possible, 42 financial institutions with a time to market earlier than January 1, 2007 are chosen to conduct our research. According to the classifications in China's stock market, these companies are divided into four subsectors- Bank, Insurance, Securities, and Others which includes five banks, two insurance companies, fifteen securities companies and twenty other diversified types of financial companies respectively. Based on the data, pre-crisis period is from September 12, 2003 to December 29, 2006, and the crisis period is from January 4, 2007 to September 26, 2008. Post-crisis period is from October 6, 2008 to May 14, 2020. For simplicity, the name of all 42 companies will not be listed in the paper. All the calculations are performed with the daily closing price of each company.

### 3.2 Results of Mean Absolute Percentage Error (MAPE)

**Table 1.** MAPE of FDG-F, FDG-CA and EV-FDG-CAF in three periods

| Period | Criteria | CA | F | EV-CAF |
|---|---|---|---|---|
| Pre-crisis | MAPE-rho | 0.114 | 0.114 | 0.114 |
| | MAPE-tau | 0.167 | 0.198 | 0.168 |
| Crisis | MAPE-rho | 0.110 | 0.110 | 0.110 |
| | MAPE-tau | 0.163 | 0.195 | 0.165 |
| Post-crisis | MAPE-rho | 0.114 | 0.114 | 0.114 |
| | MAPE-tau | 0.167 | 0.198 | 0.168 |

In Table 1, MAPE-rho and MAPE-tau were calculated with the Spearman's rho and Kendall's tau respectively by Eq. (9). Three copula families have the same value of MAPE-rho in each period whereas values of MAPE-tau vary in different families. FDG-F has the largest MAPE-tau while FDG-CA has the lowest. Thus, FDG-CA is the best copula for our data. All the results in the following parts are from this copula family.

### 3.3 Dependence Inside Each Sector

Table 2 shows the average values of Kendall's tau of each sector in three periods. Compared among these four sectors, Securities has the strongest dependence no matter in which period, and Others has the lowest dependence inside this sector. If we compare parallelly among three periods, we will find that dependence in crisis period is higher than that in the other two periods for four sectors. The highest in-sector dependence appears in Securities sector in the crisis period and the lowest is in Others sector in pre- and post-crisis periods.

**Table 2.** Dependence inside each sector in three periods

| Period | Bank | Insurance | Securities | Others |
|---|---|---|---|---|
| Pre-crisis | 0.301 | 0.312 | 0.422 | **0.246** |
| Crisis | 0.302 | 0.319 | **0.425** | 0.250 |
| Post-crisis | 0.301 | 0.312 | 0.422 | **0.246** |

Table 3 displays the average values of upper tail dependence coefficients of each sector in three periods. Like Table 2, Securities sector has the highest upper tail dependence and Others has the lowest. From parallel comparison, upper tail dependence is higher in the crisis period. In addition, upper tail dependence is the highest in Securities sector in the crisis period and the lowest in Others sector in both pre- and post-crisis periods.

**Table 3.** Upper tail dependence inside each sector in three periods

| Period | Bank | Insurance | Securities | Others |
|---|---|---|---|---|
| Pre-crisis | 0.475 | 0.488 | 0.596 | **0.406** |
| Crisis | 0.476 | 0.495 | **0.599** | 0.410 |
| Post-crisis | 0.482 | 0.488 | 0.596 | **0.406** |

### 3.4 Dependence Between Sectors

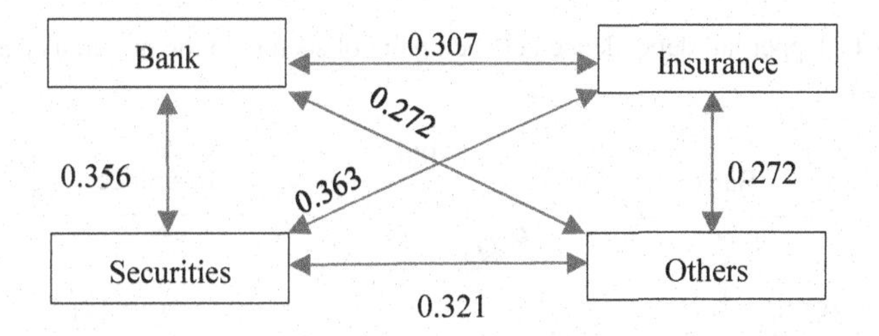


**Fig. 1.** Dependence between sectors in the pre-crisis period

Dependence between sectors in three periods are exhibited in Figs. 1, 2 and 3. In all three periods, there exist positive dependence between each pair of sectors. Dependence between Insurance sector and Securities sector is the strongest and that between Bank and Others are the weakest. And the dependence between each pair of sectors are stronger in the crisis period and weaker in the other two periods.

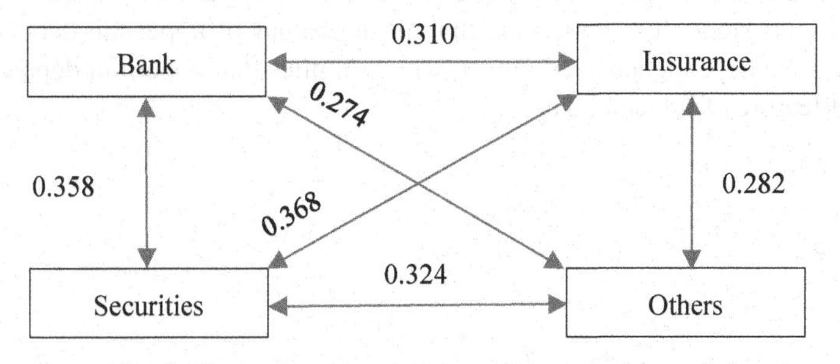


**Fig. 2.** Dependence between sectors in the crisis period

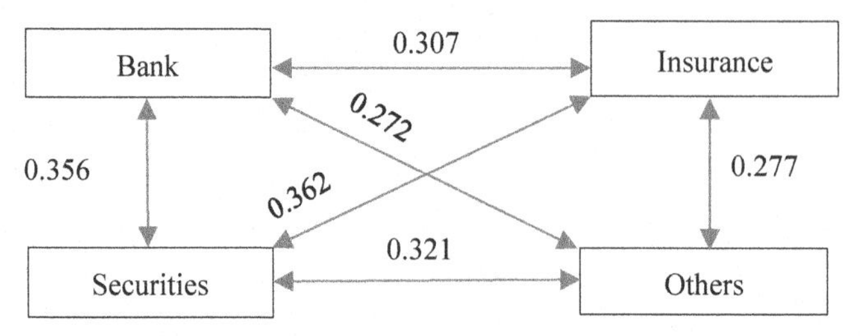


**Fig. 3.** Dependence between sectors in the post-crisis period

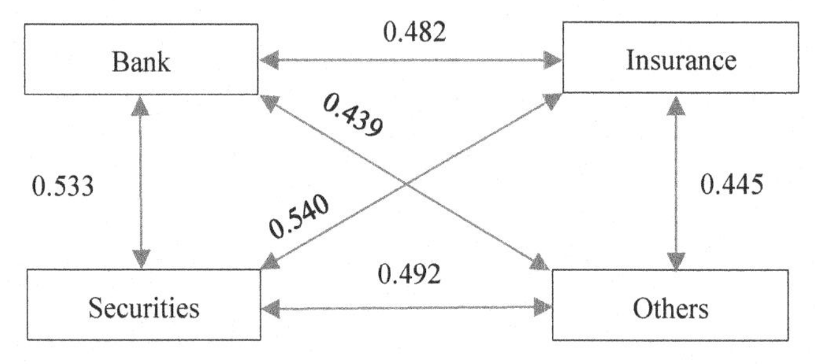


**Fig. 4.** Upper tail dependence between pairs of sectors in the pre-crisis period

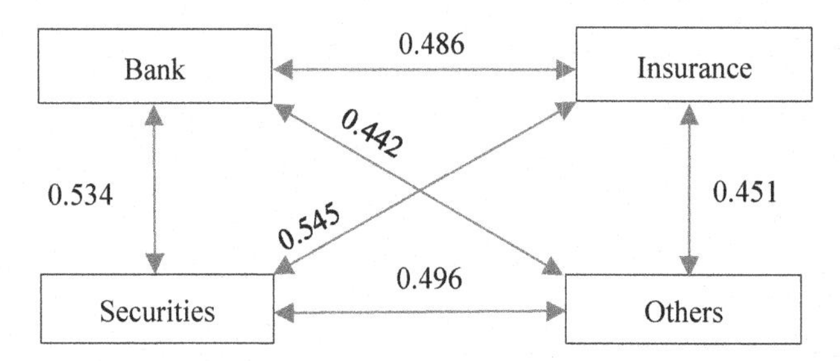


**Fig. 5.** Upper tail dependence between pairs of sectors in the crisis period

Figures 4, 5 and 6 demonstrate the upper tail dependence between sectors in three periods. Securities and Insurance still have the largest dependence and Bank and others have the smallest. The upper tail dependence in the crisis period is greater than that in the other two periods. It is noticeable that all the values of upper tail dependence are greater than the dependence coefficients, which implies that upper tail dependence is a prominent feature of financial data.

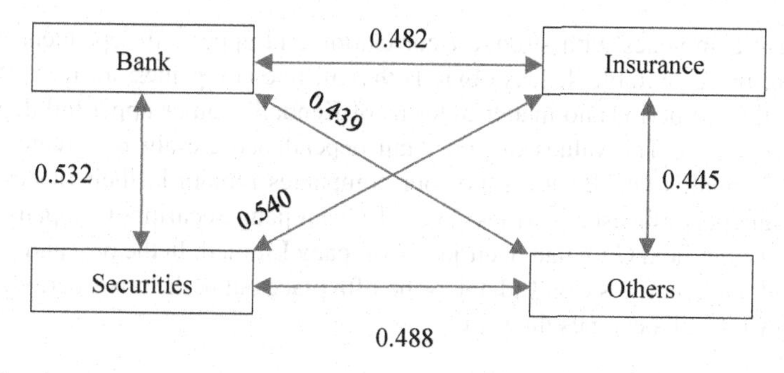


**Fig. 6.** Upper tail dependence between pairs of sectors in the post-crisis period

### 3.5 Dependence of Individual Companies

**Table 4.** Summary of top five companies with highest dependence coefficients in three periods

| Period | Index | Rank | | | | |
|---|---|---|---|---|---|---|
| | | 1 | 2 | 3 | 4 | 5 |
| Pre-crisis | Company | Northeast Securities | Changjiang Securities | CITIC Securities | Guoyuan Securities | Haitong Securities |
| | Sector | Securities | Securities | Securities | Securities | Securities |
| | tau | 0.4297 | 0.4262 | 0.4225 | 0.4142 | 0.4061 |
| | 95% CI of tau | 0.0250 | 0.0249 | 0.0247 | 0.0243 | 0.0238 |
| | utdc | 0.6032 | 0.6000 | 0.5966 | 0.5888 | 0.5811 |
| | 95% CI of utdc | 0.0237 | 0.0236 | 0.0235 | 0.0233 | 0.0231 |
| Crisis | Company | Northeast Securities | Changjiang Securities | CITIC Securities | Guoyuan Securities | Sinolink Securities |
| | Sector | Securities | Securities | Securities | Securities | Securities |
| | tau | 0.4223 | 0.4187 | 0.4153 | 0.4076 | 0.3995 |
| | 95% CI of tau | 0.0246 | 0.0244 | 0.0243 | 0.0239 | 0.0235 |
| | utdc | 0.5966 | 0.5932 | 0.5901 | 0.5828 | 0.5751 |
| | 95% CI of utdc | 0.0235 | 0.0234 | 0.0234 | 0.0232 | 0.0230 |
| Post-crisis | Company | Northeast Securities | Changjiang Securities | CITIC Securities | Guoyuan Securities | Haitong Securities |
| | Sector | Securities | Securities | Securities | Securities | Securities |
| | tau | 0.4297 | 0.4262 | 0.4225 | 0.4141 | 0.4061 |
| | 95% CI of tau | 0.0250 | 0.0248 | 0.0247 | 0.0243 | 0.0238 |
| | utdc | 0.6032 | 0.6000 | 0.5966 | 0.5888 | 0.5813 |
| | 95% CI of utdc | 0.0237 | 0.0236 | 0.0235 | 0.0233 | 0.0232 |

Top five companies with strongest correlation and upper tail dependence in three periods are listed in Table 4. It is obvious that all five companies are from Securities sector in all three periods no matter in terms of Kendall's tau or upper tail dependence coefficients (utdc). The values of upper tail dependence are always greater than the values of Kendall's tau. Besides, top four companies remain in their places in three periods, namely Northeast Securities Co., Ltd, Changjiang Securities Company Limited, CITIC Securities, and Guoyuan Securities Company Limited. In the pre- and post-crisis period Haitong Securities Co., Ltd. takes the fifth place but in the crisis period, Sinolink Securities Co., Ltd. becomes the fifth.

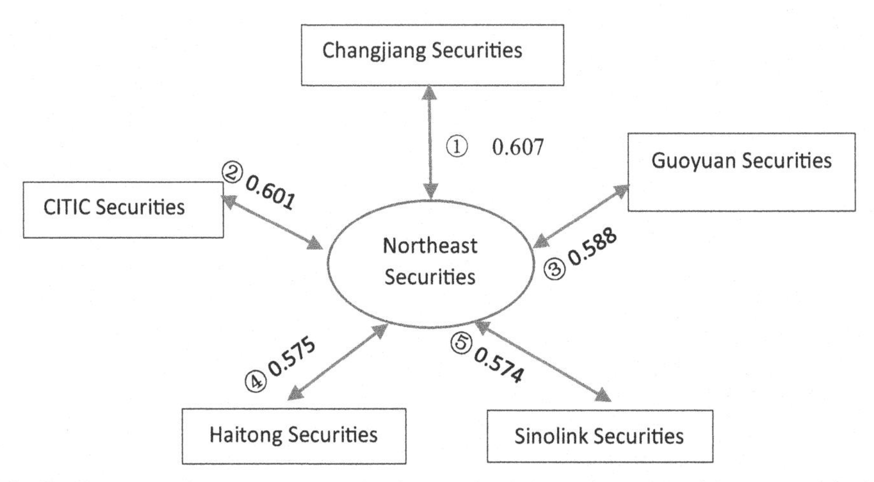


**Fig. 7.** Five companies that are most closely correlated to Northeast Securities Co., Ltd in the pre-crisis period

Securities companies listed above connect the most strongly with others in the financial sector. Conversely, for each company in our dataset, the dependence coefficients with these companies are among the highest. Let us take Northeast Securities Co., Ltd as an example. Figures 7, 8 and 9 show the top five companies that are most closely related to it (numbers next to the arrows are Kendall's tau). It is not surprising that all five companies have been listed above and the ranks are the same as that in Table 4.

To sum up, the dependence between companies and sectors in financial industry are stronger in the crisis period, implying higher risk during financial crisis. This is in accord with the fact that the crisis period witnessed greater uncertainties and larger fluctuations in financial market. Nevertheless, the differences between the crisis period with the other two periods are not huge, indicating that stock market in China is relatively stable. In addition, stronger dependence between securities companies are found. This may be attributed to the immaturity of the securities market. Problems like non-diversified services, unbalanced income structure and homogeneous competition have existed for a long time, which makes it difficult to avoid systemic risks in the stock market.

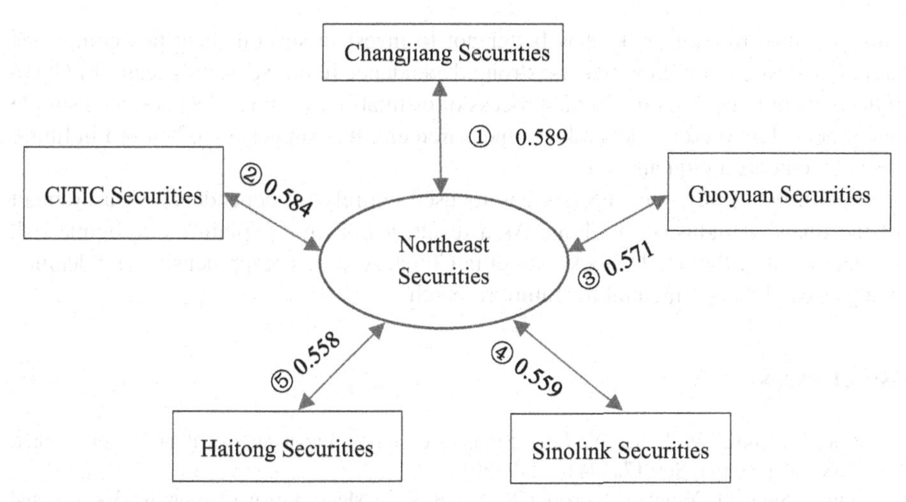


**Fig. 8.** Five companies that are most closely correlated to Northeast Securities Co., Ltd in the crisis period

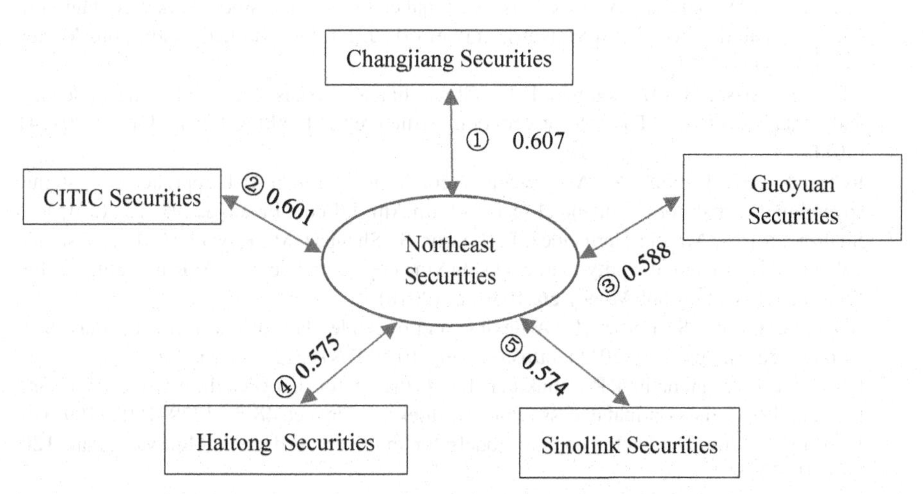


**Fig. 9.** Five companies that are most closely correlated to Northeast Securities Co., Ltd in the post-crisis period

## 4 Conclusions

This study focuses on the application of FDG copula-based ARMA-GJR-GARCH model to the data of 42 financial institutions in China over the period September 2003 to May 2020. Dependence between each pair of stocks and sectors are estimated in pre-crisis, crisis and post-crisis period, respectively. Our findings reverse the primary result from the literature to date. We show that dependence is stronger in the crisis period for all sectors and companies. We further display that the dependence coefficients of securities companies are the highest and securities sector and insurance sector are connected closely. Thus, to avoid risk, it is better not to invest in several securities companies simultaneously. For policy makers, strong dependence in the Securities sector of China reflects its immature status. In the process of estimation, we find FDG copula a simple and practical method for dependence measurement. It is suggested to be used in future research on correlation analysis.

For further study, our analysis has focused entirely on dependence measurement of the financial industry in China. As a result, it does not explain the systemic risk measurement of the whole stock market in China. A more comprehensive and detailed analysis can be implemented for future research.

## References

1. Cao, D., Long, W., Yang, W.: Sector indices correlation analysis in China's stock market. Procedia Comput. Sci. **17**, 1241–1249 (2013)
2. Liu, J., Song, Q., Yang, Q., Sanzidur, R., Songsak, S.: Measurement of systemic risk in global financial markets and its application in forecasting trading decisions. Sustainability **12**(10), 4000 (2020)
3. Kenourgios, D., Samitas, A., Paltalidis, N.: Financial crises and stock market contagion in a multivariate time-varying asymmetric framework. J. Int. Fin. Markets Institutions Money **21**(1), 92–106 (2011)
4. Aloui, R., Aïssa, M.S.B., Nguyen, D.K.: Global financial crisis, extreme interdependences, and contagion effects: The role of economic structure? J. Banking Finan. **35**(1), 130–141 (2011)
5. Reboredo, J.C., Ugolini, A.: A vine-copula conditional value-at-risk approach to systemic sovereign debt risk for the financial sector. North Am. J. Econ. Finan. **32**, 98–123 (2015)
6. Shahzad, S.J.H., Arreola-Hernandez, J., Bekiros, S., Shahbaz, M., Kayani, G.M.: A systemic risk analysis of Islamic equity markets using vine copula and delta CoVaR modeling. J. Int. Fin. Markets Institutions Money **56**, 104–127 (2018)
7. Mazo, G., Girard, S., Forbes, F.: A flexible and tractable class of one-factor copulas. Stat. Comput. **26**(5), 965–979 (2015). https://doi.org/10.1007/s11222-015-9580-7
8. Glosten, L.R., Jagannathan, R., Runkle, D.E.: On the relation between the expected value and the volatility of the nominal excess return on stocks. J. Finance **48**(5), 1779–1801 (1993)
9. Krupskii, P., Joe, H.: Factor copula models for multivariate data. J. Multivar. Anal. **120**, 85–101 (2013)
10. Durante, F.: A new class of symmetric bivariate copulas. J. Nonparametric Stat. **18**, 499–510 (2006)
11. Cuadras, C.M., Augé, J.: A continuous general multivariate distribution and its properties. Commun. Stat. Theory Methods **10**(4), 339–353 (1981)
12. Fréchet, M.: Remarques au sujet de la note précédente. CR Acad. Sci. Paris Sér. I Math. **246**, 2719–2720 (1958)
13. Joe, H., Xu, J.J.: The estimation method of inference functions for margins for multivariate models (1996)
14. Erkens, D.H., Hung, M., Matos, P.: Corporate governance in the 2007–2008 financial crisis: evidence from financial institutions worldwide. J. Corp. Finance **18**(2012), 389–411 (2012)

# Evaluation and Forecasting of Functional Port Technical Efficiency in ASEAN-4

Anuphak Saosaovaphak[1], Chukiat Chaiboonsri[2], and Satawat Wannapan[2(✉)]

[1] Faculty of Economics, Chiang Mai University, Chiang Mai, Thailand
anuphak@gmail.com

[2] Modern Quantitative Economics Research Center (MQERC), Faculty of Economics, Chiang Mai University, Chiang Mai, Thailand
chukiat1973@gmail.com, lionz1988@gmail.com

**Abstract.** The huge challenge for measuring and forecasting port efficiencies was one of the major concerns in logistics economics. This paper was aimed to deeply study the univariate calculation for the technical efficiency ratio in six major ports in Thailand, Singapore, Malaysia, and the Philippines. The annual time-series data from 2005 to 2018 was observed, including container flows, numbers of vessel arrivals, transshipments, the ranges of quay lengths, and the units of functional terminals. Observed data were categorized to be a panel. Two econometric methods such as Bootstrapping Panel Data Envelopment Analysis (BPDEA) and Bayesian Structural Time-Series Forecasting model (BSTSF) were applied for clarifying and predicting ports' bias-corrected technical efficiency ratio. The findings were used to recommend a specific policy for the uniqueness of port locational bearings.

**Keywords:** Technical efficiency · Bootstrapping Panel DEA · BSTSF · ASEAN-4

## 1 Background of Research

When the world has been being globalized, economic expansions have been still an indicator for the upcoming future, they are also considered to be the signal of better living lives. If the growth rate of economic systems is the sign of well-being images, logistics is compared as the vessel of any economy. This key playing factor refers to as supply chains. Essentially, effectiveness is involved. Toward AEC countries, logistics and supply chain management infiltrates by the master plan on ASEAN connectivity, which is the summaries of this collaborative plan consist of three milestones such as physical connectivity, people-to-people connectivity, and institutional connectivity. In terms of this connection, the goal is to achieve strategic objectives by improving logistics competitiveness, reducing cost of supply chains in each ASEAN country, and upgrading the speed and reliability of supply chains in each (Association of Southeast Asian Nations, 2019). From the point of view, the linkage between economies and supply chains should be structurally clarified.


V.-N. Huynh et al. (Eds.): IUKM 2020, LNAI 12482, pp. 297–309, 2020.
https://doi.org/10.1007/978-3-030-62509-2_25

In terms of logistics, the concept comes with two ideas; fuzzy elimination and predictable option. The former plays as a vital role for goods trading and transferring in countries locally and internationally. The latter is issued to be the tool for making a political decision to harmonize with sustainability. They are inevitably together. Real situations also support the ideas. Mass transportation brings an enormous numbers of contained products. Figure 1 illustrates that the trend of containers (the twenty-foot equivalent unit: TUEs) flowing in ASEAN countries. Interestingly, nearly nine of ten of products are moved into port terminals and sailed out from five predominant countries. Singapore, Malaysia, Thailand, Philippines, and Indonesia are the importer and exporter displayed by the red line. Almost ninety thousand million containers are arrived with various types of vessels. This makes ports become such great complexity, and the investigation for validating dynamic port effectiveness is therefore the main purpose of the research to suggest a critical policy recommendation.

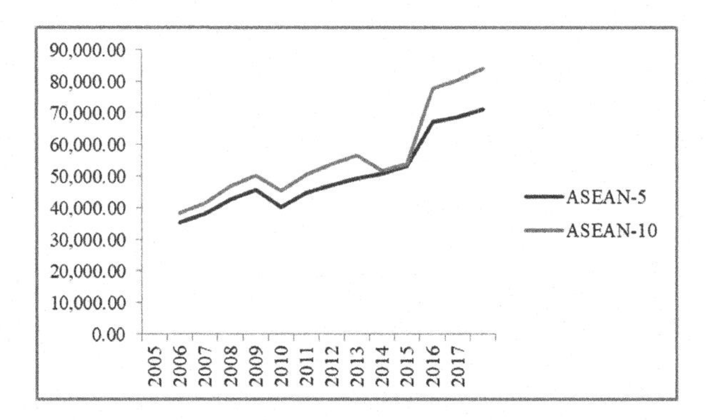


**Fig. 1.** The movement of containers through major ports in ASEAN countries

When airports are the door to open people' eyes to see the world. Container ports are the crucial gate that brings worldwide products to people' lives. In 2018, major ports in ASEAN-5 are in the top list of fifty busiest container ports (World Shipping Council, 2019). Especially, the port of Singapore is the second active port of the world ranking, which is detailed in Table 1. Consequently, this can be stated that ports are driven by the supply side, supplies need an effective management, and the direction to the point of sustainability in supply chain management eventually necessitates an econometric computation. This paper is on the account to apply some econometric approaches to clarify port efficiencies and forecast a sensible scenario for recommending a long-term policy in logistics. In this research, Bootstrapping Panel Data Envelopment Analysis (DEA) and simulation are the optimal measurement for an empirical result.

## 2 Literature Review

Logistics economics (formerly called transportation economics) is a branch of economics that deals with the allocation of resources within the transport sector. Fundamentally, in a macroeconomic view, transportation activities form a portion of a nation's total economic

**Table 1.** The situation of five complex container ports in ASEAN countries **Sources:** World shipping council.

| Worldwide rank | Port | Volume 2018 (Million TEU) | Volume 2017 (Million TEU) | Volume 2016 (Million TEU) | Volume 2015 (Million TEU) | Volume 2014 (Million TEU) |
|---|---|---|---|---|---|---|
| 2 | Singapore | 36.60 | 33.67 | 30.90 | 30.92 | 33.87 |
| 12 | Port Klang, Malaysia | 12.32 | 13.73 | 13.20 | 11.89 | 10.95 |
| 21 | Laem Chabang, Thailand | 8.07 | 7.78 | 7.22 | 6.82 | 6.58 |
| 22 | Tanjung Priok, Jakarta, Indonesia | 7.64 | 6.09 | 5.51 | 5.20 | 5.77 |
| 28 | Manila, Philippines | 5.05 | 4.82 | 4.52 | 4.23 | 3.65 |

product play a role in building or strengthening a national or regional economy. This is also included an influence in the development of land and other resources. In the sense of microeconomics, transportation involves the connectivity between firms and individual consumers. To link two theoretical concepts, data simulation management and econometrics is the key for solution. Thus, Bootstrapping Panel DEA is the non-parametric econometric tool for the future logistics efficiency, and Bayesian Structural Time Series Forecasting (BSTSF) model stands for data predictions.

It has been several years that Data Envelopment Analysis (DEA) is a role function to investigate efficiency calculations for macroeconomic and microeconomic logistics. The panel DEA was applied by Cullinane and Wang (2010) for studying the efficiencies of container port production using DEA panel data approaches. Mokhtar (2013) considered technical efficiencies of container terminal operations using the panel DEA approach for measurement of efficiency in six major container terminals of Peninsular Malaysia. Guimarãesa et al. (2014) deliberated the environmental performance of Brazilian container terminals using a DEA approach for technical efficiency measurement of ports. Continuously, Almawsheki and Shah (2015) attempted to measure the technical efficiency of container terminals in the Middle Eastern region by using a DEA approach. Cruz and Ferreira (2015) evaluated the competitiveness in Iberian ports by using a DEA approach and summarizing that those seaport efficiency scores were insignificant in its cargo throughput. Omrani and Keshavarz (2016) implemented the DEA approach to evaluate the efficiency of the shipping company in Iran. In 2017, Kutina et al. researched the relative efficiencies of ASEAN container ports using DEA analysis. The findings indicated that ports of this region need to improve the transportation network and trading competitiveness in ASEAN countries. Moreover, Intapan et al. (2019) studied the investigation of MICE tourism efficiencies. The main computational tool was the Bootstrapping Data Envelopment Analysis (Bootstrapping DEA) method comparing with the Stochastic Frontier Analysis (SFA) method by copula estimating. In agricultural economics, Somboon et al. (2018) explored the technical efficiencies of rubber productions from ASEAN countries by using the bootstrapping panel DEA analysis. In this paper, panel data is used to improve the computational ability for DEA. Applying from Simar and Wilson (1988), Nguyen et al. (2016), and Dharmapala (2018), calculations might not reveal the efficiency score because of the possibility for containing random errors. The solution is to employ the bootstrapping Panel DEA analysis.

The BSTSF model has been used in many research branches such as business, science, and engineering. In this paper, BSTSF is employed to seek simulated possible data from a forecasting period. The goal is to visualize the future observations for predicting future port efficiencies beforehand. Historically, BSTSF was applied for weekly (non-seasonally adjusted) initial claims for US unemployment benefits by Scott and Varian (2014). The authors also mentioned BSTSF is presented to be the useful for economic time series forecasting. Pinilla et al. (2018) had conducted the contribution with the BSTSF model to study how to calculate the effectiveness of a completed ban on smoking in public places, compared with a partial ban in Spain. Schmitt et al. (2018) meanwhile found the way to compare a market exposed to an advertising campaign with control markets identified through a matching procedure. In addition, Jun (2019) proposed a method of sustainable technology analysis applying a Bayesian structural time series forecasting (BSTSF) model based on time series data to provide the set of observations in the regression model. As the review, the advantage of BSTSF can be useful for exploring the complexity of port logistics time-series data.

## 3 Data Review

The paper is considered into two major issues. The first condition is the diversity of landscapes in each area. Each port needs a unique policy framework because of the different modes of local transportations for transshipments[1] in each country. The second condition is each port is nearly independent. However, there is suspicious if ports are defined as a set of absolutely identical variables as stock markets. Consequently, the annual time-series data is observed from six ports such as Laem Chabang (Thailand), Bangkok port (Thailand), Port of Singapore (Singapore), Port Klang (Malaysia), Port Kuching (Malaysia), and Port of Manila (Philippines). The time-series range is between 2005 and 2018. The locations of six selected ports are presented in Fig. 2.

[1] The transshipment is the shipment of goods or containers to an intermediate destination.

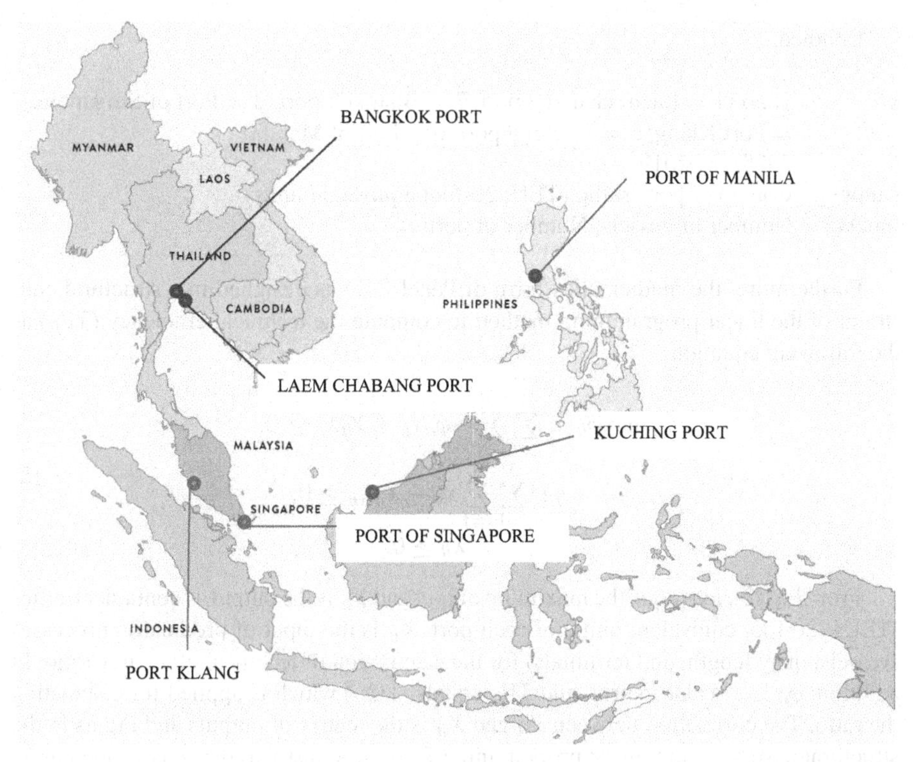


**Fig. 2.** The map of six ports in major four countries in ASEAN (Source: www.pinterest.com)

# 4 The Concept Framework and Methodology

## 4.1 Conceptual Framework of Research

### a. *The concept of efficiency estimation*

The main streams of efficiency measurement have been developed by continued studies for over 40 years, as seen in Koopmans (1951), Debreu (1957), Farrell (1975), and Coelli (1996). The importance of the efficiency measurement conceptually mentions how the firms (ports) reflects to produce the goods (containers: TEUs, 20 foot equivalent units) of ports by maximizing outputs, which are constant by using the minimum inputs. Broadly studying, the panel DEA analysis is adopted to make a better calculation for the technical efficiency of container ports, for example, Cullinane and Wang (2010); Mokhtar (2013), and Cruz and Ferreira (2015). In terms of mathematics programming for panel DEA analysis, the model can be started by Eq. (1) as

$$\text{Efficiency} = \frac{Output_{it}}{Input_{it}}, \quad (i: \text{numbers of firms}, \ t: \text{time periods}). \tag{1}$$

Denoted:

| | |
|---|---|
| *i*: | 1,..,6 (1 = Laem chabang port, 2 = Bangkok port, 3 = Port of Singapore, 4 = Port Klang, 5 = Kuching port, 6 = Port of Manila) |
| *t*: | 2005, …, 2019 |
| $Output_{it}$: | Container port traffic (TEU: 20-foot equivalent units) |
| $Input_{it}$: | Number of vessels, Number of ports. |

Furthermore, the mathematics form of Panel DEA is designed in a structural constraint of the linear programming method to compute the technical efficiency ($TE_{it}$) as the following equation.

$$\begin{aligned} \max\phi_{it}, \quad & \sum_{i=1}^{3}\sum_{t=1}^{9} -\phi_{it}y_{it} + Y_{it}\lambda_{it} \geq 0 \\ \text{s.t} \quad & \sum_{i=1}^{3}\sum_{t=1}^{9} x_{it} - X_{it}\lambda_{it} \geq 0, \\ & \lambda_{it} \geq 0. \end{aligned} \tag{2}$$

From Eq. (2), $\max\varphi_{it}$ is the maximum output and $Y_{it}$ is the output as container traffics (TEUs: 20-foot equivalent units) of each port. $X_{it}$ is the input of production processes (vessels, quay length, and terminals) for the calculation of technical efficiency ratio. In addition, $\phi_{it}$ is a scalar matrix, and $TE_{it}$ equals $(1/\phi_{it})$ which is applied for calculating the ratio. The correlation between $Y_{it}$ and $X_{it}$ is the matrix of outputs and inputs in the structural system of the linear programming form. The last parameter is formed as λ, which is a I×1 vector of constants.

**b. *Technical Efficiency (TE) bootstrapping analysis for panel data***

The disadvantage of this analysis implies that it is rarely accepted for estimation. The essential point is that DEA input/output data probably contains random errors which are abnormal (Dharmapala, 2018). Basically, the DEA analysis is ignorance for the point to calculate the technical efficiency. Therefore, Simar and Wilson (1988) considered into this problem. Their empirical results suggested the DEA bootstrapping method to resolve the problem. The bootstrapping algorithm for DEA is expressed as the following processes:

- **Step 1:** calculating the DEA from original data for each seaport to obtain $\hat{\phi}_{it}$(i = 1,2,3,4,5 (each seaport in ASEAN countries), t = 2008,..,2017).
- **Step 2:** to bootstrap samples and calculate TE $\{\phi_{it,b^*}, \phi_{it,b^*}, \ldots, \phi_{it,b^*}\}$ until *b*-th (b = 1,…,B).
- **Step 3:** constructing the smooth bootstrap samples by the process in Eq. (3)

$$\phi_{it,b}^{**} = \left\{ \begin{matrix} \phi_{it}^{*} + h\varepsilon_{it}, if\, \phi_{it}^{*} + h\varepsilon_{it} \geq 1, \quad for(i = 1, \ldots, 5, t = 2008, \ldots, 2017) \\ 2 - (\phi_{it}^{*} + h\varepsilon_{it}), \ otherwise, \ for(i = 1, .., 5, t = 2008, .., 2017) \end{matrix} \right\}. \tag{3}$$

- **Step 4:** generating the *b*-th pseudo-data by according the Eq. (4) was displayed as

$$\{(x_{it}^*, y_{it}^* = y_{it}\frac{\hat{\varphi}_{it}}{\varphi_{it}^{**}}, i = 1, \ldots, 5, t = 2008, .., 2017)\}. \tag{4}$$

- **Step 5:** using the pseudo-data set from Eq. (4) to calculate new $\hat{\phi}_{it}^*$, by the LP method which it was displayed the mathematical formula by the Eq. (2).
- **Step 6:** repeating the loop of the step (2) to step (4) in *b*-times to receive $\hat{\phi}_{it}^*$, for each i-firm at time *t*.
- **Step 7:** calculating both the bias of $\mathrm{TE_{it}}$ and bias-corrected of $\mathrm{TE_{it}}$ by bootstrapping method from formula based in Eq. (5) and Eq. (5.1) below:-

$$\hat{\mathrm{Bias}}(\mathrm{TE}_{it}) = \mathrm{Est.\ Bias}\ \hat{\phi}_{\mathrm{it}} = \frac{1}{B}\sum_{b=1}^{B}\hat{\phi}_{it}^{b} - \hat{\phi}_{\mathrm{it}} \tag{5}$$

$$Bias{-}corrected = \hat{\phi}_{\mathrm{it}} - \hat{\mathrm{Bias}}(\mathrm{TE}_{it}). \tag{5.1}$$

The finite samples of observed production units are importance for DEA bootstrapping, especially when data is the panel form.

**d.** ***Predictive data by Structural Time Series Forecasting (STSF) model***

The BSTSF model is one of nowcasting machine learning algorithms. A time series component and regression component are the crucial combination for this approach. For the structure of time-series analyzing, let $y_t$ denote observation $t$ in a real-valued time series. The structural time-series model can be explained by a pair of equations relating $y_t$ to a vector of latent state variables, $\alpha_t$ (Scott and Varian, 2014).

$$y_t = Z_t^T \alpha_t + \varepsilon_t, \quad \varepsilon_t \sim N(0, H_t) \tag{6}$$

$$\alpha_{t+1} = T_t\alpha_t + R_t\eta_t, \quad \eta_t \sim (0, Q_t). \tag{7}$$

Equation (6) is called the observation equation since it links the observed data $y_t$ with the unobserved latent state $\alpha_t$. Equation (7) is defined as the transition equation because it explains how the latent state evolves over time. $Z_T$, $T_t$, and $R_t$ are the matrices which usually contain a mix of known values (commonly using 0 and 1), and unknown parameters. The matrix $T_t$ is quadrilateral and $R_t$ is rectangular when a portion of the state transition is deterministic. The existence of the term $R_t$ allows the structural process to work with a full rank variance matrix $Q_t$. Fundamentally, any linear dependencies in the state vector can be relocated from $Q_t$ to $R_t$, and $H_t$ is a positive scalar. The model that can be estimated by Eqs. (6) and (7) is in a state space form. One useful model is obtainable by adding a regression component to the simplified 'basic structural model'. The model is described as following equations:

$$y_t = \mu_t + \tau_t + \beta^T x_t + \varepsilon_t \tag{8.1}$$

$$\mu_t = \mu_{t-1} + \delta_{t-1} + u_t \tag{8.2}$$

$$\delta_t = \delta_{t-1} + v_t \tag{8.3}$$

$$\tau_t = -\sum\nolimits_{s=1}^{S-1} \tau_{t-s} + \omega_t. \tag{8.4}$$

This model contains trend, seasonal, and regression components. The current level of the trend is $\mu_t$, the current 'slope' of the trend is $\delta_t$. The seasonal part is defined by $\tau_t$ which can be thought of as a set of $S$ dummy variables with dynamic coefficients constrained to have zero expectation over a full cycle of $S$ seasons (Scott and Varian, 2014).

Various Bayesian inference approaches have been used in economics; logistics economics is mentioned in this paper. Bayesian inference is based on the Bayes theorem as follows (Koduvely, 2015; Jun, 2019),

$$P(\theta|x) = \frac{P(x|\theta)P(\theta)}{P(x)}. \tag{9}$$

$x$ is observed information and $\theta$ is the model parameter. $P(x|\theta)$ presents the prior and likelihood functions. $P(\theta|x)$ is the posterior function. This is updated by learning the observed data $x$ given $\theta$ which is a likelihood. From Eq. (8.1), The $\tau_t$ is distributed to Gaussian and the Bayesian regression model can be expressed as follows (Gelman et al., 2013)

$$y = \beta_0 + \sum_{p=1}^{P} \beta_i x_i + e. \tag{10}$$

$e$ is the Gaussian distribution with mean = 0 and variance = $\sigma^2$. In this application, the prior of $\sigma^2$ is relied on an inverse-chi square ($Inv - \chi^2$) distribution. This can be explained as

$$\sigma^2 \sim Inv - \chi^2\left(n - p, s^2\right), \tag{11}$$

where $n$ and $p$ are data and parameter sizes, respectively. In addition, $s^2$ is given by

$$s^2 = \frac{1}{n-p}\left(y - X\hat{\beta}\right)^T\left(y - X\hat{\beta}\right). \tag{12}$$

# 5 Case Study the Efficient Logistics in ASEAN Ports

## 5.1 The Resolution for Abnormality in Data Distribution

Unlike time-series macroeconomic variables, the indexes of logistics and supply chains are unique in terms of landscapes and functional port activities. Descriptive information reported by Table 2 indicates the Jarque-Bera significant test details the critical probabilities imply the observed samples are not a normal distribution. Alternatively expressing, the box plot graphically shown in Fig. 3 suggests that the input and output variables are an abnormal distribution. This needs to calculate the technical efficiency by panel data analysis and bootstrap data repeating.

**Table 2.** Descriptive details of logistics and supply chains in ports of ASEAN-4 countries **Sources:** authors.

| | Container traffics (TUEs) | Transshipments (TMS) | Numbers of vessels (NBV) | Quay length (QL) | Port terminals (PORT) |
|---|---|---|---|---|---|
| Mean | 17517610 | 6389123 | 355126.3 | 4221.833 | 10.52381 |
| Maximum | 82238013 | 28630812 | 2799585 | 15500.00 | 23.00000 |
| Minimum | 143096 | 594.0000 | 1517.000 | 590.0000 | 1.000000 |
| Skewness | 1.536510 | 1.351553 | 2.052512 | 1.414407 | 0.196132 |
| Kurtosis | 4.245309 | 3.636630 | 5.575364 | 3.389720 | 1.664343 |
| Jarque-Bera | 38.47988 | 26.99229 | 82.19304 | 28.53925 | 6.782473 |
| Probability | 0.000000 | 0.000001 | 0.000000 | 0.000001 | 0.033667 |
| **Normality test** | NOT normality* | NOT normality* | NOT normality* | NOT normality* | NOT normality** |

Noted: */reject the null at a significance level of 1%.
**/reject the null at a significance level of 5%.

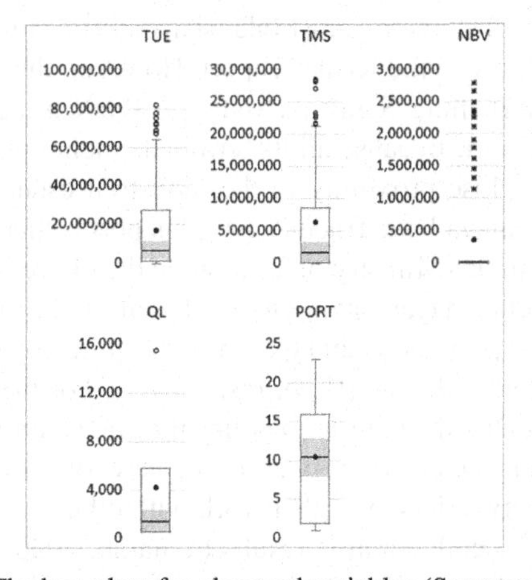


**Fig. 3.** The box plots for observed variables (Sources: authors)

## 5.2 Efficiency Scores by the Panel Bootstrapping DEA Estimation

Considering into one by one port in Table 3, the interesting point focuses on the port of Singapore. 0.554 is the quite low average bias-corrected efficiency score for the port comparing with others in the same continent. The trend of efficiency scores starts to be decreased from 0.866 to 0.448 during 2008 to 2018. Cullinane et al. (2006) stated in their suggestion for the challenge of Singapore port regarding "the landward and seaward sides". This is the chronical problem of lacking spaces causing higher generalized costs and lower profit margins for the port user. Consequently, the TE (0.554) is the evidence that the problem still occurs and it is not simply to completely solve.

The Klang port in Malaysia is also the highlight. 0.295 is the lowest average bias-corrected score when it compares with other ports' efficiencies. However, the positive

**Table 3.** Technical efficiency scores of six container ports in ASEAN-4 countries **Source:** authors.

| Seaport | | Original TE | Bias TE | Bias-corrected of TE |
|---|---|---|---|---|
| Laem Chabang Port | Average (2005–2018) | 0.941 | 0.146 | 0.795 |
| Bangkok Port | Average (2005–2018) | 0.954 | 0.148 | 0.806 |
| Port of Singapore | Average (2005–2018) | 0.588 | 0.034 | 0.554 |
| Port Kelang | Average (2005–2018) | 0.302 | 0.008 | 0.295 |
| Kuching Port | Average (2005–2018) | 0.819 | 0.123 | 0.696 |
| Port of Manila | Average (2005–2018) | 0.905 | 0.065 | 0.841 |

sign for Kelang port is the trend of the efficiency score has been continuously raising since 2005 to 2018. Jeevan et al. (2015) mentioned in the contribution regarding the obstructions to develop dry ports in Malaysia. The insufficient mode of transportation, unorganized container planning on the rail deck, highly dependent on single mode of transportation, poor recognition from the seaport community, and competition from localized seaports are key factors causing Malaysian maritime logistics to slowly expand, especially the federal port like the Klang seaport. However, the situation is different for Kuching port located in Kuching, Sarawak, Malaysia. 0.696 is the average bias-corrected efficiency score for the port. Because all its terminals including Pending Terminal and Senari Terminal are in close proximity to the industrial estates such as Demak Laut Industrial Park and Samajaya Free Trade Zone. This port is therefore the main gateway for Sarawak's external trade (Ministry of Tourism and Culture, 2016).

The situation for port management in Thailand seems to be satisfactory. During 2005 to 2018, Laem Chabang port and Bangkok port gain quite positive bias-corrected efficiency scores, which are 0.806 and 0.795, respectively. For the former port located in the east shore of the country, the long-term policy the "innovation district" implemented in Sri Racha, Chonburi is the positive effect for the new role of the port, including shipping lines and freight forwarders, etc., that work with other members in the innovation district would change the collaboration results (Senarak, 2020). Similarly, the average efficiency score for Bangkok port confirms the motivation of the country to be the center of transshipment corridors. However, the fluctuation of bias-correction efficiencies for the port is a disadvantage since the high cost of dependency on road systems (nearly indicating 90% of all transportation modes) (Deerod, 2018). Furthermore, the positive port efficiency appears on Port of Manila, Philippines. 0.841 is the highest average bias-corrected efficiency score comparing other 5 ports. This high score is backed up by the contribution conducted by Wang et al. (2019). Their conclusion states Manila port is concerned for current and future investment by Chinese terminal operators.

### 5.3 Future Overview by the BSTSF Model for Port Efficiencies

Data forecasting plays a role part for suggesting long-term policies. Also, machine learning is the vital tool for complex information. Table 4 represents the technical efficiency comparison of six selected ports. For each port, two major container ports in Thailand are predicted to be in downsizing situations. TE for Laem Chabang port is a diminished fluctuation, which starts to decrease from 0.620 to 0.393 during 2019 to 2023. The efficiency ratio for Bangkok port is additionally forecasted to be slightly reduced from 0.635 to 0.557. For Singapore, the predictive ratio is conversely reported by the continuous increment. The technical efficiency trend is predicted to be highly productive, which is increased from 0.722 to 1.021. In Malaysia, the development between Kelang and Kuching ports is reported by the contrast trend of technical efficiency ratios. Kelang port which is the main federal ocean gate of the country is predicted to have a slight improvement in port efficiency, which is from 0.415 to 0.527. On the other hand, Kuching port is computationally indicated the TE ratio continuously drops from 0.388 to 0.311 during five predictive years. For the port in Philippines, Manila port is predicted to be stable in efficiency scores. During 2019 to 2023, the TE ratio is reported by a slightly change from 0.752 and 0.725.

**Table 4.** Bias-corrected TE prediction for six ports in ASEAN **Source:** authors.

| Years | Laem Chabang Port | Bangkok Port | Port of Singapore | Port Kelang | Kuching Port | Port of Manila |
|---|---|---|---|---|---|---|
| 2016 | 0.915 | 0.899 | 0.399 | 0.399 | 0.852 | 0.729 |
| 2017 | 0.823 | 0.923 | 0.412 | 0.333 | 0.717 | 0.745 |
| 2018 | 0.480 | 0.329 | 0.448 | 0.327 | 0.336 | 0.782 |
| 2019p | 0.620 | 0.635 | 0.722 | 0.415 | 0.388 | 0.752 |
| 2020p | 0.581 | 0.611 | 0.833 | 0.437 | 0.363 | 0.749 |
| 2021p | 0.483 | 0.592 | 0.857 | 0.465 | 0.337 | 0.739 |
| 2022p | 0.509 | 0.584 | 0.857 | 0.484 | 0.328 | 0.730 |
| 2023p | 0.393 | 0.557 | 1.021 | 0.527 | 0.311 | 0.725 |

## 6 Conclusion

As evidently found that seawater logistics is the art of transportation management, this paper is conducted to investigate and predict the efficiency of functional ports in ASEAN-4 countries (Singapore, Thailand, Malaysia, and Philippines). It is obvious that container ports are the gate to massively move an enormous amount of supplies to the countries. Additionally, each country has the uniqueness of culturally structural economy. This therefore influences its major port is functional exclusive. The empirical findings show that the port efficiency calculation must be separately considered, even though mass information causes the complexity of estimations and predictions to research processes. "Transshipment containers" are the blue ocean for Singapore port (Cullinane et al., 2006). "hinterland connectivity" is the challenge for overcoming intensive competitions of functional ports in Malaysia (Chen et al. 2016). "International maritime hub" is the target for Philippines (Llanto et al., 2005; Fillone, 2016). "Blue economy" which is the operation regarding Port Safety, Health and Environ-mental Management: PSHEM on Thailand.

In conclusion, this paper is the novel way to deal with this difficult task by using panel data and a machine learning algorithm based on Bayesian thought, which is modernized for econometrics applying for logistics economics. For the upcoming future research, quantum computing can be the great potential method for better understanding how to manage, maintain, and maximize the efficiency of major container ports in the continent for shifting up the level of economic expansion and people' living standard.

## References


Almawsheki, E.S., Shah, M.Z.: Technical efficiency analysis of container terminals in the Middle Eastern region. Asian J. Shipping Logistics **31**(4), 477–486 (2015)

Association of Southeast Asian Nations: Master plan on ASEAN connectivity (2019). http://www.mfa.go.th/asean/contents/files/asean-media-center-20121203-182010-779067.pdf. Accessed 19 May 2020

Cullinane, K., Wang, T.: The efficiency analysis of container port production using DEA panel data approaches. OR Spectr. **32**(3), 717–738 (2010)

Chen, S.L., Jeevan, J., Cahoon, S.: Malaysian container seaport-hinterland connectivity: status, challenges and strategies (2016)

da Cruz, M.R.P., de Matos Ferreira, J.J.: Evaluating Iberian seaport competitiveness using an alternative DEA approach. Eur. Transp. Res. Rev. **8**(1), 1–9 (2015). https://doi.org/10.1007/s12544-015-0187-z

Coelli, T.J.: A guide to DEAP version 2.1: a data envelopment analysis (computer) program. CEPA Working Paper 96/08, Department of Econometrics, University of New England, Armidale (1996)

Cullinane, K., Yim Yap, W., Lam, J.S.L.: Chapter 13 the port of Singapore and its governance structure. Res. Transp. Econ. **17**, 285–310 (2006). https://doi.org/10.1016/s0739-8859(06)17013-4

Debreu, G.: The coefficient of resource utilization. Econometrica **9**, 273–292 (1957)

Deerod, K.: Developing port marketing strategies: a case study for Bangkok port, Thailand. World Maritime University Dissertations, 621 (2018). https://commons.wmu.se/all_dissertations/621

Dharmapala, P.S.: Bias-correction in DEA efficiency scores using simulated beta samples: an alternative view of bootstrapping in DEA. Int. J. Math. Oper. Res. **12**(4), 438–456 (2018)

Farrell, M.J.: The measurement of productive efficiency. J. Roy. Stat. Soc. **96**(3), 477–503 (1975)

Fillone, A.: Easing Port Congestion and Other Transport and Logistics Issues. In: Siar, S.V., Aranas, M.V.P. (eds.) Philippine Institute for Development Studies Publisher (2016)

Gelman, A., Carlin, J.B., Stern, H.S., Dunson, D.B., Vehtari, A., Rubin, D.B.: Bayesian Data Analysis, 3rd edn. Chapman & Hall/CRC Press, Boca Raton (2013)

Guimarãesa, A.V.D., Juniorb, I.C.L., Garcia, P.A.D.A.: Environmental performance of Brazilian container terminals: a data envelopment analysis approach. Procedia Soc. Behav. Sci. **160**, 178–187 (2014)

Jeevan, J., Chen, S.L., Lee, E.S.: The challenges of Malaysian dry ports development. Asian J. Shipping Logistics **31**(1), 109–134 (2015)

Jun, S.H.: Bayesian structural time series and regression modeling for sustainable technology management. Sustainability **11**, 4945 (2019). https://doi.org/10.3390/su11184945

Koduvely, H.M.: Learning Bayesian Models with R. Packt, Birmingham (2015)

Koopmans, T.C.: An analysis of production as an efficient combination of activities. In: Koopmans, T.C. (ed.) Activity Analysis of Production and Allocation, Cowles Commission for Research in Economics, Monograph No. 13. Wiley, New York (1951)

Intapan, C., Sriboonchitta, S., Chaiboonsri, C., Piboonrungroj, P.: Technical efficiency analysis of tourism and logistics in ASEAN: comparing bootstrapping DEA and stochastic frontier analysis based decision on copula approach. In: Kreinovich, V., Sriboonchitta, S. (eds.) TES 2019. SCI, vol. 808, pp. 389–401. Springer, Cham (2019). https://doi.org/10.1007/978-3-030-04263-9_30

Llanto, G., Basilio, E., Basilio, L.: Competition policy and regulation in ports and shipping. PIDS Discussion Paper No. 2005-02. Philippine Institute for Development Studies, Makati City (2005)

Ministry of Tourism and Culture: Tourism Malaysia (2016). https://www.tourism.gov.my/pdf/uploads/Cruise_Feb_2016.pdf. Accessed 21 May 2020

Mokhtar, K.: Technical efficiency of container terminal operations: a DEA approach. J. Oper. Supply Chain Manage. **6**(2), 1–19 (2013)

Nguyen, H.O., Nguyen, H.V., Chang, Y.T., Chin, A.T.H., Tongzon, J.: Measuring port efficiency using bootstrapped DEA: the case of Vietnamese ports. Marit. Policy Manage. **43**(5), 644–659 (2016)

Omrani, H., Keshavarz, M.: A performance evaluation model for supply chain of shipping company in Iran: an application of the relational network DEA. Marit. Policy Manage. **43**(1), 121–135 (2016)

Pinilla, J., Negrín, M., Valcárcel, B.G.L., Vázquez-Polo, F.J.: Using a Bayesian structural time–series model to infer the causal impact on cigarette sales of partial and total bans on Public Smoking. J. Econ. Stat. **238**(5), 423–439 (2018)

Schmitt, E., Tull, C., Atwater, P.: Extending Bayesian structural time-series estimates of causal impact to many-household conservation initiatives. Ann. Appl. Stat. **12**(4), 2517–2539 (2018). https://doi.org/10.1214/18-aoas1166

Scott, S.L., Varian, H.R.: Predicting the present with Bayesian structural time series. Int. J. Math. Model. Numer. Optimisation **5**(1/2), 4–23 (2014)

Senarak, C.: Shipping-collaboration model for the new generation of container port in innovation district: a case of Eastern economic corridor. Asian J. Shipping Logistics (2020, forthcoming). https://doi.org/10.1016/j.ajsl.2019.11.002

Simar, L., Wilson, P.W.: Sensitivity analysis of efficiency scores: how to bootstrap in nonparametric frontier models. Manage. Sci. **44**(1), 49–61 (1988)

Somboon, K., Chaiboonsri, C., Sriboonchitta, S.: Efficiency analysis of natural rubber production in ASEAN: the comparison of panel DEA and bootstrapping panel DEA analysis based decision on copula approach. In: Huynh, V.-N., Inuiguchi, M., Tran, D.H., Denoeux, T. (eds.) IUKM 2018. LNCS (LNAI), vol. 10758, pp. 467–476. Springer, Cham (2018). https://doi.org/10.1007/978-3-319-75429-1_39

Top 50 World Container Ports | World Shipping Council (2012). www.worldshipping.org. Archived from the original on 04 Jul 2012. Accessed 16 Oct 2019

Wang, L., Zheng, Y., Ducruet, C., Zhang, F.: Investment strategy of Chinese terminal operators along the 21st-century maritime silk road. Sustainability **11**, 2066 (2019). https://doi.org/10.3390/su11072066

# A Spatial Analysis of International Tourism Demand Model: The Exploration of ASEAN Countries

Kanchana Chokethaworn[1], Chukiat Chaiboonsri[2], and Satawat Wannapan[2(✉)]

[1] Faculty of Economics, Chiang Mai University, Chiang Mai, Thailand
kanchana@cmu.ac.th
[2] Modern Quantitative Economics Research Center (MQERC), Faculty of Economics, Chiang Mai University, Chiang Mai, Thailand
chukiat1973@gmail.com, lionz1988@gmail.com

**Abstract.** The main objective of the paper is to apply Bayesian statistics to the panel linear regression models for understanding the tourism demand function in 7 countries of South East Asia (Brunei, Indonesia, Malaysia, Singapore, Thailand, Vietnam, and the Philippines) regarding the spatial effect. The observed panel data is an annual range between 2013 and 2019. The dependent variable is the number of international tourists. The independent variables are world gross domestic products, world prices for jet fuel, domestic hotel rental prices, exchange rates, average annual temperature, and visibility. In the first methodological part, exogenous variables are investigated by the least absolute shrinkage and selection operator (LASSO) regression for validating the set of predictable variables. For the second section which is the highlight, three types of linear panel regression models such as pooled regression, spatial lag regression, and spatial errors regression are used for Bayesian approach. With comparing by deviance information criterion (DIC), the spatial lag regression (pure space-recursive model) is the most appropriate estimation can be proceeded to decide tourism policies for this equator continent.

**Keywords:** Bayesian statistics · Spatial econometrics · Panel regression model · Tourism demands · ASEAN-7

## 1 The Motivation of Research

The argument of using the null hypothesis (NP) relied on "P-Value" is the main focus in modern econometrics. To publish findings assumed to stand for the whole sample sizes is suspicious and dangerous for both statistical model developing and practical policy implementing. Inside the significant detail with $t = 2.7$ (the probabilistic value is 0.01), there is no difference in mean performance that means P-value ignores the indication of whether NP or the underlying experimental hypothesis, which is "true." Although the probability of the null hypothesis is declared as true, an alternative hypothesis cannot be told it is false. The aware of using P-Value might be the failure to control for multiple


V.-N. Huynh et al. (Eds.): IUKM 2020, LNAI 12482, pp. 310–321, 2020.
https://doi.org/10.1007/978-3-030-62509-2_26

testing, p-hacking, even misspecified tests. Additionally, the dependence on the amount of data samples is the weak point of P-Value, especially to deal with data sizes for economic effects (Harvey 2017).

To completely avoid P-value and consider into a rationally statistical inference, Bayesian statistics can be the most suitable for this issue. Using Bayesian approach has no such assumption (normality testing). This comes to involve the problem of non-sensible rejection for the null hypothesis that is the huge gap of using P-value. The strong point of this subjective statistics is "prior information", but it is ignored in classical approaches. Obviously, the Bayesian starts from the prior, which potentially consider the "hack" to the estimation. However, the prior is transparent and a skeptical reader can use whatever prior they think is appropriate [1]. Based on Bayes' rule with a multiple linear model, there are one dependent variable ($Y_{it}$) and independent variables ($X_{it}$) which are formulized as follows:

$$Y_{it} = \beta_0 + \beta_1 X_{it} + e_{it}. \tag{1}$$

$Y_{it}$ contains the matrix $n \times 1$ and $X_{it}$ has the matrix $(n \times (i+1))$. The parameter $\beta$ is the vector regression $((i+1) \times 1)$. $e_{it}$ is the vector $n \times 1$ of errors. The estimated parameter can be obtained from the formula (2),

$$\hat{\beta}_1 = \left(X_{it}^T X_{it}\right)^{-1} X_{it}^T Y_{it}. \tag{2}$$

In Bayesian linear regression, there is a prior, likelihood distribution and posterior distribution [2]. The error is assumed to be normality. Also, $\left(Y_{it} | X_{it}, \beta_1, \sigma^2\right)$ are normally distributed and probability distribution function (pdf). Based on Bayes' rule, these variables are as follows:

$$p\left(Y_{it} | X_{it}, \beta_1, \sigma^2\right) = \frac{1}{\sqrt{2\pi\sigma^2}} exp\left(-\frac{1}{2\sigma^2}(Y_{it} - X_{it}\beta_1)^T (Y_{it} - X_{it}\beta_1)\right). \tag{3}$$

The probability density function (pdf) in the formula (3) can be given the likelihood function of setting variables as follows:

$$p\left(Y_{it} | X_{it}, \beta_1, \sigma^2\right) = \left(\sigma^2\right)^{-n/2} exp\left(-\frac{1}{2\sigma^2}(Y_{it} - X_{it}\beta_1)^T (Y_{it} - X_{it}\beta_1)\right), \tag{4}$$

$$p\left(Y_{it} | X_{it}, \beta_1, \sigma^2\right) \propto \left(\sigma^2\right)^{-v/2} exp\left\{-\frac{vs^2}{2\sigma^2}\right\} \times \left(\sigma^2\right)^{-n/2}$$
$$exp\left\{-\frac{1}{2\sigma^2}(Y_{it} - X_{it}\beta_1)^T (Y_{it} - X_{it}\beta_1)\right\}. \tag{5}$$

The distribution of the conjugated prior can be used in this Bayesian approach [3]. Regression model parameters with Bayesian approach can be estimated by iterating at marginal posteriors. Posteriors can be obtained from the calculation by multiplying the prior distribution and likelihood function that is derived in the following formulas:

$$\text{Posterior} \propto \text{Likelihood} \times \text{Prior}$$

$$p\left(\beta_1, \sigma^2 | Y_{it}, X_{it}\right) \propto p\left(Y_{it} | X_{it}, \beta_1, \sigma^2\right) p\left(\sigma^2\right) p\left(\beta_1 | \sigma^2\right), \tag{6}$$

$$p\left(\beta_1, \sigma^2 | Y_{it}, X_{it}\right) \propto \left(\sigma^2\right)^{-n/2} exp\left\{-\frac{1}{2\sigma^2}(Y_{it} - X_{it}\beta_1)^T (Y_{it} - X_{it}\beta_1)\right\}$$
$$\times \left(\sigma^2\right)^{-(\frac{v}{2}+1)} exp\left\{-\frac{vs^2}{2\sigma^2}\right\} \times \left(\sigma^2\right)^{-k/2} exp\left\{-\frac{1}{2\sigma^2}(\beta_1 - \mu_{it})^T \Lambda(\beta - \mu_{it})\right\}. \tag{7}$$

The method to obtaining the estimation of parameters based on Bayesian approach can be empowered by using MCMC (Markov Chain Monte Carlo) simulation algorithm (Gibbs Sampling Algorithm). The iteration processes to estimate random parameters until the burn-in conditions are activated. For this, Bayesian approach is now challenging the potential alternative for modern econometrics.

## 2 The Bayesian Approach Contribution to Spatial Panel Estimation

In economics, distributions or samples are the random variations from sample to sample a "dance" and emphasized that variable trends can fluctuate dramatically from one random sample to the next [4]. In particular the panel data with the spatial consideration, this paper studies the matter of time presents "continuous model for spatial data". Moreover, an "identity" for each area indicates a probability distribution difference from others. This can be called "discrete model for spatial data" [5]. In common sense, spatial estimation in panel data has been widely applied since Luc Anselin proposed his book in 1988 [6]. Spatial econometrics typically deals with models related to regional and urban economics. This is continuously issued by Anselin and Hudak for a review of software options [7]. Spatial econometrics and GIS-science are merged by Anselin and Rey in 2012 [8]. In 2014, Arbia [9] contributed the handbook of spatial applications in R software because of a fast-growing field in spatial econometrics. Moreover, Bivand and Piras [10] contributed implementations of estimation methods for spatial econometrics, and Ye [11] described the definition and adoption of spatial approach for economics can be efficiently implemented by high-performance computing capacities in the cyber-infrastructure to deal with the conceptual questions raised by massive datasets and the complex dynamics underlying the existence of spatial interaction. However, as the review from previous papers, there is a rare volume that Bayesian approach is alternatively applied in the part of spatial panel estimations.

## 3 The Research Objective and Scope of Data

The main aim of this paper is to fully apply Bayesian statistics to address with the spatial panel analysis for exploring tourism demands in 7 countries in South East Asia (ASEAN) such as Brunei, Indonesia, Malaysia, Singapore, Thailand, Vietnam, and Philippines. The scope of the panel data sets is the quarterly range between 2013 and 2019. The observed variable sets are tourist arrivals (TOUR), world products (W_GDP), domestic hotel rental prices (HP), currency ratios (ERC), world jet fuel price (JET), CPI ratio by world CPI standard (RATIO_P), domestic average temperature (TEM), and capital visibility (KM). The descriptive details for observed data are presented in Table 1.

**Table 1.** Descriptive information (by growth rates)

| | G_TOUR (%) | G_W_GDP (%) | RATIO_P (%) | G_ERC (%) | G_HP (%) | G_JET (%) | G_KM (%) | G_TEM (%) |
|---|---|---|---|---|---|---|---|---|
| Mean | 6.540953 | 3.477083 | 0.981055 | 0.487549 | 5.778358 | −4.541750 | 0.778966 | 2.585753 |
| Maximum | 29.05716 | 3.900000 | 4.437323 | 6.406667 | 234.9891 | 29.32002 | 24.65753 | 18.51852 |
| Minimum | −10.66667 | 2.900000 | −0.628000 | −3.732951 | −80.51737 | −42.38147 | −12.37113 | −10.00000 |
| Std. Dev. | 8.761834 | 0.279239 | 0.987752 | 1.604175 | 43.07048 | 22.76067 | 6.475066 | 6.504838 |
| Skewness | 0.344816 | −0.857631 | 0.872806 | 1.707154 | 3.522710 | 0.016427 | 1.262621 | 0.640147 |
| Kurtosis | 3.179003 | 3.472865 | 4.517045 | 9.321016 | 19.32061 | 2.242373 | 6.263198 | 2.738926 |
| Jarque-Bera | 1.015266 | 6.331450 | 10.69717 | 103.2255 | 632.0002 | 1.150156 | 34.05061 | 3.414626 |
| Probability | 0.601919 | 0.042184 | 0.004755 | 0.000000 | 0.000000 | 0.562661 | 0.000000 | 0.181352 |

**From:** authors

## 4 Methodology

To check that the problem of useless observed independent variables is one of the major considerations of this paper. Although constraining and omitting exogenous factors can be easily simulated via MCMC approaches, the need to do this task makes inference less accessible, and approximate methods will be necessary to make these methods applicable. One such approach could be to utilize machine learning approaches [17]. Consequently, the first section of research methodology is employed "Least absolute shrinkage and selection operator (LASSO) regression" to verify predictable and unpredictable variables in data sets before using in the estimation of the Bayesian spatial panel model. For the second section, the verified set of variables is processed to estimate by three types of panel models: 1) Bayesian panel estimation (non-spatial effect), 2) Bayesian panel spatial lag model (spatial and time lags effects), and 3) Bayesian panel spatial error model (spatial and random effects). However, there is only one model that is the most sensible scenario by comparing the deviance information criterion (DIC). The overview of the methodological framework is displayed by Fig. 1.

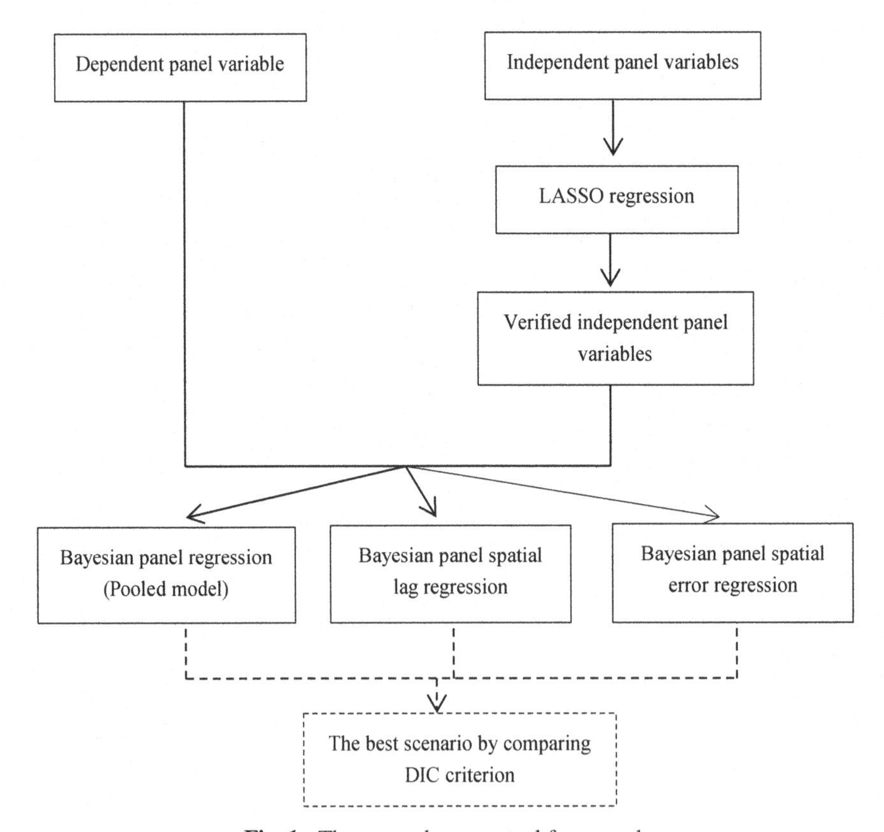


**Fig. 1.** The research conceptual framework

### 4.1 Least Absolute Shrinkage and Selection Operator (LASSO) Regression

The LASSO regression proposed by Robert Tibshirani in 1996 [12] was applied for selecting predictable variables in a regression model from observed data sets by [13]. Following those contributions, this powerful tool starts from the process of estimation by the formula as presented below:

$$SSE_{LASSO} = \sum_{i=1}^{n} (y_i - \hat{y}_i)^2 + \lambda \sum_{j}^{p} |\theta_j|. \tag{8}$$

Expressing the formula (8), the sum of squared residuals and $y_i$ are original observed data. $\hat{y}$ is the predicted value from the linear regression model estimation. $\lambda$ stands for the process to eliminate the predictable variables out of the model. Hence, all of independent variables in the right side of the demand function such as world products (W_GDP), domestic hotel rental prices (HP), currency ratios (ERC), world jet fuel price (JET), CPI ratio by world CPI standard (RATIO_P), domestic average temperature (TEM), and capital visibility (KM) are included for this screening before being estimated by the Bayesian panel regression models.

### 4.2 Bayesian Approach in Panel Regression Models

#### 4.2.1 Bayesian Pooled Panel Regression

Since it has been increasingly understood that the assumption of stationarity is unrealistic for many spatial processes [18], for example, weather forecasting, coastal structures, even tourism industrial behaviors. With respect to multidimensional covariates and spatial weights, Bayesian inference with MCMC simulations is a computationally efficient manner [19]. In Bayesian econometrics, the theoretical framework for Bayes' rule is relied on the conditional probability expressed in the formula as follows:

$$p(A, B) = p(A|B)p(B) = p(B|A)p(A). \tag{9}$$

The formula (9) can be rearranged to econometrically formalize Bayes Rule as

$$p(\theta|y) \propto p(y|\theta)p(\theta). \tag{10}$$

For the linear relationship that holds for every individual effect, the demand function model is given by

$$y_{it} = X_{it}\beta + e_{it}, \ldots i = 1, \ldots, N. \tag{11}$$

In this case, all elements of $X_{it}$ are fixed or independent of $e_{it}$, which is assumed to be the multivariate normal distribution $\left(\sigma^2 = 1/h\right)$ with a probability function $p(X_{it}|\lambda)$, where $\beta, h \neq \lambda$. With suggesting a Normal-Gamma prior ($\beta \sim \mathrm{N}\left(\beta, \mathrm{h}^{-1}V\right)$ and $h \sim \mathrm{G}\left(s^{-2}, v\right)$), the formula (11) is processed to a likelihood function of the form as follows:

$$p(y_{it}|\beta, h) = \prod_{i=1}^{N} \frac{h^{\frac{T}{2}}}{2\pi^{\frac{T}{2}}} \left[ exp\left\{ -\frac{h}{2}(y_{it} - X_{it}\beta)'(y_{it} - X_{it}\beta) \right\} \right]$$

$$= \frac{1}{2\pi^{\frac{NT}{2}}}\left\{h^{\frac{k}{2}} exp\left[-\frac{h}{2}\left(\beta-\hat{\beta}\right)' X_{it}' X_{it}\left(\beta-\hat{\beta}\right)\right]\right\}\left\{h^{\frac{v}{2}} exp\left[-\frac{hv}{2s^{-2}}\right]\right\}. \quad (12)$$

With the numerical analytics by the Markov Chain Monte Carlo (MCMC) which contains the Metropolis-Hastings algorithm and Gibbs sampling, the estimated results in $\beta, h|y \sim \text{NG}(\bar{\beta},\ \bar{V},\ \bar{s}^{-2},\ \bar{v})$ can be mathematically derived as follows:

$$\bar{V} = \left(V^{-1} + X_{it}' X_{ti}\right)^{-1}, \quad (13)$$

$$\bar{\beta} = \bar{V}\left(V^{-1}\beta + X_{it}' X_{it}\hat{\beta}\right), \quad (14)$$

$$\bar{v} = v + NT. \quad (15)$$

### 4.2.2 Bayesian Panel Spatial and Lag Regression

Interestingly, the spatial lag regression is efficient when the dependent variable is focused on the spatial interactions. This can be implied that the structure of the spatial relationship is well comprehended. For instance, the tourist arrivals in each country in ASEAN will depend on the fluctuated numbers of total international tourists travelling in the continent. In this model, the dependent variable $y$ has the spatial structure. With the consideration of spatial lags, this spatial autoregressive model includes a spatially lag dependent variable. This variable is a weighted average of its neigbors' values.

Following the concept of spatial lag mode in panel data (Panel SLM) for Bayesian approach by Luc Anselin in 2003 [14], the "pure space recursive" model, in which the dependent variable is concerned to neighboring location in a different period,

$$y_{it} = \gamma\left[Wy_{t-1}\right]_i + f(x) + e_{it}. \quad (16)$$

$\left[Wy_{t-1}\right]$ is the $i$ th element of the spatial lag vector applied to the observed samples on the dependent variable in the previous time period (using a $N \times N$ spatial weighted matrix for the cross-sectional units). $t$ is time slices of $N$ cross-sectional units. $f(x)$ is a generic designation for the regressors.

### 4.2.3 Bayesian Spatial Panel Error Regression

The spatial panel error regression, alternatively called "spatial random effect", is appropriate when the model is interested in correcting for spatial autocorrelation regarding the use of spatial data (irrespective of whether the model of the objective is spatial or not). For this case, the structure of the spatial linkage is not decisively known. Spatially correlated errors dealing with unobservable features or omitted variables associated with location. For example, foreign travellers' decisions to arrive in Brunei Darussalam may be influenced by their neighbor Islamic countries such as Malaysia or Indonesia.

In term of mathematical expressing, the spatial errors model in panel data (Panel SEM) for Bayesian approach, alternatively defined as "random space-recursive", is given by

$$y_{it} = x_{it}\beta + e_{it}. \quad (17)$$

The errors are spatially correlated by the spatial weighted matrix, which is given by

$$e_{it} = \rho W_{ij} e_{it-1} + u_{it}. \tag{18}$$

### 4.3 Deviance Information Criterion (DIC)

The deviance information criterion (DIC) was introduced in 2002 by Spiegelhalter et al. [15]. It is appropriate when using to compare the relative fit of a set of Bayesian hierarchical models. It is used extensively in many disciplines, particularly biostatistics, ecology, and econometrics [16]. In term of definition and computation, a parametric Bayesian model is defined by the prior $p(\theta)$, $\theta \in \Theta \subseteq \Re^k$, and likelihood $p(y|\theta) = \prod_{i=1}^{n} p(y_i|\theta)$ where $y = (y_1, \ldots, y_n)$ denotes the vector of observed sample. The deviance is a function of the parameter $\theta$, is defined as

$$D(\theta) = -2\, log\, p(y|\theta) + c. \tag{19}$$

$c = c(y)$ is fully specified standardizing constant that depends on the data and thus cancels out in model comparisons. With exponential family models, $E(Y) = \mu(\theta)$, setting $c(y) = 2\, log\, p(y|\mu(\theta) = y)$ for addressing with s the deviation to the saturated model.

## 5 Empirical Results

### 5.1 The Investigation of Predictable Variables for the Demands of Tourism from Observed Data

To confirm that the problem of muticolinearity in the linear regression is technically overcome, 7 independent variables (all is transformed into the growth rate) such as prices of jet fuel (g_jet), price ratios (cpiratio), rate of visibility in a capital city (g_km), rate of rental hotel prices (g_hp), world gross domestic products (g_w_gdp), rate of average annual temperature (g_temp), and rate of currencies (g_erc) are screened by the LASSO regression. The computational result shows there is only the rate of currencies (g_erc) that is defined as an unpredictable variable. The result of validated variables is also visualized in Fig. 2, and the parametric direction effects of predictable variables without the factor of currencies are represented in Table 2.

### 5.2 Spatial Panel Regressions for Bayesian Approach

Table 3 presents the result that three types of Bayesian panel regressions are compared. To validate the most appropriate model, the pure space-recursive model is indicated as the most efficient estimation, giving the lowest value of the deviance information criterion when comparing with other two scenarios, which equals 3237.015. The best acceptance rate equals 0.255. This thus implies that the spatially weighted matrix and time-lag consideration are indeed crucial. The spatial and lag panel model is chosen. Tourism demands in these 7 countries in South East Asia inevitably depend on regional

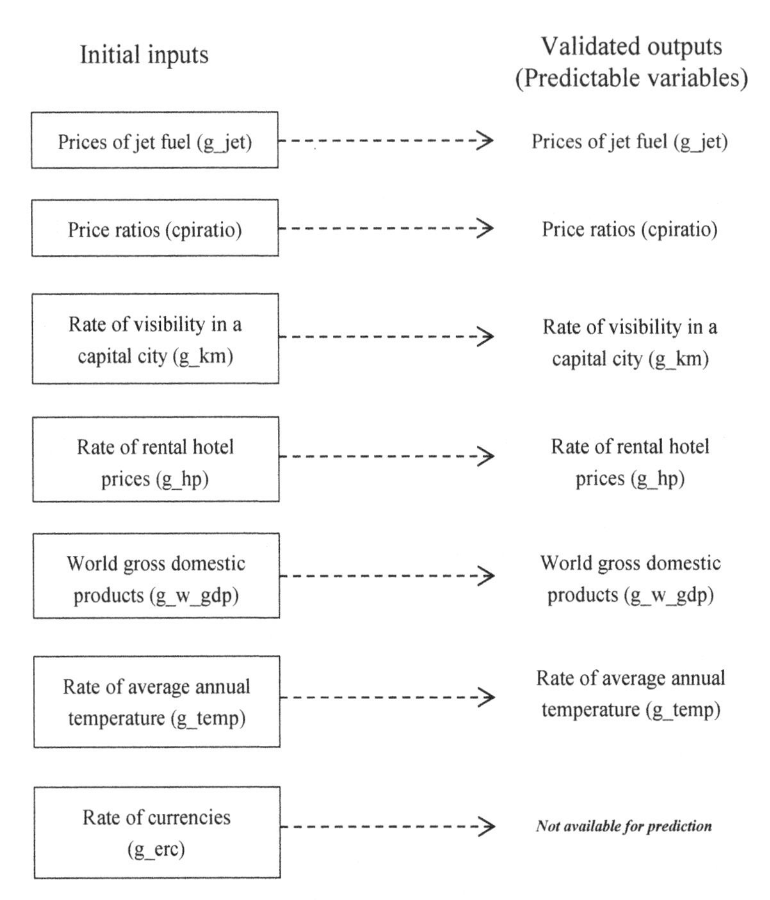


**Fig. 2.** Predictable variables validation by LASSO regression

**Table 2.** The parameters of predictable variables selected by LASSO regression

| Selected | LASSO |
|---|---|
| g_w_gdp | 2.756 |
| g_hp | 0.012 |
| g_jet | 0.045 |
| cpiratio | 1.337 |
| g_temp | −0.032 |
| g_km | −0.254 |
| Partialled-out | |
| Constant term | −3.980 |

**Source**: authors' computation

**Table 3.** Bayesian approach in three scenarios of panel regression models

| Dependent variable Y: international tourist arrivals (g_tour) | Panel pooled model | | Spatial and lag panel model | | Spatial error panel model | |
|---|---|---|---|---|---|---|
| | Mean | MCSE | Mean | MCSE | Mean | MCSE |
| Independent variable | | | | | | |
| X1: world gross domestic products (g_w_gdp) | **1.818** | 0.020 | **3.059** | 0.001 | **2.284** | 0.013 |
| X2: rental hotel prices (g_hp) | **0.014** | 0.000 | **−0.004** | 0.112 | **0.009** | 0.000 |
| X3: World prices of jet fuel (g_jet) | **0.054** | 0.000 | **0.101** | 0.000 | **0.078** | 0.000 |
| X4: Price ratios (cpiratio) | **1.384** | 0.011 | **0.959** | 0.000 | **0.821** | 0.010 |
| X5: average annual temperature | **−0.048** | 0.001 | **−0.105** | 0.010 | **−0.056** | 0.001 |
| X6: visibility in a capital city | **−0.025** | 0.001 | **−0.147** | 0.001 | **−0.170** | 0.001 |
| Constant term | **−0.692** | 0.065 | **−5.204** | 0.395 | **−1.104** | 0.044 |
| Spatial weight matrix and lag consideration ($Wy_{t-1}$) | | | **0.181** | 0.001 | | |
| $\rho$: spatially correated errors | | | | | **0.188** | 0.0001 |
| MCMC iterations | 12,500 | | 12,500 | | 12,500 | |
| Burn-in | 2,500 | | 2,500 | | 2,500 | |
| MCMC sample size | 10,000 | | 10,000 | | 10,000 | |
| Number of observations | 49 | | 49 | | 49 | |
| Acceptance rate | 0.178 | | 0.255 | | 0.232 | |
| DIC | 3382.762 | | **3237.015*****| | 3275.654 | |

**Noted**: MCSE: Monte Carlo Standard Deviation, *** stands for the most appropriate model, **Sources:** authors

attraction and historical stories. The positive signs from this indicator would be the factor that encourages the high demands to travel in this continent. It is numerically confirmed by the positive Bayesian parametric mean which equals 0.181.

Considering into the details provided by the pure space-recursive regression, it is clear that the improvement of world GDP (X1) is positive for ASEAN's tourism. The parametric equals 3.059 means if tourists' living standard trend to be developed, it is sensible that travelling in ASEAN is the parallel growth alongside this changing. The rental prices of hotels (X2) are relied on the demand theory. The increasing price will cause the improvement in travel costs. This will not be a satisfied response if the hotel price is not stable and predictable. However, the price story is different when considering the price of jet fuel (X3). The positive adjustment in this price indicates the improvement in transportation volumes. This implies world economy is expansion, and this is the positive effect for ASEAN's tourism.

Speaking to the variables representing domestic features in ASEAN countries, domestic price ratios (X4) is resulted to be positive for the growth of tourism in ASEAN. A little bit shifted up of consumer prices displays the expansion in ASEAN economy. On the other hand, the average annual temperature (X5) is a negative driven factor for tourism in ASEAN. The extreme high temperature in equator countries causes travelers to avoid coming to South East Asia. Moreover, the interesting result of visibility in capital cities in ASEAN (X6) is negative. This implies urban visibility is not the major attitude for ASEAN tourists. In other words, the case of the pure space recursive model confirms that local spatial connections are the real attraction for increasing tourist volumes. It seems only travelling and sightseeing in a capital city is not the best choice for the edge of globalization. As stated by Hassan (2008) [20], travelers of the future would likely consist of explorers and drifters who want to arrange travel independently, spatially explore their limitation, and wish to experience the lifestyle of a host community. Moreover, Bock (2015) [21] issued tourists were increasingly seeking, finding and consuming "local experiences" and the boundaries between tourists and residents become increasingly blurred since facilities for mobile accessing to information.

## 6 Conclusion

Although many economists criticize on the gap that Bayesian econometrics has a few researches for testing in parametric estimations, this paper still applies this statistical inference in the climax to estimate the demand function of tourism in ASEAN-7 countries. The strongest point is Bayesian statistics provides random parameters that can efficiently work with the panel data which is definitely random information. Additionally, the invention of spatial and time-lag effects can powerfully improve the understanding of tourism demands in ASEAN countries. Ultimately, this is obvious that tourism policies in South East Asia should be the spatial joint companion rather that individually domestic policies (locked down policy implementation).

## References

1. Harvey, C.R.: Presidential address: the scientific outlook in financial economics. J. Financ. **72**(4), 1399–1440 (2017)
2. Permai, S.D., Tanty, H.: Linear regression model using Bayesian approach for energy performance of residential building. Procedia Comput. Sci. **135**, 671–677 (2018)
3. Rubio, F.J., Genton, M.G.: Bayesian linear regression with skew-symmetric error distributions with applications to survival analysis. Stat. Med. **35**(14), 1–17 (2016)
4. Kruschke, J.K., Liddell, T.M.: The Bayesian new statistics: hypothesis testing, estimation, meta-analysis, and power analysis from a Bayesian perspective. Psychon. Bull. Rev. **25**(1), 178–206 (2017)
5. Camara, G., Carvalho, M.S.: A tutorial on spatial analysis of areas. Cardernos Saude Publica **17**(5), 1–44 (2001)
6. Anselin, L.: Spatial Econometrics: Methods and Models. 1st edn. Kluwer Academic Publishers (1988)
7. Anselin, L., Hudak, S.: Spatial econometrics in practice: a review of software options. Reg. Sci. Urban Econ. **22**, 509–536 (1992)

8. Anselin, L., Rey, S.J.: Spatial econometrics in an age of cyberGIScience. Int. J. Geogr. Inf. Sci. **26**(12), 2211–2226 (2012)
9. Arbia, G.: A Primer for Spatial Econometrics with Applications in R, 1st edn. Palgrave Macmillan, UK (2014). https://doi.org/10.1057/9781137317940
10. Bivand, R., Piras, G.: Comparing implementations of estimation methods for spatial econometrics. J. Stat. Softw. **63**(18), 1–36 (2015)
11. Ye, X.: Spatial Econometrics. The International Encyclopedia of Geography, 1st edn. Wiley, Chichester (2017)
12. Tibshirani, R.: Regression shrinkage and selection via the Lasso. J. Roy. Stat. Soc. **58**(1), 267–268 (1996)
13. Panthamit, N., Chaiboonsri, C.: China's outward foreign direct investment in the greater Mekong subregion. J. Econ. Integr. **35**(1), 129–151 (2020)
14. Anselin, L.: Spatial econometrics. In: Baltagi, B.H. (ed.) The Book of a Companion to the Theoretical Econometrics. Blackwell Publishing Ltd. (2003)
15. Spiegelhalter, D.J., Best, N.G., Carlin, B.P., Linde, A.: Bayesian measures of model complexity and model fit. J. R. Stat. Soc. Ser. B **64**, 583–639 (2002)
16. Meyer, R.: Deviance Information Criterion (DIC). In: Balakrishnan, N., Brandimarte, P., Everitt, B., Molenberghs, G., Piegorsch, W., Ruggeri, F. (eds.) Wiley StatsRef: Statistics Reference Online, 5 pages. Wiley (2016)
17. Penfold, C.A.: Bayesian parameter estimation in non-stationary semiflexible polymers from ensembles of trajectories (2018). https://doi.org/10.1101/484691
18. Yang, H.C., Bradley, J.R.: Bayesian inference for big spatial data using non-stationary spectral simulation. arXiv:2001.06477. Cornell University (2020)
19. Randell, D., Turnbull, K., Ewans, K., Jonathan, P.: Bayesian inference for non-stationary marginal extremes. Environmetrics **27**, 439–450 (2016)
20. Hassan, N.: Understanding the 'new tourist' of Asia: developing a global and local perspective. Perspect. Asian Leisure Tourism **1**(6) (2008)
21. Bock, K.: The changing nature of city tourism and its possible implications for the future of cities. Eur. J. Futures Res. **3**(1), 1–8 (2015). https://doi.org/10.1007/s40309-015-0078-5

# Measurements of the Conditional Dependence Structure Among Carbon, Fossil Energy and Renewable Energy Prices: Vine Copula Based GJR-GARCH Model

Yefan Zhou[1], Jianxu Liu[2,3], Jirakom Sirisrisakulchai[1,3](✉), and Songsak Sriboonchitta[1,3]

[1] Faculty of Economics, Chiang Mai University, Chiang Mai 50200, Thailand
sirisrisakulchai@hotmail.com
[2] School of Economics, Shandong University of Finance and Economics, Jinan 250000, China
[3] Puey Ungphakorn Center of Excellence in Econometrics, Chiang Mai University, Chiang Mai 50200, Thailand

**Abstract.** This paper explores the dependence structure among carbon prices, fossil energy prices and renewable energy price using the conditional vine copula approach. The major two contributions in our study are following. First, regarding technological innovation and development of new technologies in alternative energy sources, we consider renewable energy index into our study. Second, we simultaneously investigate the multivariate dependence among all variables so that each of them can interact with the others based on a rich variety of bivariate copula functions. We mainly find that there is a reliable and positive link between coal and oil prices, and between gas and oil prices. And we corroborate that variations in the carbon prices affect the coal price returns positively, though the association is usually found to be statistically insignificant. Moreover, carbon prices affect the renewable energy stock returns positively and strongly significant. Such findings we suggest that policy-makers could adopt effective measures and action plan to elevate carbon prices so that the emission market could provide incentives to shift from conventional fossil fuels to clean and low-carbon energy sources.

**Keywords:** Carbon price · Energy market · Renewable energy price and vine copula

## 1 Introduction

Over the past ten years, the renewable energy sector, carbon sector and energy markets have received enormous focus in order to efficiently reduce the adverse impact of greenhouse gas (GHG) emission. Establishing the stochastic relationships intertwined in carbon market, energy markets and renewable energy sector


V.-N. Huynh et al. (Eds.): IUKM 2020, LNAI 12482, pp. 322–334, 2020.
https://doi.org/10.1007/978-3-030-62509-2_27

remain a challenging task as far as it is related to the climate change objectives of carbon emission reductions, carbon efficiency, the path of transition from fossil fuels to renewable energy, and optimal energy investments. Accordingly, investments in renewable energies have increased significantly which makes this sector one of the fastest growing sectors in global energy industries.

The Kyoto Protocol implemented the targets of the United Nations Framework Convention on Climate Change (UNFCCC) to control the trend of global warming by reducing greenhouse gas concentrations. Thus, the Kyoto Protocol established three mechanisms for climate change mitigation. Among all of the three mechanisms, emission trading is a unique way for carbon reduction, as it reduces carbon emissions through a market-based mechanism, which makes carbon reduction a market activity [1,5]. The carbon product traded in the European Union Emission Trading Scheme (EU ETS) is named the European Union allowance (EUA). The EU ETS opened the first international emissions trading system in the world with the objective of reducing greenhouse gas emissions from January 1, 2005 [2]. Therefore, we selected carbon price to represent carbon trading market. Theoretically, the higher development of economic growth drives the higher energy demand then leads to carbon emissions increasing, which increases to carbon allowances prices. Accordingly, the energy market and carbon trading market seem to be connected, with causality going from the energy markets to carbon prices.

In this paper, our purpose is to investigate the cross-relationships among the renewable energy market, carbon trading market and energy markets. The first contribution deals with the choice of data. We decided to work not only with carbon market, three fossil fuels price including crude oil, natural gas and coal, but also with renewable energy market. The rationale behind this idea is that variations in renewable energy prices have significant implications on the choice of fossil energy industries and renewable energy industries, which in turn affect total carbon emissions and carbon trading prices. The fossil fuel industries, which include petroleum industries (including oil companies, petroleum refiners, fuel transport and so on), coal industries (extraction and processing) and the natural gas industries (natural gas extraction, and coal gas manufacture, as well as distribution and sales) [3]. The renewable-energy industry is the part of the energy industry that collected from renewable resources, which is focusing on emerging and renewable energy technologies, such as sunlight, wind, rain, tides, waves and geothermal heat [4]. The strategy and aim are to transition from using fossil energy to clean energy. Meanwhile, the fossil energy consumption directly impact the demand of carbon allowance. Thereby, the volatility of carbon price is affected by unpredictable the fluctuations of energy price. In addition, changing in the merit order of fossil energy and renewable energy plants is affected by the fluctuation of carbon price whereby the efficiency of the plants and the fuel costs are critical decision factors [6]. Finally, the exact relationship between energy and carbon prices could thus be biased if the dynamics of renewable energy prices is ignored.

Second, we use vine copula approach to investigate the dependence structure among renewable energy market, energy market and carbon trading market. These methodologies are flexible graphical models enabling extensions to higher dimensions using a cascade of bivariate copulas. And the methodological framework provides the advantage of modeling various dependence structures and also their potential to construct a abundant set of distributions. Thus, vine-copula GARCH models have gained increasing attention for modeling high-dimensional dependence structures.

## 2 Literature Review

Many studies have investigated the impact of carbon prices on the energy market returns and renewable energy returns. Regarding as the relationship between carbon prices and energy market returns, Chevallier et al. [6] found that there was a reliable and positive link between coal and gas prices, and between coal and oil prices, however carbon prices co-move only weakly with energy prices, and their link to oil and gas prices is negative. Ji et al. [7] investigated linkages and spillover among the carbon-energy system. Their return and volatility connectedness features that crude oil, clean energy, and coal play a pivotal role. The dynamic spillover approach has been applied recently as well by Zhang [8] and Filis et al. [9] for the relationship between the oil and stock markets. Oberndorfer [10] reported that the association between emission prices and stock returns are positively related. Additionally, the study showed that the effect of carbon market on stock returns of electricity firms is symmetric. Silva et al. [11] found that a statistically significant positive long-run impact of EU ETS on the aggregated power sector stock market return. For the relationship between the renewable energy market and carbon price, Kumar et al. [12] explored whether an increase in emission prices lead to more investments in renewable energy firms used by vector auto-regression approach. They found no significant relationship between carbon prices and clean energy stock returns. Moreover, Dutta [13] examined whether clean energy stock returns impact carbon price in European. They concluded that the carbon prices affect the renewable energy stock returns positively and their findings further indicate a significant volatility linkage between emission and European clean energy price indexes.

In terms of the research methods, different models and approaches have been developed to analyze the co-movements or interdependence among carbon market, energy market and renewable energy market. Chevallier et al. [6] made use of the vine copula models to find that carbon prices co-move only weakly with energy prices. And Lutz et al. [14] explored the regime-switching GARCH model to investigate the nonlinear relation between the carbon price and its fundamentals including energy prices, macroeconomic risk factors and weather conditions. And they found that gas, coal, and oil can be deemed to be reliable price drivers of the carbon price and that their influence depends on high- versus low-volatility regime.

# 3 Methodology and Data

## 3.1 Data

In this paper, we use the dataset of three main fossil energy, carbon and renewable energy prices from Thomson Reuters Eikon. The fossil energy prices consist of crude oil (ICE Europe Brent Crude Electronic Energy Future price), natural gas (ICE Natural Gas Future price) and the coal (ICE Natural Electronic Energy Future price). Carbon price is measured by $CO_2$ European Allowance price (ICE EUA DAILY Futures) in European Climate Exchange (ECX). Furthermore, we regard European Renewable Energy Index (ERIX) as the variable of renewable energy price. This renewable energy index tracks the price of largest alternative energy companies in the areas of renewable energy such as wind, solar, biomass and water energy. The ERIX index covers the prices of 10 largest and most liquid stocks from the list of eligible firms. It is rebalanced every quarter [15]. We obtained 2455 weekly prices that spanned the period from June 4, 2010 to November, 1, 2019. All price series were converted to US dollars using the bilateral exchange rates given by the European Central Bank and the Federal Reserve Bank of St. Louis. For our analysis, we consider the price returns at each time t as $(p_t - p_{t-1})/p_t$ where $p_t$ denotes the prices level of the corresponding series at time t. Energy prices are pictured in Fig. 1.

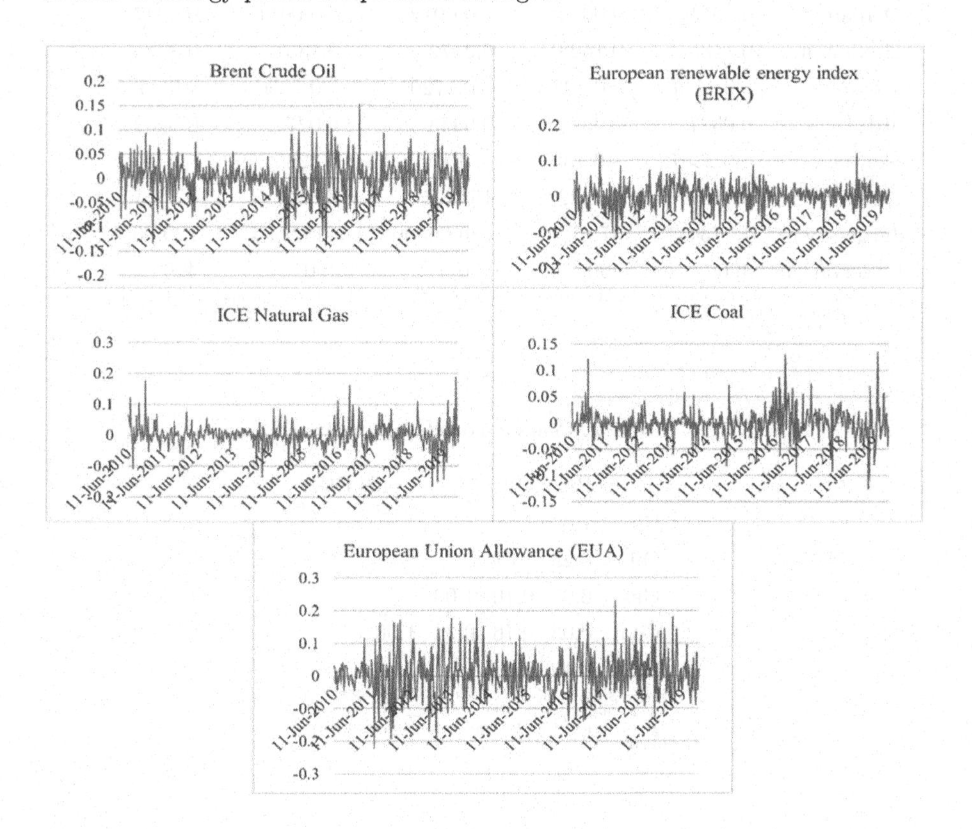


**Fig. 1.** Energy prices return for the sample period

Table 1 reports the descriptive statistics and the basic stochastic properties for renewable energy index (Re), oil price returns (Oil), gas price returns (Gas), coal price returns (Coal) and carbon price returns (EUA). The average price returns were close to zero for all series and were relatively low concerning their standard deviations. Standard deviations and differences between maximum and minimum values show that the EUA series was the greatest in all series. Coal prices display the lowest volatility in all series. Positive values for skewness are observed for the RE, Oil and Coal while RE, Oil and EUA display negative skewness. All the series exhibit excessive kurtosis, confirming the presence of fat tails in the marginal distributions or a relatively high probability of extreme observations. The results of nonnegative in Jarque-Bera test rejects the normality of the unconditional distribution for all the series.

Table 2 reports information on the Pearson linear correlation coefficient, with values indicating that all pairs display positive dependence. And all other pairs show very low or nonlinear dependence.

**Table 1.** Descriptive statistics for price returns

| | RE | Oil | Gas | Coal | EUA |
|---|---|---|---|---|---|
| Mean | 0.0018 | 0.0004 | 0.0009 | −0.0006 | 0.0034 |
| Median | 0.0051 | 0.0015 | −0.0012 | −0.0011 | 0.0047 |
| Maximum | 0.1185 | 0.1528 | 0.2872 | 0.1351 | 0.2625 |
| Minimum | −0.1117 | −0.1374 | −0.1720 | −0.1256 | −0.3413 |
| Std. Dev. | 0.0327 | 0.0390 | 0.0433 | 0.0272 | 0.0702 |
| Skewness | −0.2357 | −0.1557 | 1.7080 | 0.3226 | −0.2171 |
| Kurtosis | 3.7067 | 4.3222 | 15.3813 | 7.6447 | 5.3951 |
| Jarque-Bera | 14.7341*** | 37.6724*** | 3368.0550*** | 448.9604*** | 120.9657*** |
| Observations | 490 | 490 | 490 | 490 | 490 |

Note: Significance at the 0.01, 0.05, and 0.10 levels indicated by ***, **, *.EUA: European carbon price, RE: European Renewable Energy Index.

**Table 2.** Pearson correlation matrix.

| | RE | Oil | Gas | Coal | EUA |
|---|---|---|---|---|---|
| RE | 1.00 | | | | |
| Oil | 0.25 | 1.00 | | | |
| Gas | 0.04 | 0.10 | 1.00 | | |
| Coal | 0.03 | 0.16 | 0.34 | 1.00 | |
| EUA | 0.18 | 0.16 | 0.11 | 0.07 | 1.00 |

Note: Significance at the 0.01, 0.05, and 0.10 levels indicated by ***, **, *.EUA: European carbon price, RE: European Renewable Energy Index.

### 3.2 Methodology

In this paper, we explore multivariate dependence among the European renewable energy price, carbon allowance prices and major energy prices. Considering the dependence structure between renewable energy index, Brent crude oil, natural gas and coal prices, because this helps investigating interactions between the renewable energy market, carbon trading market and each of the major energy markets in a multivariate and integrated setting.

**Vine Copula Models.** Introduced first by Joe et al. [16] and extended by Bedford et al. [17], vine copulas are multivariate copulas that are generated using a hierarchical structure given using a cascade of bivariate copulas (pair copulas), where each bivariate pair copula captures the conditional dependence between two random variables. However, a crucial issue in this density decomposition is the choice regarding the order of the variables. The vine copulas are made of different pair copula constructions composed of d(d − 1)/2 bivariate copulas [18]. In our research, we focus on three vine models with that have been widely employed in the empirical literature, namely, C-vine, D-vine, and R-vine copulas.

A C-vine is a special case of the vine copula. The multivariate density of C-Vine copula given by:

$$f\left(x_1, x_2, ..., x_d\right) = \prod_{k=1}^{d} f_k\left(x_k\right) \prod_{h=2}^{d} c_{1,h}\left(F_1\right)\left(x_1\right), \left(F_h\right)\left(x_h\right))$$

$$\prod_{j=2}^{d-1} \prod_{i=1}^{d-j} c_{j,j+1|1,...,j-1}(F(x_j|x_1, ..., x_{j-1}), c_{j,j+1|1,...,j-1}(F(x_{j+1}|x_1, ..., x_{j-1})) \tag{1}$$

where the conditional distribution functions between variables $x_i$ and $x_j$ can be obtained [19] as

$$F_{i|j}(x_i|x_j) = c_{i|j}[F_i(x_i), F_j(x_j)] = \frac{\partial C_{i|j} F_i(x_i)}{F_j(x_j)} \tag{2}$$

with $c_{i|j}$ denoting the conditional distribution of variable i given the variable j with a joint distribution function $C_{ij}$. The dependence structure of this decomposition can be represented graphically as a hierarchical tree structure.

D-vine copula has a multivariate density given by:

$$f\left(x_1, x_2, ..., x_d\right) = \prod_{k=1}^{d} f_k\left(x_k\right) \prod_{h=1}^{d-1} c_{h,h+1}(F_h(x_h), F_{h+1}(x_{h+1}))$$

$$\prod_{j=2}^{d-1} \prod_{i=1}^{d-j} c_{i,i+j|i+1,...,i+j-1}(F(x_i|x_{i+1}, ..., x_{i+j-1}), (F(x_{i+j}|x_{i+1}, ..., x_{i+j-1}). \tag{3}$$

The specific ordering of the variables in the D-vine construction is following. The first tree (T1) models dependence between the first and second variables, the second and third variables, and so on, using bivariate copulas. The edges connecting the nodes represent this dependence relationship. The second tree (T2) models the conditional dependence between the first and the third variables, given the second variable, between the second and fourth variables, given the third variable. This process continues through different trees to the last tree, where only one bivariate dependency remains. Thus D-Vine copula have $d(d-1)/2$ pair copula models. The ordering of the variables in the first tree is crucial because it fully determines the dependence structure of the remaining trees in the D-vine. From $d!/2$ possible orderings, the best ordering is selected to capture as much dependence as possible (as measured by Kendall's tau) in the first tree [20].

According to Kurwicka and Cook [21], an R-Vine copula consists of $\mathrm{d}-1$ trees with nodes $N_i$ and $E_i$ for $i = 1, ..., d-1$. The first tree has d nodes and $E_1$ edges, whereas each tree of the remaining trees, $i = 2, ..., d-1, N_i$ has nodes $N_i = E(i-1)$. The multi variate density of the R-vine copula is given by:

$$f\left(x_1, x_2, ..., x_d\right) = \prod_{k=1}^{d} f_k\left(x_k\right) \prod_{i=1}^{d-1} \prod_{e \in E_i} c_{j(e),k(e)|D(e)}(F(x_{j(e)}|x_{D(e)}), (F(x_{k(e)}|x_{D(e)})) \tag{4}$$

where $x_D(e)$ denotes the sub-vector of $x = (x_1, ..., x_d)$ indicated by the indices contained in $D(e)$.

**Specifications of Marginal Models.** The main constructions of multivariate dependence model are the marginal densities and the bivariate copulas for vine copula. The marginal models and copula specifications used in the empirical analysis are described as follows. The marginal distribution of each energy price series was characterized (such as leverage, fat tails and asymmetries) by an ARMA(p,q)-Glosten-Jagannathan-Runkle(GJR)-GARCH(1,1) model with a disturbance term following the skewed normal distribution (snorm), skewed Student t distribution (sstd), or skewed generalized error distribution (sged) [18]. The form of the ARMA(p,q)-GJR-GARCH(1,1) is

$$r_t = \mu + \sum_{i=1}^{p} \varphi_{1i} r_{t-i} + \sum_{j=1}^{q} \varphi_{2j} \varepsilon_{t-j} + \varepsilon_t \tag{5}$$

$$\sigma_t^2 = \omega + \alpha \varepsilon_{t-1}^2 + \beta \sigma_{t-1}^2 + \gamma I[\varepsilon_{t-1<0}] \varepsilon_{t-1}^2 \tag{6}$$

where $\sum_{i=1}^{p} \phi_{1i} < 1$, $\varepsilon_t = \sigma_t \eta_t$, such that the residual $\sigma_t$ is the product of the standard variance $\eta_t$ and standardized residuals $\omega > 0$, $\alpha \geq 0$, $\beta \geq 0$, $\alpha + \gamma \geq 0$ and $\alpha + \beta + \frac{1}{2}\gamma < 1$.

Equation (5) is the mean equation, where $r_t$ is the return series of the energy price and $\phi_{1i}$ and $\phi_{2j}$ are the coefficients of the AR and MA terms. Equation (6) is the variance equation, where parameters $\alpha$ and $\beta$ are the coefficients of the ARCH term and GARCH term respectively, the leverage effect is measured by $\gamma$, and variable $I$ is a dummy variable which is only activated if the previous shock $\varepsilon_{t-1}$ is negative. The leverage effect indicates that losses have a stronger impact on future volatility than gains. If a leverage effect exists, the estimated parameter will be positive [18].

**Sequential Selection Based on Kendall's Tau Correlations.** After getting the marginal of each variable, the next step is to decide the order of the variables in the vine tree. The order of the five variables in the vine tree was decided by maximizing the sum of the absolute empirical Kendall's tau values of one variable with all the other variables. In this paper, the Coal was selected to be the first node of the first vine tree. And then renewable energy price, crude oil, carbon price and natural gas are the second, the third, the fourth and the last node respectively. In this way, the dependence among three fossil energy price, renewable energy price and carbon price could be investigated using bivariate copulas.

# 4 Results and Discussions

In this section, we present the empirical results of the marginal model and estimated the three vine copula models. The best-fitting vine copula were selected according to the BIC.

## 4.1 Results for the Marginal Distribution Models

Table 3 reports the estimation results for the marginal models described in Eq. (5) and (6). The parameters p, q were selected for different combinations of values ranging from zero to a maximum lag of two using the BIC values. From the results of the mean equation (5), the prices rate returns of the last period were found to have a significant positive influence on the returns of the current period for all the marginal models which contain AR ($\phi_1$) terms except oil price return. From the coefficients of the MA ($\phi_2$) terms, the volatility of residuals from the last week was found to have significant positive effects on the currency returns in the current period except oil price return.

The results of the variance equation indicate that the conditional volatility is quite persistent for all the series. Volatility asymmetric effects are observed for all the series which means that the volatility of price returns in the current period was positively affected by the volatility of the last period. The coefficients of the GARCH term $\beta$ were much greater than the coefficients of the ARCH term $\alpha$ which means that the effects of the volatility shocks fade away slowly. All of the return series are best captured by a skewed normal distribution according to the BIC. There is significant effect, captured by the coefficient of the GJR term $\gamma$, in Brent crude oil market and nature gas market. The leverage effects showed asymmetric volatility of the energy price returns. It implies that bad news would cause larger volatility in Brent crude oil futures markets than the influence of good news shock. Thus, if the energy price falls, the return of the energy price will tend to be more unstable and lead to higher energy risk market. The skewness parameters were estimated to be statistically significant in all return series, which further indicated that the distributions of these price return series were not symmetric. Consequently, the skewed distributions were more appropriate than the normal distribution to illustrate the skewness and kurtosis in our dataset for the estimation of the marginal models. We evaluate the sufficiency of the marginal distribution model by testing that the standardized residual is a consistent zero value. And we test whether $hat\mu$ between 0 and 1 by the Kolmogorov-Smirnov test that analyze the fitting performances of theoretical models with actual distribution. Table 3 shows that all copula sequences passed at the 5% significance level K-S test. The test results show that the copula module can be used to capture the linkage and tail dependence among main fossil energy, renewable energy and carbon prices.

## 4.2 Results for the Vine Copula Models

We first estimated the three vine copula models in Eqs. (1), (3), and (4) using the pair copulas for carbon future price, Brent crude oil future price, natural gas future price, coal future prices, and renewable energy price. Table 4 reports results for different selection criteria; the best copula fit was achieved by the D-vine copula model represented in Fig. 2, along with the corresponding information on the pair copulas. Figure 2 represents this dependence structure, along with information on the pair copulas. Pair dependence in the first tree indicates the existence of positive and strongly significant dependence between renewable energy price and Brent crude oil with Kendall's tau 0.13, Brent crude oil (O) and carbon price (E) with Kendall's tau 0.12, and between carbon prices (E) and nature gas (G) with Kendall's tau 0.24. And there is also positive but weakly significant dependence between coal (C) and renewable energy prices (R) with Kendall's tau 0.04.

In the second tree of the multivariate dependence structure, there is a positive dependence between the coal(C) and Brent oil (O) prices. The latter finding is consistent with evidence of symmetric upper and lower tail dependence (at Table 5), given that the Student-t copula is the best model in describing the Coal – Oil dependence structure. Next, there is positive and statistically significant dependence between the renewable energy price (R) and carbon price (E) with Kendall's of 0.29 which reveals that the $CO_2$ price affect the renewable energy stock returns positively. Then another positive conditional dependence between Brent crude oil (O) and natural gas (G) with the Kendall's of 0.41. This result provides evidence of co-movements between Brent crude oil (O) and natural gas (G).

**Table 3.** Parameter estimates for the marginal distribution models.

| | RE | Oil | Gas | Coal | EUA |
|---|---|---|---|---|---|
| Dis | snorm | snorm | snorm | snorm | snorm |
| *Mean equation* | | | | | |
| $\mu$ | 0.001 | −0.0004 | 0.005** | −0.002 | 0.003 |
| | (0.002) | (0.002) | (0.003) | (0.002) | (0.003) |
| $\phi_1$ | 0.945*** | −0.089 | 0.746*** | 0.717** | 0.681*** |
| | (0.092) | (0.676) | (0.155) | (0.372) | (0.144) |
| $\phi_2$ | −0.932*** | 0.138*** | −0.640*** | −0.609*** | −0.690*** |
| | (0.105) | (0.670) | (0.177) | (0.250) | (0.053) |
| *Variance equation* | | | | | |
| $\omega$ | 0.0001* | 0.0001*** | 0.0001 | 0.0001*** | 0.0004 |
| | (0.0001) | (0.0001) | (0.0001) | (0.0001) | (0.0004) |
| $\alpha$ | 0.042* | 0.000 | 0.013 | 0.083*** | 0.067* |
| | (0.028) | (0.036) | (0.009) | (0.031) | (0.055) |
| $\beta$ | 0.855*** | 0.828*** | 0.943*** | 0.853*** | 0.777*** |
| | (0.060) | (0.044) | (0.016) | (0.032) | (0.068) |
| $\gamma$ | 0.070 | 0.196*** | 0.083*** | 0.023 | 0.159 |
| | (0.061) | (0.077) | (0.032) | (0.038) | (0.125) |
| Skew | 0.890*** | 0.906*** | 1.180*** | 0.984*** | 0.973*** |
| | (0.051) | (0.051) | (0.046) | (0.044) | (0.068) |
| BIC | −3.959 | −3.685 | −3.464 | −4.410 | −2.483 |
| Likelihood | 994.765 | 927.706 | 873.470 | 1105.32 | 633.080 |
| K-S | 0.901 | 0.901 | 0.901 | 0.901 | 0.901 |

Note: RE: European Renewable Energy Index. EUA: European carbon price. Significance at the 0.01, 0.05, and 0.10 levels indicated by ***, **, *; Standard errors in parentheses. BIC: Bayesian Information Criterion.

**Table 4.** Selection criterion for vine copula models.

| | C-Vine | D-Vine | R-Vine |
|---|---|---|---|
| AIC | −142.295 | −146.637 | −137.710 |
| BIC | −96.156 | −99.190 | −94.154 |
| Log-likelihood | 82.147 | 83.664 | 76.855 |

The third tree illustrates weak conditional dependence between coal (C) and carbon (E) prices, and it is found to be statistically insignificant which indicates that no conclusion was supported for conditional dependence for coal (C) and carbon price (E). Furthermore, there is also weak dependence between renewable energy (R) and natural gas (G) prices at 10% significant level. Finally, the fourth tree indicates that conditional dependence between coal and the natural gas variable is positive and highly significant that given Student-t copula is the best fitting model, which is indicated by an associated Kendall's tau of 0.16 with symmetric upper and lower tail dependence.

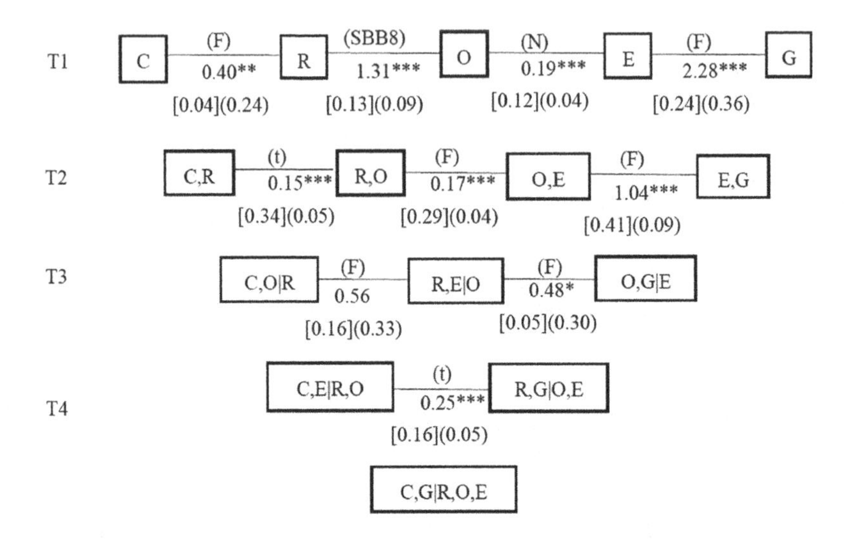


**Fig. 2.** D–vine copula tree for Coal future price (C), Renewable energy price index (R), Brent crude oil future price (O), European carbon price (E) (4), and Nature gas price future price (G). Note: For each edge, we have indicated the name of the best pair copula (Gaussian (N), Student-t (t), and Frank (F), survival BB8 (SBB8)) and its estimated parameter value with Kendall's tau (in brackets) and the corresponding standard error (in parenthesis); Significance at the 0.01, 0.05, and 0.10 levels indicated by ***, **, *.

**Table 5.** Estimation results of tail dependence from the D-vine copulas.

| | Pairs | Copulas | Tail dep | Symmetric |
|---|---|---|---|---|
| Tree1 | C-R | Frank | 0.000 | √ |
| | R-O | SBB8 | 0.000 | – |
| | O-E | Gaussian | 0.000 | √ |
| | E-G | Frank | 0.000 | √ |
| Tree2 | C, R-R, O | Student-t | 0.023[U] 0.023[L] | √ |
| | R, O-O, E | Frank | 0.000 | √ |
| | O, E-E, G | Frank | 0.000 | √ |
| Tree3 | $C, O\|R$-$R, E\|0$ | Frank | 0.000 | √ |
| | $R, E\|O$-$O, G\|E$ | Frank | 0.000 | √ |
| Tree4 | $C, E\|R, O$-$R, G\|O, E$ | Student-t | 0.016[U] 0.016[L] | √ |

Note: Coal price future price (C), European Renewable Energy Index (R), Brent crude oil future price (O), European carbon price (E), and Nature gas price future price (G). Survival BB8 (SBB8). Significance at the 0.01, 0.05, and 0.10 levels indicated by ***, **, and *; The symbol "[U]" represents upper tail dependence, "[L]" represents lower tail dependence.

## 5 Conclusion

Global warming and climate change have seriously affected environmental quality, human health, and economic activity, and it requires coordinated policies from countries around the world. The core of the threats for climate changing is the increasing tendency of global CO2 emissions. And increasing the supply of renewable energy would allow us to replace carbon-intensive energy sources and significantly reduce global warming emissions. Hence, in this study we use vine copula to investigate the multivariate dependence patterns between carbon prices and energy prices, including renewable energy prices. Moreover, the vine copula-based approach allows to capture a wide range of multidimensional dependence characteristics (such as average dependence, upper and lower tail dependence, and asymmetric dependence) of the price variables under consideration for the changing market states through time.

Our results mainly show evidence of a strong and positive link between oil and gas prices, between coal and oil prices, and between and renewable energy and carbon prices. It reveals that the carbon price affects the renewable energy price returns positively. And another finding that is a weak and positive link is found between renewable energy and natural gas prices, which typically reflects the competition between renewable energy and gas in both consumption and production processes in European countries. Due to the substitutability between the renewable energy and natural gas, if one of them price rises dramatically, then the end consumers and producers will move to another source of energy, causing its price to rise.

Finally, our empirical results find the high carbon price could push renewable energy price. Accordingly, we suggest that policymakers could take effective measures and action plan to assign carbon prices so that the emission market could provide incentives to transfer from traditional energy sources to renewable and clean energy sources. The ultimate goal of the European emissions trading system (EU ETS) is to help European Union member states effectively reduce greenhouse gas emissions and offer incentives for innovation and investment in clean and renewable energies. Therefore, increasing carbon price will play a pivotal role in the purpose of improving environmental quality, human health, and economic activity.

## References


1. Chen, H., et al.: The linkages of carbon spot-futures: evidence from EU-ETS in the third phase. Sustainability **12**(6), 2517 (2020)
2. EU Emissions Trading System—Climate Change (2020)
3. Motoaki, S.: Thermochemistry of the formation of fossil fuels. In: Fluid-Mineral Interactions: A Tribute to H. P. Eugster, pp. 271–283 (1990)
4. Ellabban, O., et al.: Renewable energy resources: current status, future prospects and their enabling technology. Renew. Sustain. Energy Rev. **39**, 748–764 (2014)
5. Jiménez-Rodríguez, R.: What happens to the relationship between EU allowances prices and stock market indices in Europe? Energy Econ. **81**, 13–24 (2019)

6. Chevallier, J., Nguyen, D.K., Reboredo, J.C.: A conditional dependence approach to $CO_2$-energy price relationships. Energy Econ. **81**, 812–821 (2019)
7. Ji, Q., Zhang, D., Geng, J.-B.: Information linkage, dynamic spillovers in prices and volatility between the carbon and energy markets. J. Clean. Prod. **198**, 972–978 (2018)
8. Zhang, D., Broadstock, D.C.: Global financial crisis and rising connectedness in the international commodity markets. Int. Rev. Financ. Anal. **68**, 101239 (2020)
9. Filis, G., Degiannakis, S., Floros, C.: Dynamic correlation between stock market and oil prices: the case of oil-importing and oil-exporting countries. Int. Rev. Financ. Anal. **20**(3), 152–164 (2011)
10. Oberndorfer, U.: EU emission allowances and the stock market: evidence from the electricity industry. Ecol. Econ. **68**(4), 1116–1126 (2009)
11. Silva, P.P., Moreno, B., Figueiredo, N.C.: Firm-specific impacts of $CO_2$ prices on the stock market value of the Spanish power industry. Energy Policy **94**, 492–501 (2016)
12. Kumar, S., Managi, S., Jain, R.K.: $CO_2$ mitigation policy for Indian thermal power sector: potential gains from emission trading. Energy Econ. **86**, 104653 (2020)
13. Dutta, A., Bouri, E., Noor, M.H.: Return and volatility linkages between $CO_2$ emission and clean energy stock prices. Energy **164**, 803–810 (2018)
14. Lutz, B.J., Pigorsch, U., Rotfuß, W.: Nonlinearity in cap-and-trade systems: the EUA price and its fundamentals. Energy Econ. **40**, 222–232 (2013)
15. Song, Y., et al.: The dynamic dependence of fossil energy, investor sentiment and renewable energy stock markets. Energy Econ. **84**, 104564 (2019)
16. Joe, H., Xu, J.: The estimation method of inference functions for margins for multivariate models. Technical Report No. 166, Department of Statistics, University of British Columbia (1996)
17. Bedford, T., Cooke, R.M.: Probability density decomposition for conditionally dependent random variables modeled by vines. Ann. Math. Artif. Intell. **32**, 245–268 (2001). https://doi.org/10.1023/A:1016725902970
18. Liu, J., Wang, M., Sriboonchitta, S.: Examining the interdependence between the exchange rates of China and ASEAN countries: a canonical vine copula approach. Sustainability **11**, 5487–5505 (2019)
19. Joe, H.: Multivariate Models and Dependence Concepts. Chapman & Hall, London (1997)
20. Song, Q., Liu, J., Sriboonchitta, S.: Risk measurement of stock markets in BRICS, G7, and G20: vine copulas versus factor copulas. Mathematics **7**(3), 274 (2019)
21. Kurowicka, D., Cooke, R.M.: Uncertainty Analysis with High Dimensional Dependence Modelling. Wiley, Chichester (2006)

# Prediction of Closing Stock Prices Using the Artificial Neural Network in the Market for Alternative Investment (MAI) of the Stock Exchange of Thailand (SET)

Rujira Chaysiri(✉) and Chanrathanak Ngauv

School of Management Technology, Sirindhorn International Institute of Technology, Thammasat University, Pathum Thani 12120, Thailand
rchaysiri@siit.tu.ac.th

**Abstract.** Forecasting stock prices has been a challenging problem due to the involvement of many variables and indicators. This study investigates daily closing stock prices in the Market for Alternative Investment (MAI) of the Stock Exchange of Thailand (SET). The Artificial Neural Network (ANN) is applied to predict daily closing stock prices of the most active stocks in MAI. This paper aims to investigate whether day traders are capable or not of capturing profit by using ANN in MAI and comparing day trading strategies with the buy and hold strategy. The result shows that day traders have the potential to generate profit when they trade the most active stocks in MAI and the profit from day trading strategies is statistically significantly higher than the profit from the buy and hold strategy.

**Keywords:** Stock market · Artificial Neural Network (ANN) · Market for Alternative Investment (MAI) · Stock Exchange of Thailand (SET)

## 1 Introduction

People who invest in stock markets could get rich quick or could go bankrupt due to the volatility over time both during crises and non-crises [1]. For example, the Global Financial Crisis in 2007–2008 affected the financial sector around the world. After the Global Financial Crisis, there was negative growth of GDP in the US and EU of around 2–3% (Eurostat) [2]. Therefore, some investors would gain profits while some would lose their investments. Investors are mostly concerned with how to pick the right stocks at the right time. Buying stocks that are going up and selling them for a profit is probably the most obvious strategy to generate profit. Short selling is another strategy whereby investors borrow stocks from a broker then sell them back. The aim is to buy stocks back at a lower price and return them to the broker [3]. Thus, there are opportunities for investors to gain profit from stock markets in both bull and bear markets. However, the formulation of a good prediction model is a complex task for investors in the financial sector because data on stock markets is nonlinear, dynamic, non-parametric, and disorganized in nature.


V.-N. Huynh et al. (Eds.): IUKM 2020, LNAI 12482, pp. 335–345, 2020.
https://doi.org/10.1007/978-3-030-62509-2_28

Day traders and investors use different methods to earn their profit on stock markets. Day traders usually do not hold their position overnight. Furthermore, they use historical data to learn the trends of the market and aim to gain profit in a short period of time. Day traders or speculators would bear higher risk than investors [4]. Unlike day traders, the goal of investors is to gain profit over an extended period of time. Investors often hold their positions for a long period of time such as months and years [5]. Both day traders and investors want to gain profits when they invest in financial sectors, but they use different techniques to gain their profit.

Machine Learning techniques are important for modern stock traders in order to gain their profit and the techniques will continue to grow in the future. There are number of techniques that can analyze large datasets and make trade decisions instantly. The Artificial Neural Network (ANN) is a commonly used method in stock price predictions [6]. Many studies used the ANN model to predict stock prices that have surfaced in literature [6, 7]. An ANN model has its strength over traditional methods because it can have broad estimation capabilities, can identify data by itself, and make quick adaptations. Many stock price prediction papers applied ANN in their studies with different stock markets. For example, the prediction of the daily stock exchange rates of the NASDAQ [6], the prediction of the stock price of companies listed under the National Stock Exchange of India Ltd (NSE India) [8], and the prediction using five different companies such as Nike, Goldman Sachs, Johnson, & Johnson, JP Morgan and Pfizer [9]. ANN has the ability to deal with non-linear data such as historical data of stock markets, and the model can perform well for both regression and classification. It can possibly be combined with other methods to improve the prediction performance [10]. However, there are some limitations for ANN which requires large datasets and it is an unstable learning method [11]. In addition, the prediction accuracy of ANN models would decrease when analyzing complex networks or overtraining [10]. The number of neurons in each hidden layer, which are adjustable, is critical for the ANN models [12]. Researchers could use a trial and error method for constructing the ANN models in order to cope with some noise as well as the incapability of differentiating between complex patterns. Otherwise, ANN models could be constructed by using factorial design in Design of Experiment (DOE) to obtain a good model.

The purpose of this study aims to predict daily closing stock prices of the most active stocks in the Market for Alternative Investment (MAI) after the market opens. This research applies the Artificial Neural Network to build prediction models by using RapidMiner Studio. This study could illustrate that day trading strategies produced by ANN models could outperform the buy and hold strategy in MAI.

This paper is organized as follows: A literature review is presented in Sect. 2; Methodology is described in Sect. 3; Sect. 4 shows the results and discussion of our research; and Sect. 5 draws conclusions and explains the future directions of our study.

## 2 Literature Review

Different Neural Network techniques have been applied to predict daily stock prices. Many studies used the ANN model to apply to stock market predictions with different types of datasets [13–15]. Lasfer et al. [13] constructed the best performing ANN model using factorial design in Design of Experiment (DOE) to predict the closing prices of the United Arab Emirates MSCI Index with historical data between August 2002 to August 2012. The proposed idea from this study helps to reduce the number of random trial and error experiments when constructing Artificial Neural Network models. Jeenanuta et al. [14] proposed Recurrent Neural Network (RNN) with Long Short-Term Memory (LSTM) and the Deep Belief Network (DBN) with test data of PTT, SCC, KBANK, SCB, and CPALL. This study used six variables such as High Price, Low Price, Open Price, Close Price, Volume, and a three-day Simple Moving Average in their study. The result shows that SCC and CPALL give the lowest median value of MAPE for the comparison of one-year training data. SCC gives the lowest median value of MAPE in both three-year and five-year training data. The second experiment used DBN to compare with LSTM to then apply it to predict high, low, open, and close prices. The result shows that LSTM outperforms for low volatility stocks and DBN for high volatility stocks. Hasan and Rasel used an ANN model to predict the closing price of Wal-Mart with one-day ahead, five-days ahead and 10-days ahead by using five attributes: Date, Open Price, Close Price, High Price and Low Price [15]. The MAPE results from ANN models show that the prediction accuracy of one-day ahead has an error of 0.75%, five-days ahead and 10-days ahead have errors of 3.28% and 2.01%, respectively.

Furthermore, many studies used ANN to compare it with other methods. For example, Simon and Raoot constructed an ANN model to compare with other traditional methods such as Technical Analysis and Fundamental Analysis [16]. The result shows that the ANN model has better performance than traditional models. The combination of an ANN model with other algorithms can also improve prediction accuracy by combining the process of the technical and fundamental analyses, choosing the useful inputs, developing methodologies for processing stock market data, using the numerical techniques to prepare input, selecting the best model components, and applying more than one data mining technique. Kara et al. [17] used ANN to compare with Support Vector Machine to predict the stock price index movements on the Istanbul Stock Exchange with 10 years of historical data. The result shows that the ANN model provides a forecast accuracy of 75.74%, while SVM has the prediction accuracy of 71.52%. Şenol and Ozturan used an ANN model to compare with the Logistic Regression model to predict the stock price direction on the Turkey Stock Market (ISE-30) [18]. The result illustrates that the ANN model has a better prediction accuracy compare to the Logistic Regression model with the average prediction accuracy rate of 78.47% and 64.51%, respectively. Bustos et al. [19] investigated the comparison of ANN with the SVM model to forecast the Colombian stock market from January 2015 to December 2016. This paper used variables such as the Simple Day Moving average, Weighted Day Moving average, Stochastic Oscillators (K and D), Momentum, Relative Strength Index, Moving Average Convergence Divergence, Accumulation/Distribution oscillator and Commodity Channel Index in both methods. SVM has a prediction accuracy of 0.78 while ANN has a prediction accuracy of 0.76. Bouktif et al. [20] forecasted stock market movement direction on the AMZN NASDAQ by using historical data of Open, High, Low, Close and Volume. This study used ANN to compare with the Logistic Regression, Support Vector Machine, Random Forest, and Extreme Gradient Boosting. The prediction accuracy of Random Forecast is the best at 0.627.

In the literature, we can recognize that using the Artificial Neural Network model to predict the stock price is convincingly useful among the existing works. Nevertheless, some papers brought up some weakness of using ANN to predict stock prices in comparison to other techniques due to the different dataset used in the study. In this research, we use ANN for the prediction of stock price movements in the Market for Alternative Investment (MAI) on the Stock Exchange of Thailand (SET).

# 3 Methodology

We constructed an ANN model in order to predict the closing price of each stock after the market opened. In this study, the closing price is used as an output target in the model after we have obtained the predicted closing price, then we take those values to evaluate our position on whether to buy or to sell. For input value, we use today's Open prices, yesterday's Open prices, High prices, Low prices, Volume, Return-rates, and 20-day Moving Average. The 20-day Moving average is used as a technical parameter because we would like to add more information about the trends of stock prices into our inputs.

We constructed an Artificial Neural Network model in the form of trial and error which randomly created the hidden layers and hidden nodes with seven variables for the input layer as mentioned above and the output represents the result of the stock price movement to buy or to sell after the market opens on the current day. To construct the ANN model, we set up from one hidden layer to three hidden layers with hidden nodes from two to 10 hidden nodes. We use the cross-validation operator in RapidMiner with 10-fold cross-validation because of a limited data sample.

## 3.1 Data Collection

The Market for Alternative Investment (MAI) is the index of the Stock Exchange of Thailand and was established on 11 November 1998. It enacted the first trade on 17 September 2001 [21]. The MAI Index comprises 166 companies with 220,312.13 million baht [22]. MAI was established to create new fund-raising opportunities for businesses which have high growth potential. It can support small to large enterprises which have a minimum market capitalization from 20 million Thai baht while the SET can only support businesses that have at least 300 million Thai baht.

Table 1 shows the industrial groups which are classified by types of business. There are eight industry groups and 28 sectors of different businesses which follow this classification [22].

In SET, there are four different types of stock indexes: SET index, SET50 index, SET100 index and MAI. Figure 1 shows that MAI has the highest volatility of 0.061 among other indexes. It has a great potential for day traders to generate their profit.

In this paper, we investigate the most active stocks which have an average daily trading stock volume of more than 10 million trading stock volumes in 2019. There are 166 companies registered in MAI. CHAYO, ECF, META, MORE, NEWS, PSTC and UWC are selected due to the average daily trading stock volume and that they started to trade before 2019.

**Table 1.** MAI Industrial Group.

| | Industries | Symbol |
|---|---|---|
| 1 | Agricultural and Food Industry | AGRO |
| 2 | Consumer Products | CONSUMP |
| 3 | Financial | FINCIAL |
| 4 | Industrials | INDUS |
| 5 | Properties and Construction | PROPCON |
| 6 | Resource | RESOURC |
| 7 | Services | SERVICE |
| 8 | Technology | TECH |

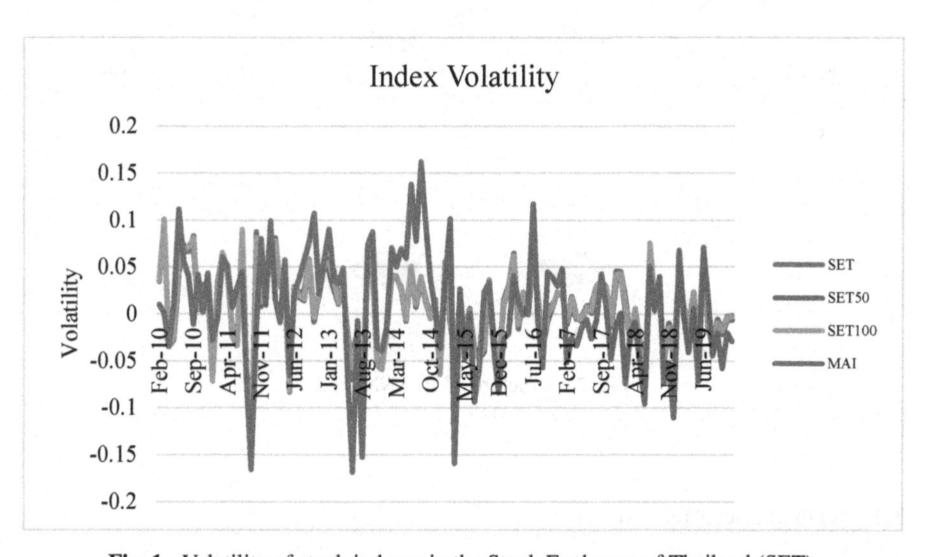


**Fig. 1.** Volatility of stock indexes in the Stock Exchange of Thailand (SET).

### 3.2 Artificial Neural Network (ANN)

It is recognized that the human brain computes in a definitively opposite way from the functioning of a computer, from which ANN has been inspired from its initiation. ANN is the method which provides a significant result of the prediction and it is used in numerous sectors, for example the medical sector [23] and weather forecasting [10], and the financial sector [15]. ANN is an effective tool which aids the provision of the efficient outcome for scientific experiments. The functions of the neural network are used for prediction and classification that can usually give more precise results than simple classification methods [15].

There are three important elements in the ANN model structure: an input neuron, hidden neurons, and output neurons, as shown in Fig. 2. An input neuron contains multiple variables, $x_i$, as the data information then calculates information and transfers to the next layer. A weight, $w_{ij}$, connects with hidden neurons, which are the middle layers. Hidden neurons could have one layer or more than one layer with multiple hidden nodes. A hidden neuron is a black box that is in the prediction process. A weight is a strength connection between each neuron and a small change of weight can result in a small change of weight in output, giving the artificial neuron their desired functionality [24]. The bias term is also weighted and added to the sum of the weighted input which serves as a threshold value [25]. An output neuron, $y_j$, is the result of prediction that is transformed from the input neurons.

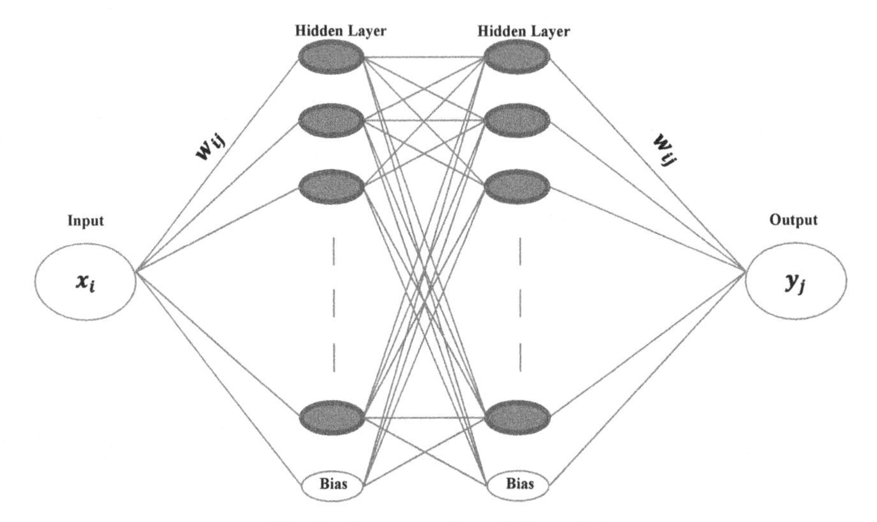


**Fig. 2.** Structure of an ANN model.

### 3.3 Performance Measure

This study used Mean Absolute Percentage Error (MAPE) as a performance measure for ANN models. MAPE is used to evaluate prediction accuracy. It shows how many percent that the prediction value differs from the actual value.

The function of MAPE is shown as follows [26]:

$$MAPE = \frac{1}{n}\sum\nolimits_{t=1}^{N}\left|\frac{actual_t - predict_t}{actual_t}\right| \quad (1)$$

### 3.4 Trading Strategies

After we obtained the most effective structure of the ANN model to predict the daily closing stock prices, we use the results of prediction to calculate profit using day trading strategies. Then we compare the profit from day trading strategies with the profit from the Buy and Hold strategy while using a month as a time horizon.

**Day Trading Strategy.** We have four types of day trading strategies that we consider. Each strategy is constructed by changing the percentage difference (2%, 3%, 4% and 5%) between the forecasted closing prices from the ANN model and the open prices. We use the predicted closing price to generate Buy and Sell signals. The basic structure of each day trading rule is defined as follows:

1. Buy when the predicted closing price is greater than the open price by 2%, 3%, 4% and 5%.
2. Sell when the predicted closing price is lower than the open price by 2%, 3%, 4% and 5%.
3. Close any position at the end of the day.

After Buy and Sell signals are generated, we calculated the profit/loss for each month.

**Buy and Hold Strategy.** The Buy and Hold strategy is defined as follow:

1. Buy on the first trading period of each month.
2. Sell at the end of each month.

The profit/loss of buy and hold strategy is calculated for each month.

### 3.5 Tukey's Test

We use a one-way ANOVA test to compare the p-value of the sample means with the significance level $\alpha = 0.05$. When the p-value is smaller than significant level, the null hypothesis is rejected. That means at least one of the sample means is statistically significant different from others. Otherwise, if there is no difference between the sample means, then the p-value will be greater than the significant level.

We use the Tukey's test to compare the difference among day trading strategies and the Buy and Hold strategy at the significance level $\alpha = 0.05$. Tukey's test computes statistically significant differences in the average profit/loss of day trading strategies and the Buy and Hold strategy in each month. Tukey's test also groups the sample means which are not statistically significantly different.

## 4 Result and Discussion

ANN models are built to forecast daily closing stock prices of CHAYO, ECF, META, MORE, NEWS, PSTC and UWC. The best structure of ANN models was identified based on the average MAPE from each structure of ANN models that were applied to the selected stocks. Figure 3 shows the average MAPE of various structures of ANN models and it is found that the best structure has the lowest MAPE at 7.06%. The best structure has three hidden layers which has five nodes in the first layer, 10 nodes in the second layer and 10 nodes in the third layer.

Figure 4 shows the result from the comparison among the day trading strategies that used the best structure of the ANN model with the Buy and Hold strategy. The result

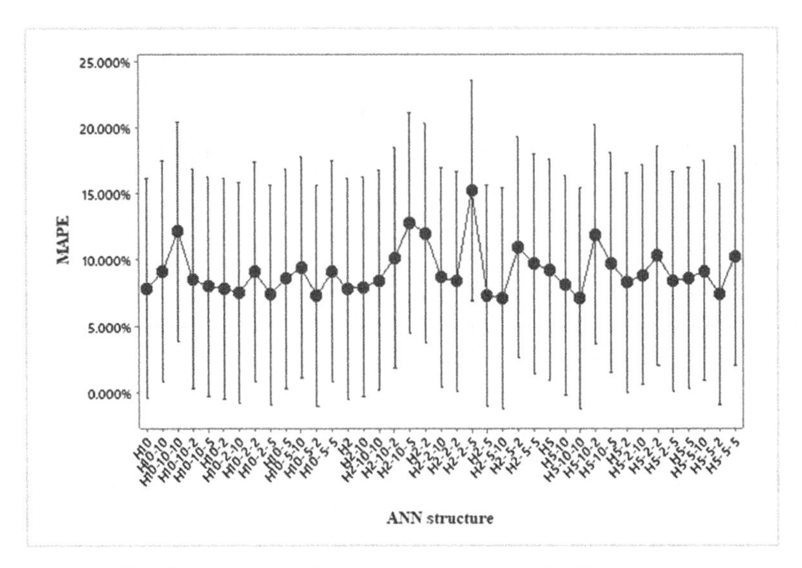


**Fig. 3.** Average MAPE of various structure of ANN models.

illustrates that the day traders who follow the ANN prediction model gain more profit than the Buy and Hold strategy. Day traders can gain an average profit of more than 10% per month, but investors who follow the Buy and Hold strategy can only gain a profit of less than 1%. In this paper, the transaction cost[1] of 0.207% and VAT 7% are also included of every buy and sell of a share of stocks.

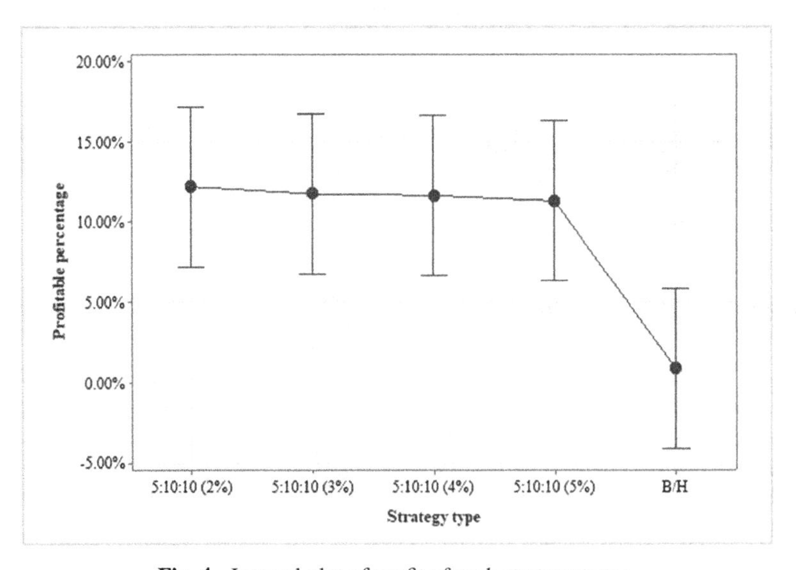


**Fig. 4.** Interval plot of profit of each strategy types.

[1] https://wwwa1.settrade.com/brokerpage/023/StaticPage/home/attachfile/equitycomm.html.

Table 2 is the result of comparing the mean profit of all strategy types using analysis of variance (ANOVA) to illustrate which one of the trading strategies is different from another. It shows that the sample means are not equal because the p-value of the population mean is less than significant level ($p\text{-}value = 0.006 < \alpha = 0.05$). Therefore, there is a significant difference among the population mean.

**Table 2.** ANOVA table of trading strategies.

| Source | SS | Df | MS | F-value | p-value |
|---|---|---|---|---|---|
| Type | 4.963 | 4 | 1.2407 | 3.66 | 0.006 |
| Error | 888.36 | 2620 | 0.3391 | | |
| Total | 893.327 | 2624 | | | |

We would like to investigate further where the difference in means is. Tukey's test is performed. Table 3 shows that the average profit from the Buy and Hold strategy is statistically significantly different from all day trading strategies. All day trading strategies are placed in the same group "A" while the Buy and Hold strategy is placed in the different group "B".

**Table 3.** Grouping using Tukey's test ($n = 525$, $\alpha = 0.05$) for trading strategies.

| Type | Mean | Standard deviation | Grouping |
|---|---|---|---|
| 5:10:10 (2%) | 12.17% | 0.6325 | A |
| 5:10:10 (3%) | 11.73% | 0.6291 | A |
| 5:10:10 (4%) | 11.61% | 0.6353 | A |
| 5:10:10 (5%) | 11.28% | 0.6379 | A |
| B/H | 0.85% | 0.2983 | B |

Our results show that day trading strategies from the ANN models can generate more profit than the Buy and Hold strategy on the most active stocks in MAI. In addition, a day trading strategy that follows a 2% difference from the Open price has the highest average profit at 12.17% per month.

## 5 Conclusion

This paper applies an Artificial Neural Network (ANN) for the prediction of stock closing prices of the Market for Alternative Investment (MAI) on the Stock Exchange of Thailand (SET). Our study focuses on the most active stocks which have an average daily trading stock volume more than 10 million trading stock volume and those stocks that have been traded before 2019. The result shows that ANN performs well for the prediction of closing prices in MAI. Moreover, using the best structure of ANN models to generate day trading strategies is more profitable than using the Buy and Hold strategy.

In future studies, we will use our results to compare with the performance of other prediction techniques such as other Machine Learning algorithm techniques and Technical Analysis. Design of Experiment (DOE) will be applied in the future studies for factorial designs to construct an ANN model more effectively.

## References

1. Banchit, A., Abidin, S., Wu, J.: Are shares more volatile during the global financial crisis? In: 6th International Research Symposium in Service Management, Kuching (2016)
2. Tangpornpaiboon, S., Puttapong, N.: Financial contagion of the global financial crisis from the US to other developed countries. J. Adm. Bus. Stud. **2**(1), 49–55 (2016)
3. Grünewald, S.N., Wagner, A.F., Weber, R.H.: Short selling regulation after the financial crisis – first principles revisited. Int. J. Discl. Gov. **7**(2010), 108–135 (2010)
4. Cheng, T.Y., Lin, C.H., Li, H., Lai, S., Watkins, K.A.: Day trader behavior and performance: evidence from Taiwan futures market. Emerg. Markets Financ. Trade **52**(11), 2495–2511 (2016)
5. Kuo, W.-Y., Lin, T.-C.: Overconfident individual day traders: evidence from the Taiwan futures market. J. Bank. Finance **37**(9), 3548–3561 (2013)
6. Moghaddam, A.H., Moghaddam, M.H., Esfandyari, M.: Stock market index prediction using artificial neural network. J. Econ. Financ. Adm. Sci. **21**(41), 89–93 (2016)
7. Vui, C.S., Soon, G.K., On, C.K., Alfred, R., Anthony, P.: A review of stock market prediction with Artificial neural network (ANN). In: 2013 IEEE International Conference on Control System, Computing and Engineering, Mindeb (2013)
8. Modi, C., Pasha, S.K., Devi, M.: Stock price prediction using artificial neural network. Int. J. Eng. Res. Technol. (IJERT) **4**(2016), 1–5 (2016)
9. Vijh, M., Chandola, D., Tikkiwal, V.A., Kumar, A.: Stock closing price prediction using machine learning technique. Procedia Comput. Sci. **167**(2020), 599–606 (2020)
10. Sureshkumar, K., Elango, N.: Performance analysis of stock price prediction using artificial neural network. Global J. Comput. Sci. Technol. **12**(1), 19–26 (2012)
11. Kumar, A., Murugan, S.: Performance analysis of indian stock market index using neural network time series model. In: 2013 International Conference on Pattern Recognition, Informatics and Mobile Engineering, Salem (2013)
12. Göcken, M., Özcalici, M., Boru, A., Dosdogru, A.T.: Integrating metaheuristics and Artificial Neural Networks for improved. Expert Syst. Appl. **44**(2016), 320–331 (2016)
13. Lasfer, A., El-Baz, H., Zualkernan, I.: Neural network design parameters for forecasting financial time series. In: 2013 5th International Conference on Modeling, Simulation and Applied Optimization (ICMSAO), Hammamet (2013)
14. Jeenanunta, C., Chaysiri, R., Thong, L.: Stock price prediction with long short-term memory recurrent neural network. In: 2018 International Conference on Embedded Systems and Intelligent Technology & International Conference on Information and Communication Technology for Embedded Systems (ICESIT-ICICTES), Khon Kaen, Thailand (2018)
15. Hasan, N., Rasel, R.I.: Artificial neural network approach for stock price and trend prediction. In: International Conference on Advanced Information & Communication Technology, Bangladesh (2016)
16. Simon, S., Raoot, A.: Accuracy driven Artificial Neural Networks in Stock Market Prediction. Int. J. Soft Comput. (IJSC) **3**(2), 35–44 (2012)

17. Kara, Y., Moyacioglu, M.A., Baykan, Ö.K.: Predicting direction of stock price index movement using artificial neural networks and support vector machines: the sample of the Istanbul Stock Exchange. Expert Syst. Appl. **38**, 5311–5319 (2011)
18. Şenol, D., Ozturan, M.: Stock prediction direction prediction using artificial neural network approach: the case of turkey. J. Artif. Intell. **2**(1), 70–77 (2008)
19. Bustos, O., Pomares, A., Gonzalez, E.: A comparison between SVM and multilayer perceptron in predicting an emerging financial market: Colombian stock market. In: 2017 Congreso Internacional de Innovacion y Tendencias en Ingenieria (CONIITI), Bogota (2017)
20. Bouktif, S., Fiaz, A., Awad, M.: Stock market movement prediction using disparate text features with machine learning. In: 2019 Third International Conference on Intelligent Computing in Data Sciences (ICDS), Marrakech (2019)
21. Chorruk, J., Worthington, A.C.: The pricing and performance of IPOs for small- and medium-sized enterprises: evidence from Thailand. J. Asia Pac. Econ. **18**(4), 543–559 (2013)
22. The Stock Exchange of Thailand (2019). https://www.set.or.th/en/market/market_statistics.html
23. Haglin, J.M., Jimenez, G., Eltorai, A.E.: Artificial neural networks in medicine. Health Technol. **9**, 1–6 (2019)
24. Kar, A.: Stock Prediction Using Artificial Neural Networks. Department of Computer Science and Engineering, IIT Kanpur (1990)
25. Kantardzic, M.: Data Mining: Concepts, Models, Methods, and Algorithms, 2nd edn. Wiley, Hoboken (2011)
26. Yousif, J.H., Kazem, H.A., Alattar, N.N., Elhassan, I.I.: A comparison study based on artificial neural network for assessing PV/T solar energy production. Case Stud. Therm. Eng. **13**(2019), 1–13 (2019)

# Estimating Fish Dispersal Using Interval Estimations for the Single Variance of a Delta-Lognormal Distribution

Patcharee Maneerat[1](✉), Sa-Aat Niwitpong[2], and Pisit Nakjai[3]

[1] Department of Mathematics, Uttaradit Rajabhat University, Uttaradit 53000, Thailand
m.patcharee@uru.ac.th

[2] Department of Applied Statistics, King MongKut's University of Technology North Bangkok, Bangkok 10800, Thailand
sa-aat.n@sci.kmutnb.ac.th

[3] Department of Computer Science, Uttaradit Rajabhat University, Uttaradit 53000, Thailand
mynameisbee@uru.ac.th

**Abstract.** Fish dispersal can be used to indicate the abundance of fish populations in a given local environment, and it is also important for understanding the ecology of fish species. The motivation of this research is how to estimate the variation in fish abundance from trawl surveys using interval estimations for the single variance of a delta-lognormal distribution. Our proposed methods are the normal approximation (NA), fiducial generalized confidence interval (FGCI) and the method of variance estimates recovery (MOVER). Monte Carlo simulation was used to assess the performance of the three methods in terms of coverage rate and average width. The findings from the simulation study show that NA and FGCI worked well for small and large variances, respectively. Data on the densities of red cod from trawl surveys were used to illustrate the efficacy of our methods.

**Keywords:** Delta-lognormal distribution · FGCI · Fish dispersal · Normal approximation · Variance

## 1 Introduction

In marine surveys, fish density can be expressed as the fish abundance of a particular species compared to all species as representative of the total fish community. The spatial variation of abundance depends on the local environmental conditions that directly affect both the local fish population dynamics and dispersal [14]. Mullen [14] stated that the dispersion of fish is a measure of diffusion. Smith et al. [18] noted that changes in catchability can be considered from the variation caused by the impacts of environmental variables on fish distribution. Furthermore, Fletcher [5], and Wu and Hsieh [22] estimated the mean density


V.-N. Huynh et al. (Eds.): IUKM 2020, LNAI 12482, pp. 346–357, 2020.
https://doi.org/10.1007/978-3-030-62509-2_29

of red cod (*Pseudophycis bachus*) population taken from a trawl survey in the National Institute of Water and Atmospheric Research in New Zealand. Motivated by these studies, the fish dispersal for each species can be utilized to not only indicate the abundance dispersion but also the ecology of the fish species in the local environment. Importantly, the red cod followed the assumptions of a delta-lognormal distribution, as evidenced by histogram, normal Q-Q plot and Akaike and Bayesian information criterion (AIC and BIC) results.

It is well-known that a lognormal distribution of right-skewed positive data combined with zero observations fitting a binomial distribution forms the delta-lognormal distribution, defined by Aitchison [1]. The delta-lognormal distribution has been applied to real-world applications in several research areas. For example, airborne chlorine concentration to measure air contaminant at an industrial site in the US [8,17,20], medical cost for the treatment of some diseases [20,23], and the average and dispersion of rainfall data [7,11].

In applied statistics, the variance is commonly used to measure dispersion, the standard deviation is obtained as the square root of the variance. In probability and statistical inference, the variance is also defined as the second central moment [4]. It is one of the parameters of interest that depends on its distribution function. When estimating a parameter, interval estimation can be a confidence interval (CI) that possibly provides more information about data than point estimation. There are a few researchers who have investigated interval estimation for the variance, as well as functions of the variance (the ratio and the difference between variances). For example, Bebu and Methew [3] presented the modified signed loglikelihood ratio and generalized confidence interval (GCI) for the ratio of variances of bivariate lognormal distributions. Later, Mathew and Webb [13] proposed the GCI for the variance components in mixed models. Niwitpong [16] studied the GCIs for a single variance and the ratio and the difference between two variances of lognormal data. Finally, Maneerat et al. [12] proposed a Bayesian credible interval in the form of the highest posterior density interval for the ratio of delta-lognormal variances.

For applying to and estimating fish dispersal, the aim of this study is to construct CIs for the single variance of a delta-lognormal distribution using normal approximation (NA), the fiducial generalized confidence interval (FGCI) and the method of variance estimates recovery (MOVER). Both FGCI and MOVER based on Li et al. [10] and Hasan and Krishnamoorthy [8], respectively, are detailed in Sect. 2. The simulation procedure and numerical results are revealed in Sect. 3. In Sect. 4, data from red cod densities are used to demonstrate the efficacies of our methods for analyzing fish dispersal. This article is ended with conclusions in Sect. 5.

## 2 Notation and Methods

Let $W = (W_1, W_2, ..., W_n)$ be a non-negative random sample draw from a delta-lognormal distribution, denoted as $\Delta \sim N(\mu, \sigma^2, 1-\delta')$ where $\delta'$ and $\delta = 1-\delta'$ be the proportions of non-zero and zero observations, respectively. For $w = 0$, the probability density function (pdf) is $h(w;\mu,\sigma^2,\delta') = 1-\delta'$. For $w>0$, the pdf of $W$ is given by

$$h(w;\mu,\sigma^2,\delta') = (1-\delta') + \delta' \left\{ \frac{1}{w\sqrt{2\pi\sigma^2}} \exp\left[-\frac{1}{2\sigma^2}(\ln w - \mu)^2\right] \right\} \quad (1)$$

where $\ln W \sim N(\mu,\sigma^2)$ and the number of zero values has a binomial distribution, denoted as $n_{i(0)} = \#\{i : w_i = 0\} \sim Bi(n, 1-\delta')$. The unbiased estimates of $\mu$, $\sigma^2$, $\delta'$ are $\bar{w} = n_{(1)}^{-1} \sum_{i:w_i>0} \ln w_i$, $s^2 = (n_{(1)}-1)^{-1} \sum_{i:w_i>0} (\ln w_i - \bar{w})^2$ and $\hat{\delta}' = n_{(1)}/n$ where $n_{(1)} = \#\{i : w_i > 0\}$ and $n = n_{(0)} + n_{(1)}$. The population variance of $W$ is $\lambda = \delta' \exp(2\mu+\sigma^2)[\exp(\sigma^2) - \delta']$ log-transformed as

$$\theta = \ln(\delta') + (2\mu+\sigma^2) + \ln\left[\exp(\sigma^2) - \delta'\right] \quad (2)$$

which is the parameter of interest in this study. The following methods are explained to construct the CIs for $\theta$.

### 2.1 Normal Approximation

The idea of this method is obtained from the central limit theorem (Theorem 1) which is related with the sample mean and its limiting distribution (standard normal distribution) when the sample size increases. Generally, this theorem is enough for almost all statistic purposes [4]. It can be seen that there is a finite variance which is only the assumption on the parent distributions.

**Theorem 1.** *Let $X_1, X_2, ...$ be a sequence of independent and identically distributed random variables with the mean $E(X_i) = \mu$ and variance $Var(X_i) = \sigma^2 > 0$. Define $\bar{X}_n = n^{-1}\sum_{i=1}^{n} X_i$. Let $G_n$ denote the cumulative distribution function of $Z = \sqrt{n}(\bar{X}_n - \mu)/\sigma$. Then, for $-\infty < x < \infty$,*

$$\lim_{n\to\infty} \int_{-\infty}^{x} \frac{1}{2\pi} \exp(-z^2/2)dz \quad (3)$$

*that is, $Z$ has a limiting standard normal distribution.*

Here $W \sim \Delta(\mu,\sigma^2,1-\delta')$, and the logarithm of variance can be written as

$$\theta = \ln(\delta') + 2(\mu+\sigma^2) + \ln\left(1 - \frac{\delta'}{\exp(\sigma^2)}\right) \quad (4)$$

It can be implied that $\lim_{\sigma\to\infty} \ln\left(1 - \frac{\delta'}{\exp(\sigma^2)}\right) = 0$ such that $\theta \approx \ln(\delta') + 2(\mu+\sigma^2)$. Replacing the unbiased estimates $\bar{w}$, $s^2$ and $\hat{\delta}'$, the estimated delta-lognormal variance becomes

$$\hat{\theta} \approx \ln(\hat{\delta}') + 2(\bar{w} + s^2) \quad (5)$$

By the delta method, the variance of $\hat{\theta}$ is

$$Var(\hat{\theta}) \approx Var\left[\ln(\hat{\delta}') + 2(\bar{w} + s^2)\right] \quad (6)$$

$$\approx \frac{1-\delta'}{n\delta'} + \frac{4\sigma^2}{n\delta'} + \frac{8\sigma^4}{n\delta' - 1} \quad (7)$$

Apply Theorem 1, the random variable is

$$Z = \frac{\hat{\theta} - \theta}{\sqrt{Var(\hat{\theta})}} \sim N(0,1) \quad (8)$$

as $n \to \infty$. Therefore, the $100(1-\alpha)\%$ CI-based NA is given by

$$NA_{\theta} = \hat{\theta} \pm z_{1-\alpha/2}\sqrt{\widehat{Var}(\hat{\theta})} \quad (9)$$

where $z_{\alpha}$ denotes the $\alpha^{th}$ percentile of standard normal distribution. The estimated variance $\widehat{Var}(\hat{\theta})$ is obtained using $\bar{w}$ $s^2$ and $\hat{\delta}'$ from the samples.

**Algorithm 1: NA**

1) Compute the estimated delta-lognormal variance $\hat{\theta}$.
2) Compute the approximated variance of $\hat{\theta}$, denoted as $\widehat{Var}(\hat{\theta})$.
3) Compute the $100(1-\alpha)\%$ of NA in Eq. (9).

## 2.2 Fiducial Generalized Confidence Interval

This method is constructed based on a pivotal quantity which is a random variable whose distribution does not depend on the parameter [4]. Hannig et al. [6] have shown that the generalized pivotal quantity (GPQ) can be turned out a subclass of GPQs in the (GPQ2) condition (the observed value of GPQ depends on the parameter of interest), defined by Weeranhadi [21]. Specially, the subclass of GPQs can be connected with fiducial inference, called a fiducial generalized pivotal quantity (FGPQ) in Definition 1.

**Definition 1.** *Let $W \in \mathbb{R}^k$ be a random variable with the distribution function $F_W(w;\gamma)$; $\gamma \in \mathbb{R}^p$ be a (possibly vector) parameter. Suppose that $\eta = \pi(\gamma) \in \mathbb{R}^q$ be a parameter of interest; $q \leq 1$. Let $W^*$ be an independent copy of $W$. Given an observation, let $w$ and $w^*$ be the values of $W$ and $W^*$. A GPQ for $\eta$, denoted by $R_\eta(W, W^*, \gamma)$ is called a fiducial generalized pivotal quantity (FGPQ) if it satisfies the following conditions:*

*(FGPQ1) For given $W = w$, the distribution of $R_\eta(W, W^*, \gamma)$ is free from all unknown parameters.*

*(FGPQ2) For every allowable $w \in \mathbb{R}^k$, $R_\eta(w, w^*, \gamma) = \eta$.*

The difference between the GPQ and FGPQ is that the condition (FGPQ2) is more powerful than the (GPQ2). The $100(1-\alpha)\%$ FGCI for $\eta$ is $[R_\eta(\alpha/2), R_{1-\alpha/2}]$ where $R_\eta(\alpha)$ stands for the $\alpha^{th}$ quantile of $R_\eta(W, W^*, \gamma)$. Recall that $\theta = \ln(\delta') + (2\mu + \sigma^2) + \ln[\exp(\sigma^2) - \delta']$ be a parameter of interest. By Definition 1, the FGPQs of $\mu$, $\sigma^2$ and $\delta$ are considered. Motivated by Li et al. [10], the FGPQ of $\delta'$ is developed using the weighted beta distributions as

$$R_{\delta'} = [beta(n_{(1)}, n_{(0)} + 1) + beta(n_{(1)} + 1, n_{(0)})]/2 \tag{10}$$

According to Hasan and Krishnamoorthy [8], the FGPQs of $\mu$ and $\sigma^2$ are given by $R_\mu = \bar{w} - T\sqrt{R_{\sigma^2}/n_{(1)}}$ and $R_{\sigma^2} = [n_{(1)} - 1]s^2/U$ where $T$ and $U$ be independent random variables of standard normal $N(0,1)$ and chi-square $\chi^2_{n_{(1)}-1}$ distributions, respectively. By the pivots $R_\mu$, $R_{\sigma^2}$ and $R_\delta$, the FGPQ of $\theta$ is

$$R_\theta = \ln(R_{\delta'}) + (2R_\mu + R_{\sigma^2}) + \ln[\exp(R_{\sigma^2}) - R_{\delta'}] \tag{11}$$

Thus, the $100(1-\alpha)\%$ FGCI for $\theta$ is

$$FGCI_\theta = [R_\theta(\alpha/2), R_\theta(1-\alpha/2)] \tag{12}$$

where $R_\theta(\alpha)$ stands for the $\alpha^{th}$ quantile of $R_\theta$.

**Algorithm 2: FGCI**

1) Generate $T \sim N(0,1)$ and $U \sim \chi^2_{n_{(1)}-1}$ are independent random variables.
2) Compute the FGPQs: $R_\mu$, $R_{\sigma^2}$ and $R_{\delta'}$.
3) Compute $R_\theta$.
4) Repeat steps 1–3, a large number of times, m = 2500, compute the $(\alpha/2)^{th}$ and $(1-\alpha/2)^{th}$ percentiles of $R_\theta$ becomes the $100(1-\alpha)\%$ FGCI for $\theta$.

### 2.3 Method of Variance Estimates Recovery

For given CI for $\theta_r$, the simple closed-form CI for a linear combination of parameters $\theta_r$; $r = 1, 2, ..., k$ is obtained from the MOVER idea, presented by Krishnamoorthy and Oral [9]. Let $(\hat{\theta}_1, \hat{\theta}_2, ..., \hat{\theta}_k)$ be independent and unbiased estimates of $(\theta_1, \theta_2, ..., \theta_k)$. Also, let $(l_{\theta_r}, u_{\theta_r})$ be the CI for $\theta_r$. The $100(1-\alpha)\%$ MOVER for $\sum_{r=1}^{k} c_r\theta_r$ can be written as

$$MOVER_\theta = \left[\sum_{r=1}^{k} c_r\hat{\theta}_r - \left\{\sum_{r=1}^{k} c_r^2(\hat{\theta}_r - l_r^*)^2\right\}^{1/2}, \sum_{r=1}^{k} c_r\hat{\theta}_r + \left\{\sum_{r=1}^{k} c_r^2(\hat{\theta}_r - u_r^*)^2\right\}^{1/2}\right] \tag{13}$$

where $l_r^* = l_r$ if $c_r > 0$ and $u_r$ if $c_r < 0$, and $u_r^* = u_r$ if $c_r > 0$ and $l_r$ and $c_r$ if $c_r < 0$. Here, we let

$$\theta = \theta_1 + \theta_2 + \theta_3 = \ln(\delta') + (2\mu + \sigma^2) + \ln[\exp(\sigma^2) - \delta'] \tag{14}$$

According to Hasan and Krishnamoorthy [8], the Wilson interval for $\delta'$ is given by

$$[l_{\theta_1}, u_{\theta_1}] = \ln\left[\frac{n_{(1)} + 0.5V^2_{\alpha/2}}{n + V^2_{\alpha/2}} \pm \frac{V_{1-\alpha/2}}{n + V^2_{\alpha/2}}\left\{\frac{n_{(0)}n_{(1)}}{n} + \frac{V^2_{\alpha/2}}{4}\right\}^{1/2}\right] \tag{15}$$

where $V = [n_{(1)} - n\delta']/\sqrt{n\delta'(1-\delta')} \sim N(0,1)$. Next, the CI for $\theta_2$ can be expressed as

$$\begin{aligned}[l_{\theta_2}, u_{\theta_2}] = &\left[2\bar{w} + s^2 - \left\{\frac{4Q^2_{1-\alpha/2}s^2}{n_1} + s^4\left(1 - \frac{n_{(1)} - 1}{\chi^2_{1-\alpha/2,n_{(1)}-1}}\right)^2\right\}^{1/2},\right. \\ &\left. 2\bar{w} + s^2 + \left\{\frac{4Q^2_{1-\alpha/2}s^2}{n_1} + s^4\left(\frac{n_{(1)} - 1}{\chi^2_{\alpha/2,n_{(1)}-1}} - 1\right)^2\right\}^{1/2}\right]\end{aligned} \tag{16}$$

where $Q = (\bar{w} - \mu)/\sqrt{\sigma^2/n_{(1)}} \sim N(0,1)$ and $\chi^2_{n_{(1)}-1}$ stands for chi-square distribution with degree of freedom $n_{(1)} - 1$. We develop the CI for $\theta_3$ that is

$$\begin{aligned}[l_{\theta_3}, u_{\theta_3}] = \ln &\left[\exp(s^2) - \delta' - \left\{\exp\left(s^4\left[1 - \frac{n_{(1)} - 1}{\chi^2_{1-\alpha/2,n_{(1)}-1}}\right]^2\right) + \frac{V^2_{1-\alpha/2}}{(n + V^2_{\alpha/2})^2}\left(\frac{n_{(0)}n_{(1)}}{n} + \frac{V^2_{\alpha/2}}{4}\right)\right\}^{1/2}\right. \\ &\left.\exp(s^2) - \delta' + \left\{\exp\left(s^4\left[\frac{n_{(1)} - 1}{\chi^2_{\alpha/2,n_{(1)}-1}} - 1\right]^2\right) + \frac{V^2_{1-\alpha/2}}{(n + V^2_{\alpha/2})^2}\left(\frac{n_{(0)}n_{(1)}}{n} + \frac{V^2_{\alpha/2}}{4}\right)\right\}^{1/2}\right]\end{aligned} \tag{17}$$

The estimate $\hat{\theta} = \hat{\theta}_1 + \hat{\theta}_2 + \hat{\theta}_3$ are obtained; $\hat{\theta}_1 = \ln(\hat{\delta}')$, $\hat{\theta}_2 = 2\bar{w} + s^2$ and $\hat{\theta}_3 = \ln[\exp(s^2) - \hat{\delta}']$. From Eq. (13), the $100(1-\alpha)\%$ MOVER interval for $\theta$ is

$$\begin{aligned}MOVER_\theta = &\left[\hat{\theta} - \left\{(\hat{\theta}_1 - l_{\theta_1})^2 + (\hat{\theta}_2 - l_{\theta_2})^2 + (\hat{\theta}_3 - l_{\theta_3})^2\right\}^{1/2},\right. \\ &\left.\hat{\theta} + \left\{(\hat{\theta}_1 - u_{\theta_1})^2 + (\hat{\theta}_2 - u_{\theta_2})^2 + (\hat{\theta}_3 - u_{\theta_3})^2\right\}^{1/2}\right]\end{aligned} \tag{18}$$

**Algorithm 3: MOVER**

1) Generate $V, Q \sim N(0,1)$ and $U \sim \chi^2_{n_{(1)}-1}$ are independent.
2) Compute the CIs for $\theta_1$, $\theta_2$ and $\theta_3$ in Eq. (15), (16) and (17), respectively.
3) Compute the $100(1-\alpha)\%$ MOVER for $\theta$ in Eq. (18).

## 3 Simulation Studies and Results

In this study, computer simulations were used to investigate the performance measures (the coverage rate (CR) and the average width (AW)) of the proposed methods: NA, FGCI and MOVER for constructing CIs for the delta-lognormal variance. The CR is defined as the probability that the parameter falls within the interval, while the AW is the mean of widths of the simulated intervals. In the comparison, the criteria for a recommended method is that the CI provides a CR close to or greater than the nominal confidence level ($1 - \alpha = 95\%$) and also has the shortest AW. Throughout the simulation study, the parameter settings are as follows: sample sizes $n = 25, 50, 75, 100, 200$; the proportions of zero observations $\delta = 20\%, 50\%, 70\%$; the variance $\sigma^2 = 1.25, 2.0, 3.0$; and the mean $\mu = -\sigma^2/2$. Algorithm 4 shows the steps to compute the performances of the methods via Monte Carlo simulation.

**Algorithm 4:**

1) Generate random samples from delta-lognormal distribution $W \sim \Delta(\mu, \sigma^2, \delta)$.
2) Compute the unbiased estimates $\bar{w}$, $s^2$ and $\hat{\delta}'$.
3) Compute the NA, FGCI, MOVER for $\theta$ in Algorithms 1, 2 and 3, respectively.
4) Repeat steps 1–3, a large number of times, $\mathrm{M} = 5000$, the CRs and AWs of all proposed methods are obtained at the 95% nominal level.

From Table 1, numerical evaluations show that NA was one of the methods with CRs more closed to the nominal confidence level, and it also produced shorter interval than other methods for small variance $\sigma^2 = 1.25$ as well as $\sigma^2 = 2$ and large sample size. For large variance, FGCI gave the best CR with the smallest AW. On the contrary, MOVER tended to underestimate the CR in all situations.

## 4 An Empirical Study

Fletcher [5], and Wu and Hsieh [22] estimated the density (kg/km$^2$) of red cod for a sample of 64 trawls (13 of the 64 for produced empty hauls) in New Zealand. For these situations, the variance can be used to estimate the fish dispersion to explain variation in the abundance of fish species. There are two important assumptions of the delta-lognormal distribution: the zero observations are contained with the probability $0 < \delta < 1$, and the positive observations with the remaining probability $1-\delta$ have a lognormal distribution. The histogram (Fig. 1) and normal Q-Q plots (Fig. 2) show that the fish density possibly follows with the assumptions for a delta-lognormal distribution. Nguyen [15] have argued that it is inadequate to use probability value (p-value) alone to make decisions in statistical hypothesis testing. This is a way to avoid using it for evaluating model, so the AIC and BIC are used to check the fitting distribution instead. Both are methods for scoring and selecting a model derived from frequentist and

**Table 1.** The coverage rate (CR) and average width (AW) of 95% CIs for $\theta$: the variances $\sigma^2 = 1.25, 2.0, 3.0$.

| $n$ | $\delta$ | $\sigma^2 = 1.25$ | | | | | | $\sigma^2 = 2.0$ | | | | | | $\sigma^2 = 3.0$ | | | | | |
|---|---|---|---|---|---|---|---|---|---|---|---|---|---|---|---|---|---|---|---|
| | | CR | | | AW | | | CR | | | AW | | | CR | | | AW | | |
| | | NA | FG | MO | NA | FG | MO | NA | FG | MO | NA | FG | MO | NA | FG | MO | NA | FG | MO |
| 25 | 0.2 | 97.32 | 96.24 | 85.52 | 3.712 | 4.720 | 4.082 | 93.08 | 94.38 | 79.98 | 5.507 | 6.773 | 5.419 | 92.20 | 94.88 | 79.14 | 7.941 | 9.603 | 8.029 |
| | 0.5 | 95.82 | 95.34 | 86.08 | 4.546 | 6.050 | 5.668 | 91.68 | 94.20 | 81.58 | 6.814 | 9.016 | 8.666 | 91.92 | 95.12 | 81.28 | 9.822 | 12.952 | 14.959 |
| | 0.7 | 94.38 | 95.42 | 88.34 | 5.780 | 8.976 | 10.639 | 90.54 | 94.54 | 84.62 | 8.566 | 13.480 | 20.387 | 88.76 | 96.18 | 84.18 | 12.102 | 19.188 | 37.958 |
| 50 | 0.2 | 96.10 | 95.12 | 84.44 | 2.571 | 3.044 | 2.766 | 95.06 | 95.08 | 80.30 | 3.890 | 4.398 | 3.476 | 93.46 | 94.72 | 78.80 | 5.602 | 6.191 | 4.679 |
| | 0.5 | 95.34 | 94.62 | 86.24 | 3.222 | 3.798 | 3.676 | 94.06 | 94.94 | 81.56 | 4.838 | 5.621 | 4.783 | 92.94 | 94.86 | 80.60 | 6.954 | 8.009 | 6.497 |
| | 0.7 | 94.76 | 95.08 | 88.22 | 4.123 | 5.139 | 5.165 | 93.00 | 94.02 | 83.18 | 6.212 | 7.767 | 7.326 | 92.34 | 94.62 | 81.16 | 8.955 | 11.224 | 11.517 |
| 75 | 0.2 | 95.02 | 95.10 | 85.88 | 2.094 | 2.422 | 2.275 | 94.84 | 94.90 | 79.98 | 3.164 | 3.486 | 2.789 | 94.58 | 94.96 | 78.46 | 4.583 | 4.924 | 3.748 |
| | 0.5 | 95.44 | 94.06 | 86.36 | 2.631 | 2.982 | 3.033 | 94.58 | 94.66 | 83.32 | 3.932 | 4.374 | 3.815 | 93.86 | 94.86 | 79.54 | 5.674 | 6.245 | 5.056 |
| | 0.7 | 95.12 | 93.74 | 86.92 | 3.344 | 3.881 | 4.140 | 93.14 | 93.78 | 82.78 | 5.025 | 5.829 | 5.370 | 93.28 | 94.70 | 80.92 | 7.262 | 8.404 | 7.317 |
| 100 | 0.2 | 94.00 | 94.92 | 87.14 | 1.816 | 2.074 | 2.013 | 95.04 | 94.48 | 79.88 | 2.740 | 2.981 | 2.411 | 95.16 | 95.14 | 78.32 | 3.985 | 4.219 | 3.238 |
| | 0.5 | 95.32 | 94.50 | 87.72 | 2.286 | 2.542 | 2.704 | 95.12 | 95.08 | 83.40 | 3.431 | 3.740 | 3.360 | 94.44 | 94.54 | 79.66 | 4.978 | 5.361 | 4.429 |
| | 0.7 | 95.36 | 93.98 | 88.56 | 2.921 | 3.283 | 3.707 | 93.64 | 94.86 | 84.76 | 4.351 | 4.881 | 4.630 | 93.52 | 94.66 | 80.34 | 6.363 | 7.111 | 6.155 |
| 200 | 0.2 | 89.92 | 95.02 | 90.98 | 1.289 | 1.447 | 1.567 | 95.34 | 95.08 | 81.50 | 1.944 | 2.074 | 1.742 | 94.66 | 94.78 | 78.40 | 2.812 | 2.914 | 2.278 |
| | 0.5 | 94.28 | 94.16 | 90.44 | 1.627 | 1.760 | 2.136 | 95.36 | 94.96 | 84.32 | 2.438 | 2.576 | 2.492 | 94.72 | 94.44 | 80.84 | 3.530 | 3.678 | 3.203 |
| | 0.7 | 95.36 | 94.52 | 90.64 | 2.093 | 2.241 | 2.929 | 94.84 | 94.74 | 85.62 | 3.121 | 3.323 | 3.486 | 93.90 | 94.06 | 81.28 | 4.531 | 4.797 | 4.436 |

Remark: NA, FG and MO are abbreviated in normal approximation, fiducial generalized confidence interval and method of variance estimates recovery, respectively.

Bayesian probabilities, respectively. Akaike [2] and Stone [19] defined the AIC and BIC of a model, which can be calculated as

$$\mathrm{AIC} = -2\ln L + 2k \tag{19}$$

$$\mathrm{BIC} = -2\ln L + 2k\ln(n) \tag{20}$$

where $L$ be the likelihood function, $k$ be the number of parameters in the model, and $n$ be the number of recorded measurements. It is possible that the positive density of fish might be fitted with other distributions such as Cauchy, exponential, gamma, logistic, lognormal, normal, t and Weibull distributions. The reason is that the random samples $W = (W_1, W_2, ..., W_n)$ of these distributions are in the same range as $W > 0$. Although, a model is suitable for the data which is the one that has minimum AIC and BIC among a set of candidate models. It can be indicated the positive density of fish fitting for lognormal distribution, as shown in Table 2. Some trawls are also empty. Thus, it can be concluded that the distribution of fish density is delta-lognormal distribution which is in line with Fletcher [5], and Wu and Hsieh [22].

**Table 2.** AIC and BIC results of positive fish densities.

| Distributions | AIC | BIC | Distributions | AIC | BIC |
|---|---|---|---|---|---|
| Cauchy | 734.7968 | 738.7748 | Lognormal | 702.8736 | 706.8516 |
| Exponential | 722.5387 | 724.5277 | Normal | 826.5334 | 830.5113 |
| Gamma | 720.9743 | 724.9523 | T | 734.6301 | 740.5970 |
| Logistic | 796.7544 | 800.7324 | Weibull | 716.9205 | 720.8985 |

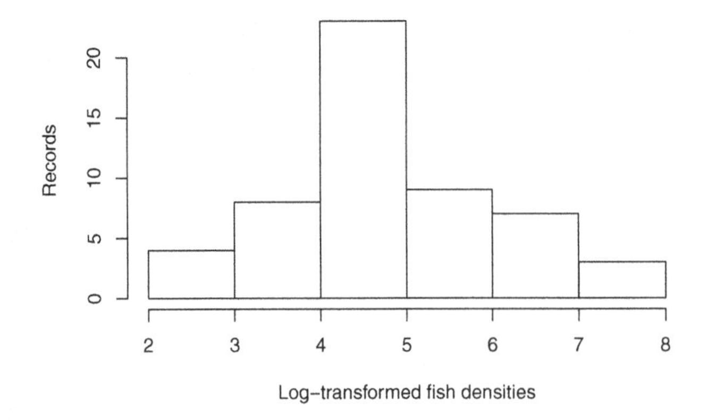


**Fig. 1.** Histogram plot of log-transformed fish densities.

**Table 3.** 95% CIs for $\theta$

| CIs | NA | FGCI | MOVER |
|---|---|---|---|
| Lower limits | 11.1724 | 11.1533 | 11.4049 |
| Upper limits | 13.7921 | 14.1077 | 24.7686 |
| Lengths | 2.6197 | 2.9544 | 13.3637 |

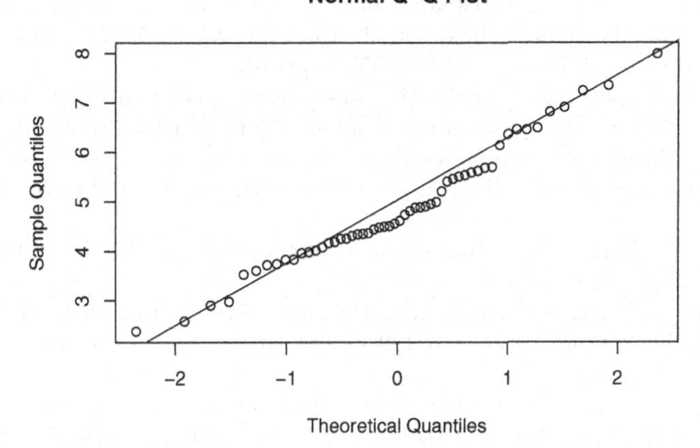


**Fig. 2.** Normal Q-Q plot of log-transformed fish densities.

On the basis of the log-transformed data from 57 trawls, the sample mean $\bar{w} = 4.8636$ kg/km$^2$ and the sample variance $s^2 = 1.4854$. The proportion of empty hauls is approximately 20.31% of the trawl total. The variance of fish density was estimated as $\exp(\hat{\theta}) = \exp(12.2808) = 215514.5$. The 95% NA, FGCI and MOVER for $\hat{\theta}$ are reported in Table 3. It can be interpreted that the red cod densities had a large dispersion, indicating that the dispersal of local fish abundance was high in New Zealand. These results are accordance with the simulation results for $(n, s^2, \hat{\delta}) = (75, 2.0, 20\%)$ in the previous section.

## 5 Conclusions

In statistical inference, the delta-lognormal variance can be used to estimate and explain the statistical dispersion of data that fit a right-skewed distribution and contained some zero observations. Real-world data on fish abundance can offer information on fish dispersion. In this article, the NA, FGCI and MOVER methods to construct CIs for the delta-lognormal variance are offered. In conclusion, research results recommend that the following methods performed quite well in the different situations: NA for small $\sigma^2$ and FGCI for large $\sigma^2$. Although, MOVER is not recommended for estimating the delta-lognormal variance.

It can be seen that NA obtained from the estimated delta-lognormal variance $\hat{\theta}$ and its estimated variance $\widehat{Var}(\hat{\theta})$ is simple and straightforward, although its limitation is that it is unsuitable for situations with large $\sigma^2$. When considering FGCI, the FGPQs of $\sigma^2$ and $\delta'$ are more powerful than the other methods, thereby making it suitable for practical problems involving the delta-lognormal distribution with large $\sigma^2$.

## References

1. Aitchison, J.: On the distribution of a positive random variable having a discrete probability mass at the origin. J. Am. Stat. Assoc. **50**(271), 901–908 (1955). https://doi.org/10.1080/01621459.1955.10501976
2. Akaike, H.: A new look at the statistical model identification. IEEE Trans. Autom. Control **19**, 716–723 (1974). https://doi.org/10.1109/TAC.1974.1100705
3. Bebu, I., Mathew, T.: Comparing the means and variances of a bivariate log-normal distribution. Stat. Med. **27**(14), 2684–2696 (2008). https://doi.org/10.1002/sim.3080
4. Casella, G., Berger, R.L.: Statistical Inference, 2nd edn. Duxbury, Pacific Grove (2002)
5. Fletcher, D.: Confidence intervals for the mean of the delta-lognormal distribution. Environ. Ecol. Stat. **15**(2), 175–189 (2008). https://doi.org/10.1007/s10651-007-0046-8
6. Hannig, J., Iyer, H., Patterson, P.: Fiducial generalized confidence intervals. J. Am. Stat. Assoc. **101**(473), 254–269 (2006). https://doi.org/10.1198/016214505000000736
7. Harvey, J., Merwe, A.J.: Bayesian confidence intervals for means and variances of lognormal and bivariate lognormal distributions. J. Stat. Plan. Infer. **142**(6), 1294–1309 (2012). https://doi.org/10.1016/j.jspi.2011.12.006
8. Hasan, M.S., Krishnamoorthy, K.: Confidence intervals for the mean and a percentile based on zero-inflated lognormal data. J. Stat. Comput. Simul. **88**(8), 1499–1514 (2018). https://doi.org/10.1080/00949655.2018.1439033
9. Krishnamoorthy, K., Oral, E.: Standardized likelihood ratio test for comparing several log-normal means and confidence interval for the common mean. Stat. Methods Med. Res. 1–23 (2015). https://doi.org/10.1177/0962280215615160
10. Li, X., Zhou, X., Tian, L.: Interval estimation for the mean of lognormal data with excess zeros. Stat. Probab. Lett. **83**(11), 2447–2453 (2013). https://doi.org/10.1016/j.spl.2013.07.004
11. Maneerat, P., Niwitpong, S.A., Niwitpong, S.: Bayesian confidence intervals for a single mean and the difference between two means of delta-lognormal distributions. Commun. Stat. Simul. Comput. 1–29 (2019). https://doi.org/10.1080/03610918.2019.1616095
12. Maneerat, P., Niwitpong, S.A., Niwitpong, S.: A Bayesian approach to construct confidence intervals for comparing the rainfall dispersion in Thailand. PeerJ **8**, e8502 (2020). https://doi.org/10.7717/peerj.8502
13. Mathew, T., Webb, D.W.: Generalized p values and confidence intervals for variance components: applications to army test and evaluation. Technometrics **47**(3), 312–322 (2005). https://doi.org/10.1198/004017005000000265
14. Mullen, A.J.: Aggregation of fish through variable diffusivity. Fish. Bull. **88**(2), 353–362 (1989)
15. Nguyen, H.T.: How to test without p-values? Thai. Stat. **17**(2), i–x (2019). https://ph02.tci-thaijo.org/index.php/thaistat/article/view/202439

16. Niwitpong, S.: Generalized confidence intervals for function of variances of log-normal distribution. Adv. Appl. Stat. **51**(2), 151–163 (2017). https://doi.org/10.17654/AS051020151
17. Owen, W.J., DeRouen, T.A.: Estimation of the mean for lognormal data containing zeroes and left- censored values, with applications to the measurement of worker exposure to air contaminants. Biometrics **36**(4), 707–719 (1980). https://doi.org/10.2307/2556125
18. Smith, S.J., Perry, R.I., Fanning, L.P.: Relationships between water mass characteristics and estimates of fish population abundance from trawl surveys. Environ. Monit. Assess. **17**(2–3), 227–245 (1991). https://doi.org/10.1007/BF00399305
19. Stone, M.: Comments on model selection criteria of Akaike and Schwarz. J. Roy. Stat. Soc. **41**(2), 276–278 (1979). https://www.jstor.org/stable/2985044
20. Tian, L.: Inferences on the mean of zero-inflated lognormal data: the generalized variable approach. Stat. Med. **24**(20), 3223–3232 (2005). https://doi.org/10.1002/sim.2169
21. Weerahandi, S.: Generalized confidence intervals. J. Am. Stat. Assoc. **88**(423), 899–905 (1993). https://doi.org/10.2307/2290779
22. Wu, W.H., Hsieh, H.N.: Generalized confidence interval estimation for the mean of delta-lognormal distribution: an application to New Zealand trawl survey data. J. Appl. Stat. **41**(7), 1471–1485 (2014). https://doi.org/10.1080/02664763.2014.881780
23. Zhou, X.H., Tu, W.: Comparison of several independent population means when their samples contain log-normal and possibly zero observations. Biometrics **55**(2), 645–651 (1999). https://doi.org/10.1111/j.0006-341X.1999.00645.x

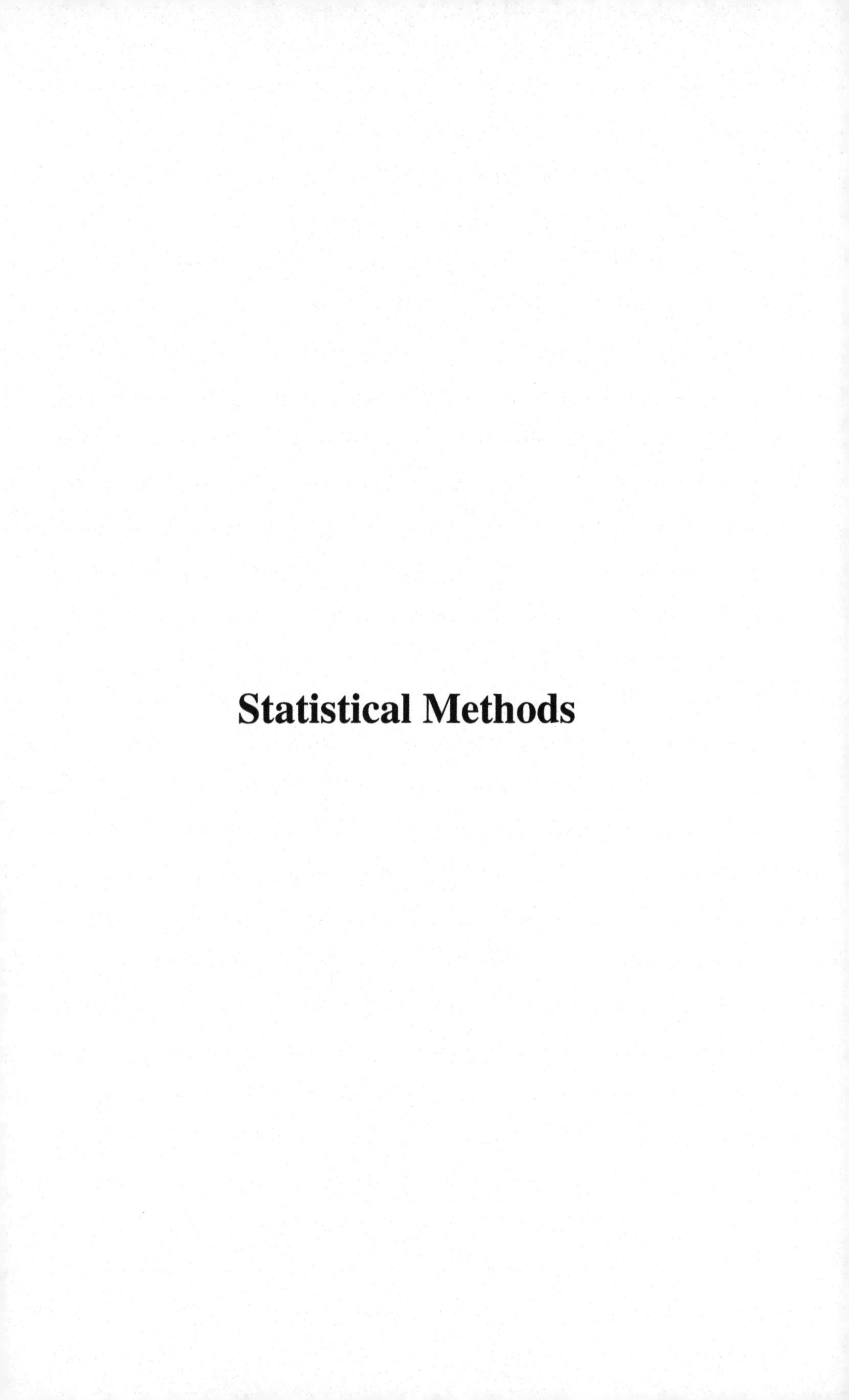

# Statistical Methods

# Bayesian Confidence Intervals for Means of Normal Distributions with Unknown Coefficients of Variation

Warisa Thangjai[1(✉)], Sa-Aat Niwitpong[2], and Suparat Niwitpong[2]

[1] Department of Statistics, Faculty of Science, Ramkhamhaeng University, Bangkok 10240, Thailand
wthangjai@yahoo.com

[2] Department of Applied Statistics, Faculty of Applied Science, King Mongkut's University of Technology North Bangkok, Bangkok 10800, Thailand
{sa-aat.n,suparat.n}@sci.kmutnb.ac.th

**Abstract.** There are two main approaches to statistical machine learning. The approaches are frequentist approach and Bayesian approach. The frequentist approach is commonly discussed in statistical inference. This approach assumes that unknown parameters are fixed constants. Frequentist approach uses maximum likelihood estimates of unknown parameters to predict new data points. Bayesian approach treats parameters as random variables. Bayesian approach is an important technique in statistics and mathematical statistics. Bayesian inference has found application in science, medicine, philosophy, and engineering. In this paper, the Bayesian confidence intervals for single mean and difference of two means with unknown coefficients of variation (CVs) of normal distributions are introduced. An evaluation of the performance of the Bayesian confidence intervals compared to frequentist confidence intervals of Thangjai et al. [1] for mean and difference of means with unknown CVs using Monte Carlo simulations is conducted. Simulations showed that the Bayesian approach performs better than the existing approaches when sample sizes are moderate and large. The Bayesian approach and other existing approaches are illustrated using two examples.

**Keywords:** Bayesian · CV · Mean · MOVER · Simulation

## 1 Introduction

Normal distribution is the most important probability distribution. This is because this distribution fits many natural phenomena. For instance, weight, height, and IQ score etc. Normal distribution has two parameters. The parameter $\mu$ is mean and the parameter $\sigma^2$ is variance. It is well known that $\bar{X}$ is sample mean which is unbiased estimator of $\mu$. Also, the sample variance $S^2$ is unbiased estimator of $\sigma^2$. In statistics, the CV is a measure of dispersion of a probability distribution. It is defined by $\tau = \sigma/\mu$. The CV needs to be estimated because it


V.-N. Huynh et al. (Eds.): IUKM 2020, LNAI 12482, pp. 361–371, 2020.
https://doi.org/10.1007/978-3-030-62509-2_30

is unknown. Therefore, Srivastava [2] introduced the estimator $\hat{\theta}$ which is uniformly minimum variance unbiased estimator of mean with unknown CV. Next, Sahai [3] proposed the estimator $\hat{\theta}^*$ which estimator of mean with unknown CV.

For single mean, it is generally accepted that the standard confidence intervals for mean of normal distribution are the confidence interval based on the Student's $t$-distribution and the $z$-distribution. Thangjai et al. [1] proposed large sample approach to estimate the confidence intervals for mean with unknown CV. Therefore, this paper will introduce the Bayesian approach for the confidence intervals.

For two means, the Welch-Satterthwaite approach is used to estimate the confidence interval for difference between two means of normal distributions. According to Thangjai et al. [1], the confidence intervals for difference between two means of normal distributions with unknown CVs were constructed using the large sample approach and the method of variance estimates recovery (MOVER) approach. Furthermore, Thangjai et al. [4] proposed the confidence interval for single inverse mean and difference of inverse means of normal distributions with unknown CVs. Therefore, the Bayesian approach was proposed to construct the confidence intervals for difference between two means of normal distributions with unknown CVs in this paper.

## 2 Bayesian Confidence Intervals for Mean with Unknown CV

Let $X_1, X_2, ..., X_n$ be a sample of size $n$ taken from normal distribution. Let $\mu$ and $\sigma^2$ be mean and variance. Let $\bar{X}$ and $S^2$ be estimators of $\mu$ and $\sigma^2$, respectively. Srivastava [2] and Sahai [3] introduced estimators of mean with unknown CV. The estimators of Srivastava [2] and Sahai [3] are

$$\hat{\theta} = n\bar{X}/(n + (S^2/\bar{X}^2)) \quad and \quad \hat{\theta}^* = n\bar{X}/(n - (S^2/\bar{X}^2)). \tag{1}$$

Tongmol et al. [5] suggested the independence Jeffreys prior distribution using the Fisher information matrix. The conditional posterior distribution $\mu|\sigma^2, x$ is the normal distribution with mean $\hat{\mu}$ and variance $\sigma^2/n$. Moreover, the posterior distribution $\sigma^2|x$ is the inverse gamma distribution with shape parameter $(n-1)/2$ and scale parameter $(n-1)s^2/2$. The mean with unknown CV based on the estimators of Srivastava [2] and Sahai [3] are

$$\hat{\theta}_{BS} = n\mu/(n + (\sigma^2/\mu^2)) \quad and \quad \hat{\theta}^*_{BS} = n\mu/(n - (\sigma^2/\mu^2)), \tag{2}$$

where $\mu$ and $\sigma^2$ are simulated from the posterior distributions.

The posterior distribution is used to construct the Bayesian confidence interval using the highest posterior density interval. Therefore, the $100(1-\alpha)\%$ two-sided confidence intervals for the single mean with unknown CV based on the Bayesian approach using the estimators of Srivastava [2] and Sahai [3] are obtained by

$$CI_{BS.\theta} = [L_{BS.\theta}, U_{BS.\theta}] \quad and \quad CI_{BS.\theta^*} = [L_{BS.\theta^*}, U_{BS.\theta^*}], \tag{3}$$

where $L_{BS.\theta}$, $U_{BS.\theta}$, $L_{BS.\theta^*}$, and $U_{BS.\theta^*}$ are the lower limit and the upper limit of the shortest $100(1-\alpha)\%$ highest posterior density intervals of $\hat{\theta}_{BS}$ and $\hat{\theta}^*_{BS}$, respectively.

# 3 Bayesian Confidence Intervals for Difference Between Means with Unknown CVs

Let $X_1, X_2, ..., X_n$ be random variables of size $n$ taken from normal distribution with mean $\mu_X$ and variance $\sigma_X^2$. Let $Y_1, Y_2, ..., Y_m$ be the random variables of size $m$ taken from normal distribution with mean $\mu_Y$ and variance $\sigma_Y^2$. Let $\bar{X}$ and $\bar{Y}$ be the sample means and let $S_X^2$ and $S_Y^2$ be the sample variances. The difference of means with unknown CVs based on the estimators of Srivastava [2] and Sahai [3] are

$$\hat{\delta}_{BS} = (n\mu_X/(n + (\sigma_X^2/\mu_X^2))) - (m\mu_Y/(m + (\sigma_Y^2/\mu_Y^2))) \tag{4}$$

and

$$\hat{\delta}^*_{BS} = (n\mu_X/(n - (\sigma_X^2/\mu_X^2))) - (m\mu_Y/(m - (\sigma_Y^2/\mu_Y^2))), \tag{5}$$

where $\mu$ and $\sigma^2$ are simulated from the posterior distributions.

The Bayesian confidence interval is constructed based on the highest posterior density interval. Therefore, the $100(1-\alpha)\%$ two-sided confidence intervals for the difference of means with unknown CVs based on the Bayesian approach using the estimators of Srivastava [2] and Sahai [3] are obtained by

$$CI_{BS.\delta} = [L_{BS.\delta}, U_{BS.\delta}] \quad and \quad CI_{BS.\delta^*} = [L_{BS.\delta^*}, U_{BS.\delta^*}], \tag{6}$$

where $L_{BS.\delta}$, $U_{BS.\delta}$, $L_{BS.\delta^*}$, and $U_{BS.\delta^*}$ are the lower limit and the upper limit of the shortest $100(1-\alpha)\%$ highest posterior density intervals of $\hat{\delta}_{BS}$ and $\hat{\delta}^*_{BS}$, respectively.

# 4 Simulation Studies

Two simulations investigated the coverage probability and average length. Generally, the confidence interval was chosen when the coverage probability greater than or close to the nominal confidence level $1-\alpha = 0.95$ and has the shortest average length. First simulation, the standard confidence interval for single mean of normal distribution was estimated using the Student's $t$-distribution and the $z$-distribution. The standard confidence interval was defined as $CI_\mu$. Thangjai et al. [1] proposed the confidence intervals for single mean with unknown CV based on large confidence intervals ($CI_{LS.\theta}$ and $CI_{LS.\theta^*}$). The standard confidence interval and the confidence intervals of Thangjai et al. [1] were compared with the Bayesian confidence intervals for single mean with unknown CV ($CI_{BS.\theta}$ and $CI_{BS.\theta^*}$). The data were generated from normal distribution $N(\mu, \sigma^2)$. The parameters and sample size configurations were the same as in Thangjai et al.

[1]. The number of simulation runs was 5000. For the Bayesian confidence interval, 2500 posterior distributions were simulated in each of 5000 simulation. From Tables 1 and 2, the result showed that the Bayesian approaches are better than the large sample approach in term of the coverage probabilities. Considering the mean with unknown CV, the three confidence intervals based on estimators of Srivastava [2] are better than the three confidence intervals based on estimators of Sahai [3] in term of the coverage probabilities. This is because the coverage probabilities of the confidence intervals based on estimators of Sahai [3] are close to 1.00 when $\sigma$ increases. The Bayesian confidence intervals based on estimator of Srivastava [2] is better than others in terms of coverage probability and average length.

Second simulation, the Welch-Satterthwaite confidence interval for difference of means of normal distributions were constructed to compare with the confidence intervals for difference of means of normal distributions with unknown CVs. The Welch-Satterthwaite confidence interval was defined as $CI_{\mu_X-\mu_Y}$. Thangjai et al. [1] introduced the confidence intervals for difference of means of normal distributions with unknown CVs based on large sample confidence intervals ($CI_{LS.\delta}$ and $CI_{LS.\delta^*}$) and MOVER confidence intervals ($CI_{MOVER.\delta}$ and $CI_{MOVER.\delta^*}$). The Welch-Satterthwaite confidence interval and the confidence intervals of Thangjai et al. [1] were compared with the Bayesian confidence intervals for difference of means of normal distributions with unknown CVs ($CI_{BS.\delta}$ and $CI_{BS.\delta^*}$). The first data were generated $n$ sample size from normal distribution $N(\mu_X, \sigma_X^2)$ and the second data were generated $m$ sample size from normal distribution $N(\mu_Y, \sigma_Y^2)$. The parameters and sample sizes were defined the same as in Thangjai et al. [1]. The posterior distributions of the Bayesian confidence intervals were evaluated with 2500 runs. The coverage probability and average length of all confidence intervals were computed using Monte Carlo method with 5000 runs. From Tables 3 and 4, the results indicated that the performances of the Bayesian confidence intervals are better than the performance of the Welch-Satterthwaite confidence interval. The confidence intervals based on estimator of Srivastava [2] are better than the confidence intervals based on estimator of Sahai [3] in terms of coverage probability and average length. For example, $CI_{BS.\delta}$ is better than $CI_{BS.\delta^*}$.

## 5 Empirical Applications

The simulation results indicate that the Bayesian confidence intervals give a good performance. The applicability is examined by re-analyzing two real data sets.

**Example 1.** The 15 participants were given 8 weeks of training to truly reduce cholesterol level. Niwitpong [6] studied the cholesterol level of the participants. The cholesterol levels were 129, 131, 154, 172, 115, 126, 175, 191, 122, 238, 159, 156, 176, 175, and 126. The sample mean and sample standard deviation were 156.3333 and 33.09006, respectively. The sample mean and sample variance were 156.3333 and 1094.9547, respectively. The confidence intervals for

**Table 1.** The coverage probabilities of 95% of two-sided confidence intervals for mean of normal distribution with unknown CV.

| $n$ | $\sigma$ | $CI_{LS.\theta}$ | $CI_{BS.\theta}$ | $CI_{LS.\theta^*}$ | $CI_{BS.\theta^*}$ | $CI_{\mu}$ |
|---|---|---|---|---|---|---|
| 10 | 0.3 | 0.9208 | 0.9484 | 0.9266 | 0.9472 | 0.9512 |
| | 0.5 | 0.9108 | 0.9470 | 0.9334 | 0.9488 | 0.9492 |
| | 0.7 | 0.9010 | 0.9390 | 0.9496 | 0.9568 | 0.9464 |
| | 0.9 | 0.8912 | 0.9394 | 0.9760 | 0.9708 | 0.9534 |
| | 1.0 | 0.8772 | 0.9364 | 0.9736 | 0.9688 | 0.9478 |
| | 1.1 | 0.8676 | 0.9356 | 0.9624 | 0.9684 | 0.9442 |
| | 1.3 | 0.8518 | 0.9452 | 0.9306 | 0.9728 | 0.9526 |
| | 1.5 | 0.8340 | 0.9460 | 0.8764 | 0.9756 | 0.9496 |
| | 1.7 | 0.8156 | 0.9358 | 0.8250 | 0.9764 | 0.9424 |
| | 2.0 | 0.7958 | 0.9398 | 0.7396 | 0.9920 | 0.9442 |
| 20 | 0.3 | 0.9354 | 0.9480 | 0.9388 | 0.9476 | 0.9498 |
| | 0.5 | 0.9266 | 0.9452 | 0.9380 | 0.9452 | 0.9480 |
| | 0.7 | 0.9300 | 0.9498 | 0.9564 | 0.9478 | 0.9510 |
| | 0.9 | 0.9310 | 0.9438 | 0.9742 | 0.9604 | 0.9524 |
| | 1.0 | 0.9308 | 0.9382 | 0.9824 | 0.9650 | 0.9496 |
| | 1.1 | 0.9284 | 0.9308 | 0.9860 | 0.9658 | 0.9454 |
| | 1.3 | 0.9230 | 0.9362 | 0.9906 | 0.9716 | 0.9502 |
| | 1.5 | 0.8980 | 0.9320 | 0.9698 | 0.9678 | 0.9420 |
| | 1.7 | 0.8784 | 0.9402 | 0.9450 | 0.9682 | 0.9474 |
| | 2.0 | 0.8358 | 0.9452 | 0.8846 | 0.9704 | 0.9488 |
| 30 | 0.3 | 0.9392 | 0.9492 | 0.9418 | 0.9476 | 0.9404 |
| | 0.5 | 0.9352 | 0.9462 | 0.9438 | 0.9466 | 0.9384 |
| | 0.7 | 0.9362 | 0.9498 | 0.9548 | 0.9474 | 0.9420 |
| | 0.9 | 0.9390 | 0.9474 | 0.9692 | 0.9436 | 0.9378 |
| | 1.0 | 0.9460 | 0.9492 | 0.9804 | 0.9516 | 0.9428 |
| | 1.1 | 0.9490 | 0.9428 | 0.9838 | 0.9536 | 0.9392 |
| | 1.3 | 0.9506 | 0.9384 | 0.9954 | 0.9698 | 0.9398 |
| | 1.5 | 0.9380 | 0.9318 | 0.9914 | 0.9722 | 0.9358 |
| | 1.7 | 0.9166 | 0.9384 | 0.9862 | 0.9748 | 0.9422 |
| | 2.0 | 0.8744 | 0.9418 | 0.9564 | 0.9690 | 0.9386 |
| 50 | 0.3 | 0.9458 | 0.9504 | 0.9482 | 0.9512 | 0.9476 |
| | 0.5 | 0.9420 | 0.9466 | 0.9466 | 0.9458 | 0.9440 |
| | 0.7 | 0.9400 | 0.9466 | 0.9474 | 0.9440 | 0.9416 |
| | 0.9 | 0.9488 | 0.9514 | 0.9656 | 0.9490 | 0.9466 |
| | 1.0 | 0.9512 | 0.9498 | 0.9728 | 0.9472 | 0.9470 |
| | 1.1 | 0.9508 | 0.9448 | 0.9782 | 0.9386 | 0.9394 |
| | 1.3 | 0.9600 | 0.9428 | 0.9916 | 0.9464 | 0.9424 |
| | 1.5 | 0.9660 | 0.9432 | 0.9992 | 0.9680 | 0.9480 |
| | 1.7 | 0.9538 | 0.9264 | 0.9996 | 0.9696 | 0.9414 |
| | 2.0 | 0.9318 | 0.9354 | 0.9950 | 0.9758 | 0.9454 |
| 100 | 0.3 | 0.9430 | 0.9446 | 0.9444 | 0.9440 | 0.9436 |
| | 0.5 | 0.9474 | 0.9478 | 0.9498 | 0.9488 | 0.9490 |
| | 0.7 | 0.9430 | 0.9432 | 0.9478 | 0.9432 | 0.9440 |
| | 0.9 | 0.9510 | 0.9518 | 0.9580 | 0.9524 | 0.9504 |
| | 1.0 | 0.9546 | 0.9502 | 0.9638 | 0.9498 | 0.9506 |
| | 1.1 | 0.9532 | 0.9474 | 0.9676 | 0.9458 | 0.9468 |
| | 1.3 | 0.9664 | 0.9526 | 0.9858 | 0.9500 | 0.9504 |
| | 1.5 | 0.9768 | 0.9482 | 0.9966 | 0.9446 | 0.9458 |
| | 1.7 | 0.9794 | 0.9496 | 0.9990 | 0.9422 | 0.9478 |
| | 2.0 | 0.9702 | 0.9438 | 1.0000 | 0.9588 | 0.9456 |

* $CI_{BS.\theta}$ is the shortest highest posterior density interval of $\hat{\theta}_{BS}$

** $CI_{BS.\theta^*}$ is the shortest highest posterior density interval of $\hat{\theta}^*_{BS}$

**Table 2.** The average lengths of 95% of two-sided confidence intervals for mean of normal distribution with unknown CV.

| $n$ | $\sigma$ | $CI_{LS.\theta}$ | $CI_{BS.\theta}$ | $CI_{LS.\theta^*}$ | $CI_{BS.\theta^*}$ | $CI_{\mu}$ |
|---|---|---|---|---|---|---|
| 10 | 0.3 | 0.3579 | 0.4195 | 0.3657 | 0.4035 | 0.4152 |
| | 0.5 | 0.5969 | 0.7176 | 0.6442 | 0.6492 | 0.6923 |
| | 0.7 | 0.8603 | 1.0106 | 1.8084 | 1.0661 | 0.9715 |
| | 0.9 | 1.1644 | 1.2557 | 15.8671 | 1.9163 | 1.2552 |
| | 1.0 | 1.3031 | 1.3532 | 28.7307 | 2.4151 | 1.3830 |
| | 1.1 | 1.4500 | 1.4560 | 25.5203 | 3.0735 | 1.5288 |
| | 1.3 | 1.7200 | 1.6510 | 30.9951 | 4.3496 | 1.8045 |
| | 1.5 | 2.0826 | 1.8464 | 30.4435 | 5.8251 | 2.0945 |
| | 1.7 | 2.2147 | 2.0249 | 36.0163 | 7.1355 | 2.3555 |
| | 2.0 | 2.6573 | 2.3284 | 63.6102 | 9.3794 | 2.7888 |
| 20 | 0.3 | 0.2584 | 0.2765 | 0.2610 | 0.2729 | 0.2769 |
| | 0.5 | 0.4299 | 0.4664 | 0.4427 | 0.4489 | 0.4616 |
| | 0.7 | 0.6111 | 0.6634 | 0.6584 | 0.6082 | 0.6461 |
| | 0.9 | 0.8270 | 0.8638 | 1.1785 | 0.7819 | 0.8304 |
| | 1.0 | 0.9486 | 0.9604 | 2.3401 | 0.9042 | 0.9215 |
| | 1.1 | 1.0886 | 1.0574 | 5.7519 | 1.1075 | 1.0201 |
| | 1.3 | 1.3386 | 1.2175 | 16.8972 | 1.6142 | 1.1979 |
| | 1.5 | 1.5458 | 1.3684 | 102.4481 | 2.3788 | 1.3878 |
| | 1.7 | 1.7157 | 1.5002 | 29.9521 | 3.2196 | 1.5700 |
| | 2.0 | 4.9440 | 1.8307 | 1.6973 | 4.6490 | 1.8504 |
| 30 | 0.3 | 0.2125 | 0.2215 | 0.2139 | 0.2198 | 0.2131 |
| | 0.5 | 0.3522 | 0.3702 | 0.3588 | 0.3622 | 0.3538 |
| | 0.7 | 0.4980 | 0.5239 | 0.5191 | 0.5002 | 0.4957 |
| | 0.9 | 0.6678 | 0.6837 | 0.7483 | 0.6258 | 0.6380 |
| | 1.0 | 0.7710 | 0.7666 | 0.9921 | 0.6855 | 0.7104 |
| | 1.1 | 0.8802 | 0.8460 | 1.8413 | 0.7501 | 0.7799 |
| | 1.3 | 1.1157 | 1.0024 | 16.3304 | 0.9587 | 0.9228 |
| | 1.5 | 1.3237 | 1.1408 | 28.2576 | 1.3204 | 1.0636 |
| | 1.7 | 1.4914 | 1.2763 | 2326.3080 | 1.8296 | 1.2127 |
| | 2.0 | 1.6103 | 1.4365 | 13.2521 | 2.7858 | 1.4168 |
| 50 | 0.3 | 0.1653 | 0.1688 | 0.1659 | 0.1681 | 0.1656 |
| | 0.5 | 0.2753 | 0.2827 | 0.2782 | 0.2794 | 0.2761 |
| | 0.7 | 0.3861 | 0.3967 | 0.3949 | 0.3873 | 0.3855 |
| | 0.9 | 0.5100 | 0.5141 | 0.5350 | 0.4930 | 0.4958 |
| | 1.0 | 0.5829 | 0.5747 | 0.6309 | 0.5439 | 0.5515 |
| | 1.1 | 0.6645 | 0.6349 | 0.7640 | 0.5920 | 0.6063 |
| | 1.3 | 0.8548 | 0.7590 | 2.5003 | 0.6812 | 0.7174 |
| | 1.5 | 1.0538 | 0.8848 | 12.2797 | 0.7875 | 0.8296 |
| | 1.7 | 1.2158 | 0.9985 | 48.1625 | 0.9444 | 0.9357 |
| | 2.0 | 1.3719 | 1.1614 | 13.1509 | 1.3562 | 1.1030 |
| 100 | 0.3 | 0.1171 | 0.1178 | 0.1173 | 0.1176 | 0.1172 |
| | 0.5 | 0.1949 | 0.1966 | 0.1959 | 0.1956 | 0.1952 |
| | 0.7 | 0.2742 | 0.2768 | 0.2771 | 0.2738 | 0.2741 |
| | 0.9 | 0.3568 | 0.3565 | 0.3637 | 0.3500 | 0.3518 |
| | 1.0 | 0.4034 | 0.3976 | 0.4148 | 0.3887 | 0.3915 |
| | 1.1 | 0.4531 | 0.4375 | 0.4730 | 0.4253 | 0.4297 |
| | 1.3 | 0.5764 | 0.5208 | 0.6571 | 0.4997 | 0.5087 |
| | 1.5 | 0.7222 | 0.6042 | 1.2491 | 0.5694 | 0.5865 |
| | 1.7 | 0.8755 | 0.6912 | 8.9144 | 0.6360 | 0.6661 |
| | 2.0 | 1.0511 | 0.8192 | 18.9046 | 0.7305 | 0.7816 |

* $CI_{BS.\theta}$ is the shortest highest posterior density interval of $\hat{\theta}_{BS}$
** $CI_{BS.\theta^*}$ is the shortest highest posterior density interval of $\hat{\theta}^*_{BS}$

**Table 3.** The coverage probabilities of 95% of two-sided confidence intervals for difference between means of normal distributions with unknown CVs.

| $n$ | $m$ | $\sigma_X/\sigma_Y$ | $CI_{LS.\delta}$ | $CI_{MOVER.\delta}$ | $CI_{BS.\delta}$ | $CI_{LS.\delta^*}$ | $CI_{MOVER.\delta^*}$ | $CI_{BS.\delta^*}$ | $CI_{\mu_X-\mu_Y}$ |
|---|---|---|---|---|---|---|---|---|---|
| 10 | 10 | 0.3 | 0.8928 | 0.9228 | 0.9448 | 0.9718 | 0.9814 | 0.9732 | 0.9500 |
| | | 0.5 | 0.8984 | 0.9302 | 0.9472 | 0.9730 | 0.9826 | 0.9726 | 0.9434 |
| | | 0.7 | 0.9164 | 0.9450 | 0.9562 | 0.9816 | 0.9902 | 0.9852 | 0.9504 |
| | | 0.9 | 0.9214 | 0.9512 | 0.9590 | 0.9858 | 0.9930 | 0.9926 | 0.9540 |
| | | 1.0 | 0.9186 | 0.9464 | 0.9536 | 0.9884 | 0.9932 | 0.9918 | 0.9502 |
| | | 1.1 | 0.9240 | 0.9474 | 0.9572 | 0.9822 | 0.9882 | 0.9926 | 0.9560 |
| | | 1.3 | 0.9180 | 0.9440 | 0.9582 | 0.9640 | 0.9742 | 0.9954 | 0.9542 |
| | | 1.5 | 0.9192 | 0.9452 | 0.9572 | 0.9380 | 0.9570 | 0.9964 | 0.9520 |
| | | 1.7 | 0.9122 | 0.9426 | 0.9534 | 0.9082 | 0.9358 | 0.9954 | 0.9464 |
| | | 2.0 | 0.9154 | 0.9440 | 0.9588 | 0.8452 | 0.8806 | 0.9974 | 0.9526 |
| 10 | 20 | 0.3 | 0.9396 | 0.9578 | 0.9546 | 0.9858 | 0.9894 | 0.9720 | 0.9546 |
| | | 0.5 | 0.9428 | 0.9586 | 0.9586 | 0.9838 | 0.9896 | 0.9722 | 0.9540 |
| | | 0.7 | 0.9394 | 0.9590 | 0.9580 | 0.9862 | 0.9916 | 0.9796 | 0.9548 |
| | | 0.9 | 0.9276 | 0.9502 | 0.9546 | 0.9884 | 0.9942 | 0.9812 | 0.9502 |
| | | 1.0 | 0.9162 | 0.9390 | 0.9458 | 0.9858 | 0.9910 | 0.982 | 0.9442 |
| | | 1.1 | 0.9012 | 0.9284 | 0.9470 | 0.9764 | 0.9856 | 0.9856 | 0.9460 |
| | | 1.3 | 0.9052 | 0.9260 | 0.9480 | 0.9534 | 0.9684 | 0.9840 | 0.9504 |
| | | 1.5 | 0.8920 | 0.9162 | 0.9490 | 0.9152 | 0.9434 | 0.9862 | 0.9470 |
| | | 1.7 | 0.8958 | 0.9180 | 0.9508 | 0.8598 | 0.8990 | 0.9886 | 0.9502 |
| | | 2.0 | 0.8792 | 0.9092 | 0.9512 | 0.7572 | 0.7968 | 0.9928 | 0.9486 |
| 30 | 30 | 0.3 | 0.9482 | 0.9568 | 0.9492 | 0.9752 | 0.9808 | 0.9546 | 0.9536 |
| | | 0.5 | 0.9450 | 0.9530 | 0.9478 | 0.9770 | 0.9822 | 0.9532 | 0.9492 |
| | | 0.7 | 0.9526 | 0.9612 | 0.9592 | 0.9796 | 0.9838 | 0.9594 | 0.9562 |
| | | 0.9 | 0.9516 | 0.9810 | 0.9846 | 0.9606 | 0.9538 | 0.9578 | 0.9516 |
| | | 1.0 | 0.9510 | 0.9614 | 0.9524 | 0.9854 | 0.9892 | 0.9610 | 0.9502 |
| | | 1.1 | 0.9538 | 0.9620 | 0.9520 | 0.9900 | 0.9924 | 0.9598 | 0.9476 |
| | | 1.3 | 0.9582 | 0.9654 | 0.9502 | 0.9960 | 0.9970 | 0.9738 | 0.9516 |
| | | 1.5 | 0.9518 | 0.9564 | 0.9494 | 0.9952 | 0.9952 | 0.9802 | 0.9516 |
| | | 1.7 | 0.9414 | 0.9486 | 0.9482 | 0.9888 | 0.9890 | 0.9764 | 0.9502 |
| | | 2.0 | 0.9140 | 0.9238 | 0.9442 | 0.9586 | 0.9592 | 0.9734 | 0.9522 |
| 20 | 30 | 0.3 | 0.9444 | 0.9530 | 0.9488 | 0.9756 | 0.9812 | 0.9528 | 0.9492 |
| | | 0.5 | 0.9464 | 0.9568 | 0.9516 | 0.9742 | 0.9818 | 0.9564 | 0.9510 |
| | | 0.7 | 0.9484 | 0.9584 | 0.9552 | 0.9762 | 0.9830 | 0.9578 | 0.9532 |
| | | 0.9 | 0.9472 | 0.9598 | 0.9516 | 0.9866 | 0.9906 | 0.9640 | 0.9536 |
| | | 1.0 | 0.9502 | 0.9598 | 0.9540 | 0.9908 | 0.9930 | 0.9690 | 0.9520 |
| | | 1.1 | 0.9442 | 0.9530 | 0.9470 | 0.9906 | 0.9940 | 0.9688 | 0.9474 |
| | | 1.3 | 0.9382 | 0.9486 | 0.9474 | 0.9894 | 0.9904 | 0.9776 | 0.9522 |
| | | 1.5 | 0.9226 | 0.9324 | 0.9472 | 0.9710 | 0.9722 | 0.9758 | 0.9524 |
| | | 1.7 | 0.9072 | 0.9138 | 0.9426 | 0.9486 | 0.9504 | 0.9742 | 0.9462 |
| | | 2.0 | 0.8740 | 0.8854 | 0.9470 | 0.8908 | 0.8940 | 0.9770 | 0.9506 |
| 50 | 50 | 0.3 | 0.9528 | 0.9556 | 0.9510 | 0.9672 | 0.9702 | 0.9498 | 0.9528 |
| | | 0.5 | 0.9520 | 0.9576 | 0.9526 | 0.9702 | 0.9746 | 0.9478 | 0.9530 |
| | | 0.7 | 0.9500 | 0.9548 | 0.9494 | 0.9708 | 0.9730 | 0.9466 | 0.9502 |
| | | 0.9 | 0.9574 | 0.9634 | 0.9568 | 0.9756 | 0.9798 | 0.9542 | 0.9554 |
| | | 1.0 | 0.9518 | 0.9574 | 0.9464 | 0.9724 | 0.9770 | 0.9440 | 0.9456 |
| | | 1.1 | 0.9590 | 0.9644 | 0.9530 | 0.9826 | 0.9846 | 0.9514 | 0.9510 |
| | | 1.3 | 0.9660 | 0.9694 | 0.9500 | 0.9916 | 0.9922 | 0.9528 | 0.9496 |
| | | 1.5 | 0.9662 | 0.9698 | 0.9506 | 0.9988 | 0.9992 | 0.9638 | 0.9496 |
| | | 1.7 | 0.9552 | 0.9594 | 0.9404 | 0.9990 | 0.9990 | 0.9664 | 0.9446 |
| | | 2.0 | 0.9468 | 0.9494 | 0.9448 | 0.9934 | 0.9934 | 0.9750 | 0.9530 |

(*continued*)

**Table 3.** (*continued*)

| $n$ | $m$ | $\sigma_X/\sigma_Y$ | $CI_{LS.\delta}$ | $CI_{MOVER.\delta}$ | $CI_{BS.\delta}$ | $CI_{LS.\delta^*}$ | $CI_{MOVER.\delta^*}$ | $CI_{BS.\delta^*}$ | $CI_{\mu_X-\mu_Y}$ |
|---|---|---|---|---|---|---|---|---|---|
| 30 | 50 | 0.3 | 0.9484 | 0.9550 | 0.9480 | 0.9670 | 0.9722 | 0.9452 | 0.9488 |
| | | 0.5 | 0.9484 | 0.9550 | 0.9512 | 0.9678 | 0.9706 | 0.9520 | 0.9508 |
| | | 0.7 | 0.9444 | 0.9532 | 0.9496 | 0.9674 | 0.9724 | 0.9460 | 0.9464 |
| | | 0.9 | 0.9450 | 0.9540 | 0.9460 | 0.9734 | 0.9786 | 0.9490 | 0.9472 |
| | | 1.0 | 0.9508 | 0.9594 | 0.9534 | 0.9820 | 0.9864 | 0.9562 | 0.9506 |
| | | 1.1 | 0.9556 | 0.9626 | 0.9506 | 0.9866 | 0.9880 | 0.9622 | 0.9514 |
| | | 1.3 | 0.9534 | 0.9574 | 0.9430 | 0.9936 | 0.9952 | 0.9668 | 0.9466 |
| | | 1.5 | 0.9510 | 0.9562 | 0.9468 | 0.9956 | 0.9958 | 0.9728 | 0.9516 |
| | | 1.7 | 0.9330 | 0.9376 | 0.9412 | 0.9844 | 0.9844 | 0.9708 | 0.9456 |
| | | 2.0 | 0.9072 | 0.9140 | 0.9376 | 0.9560 | 0.9562 | 0.9676 | 0.9452 |
| 100 | 100 | 0.3 | 0.9522 | 0.9554 | 0.9508 | 0.9612 | 0.9634 | 0.9482 | 0.9518 |
| | | 0.5 | 0.9562 | 0.9586 | 0.9544 | 0.9646 | 0.9672 | 0.9530 | 0.9556 |
| | | 0.7 | 0.9544 | 0.9576 | 0.9542 | 0.9638 | 0.9652 | 0.9524 | 0.9538 |
| | | 0.9 | 0.9492 | 0.9508 | 0.9464 | 0.9592 | 0.9620 | 0.9476 | 0.9478 |
| | | 1.0 | 0.9546 | 0.9576 | 0.9488 | 0.9642 | 0.9664 | 0.9490 | 0.9522 |
| | | 1.1 | 0.9532 | 0.9574 | 0.9484 | 0.9652 | 0.9670 | 0.9476 | 0.9500 |
| | | 1.3 | 0.9654 | 0.9670 | 0.9504 | 0.9844 | 0.9854 | 0.9486 | 0.9518 |
| | | 1.5 | 0.9736 | 0.9752 | 0.9502 | 0.9964 | 0.9964 | 0.9478 | 0.9508 |
| | | 1.7 | 0.9758 | 0.9768 | 0.9482 | 0.9994 | 0.9996 | 0.9478 | 0.9516 |
| | | 2.0 | 0.9726 | 0.9746 | 0.9456 | 1.0000 | 1.0000 | 0.9624 | 0.9492 |
| 50 | 100 | 0.3 | 0.9486 | 0.9506 | 0.9462 | 0.9574 | 0.9588 | 0.9476 | 0.9476 |
| | | 0.5 | 0.9474 | 0.9504 | 0.9458 | 0.9548 | 0.9576 | 0.9444 | 0.9462 |
| | | 0.7 | 0.9550 | 0.9584 | 0.9528 | 0.9636 | 0.9656 | 0.9534 | 0.9556 |
| | | 0.9 | 0.9472 | 0.9524 | 0.9452 | 0.9616 | 0.9650 | 0.9440 | 0.9472 |
| | | 1.0 | 0.9528 | 0.9570 | 0.9488 | 0.9702 | 0.9720 | 0.9478 | 0.9504 |
| | | 1.1 | 0.9508 | 0.9550 | 0.9474 | 0.9774 | 0.9798 | 0.9438 | 0.9450 |
| | | 1.3 | 0.9648 | 0.9682 | 0.9512 | 0.9922 | 0.9938 | 0.9534 | 0.9512 |
| | | 1.5 | 0.9666 | 0.9688 | 0.9484 | 0.9978 | 0.9980 | 0.9634 | 0.9504 |
| | | 1.7 | 0.9574 | 0.9606 | 0.9390 | 0.9986 | 0.9986 | 0.9686 | 0.9462 |
| | | 2.0 | 0.9352 | 0.9398 | 0.9364 | 0.9946 | 0.9946 | 0.9718 | 0.9458 |

* $CI_{BS.\delta}$ is the shortest highest posterior density interval of $\hat{\delta}_{BS}$
** $CI_{BS.\delta^*}$ is the shortest highest posterior density interval of $\hat{\delta}^*_{BS}$

the single mean with unknown CV using the estimators of Srivastava [2] and Sahai [3] were constructed. The large sample confidence intervals were $CI_{LS.\theta} = [-4679.6220, 4991.3580]$ and $CI_{LS.\theta^*} = [-4709.2810, 5022.8840]$ with interval lengths of 9670.9800 and 9732.1650, respectively. The Bayesian confidence intervals were $CI_{BS.\theta} = [138.2233, 174.1105]$ and $CI_{BS.\theta^*} = [140.1681, 175.7842]$ with interval lengths of 35.8872 and 35.6161, respectively. The confidence interval for the single mean was $CI_{\mu} = [138.0087, 174.6580]$ with interval length of 36.6493. The interval lengths of the Bayesian confidence interval were shorter than interval lengths of other confidence intervals.

**Example 2.** The data were presented in Fung and Tsang [7] and Tian [8]. The Hong Kong Medical Technology Association provided the data about measurement of the hemoglobin from 1995 and 1996 surveys. The sample sizes were $n = 63$ and $m = 72$ for 1995 and 1996 surveys, respectively. For 1995 survey, the sample mean and sample variance were $\bar{x} = 84.13$ and $s_X^2 = 3.390$. For 1996 survey, the sample mean and sample variance were $\bar{y} = 85.68$ and

**Table 4.** The average lengths of 95% of two-sided confidence intervals for difference between means of normal distributions with unknown CVs.

| $n$ | $m$ | $\sigma_X/\sigma_Y$ | $CI_{LS.\delta}$ | $CI_{MOVER.\delta}$ | $CI_{BS.\delta}$ | $CI_{LS.\delta^*}$ | $CI_{MOVER.\delta^*}$ | $CI_{BS.\delta^*}$ | $CI_{\mu_X-\mu_Y}$ |
|---|---|---|---|---|---|---|---|---|---|
| 10 | 10 | 0.3 | 1.3601 | 1.5699 | 1.4420 | 13.7186 | 15.8337 | 2.5301 | 1.4229 |
| | | 0.5 | 1.4754 | 1.7029 | 1.5652 | 18.9670 | 21.8914 | 2.6446 | 1.4947 |
| | | 0.7 | 1.6092 | 1.8573 | 1.7348 | 17.9090 | 20.6702 | 3.0682 | 1.6230 |
| | | 0.9 | 1.8028 | 2.0807 | 1.9068 | 26.7271 | 30.8479 | 3.9131 | 1.7821 |
| | | 1.0 | 1.8986 | 2.1913 | 1.9794 | 35.4492 | 40.9149 | 4.4303 | 1.8636 |
| | | 1.1 | 2.0322 | 2.3456 | 2.0680 | 43.3879 | 50.0776 | 4.9437 | 1.9599 |
| | | 1.3 | 2.2504 | 2.5974 | 2.2276 | 45.1086 | 52.0636 | 6.3913 | 2.1763 |
| | | 1.5 | 2.4526 | 2.8308 | 2.3741 | 47.9796 | 55.3772 | 7.7441 | 2.3852 |
| | | 1.7 | 2.5800 | 2.9778 | 2.5451 | 74.9159 | 86.4666 | 9.1243 | 2.6258 |
| | | 2.0 | 2.7841 | 3.2133 | 2.8047 | 51.2322 | 59.1314 | 11.2934 | 3.0136 |
| 10 | 20 | 0.3 | 1.0257 | 1.1076 | 1.0575 | 2.1070 | 2.2601 | 1.0159 | 0.9931 |
| | | 0.5 | 1.1388 | 1.2456 | 1.2098 | 2.8945 | 3.1182 | 1.1773 | 1.1148 |
| | | 0.7 | 1.3046 | 1.4453 | 1.4074 | 3.9599 | 4.3598 | 1.5645 | 1.2884 |
| | | 0.9 | 1.5232 | 1.7045 | 1.5987 | 9.3249 | 10.6246 | 2.3371 | 1.4913 |
| | | 1.0 | 1.6448 | 1.8488 | 1.6919 | 18.2168 | 20.8889 | 2.9039 | 1.6049 |
| | | 1.1 | 1.7604 | 1.9849 | 1.7828 | 20.0674 | 22.9999 | 3.5049 | 1.7231 |
| | | 1.3 | 2.0294 | 2.2999 | 1.9679 | 39.6520 | 45.5056 | 4.7720 | 1.9726 |
| | | 1.5 | 2.2239 | 2.5263 | 2.1384 | 55.2958 | 63.7146 | 6.1119 | 2.2212 |
| | | 1.7 | 2.4597 | 2.8000 | 2.3045 | 32.1353 | 36.9879 | 7.4674 | 2.4724 |
| | | 2.0 | 2.5548 | 2.9105 | 2.5719 | 32.4171 | 37.3181 | 9.6419 | 2.8765 |
| 30 | 30 | 0.3 | 0.7992 | 0.8339 | 0.7959 | 1.0571 | 1.1031 | 0.7227 | 0.7682 |
| | | 0.5 | 0.8495 | 0.8864 | 0.8506 | 1.0495 | 1.0952 | 0.7818 | 0.8184 |
| | | 0.7 | 0.9224 | 0.9625 | 0.9287 | 1.1176 | 1.1662 | 0.8568 | 0.8902 |
| | | 0.9 | 1.0263 | 1.0709 | 1.0274 | 1.2765 | 1.3321 | 0.9427 | 0.9796 |
| | | 1.0 | 1.0988 | 1.1466 | 1.0856 | 1.4638 | 1.5274 | 0.9924 | 1.0312 |
| | | 1.1 | 1.1797 | 1.2311 | 1.1438 | 4.8939 | 5.1068 | 1.0558 | 1.0838 |
| | | 1.3 | 1.3698 | 1.4294 | 1.2676 | 11.1750 | 11.6611 | 1.2568 | 1.1989 |
| | | 1.5 | 1.5449 | 1.6121 | 1.3847 | 21.3674 | 22.2970 | 1.5754 | 1.3160 |
| | | 1.7 | 1.6828 | 1.7560 | 1.4994 | 50.8189 | 53.0298 | 2.0302 | 1.4393 |
| | | 2.0 | 1.8140 | 1.8929 | 1.6634 | 15.2829 | 15.9477 | 2.9408 | 1.6364 |
| 20 | 30 | 0.3 | 0.8126 | 0.8500 | 0.8141 | 1.0935 | 1.1430 | 0.7409 | 0.7826 |
| | | 0.5 | 0.8833 | 0.9271 | 0.8958 | 1.1110 | 1.1643 | 0.8263 | 0.8567 |
| | | 0.7 | 0.9905 | 1.0431 | 1.0161 | 1.2142 | 1.2767 | 0.9305 | 0.9634 |
| | | 0.9 | 1.1422 | 1.2070 | 1.1588 | 1.5796 | 1.6698 | 1.0743 | 1.0912 |
| | | 1.0 | 1.2348 | 1.3065 | 1.2311 | 3.1487 | 3.3464 | 1.1913 | 1.1584 |
| | | 1.1 | 1.3365 | 1.4161 | 1.3068 | 7.2226 | 7.6973 | 1.3684 | 1.2310 |
| | | 1.3 | 1.5600 | 1.6561 | 1.4578 | 19.4081 | 20.7178 | 1.8831 | 1.3897 |
| | | 1.5 | 1.7565 | 1.8668 | 1.5940 | 108.2921 | 115.6373 | 2.5903 | 1.5495 |
| | | 1.7 | 1.8851 | 2.0046 | 1.7211 | 70.5853 | 75.3712 | 3.3791 | 1.7104 |
| | | 2.0 | 1.9933 | 2.1203 | 1.9044 | 15.3269 | 16.3618 | 4.7565 | 1.9654 |
| 50 | 50 | 0.3 | 0.6066 | 0.6219 | 0.5984 | 0.6520 | 0.6685 | 0.5695 | 0.5883 |
| | | 0.5 | 0.6474 | 0.6638 | 0.6415 | 0.6939 | 0.7114 | 0.6124 | 0.6291 |
| | | 0.7 | 0.7023 | 0.7200 | 0.6996 | 0.7483 | 0.7673 | 0.6694 | 0.6846 |
| | | 0.9 | 0.7789 | 0.7986 | 0.7729 | 0.8338 | 0.8549 | 0.7359 | 0.7539 |
| | | 1.0 | 0.8305 | 0.8515 | 0.8147 | 0.9027 | 0.9256 | 0.7714 | 0.7933 |
| | | 1.1 | 0.8876 | 0.9101 | 0.8571 | 1.0084 | 1.0339 | 0.8056 | 0.8319 |
| | | 1.3 | 1.0378 | 1.0641 | 0.9517 | 2.3948 | 2.4554 | 0.8798 | 0.9184 |
| | | 1.5 | 1.2052 | 1.2357 | 1.0525 | 20.6228 | 21.1448 | 0.9781 | 1.0106 |
| | | 1.7 | 1.3544 | 1.3887 | 1.1555 | 37.2831 | 38.2268 | 1.1397 | 1.1073 |
| | | 2.0 | 1.4989 | 1.5368 | 1.3058 | 13.6467 | 13.9921 | 1.5152 | 1.2565 |

(continued)

**Table 4.** (*continued*)

| $n$ | $m$ | $\sigma_X/\sigma_Y$ | $CI_{LS.\delta}$ | $CI_{MOVER.\delta}$ | $CI_{BS.\delta}$ | $CI_{LS.\delta^*}$ | $CI_{MOVER.\delta^*}$ | $CI_{BS.\delta^*}$ | $CI_{\mu_X-\mu_Y}$ |
|---|---|---|---|---|---|---|---|---|---|
| 30 | 50 | 0.3 | 0.6198 | 0.6368 | 0.6143 | 0.6646 | 0.6828 | 0.5857 | 0.6023 |
| | | 0.5 | 0.6830 | 0.7037 | 0.6836 | 0.7279 | 0.7497 | 0.6540 | 0.6675 |
| | | 0.7 | 0.7732 | 0.7988 | 0.7818 | 0.8263 | 0.8534 | 0.7426 | 0.7590 |
| | | 0.9 | 0.8938 | 0.9257 | 0.8961 | 0.9929 | 1.0286 | 0.8342 | 0.8639 |
| | | 1.0 | 0.9712 | 1.0070 | 0.9582 | 1.3896 | 1.4434 | 0.8844 | 0.9204 |
| | | 1.1 | 1.0606 | 1.1007 | 1.0231 | 1.9628 | 2.0426 | 0.9432 | 0.9788 |
| | | 1.3 | 1.2692 | 1.3193 | 1.1585 | 9.2391 | 9.6376 | 1.1409 | 1.1051 |
| | | 1.5 | 1.4615 | 1.5207 | 1.2907 | 29.3267 | 30.6005 | 1.4705 | 1.2367 |
| | | 1.7 | 1.6040 | 1.6696 | 1.4066 | 18.8086 | 19.6252 | 1.9267 | 1.3638 |
| | | 2.0 | 1.7399 | 1.8117 | 1.5814 | 13.3950 | 13.9759 | 2.8277 | 1.5736 |
| 100 | 100 | 0.3 | 0.4191 | 0.4243 | 0.4137 | 0.4301 | 0.4354 | 0.4050 | 0.4122 |
| | | 0.5 | 0.4477 | 0.4532 | 0.4432 | 0.4583 | 0.4640 | 0.4348 | 0.4410 |
| | | 0.7 | 0.4879 | 0.4940 | 0.4843 | 0.4990 | 0.5052 | 0.4752 | 0.4813 |
| | | 0.9 | 0.5398 | 0.5465 | 0.5347 | 0.5532 | 0.5601 | 0.5237 | 0.5307 |
| | | 1.0 | 0.5695 | 0.5766 | 0.5607 | 0.5857 | 0.5929 | 0.5480 | 0.5562 |
| | | 1.1 | 0.6084 | 0.6159 | 0.5919 | 0.6313 | 0.6392 | 0.5768 | 0.5861 |
| | | 1.3 | 0.7037 | 0.7125 | 0.6548 | 0.7789 | 0.7885 | 0.6325 | 0.6463 |
| | | 1.5 | 0.8276 | 0.8379 | 0.7237 | 1.3839 | 1.4010 | 0.6897 | 0.7105 |
| | | 1.7 | 0.9605 | 0.9723 | 0.7949 | 6.1894 | 6.2660 | 0.7445 | 0.7768 |
| | | 2.0 | 1.1265 | 1.1404 | 0.9099 | 22.8393 | 23.1219 | 0.8366 | 0.8817 |
| 50 | 100 | 0.3 | 0.4355 | 0.4417 | 0.4314 | 0.4462 | 0.4526 | 0.4228 | 0.4288 |
| | | 0.5 | 0.4881 | 0.4961 | 0.4868 | 0.4992 | 0.5074 | 0.4775 | 0.4828 |
| | | 0.7 | 0.5597 | 0.5701 | 0.5620 | 0.5741 | 0.5847 | 0.5490 | 0.5557 |
| | | 0.9 | 0.6515 | 0.6647 | 0.6503 | 0.6786 | 0.6925 | 0.6280 | 0.6397 |
| | | 1.0 | 0.7109 | 0.7259 | 0.6991 | 0.7579 | 0.7741 | 0.6691 | 0.6859 |
| | | 1.1 | 0.7792 | 0.7962 | 0.7499 | 0.8800 | 0.8996 | 0.7089 | 0.7327 |
| | | 1.3 | 0.9453 | 0.9670 | 0.8550 | 1.8884 | 1.9344 | 0.7897 | 0.8295 |
| | | 1.5 | 1.1277 | 1.1544 | 0.9677 | 12.5507 | 12.8674 | 0.8927 | 0.9316 |
| | | 1.7 | 1.2854 | 1.3162 | 1.0791 | 47.7730 | 48.9817 | 1.0482 | 1.0363 |
| | | 2.0 | 1.4328 | 1.4675 | 1.2353 | 18.6640 | 19.1358 | 1.4490 | 1.1947 |

* $CI_{BS.\delta}$ is the shortest highest posterior density interval of $\hat{\delta}_{BS}$
** $CI_{BS.\delta^*}$ is the shortest highest posterior density interval of $\hat{\delta}^*_{BS}$

$s_Y^2 = 2.946$. The confidence intervals for difference between means of normal distributions with unknown CVs were constructed based on the large sample, MOVER, and Bayesian approaches. The large sample confidence intervals were $CI_{LS.\delta} = [-52.8629, 49.7626]$ and $CI_{LS.\delta^*} = [-52.8795, 49.7798]$ with interval lengths of 102.6255 and 102.6593, respectively. In addition, the MOVER confidence intervals were $CI_{MOVER.\delta} = [-53.8431, 50.7428]$ and $CI_{MOVER.\delta^*} = [-53.8600, 50.7603]$ with interval lengths of 104.5859 and 104.6203, respectively. Furthermore, the Bayesian confidence intervals were $CI_{BS.\delta} = [-2.1085, -0.9213]$ and $CI_{BS.\delta^*} = [-2.1076, -0.9208]$ with interval lengths of 1.1872 and 1.1868, respectively. The confidence interval for difference between means of normal distributions was $CI_{\mu_X-\mu_Y} = [-2.1590, -0.9410]$ with interval length of 1.2180. Therefore, the Bayesian approach provided shortest interval length.

## 6 Discussion and Conclusions

For one population, Thangjai et al. [1] provided large sample approach for the confidence intervals for the mean of normal distribution with unknown CV. Then compared with the Student's $t$-distribution and the $z$-distribution for the confidence intervals for the mean of normal distribution. In this paper, the confidence intervals for the mean of normal distribution with unknown CV based on the Bayesian approach were proposed. The results indicated that the Bayesian confidence intervals which were new approach were better than the existing confidence intervals.

For two populations, Thangjai et al. [1] proposed the confidence intervals for difference between means of normal distributions with unknown CVs based on large sample and method of variance estimates recovery (MOVER) approach. In this paper, the Bayesian approach was presented to construct the confidence intervals for difference between means of normal distributions with unknown CVs. The Bayesian approach was the best in term of interval length.

## References

1. Thangjai, W., Niwitpong, S., Niwitpong, S.: Confidence intervals for mean and difference of means of normal distributions with unknown coefficients of variation. Mathematics **5**, 1–23 (2017). https://doi.org/10.3390/math5030039
2. Srivastava, V.K.: A note on the estimation of mean in normal population. Metrika **27**(1), 99–102 (1980). https://doi.org/10.1007/BF01893580
3. Sahai, A.: On an estimator of normal population mean and UMVU estimation of its relative efficiency. Appl. Math. Comput. **152**, 701–708 (2004). https://doi.org/10.1016/S0096-3003(03)00588-5
4. Thangjai, W., Niwitpong, S., Niwitpong, S.: Confidence intervals for the inverse mean and difference of inverse means of normal distributions with unknown coefficients of variation. Stud. Comput. Intell. **808**, 245–263 (2019). https://doi.org/10.1007/978-3-030-04263-9_19
5. Tongmol, N., Srisodaphol, W., Boonyued, A.: A Bayesian approach to the one way ANOVA under unequal variance. Sains Malaysiana **45**, 1565–1572 (2016)
6. Niwitpong, S.: Confidence intervals for the normal mean with a known coefficient of variation. Far East J. Math. Sci. **97**, 711–727 (2015). https://doi.org/10.17654/FJMSJul2015_711_727
7. Fung, W.K., Tsang, T.S.: A simulation study comparing tests for the equality of coefficients of variation. Stat. Med. **17**, 2003–2014 (1998). https://doi.org/10.1002/(SICI)1097-0258(19980915)17:17⟨2003::AID-SIM889⟩3.0.CO;2-I
8. Tian, L.: Inferences on the common coefficient of variation. Stat. Med. **24**, 2213–2220 (2005). https://doi.org/10.1002/sim.2088

# Confidence Intervals for the Difference Between the Coefficients of Variation of Inverse Gaussian Distributions

Wasana Chankham(✉), Sa-Aat Niwitpong, and Suparat Niwitpong

Department of Applied Statistics, Faculty of Applied Science,
King Mongkut's University of Technology North Bangkok, Bangkok, Thailand
wasana.ch.kh@gmail.com, {sa-aat.n,suparat.n}@sci.kmutnb.ac.th

**Abstract.** The aim of this study is to propose confidence intervals for the difference between the coefficients of variation of inverse Gaussian distributions based on the generalized confidence interval (GCI), the adjusted generalized confidence interval (AGCI), the bootstrap percentile confidence interval (BPCI), and the method of variance estimates recovery (MOVER). The performances of the proposed confidence intervals were evaluated using coverage probabilities and average lengths via Monte Carlo simulation. The results showed that the GCI and AGCI methods were higher than or close to the nominal level in all cases. For small sample sizes, MOVER was better than the other methods because it provided the narrowest average length. The performances of all the approaches were illustrated using two real data examples.

**Keywords:** Bootstrap · Coefficients of variation · Generalized confidence interval · Inverse Gaussian distribution · Method of variance estimates recovery

## 1 Introduction

The inverse Gaussian distribution is used to describe and analyze positive and right-skewed data and has been applied in useful applications in a variety of fields, such as cardiology, pharmacokinetics, economics, medicine, and finance. The inverse Gaussian distribution was first derived by Schordinger [1] for the first passage of time of a Wiener process to an absorbing barrier and has been used to describe the cycle time distribution of particles in the blood [2]. Liu et al. [3] used data sets from an inverse Gaussian distribution applied to a lifetime model for reliability analysis. Banerjee and Bhattacharyya [4] applied this distribution in a study of market incidence models, while Lancaster [5] used it as a model for the duration of strikes, and Sheppard [6] proposed applying it for the time duration of injected labeled substances called tracers in a biological system. For more informations and applications, Chhikara and Folks [7], Krishnamoorthy and Lu [8], Tian and Wu [9], Lin et al. [10], and Ye et al. [11].


V.-N. Huynh et al. (Eds.): IUKM 2020, LNAI 12482, pp. 372–383, 2020.
https://doi.org/10.1007/978-3-030-62509-2_31

The coefficient of variation, which is the ratio of the standard deviation to the mean, can be used to measure data dispersion with non-homogeneous units. It has been used in many fields, such as biology, economics, medicine, agriculture, and finance. Many researchers have proposed confidence intervals for the coefficient of variation. Wongkhao et al. [12] presented confidence intervals for the ratio of two independent coefficients of variation of normal distributions based on the concept of the general confidence interval (GCI) and the method of variance estimates recovery (MOVER). Banik and Kibria [13] estimated the population coefficient of variation and compared it with bootstrap interval estimators. Mahmoudvand and Hassani [14] proposed an unbiased estimator to construct confidence intervals for the population coefficient of variation of normal distribution. Hasan and Krishnamoorthy [15] proposed two new approximate confidence intervals for the ratio of the coefficients of variation of lognormal populations using MOVER and the fiducial confidence intervals method. Sangnawakij et al. [16] proposed two confidence intervals for the ratio of coefficients of variation of gamma distributions based on MOVER with the methods of Score and Wald. Thangjai and Niwitpong [17] constructed confidence intervals for the weighted coefficient of constructed of two-parameter exponential distributions based on the adjusted method of variance estimates recovery method (adjusted MOVER) and compared it with the general confidence interval (GCI) and the large sample method. Yosboonruang et al. [18] proposed confidence intervals for the coefficient of variation for a delta-lognormal distribution based on GCI and the modified Fletcher method.

As mentioned previously, there have been many comprehensive studies on confidence intervals for the difference between the coefficients of variation of inverse Gaussian distributions. However, few researchers have investigated confidence intervals for the parameters of inverse Gaussian distributions. For instance, Ye et al. [11] proposed confidence intervals for the common mean of several inverse Gaussian populations when the scalar parameters are unknown and unequal. Tian and Wilding [19] presented confidence intervals for the ratio of the means of two independent inverse Gaussian distribution. Krishnamoorthy and Tian [20] developed confidence intervals for the difference between and ratio of the means of two inverse Gaussian distributions based on GCI. The purpose of the current study is to establish new confidence intervals for the difference between the coefficients of variation of inverse Gaussian distributions based on GCI, the adjusted generalized confidence interval (AGCI), the bootstrap percentile confidence interval (BPCI), and MOVER.

The organization of this paper are as follows. Section 2 provides preliminaries for the difference between the coefficients of variation of inverse Gaussian distributions. Simulation studies are presented in Sect. 3. Section 4 presents empirical studies. Finally, concluding remarks are summarized in Sect. 5.

## 2 Confidence Intervals for the Difference Between the Coefficients of Variation of Inverse Gaussian Distributions

Let X = ($X_1$, $X_2$,..., $X_n$) be an independent random sample of size $n$ from the two-parameter inverse Gaussian distribution, IG($\mu$, $\lambda$), is defined as

$$f(x,\mu,\lambda) = \left(\frac{\lambda}{2\pi x^3}\right)^{\frac{1}{2}} exp\left\{-\frac{\lambda\,(x-\mu)^2}{2\mu^2 x}\right\}, x>0, \mu>0, \lambda>0, \quad (1)$$

where $\mu$ and $\lambda$ are the mean parameter and the scale parameter. The population mean and variance of X are define as

$$E(X) = \mu \quad (2)$$

and

$$Var(X) = \mu^3/\lambda. \quad (3)$$

Then, the coefficient of variation of X is expressed by

$$\omega = CV(X) = \sqrt{\frac{\mu}{\lambda}} \quad (4)$$

and the difference between of coefficients of variation is defined as

$$\eta = \omega_X - \omega_Y = \sqrt{\frac{\mu_X}{\lambda_X}} - \sqrt{\frac{\mu_Y}{\lambda_Y}}. \quad (5)$$

### 2.1 The Generalized Confidence Interval (GCI) Method

Weerahandi [21] introduced the generalized confidence interval (GCI) method. The concept of this method is to the definition of generalized pivotal quantity (GPQ). Suppose that X is a random sample from a distribution having the parameter $(\theta, \delta)$, where $\theta$ is the parameter of interest and $\delta$ is the nuisance parameter.

**Definition 1.** *Let* $X = (X_1, X_2, ..., X_n)$ *be the observed value of* $X$ *and the probability density function of* $(X; x, \theta, \delta)$ *is* $R(X; x, \theta, \delta)$. *The generalized pivotal quantity* $R(X; x, \theta, \delta)$ *satisfies the following two properties:*

*(a) The probability distribution of the function* $R(X; x, \theta, \delta)$ *is independent of unknown parameters.*
*(b) The observed value of* $R(X; x, \theta, \delta)$, $X = x$, *does not depend on nuisance parameters.*

Therefore, the $100\,(1-\alpha)\,\%$ two-sided GCI for the parameter of interest is defined as $[R(\alpha/2), R(1-\alpha/2)]$, where $R(\alpha/2)$ and $R(1-\alpha/2)$ are the $100(\alpha/2)-th$ and $100(1-\alpha/2)-th$ percentile of $R(X; x, \theta, \delta)$.

Ye et al. [11] developed the generalized confidence interval based on general pivotal quantities for the mean and scale parameter of the inverse Gaussian distribution. Suppose that $k$ is independent populations of the inverse Gaussian distributions with mean parameters $\mu_i$ and scale parameters $\lambda_i$, $i = 1, 2, .., k$. Let $X_{i1}, X_{i2}, ..., X_{in_i}$ be the random sample from $IG(\mu_i, \lambda_i)$, $i = 1, ..., k$. Form the $i$th population, the maximum likelihood estimators (MLEs) of $\mu_i$ and $\lambda_i$ can be found as

$$\hat{\mu} = \bar{X}_i, \qquad \hat{\lambda}_i^{-1} = \frac{1}{n_i}\sum_{j=1}^{n_i}(X_{ij}^{-1} - \bar{X}_i^{-1}), \tag{6}$$

where $\bar{X}_i = \sum_{j=1}^{n_i} X_{ij}/n_i$. Let $V_i = \hat{\lambda}_i^{-1}$, it is well known that $\bar{X}_i$ and $V_i$ are mutually independent random variables. Note that

$$\bar{X}_i \sim IG(\mu, n_i\lambda_i), \qquad n_i\lambda_i V_i \sim \chi^2_{n_i-1}, i = 1, .., k, \tag{7}$$

where $\chi^2_m$ denotes as the Chi-square distribution with $m$ degrees of freedom. It is easily proved that $(\bar{X}_i, V_i)$ forms a set complete sufficient statistics for $(\mu_i, \lambda_i)$. From Eq. (7), the generalized pivotal quantity $R_{\lambda_i}$ for $\lambda_i$ based on the $i$th sample is defined as

$$R_{\lambda_i} = \frac{n_i\lambda_i V_i}{n_i v_i} \sim \frac{\chi^2_{n_i-1}}{n_i v_i}, i = 1, ..., k, \tag{8}$$

where $v_i$ denotes as the observed value of $V_i$. The distribution of $R_{\lambda_i}$ is free of any unknown parameters and the observed values only relate to the parameter $\lambda_i$. Therefore, $R_{\lambda_i}$ is the generalized pivotal quantity for$\lambda_i$.

According to Ye et al. [11], the generalized pivotal quantity $R_{\mu_i}$ for $\mu_i$ based on the $i$th sample is defined as

$$R_{\mu_i} = \frac{\bar{x}_i}{\left|1 + \frac{\sqrt{n_i\lambda_i}(\bar{x}_i - \mu)}{\mu\sqrt{\bar{x}_i}}\sqrt{\frac{\bar{x}_i}{n_i R_{\lambda_i}}}\right|} \sim \frac{\bar{x}_i}{\left|1 + Z_i\sqrt{\frac{\bar{x}_i}{n_i R_{\lambda_i}}}\right|}, \tag{9}$$

where $\sim$ denotes as "approximately distributed" and $Z_i \sim N(0, 1)$. $\bar{x}_i$ and $v_i$ are the observed values of $\bar{X}_i$ and $V_i$. The approximations in Eq. (9) are derived by using Theorem 2.1 given by Chhikara and Folks [7]. Using the moment matching method, it can be shown that$\sqrt{n_i\lambda_i}(\bar{X}_i - \mu)/\mu\sqrt{\bar{x}_i}$ has a limiting distribution of $Z_i \sim N(0, 1)$. Note that the observed value of $R_{\mu_i}$ is $\mu_i$. Therefore, $R_{\mu_i}$ satisfies conditions (a) and (b) in Definition 1. However, $R_{\mu_i}$ is an approximate generalized pivotal quantity for $\mu_i$ based on the $i$th sample.

Therefore, the generalized pivotal quantities for difference between the coefficients of variation are given by

$$R_\eta = R_{\omega_X} - R_{\omega_Y} = \sqrt{\frac{R_{\mu_X}}{R_{\lambda_X}}} - \sqrt{\frac{R_{\mu_Y}}{R_{\lambda_Y}}}. \tag{10}$$

Then, the $100(1-\alpha)\%$ two-sided confident interval for the difference between the coefficients of variation based on generalized confidence interval method is given by

$$CI_{(GCI)} = (L_{(GCI)}, U_{(GCI)}) = (R_\eta(\alpha/2), R_\eta(1-\alpha/2)), \tag{11}$$

where $R_\eta(\alpha/2)$ and $R_\eta(1-\alpha/2)$ are the $100(\alpha/2)\%$ and $100(1-\alpha/2)\%$ percentiles of the distribution of $R = R(X; x, \theta, \delta)$, respectively.

The following algorithm is used to construct the generalized confidence interval:

Algorithm 1.

Step 1. Generate $x_j$ and $y_j$ , $j = 1, 2, ..., n_i$ from the inverse Gaussian distribution.

Step 2. Compute $\bar{x}_j$, $\bar{y_j}$, $\hat{v}_{x_j}$ and $\hat{v}_{y_j}$.

Step 3. For $t = 1$ to $T$.

Step 4. Generate $\chi^2_{n_i-1}$ from chi-square distribution and $Z \sim N(0, 1)$.

Step 5. Compute $R_{\lambda_i}$ from Eq. (8).

Step 6. Compute $R_{\mu_i}$ from Eq. (9).

Step 7. Compute $R_\eta$ from Eq. (10).

Step 8. End $t$ loop.

Step 9. Compute $R_\eta(\alpha/2)$ and $R_\eta(1-\alpha/2)$.

## 2.2 The Adjusted Generalized Confidence Interval (AGCI) Method

According to Ye et al. [11], We can use a similar method in GCI method for calculating the difference between the coefficients of variation $\eta$. $R_{\tilde{\lambda}_i}$ can be computed by the generalized pivotal quantity for $\tilde{\lambda}$ based on $i$th sample shown in Eq. (8). Then, Krishnomoorthy and Tian [20] presented an approximate generalized pivotal quantity for $R_{\tilde{\mu_i}}$ based on the $i$th sample, as follows:

$$R_{\tilde{\mu_i}} = \frac{\bar{x}_i}{\max\left\{0, t_{n_i-1}\sqrt{\frac{\bar{x}_i v_i}{n_i-1}}\right\}}, \tag{12}$$

where $t_{n_i-1}$ denotes the t distribution with $n_i - 1$ degrees of freedom. Therefore, the denominator in Eq. (12) may be zero when $t_{n_i-1}$ obtains the negative value. $R_{\tilde{\mu_i}}$ is an approximate generalized pivotal quantity.

Therefore, the generalized pivotal quantities for the difference between the coefficients of variation are given by

$$R_{\tilde{\eta}} = R_{\tilde{\omega}_X} - R_{\tilde{\omega}_Y} = \sqrt{\frac{R_{\tilde{\mu}_X}}{R_{\tilde{\lambda}_X}}} - \sqrt{\frac{R_{\tilde{\mu}_Y}}{R_{\tilde{\lambda}_Y}}}. \tag{13}$$

Then, the $100(1-\alpha)\%$ two-sided confident interval for the difference between the coefficients of variation based on generalized confidence interval method is given by

$$CI_{(AGCI)} = (L_{(AGCI)}, U_{(AGCI)}) = (R_{\tilde{\eta}}(\alpha/2), R_{\tilde{\eta}}(1-\alpha/2)), \tag{14}$$

where $R_{\tilde{\eta}}(\alpha/2)$ and $R_{\tilde{\eta}}(1-\alpha/2)$ which are the $100(\alpha/2)\%$ and $100(1-\alpha/2)\%$ percentiles of the distribution of $R = R(X; x, \theta, \delta)$ can be obtained from the concept of Algorithm 1.

### 2.3 The Bootstrap Percentile Confidence Interval (BPCI) Method

Efron and Tibshirani [22] introduced the bootstrap percentile method. The bootstrap is a re-sampling method for assigning measures of accuracy to statistical estimate a random selection of resamples from the original sample with replacement. Let $x$ be a random sample of size $n$ from the inverse Gaussian distribution. Suppose that $x = x_1, x_2, ..., x_n$ is a random sample of size $n$ from the inverse Gaussian distribution. Sampling is replaced by $x^* = x_1^*, x_2^*, ..., x_n^*$, which can be obtained by the bootstrap sample with $B$ times. When the re-sampling bootstrap sample is operated, the difference between coefficients of variation is then calculated.

The $100(1-\alpha)\%$ two-sided confidence interval for the difference between the coefficients of variation based on the bootstrap percentile confidence interval is defined by

$$CI_{(BPCI)} = (L_{(BPCI)}, U_{(BPCI)}) = (\eta^*(\alpha/2), \eta^*(1-\alpha/2)), \tag{15}$$

where $\eta^*(\alpha/2)$ and $\eta^*(1-\alpha/2)$ are the $100(\alpha/2)\%$ and $100(1-\alpha/2)\%$ percentiles of the distribution.

Algorithm 2.
Step 1. Generate$X_1, X_2, ...X_n$ from the inverse Gaussian distribution
Step 2. Obtain a bootstrap sample $X^* = X_1^*, X_2^*, ..., X_n^*$ from Step 1.
Step 3. Compute $\eta^*$
Step 4. Repeat Steps 2 and 3, B times.
Step 5. Compute $\eta^*(\alpha/2)$ and $\eta^*(1-\alpha/2)$

### 2.4 The Method of Variance Estimates Recovery (MOVER)

Gulhar et al. [23] proposed the confidence interval for a coefficient of variation of X which is

$$(l_x, u_x) = \left( \frac{\sqrt{n-1}(\hat{\omega_x})}{\sqrt{\chi^2_{1-\alpha/2,n-1}}}, \frac{\sqrt{n-1}(\hat{\omega_x})}{\sqrt{\chi^2_{\alpha/2,n-1}}} \right) \tag{16}$$

and the confidence interval for a coefficient of variation of Y is defined as

$$(l_y, u_y) = \left( \frac{\sqrt{n-1}(\hat{\omega_y})}{\sqrt{\chi^2_{1-\alpha/2,n-1}}}, \frac{\sqrt{n-1}(\hat{\omega_y})}{\sqrt{\chi^2_{\alpha/2,n-1}}} \right), \tag{17}$$

where $\chi^2_{1-\alpha/2,n-1}$ and $\chi^2_{\alpha/2,n-1}$ are respectively the $100(\alpha)\%$ -th and $100(1-\alpha)\%$ -th percentile of the chi-square distribution with $n-1$ degrees of freedom.

Donner and Zou [24] introduced the confidence interval estimation for the difference of parameters of interest using by MOVER. The lower limit and upper limit are given by

$$L_\eta = \hat{\omega}_x - \hat{\omega}_y - \sqrt{(\hat{\omega}_x - l_x)^2 + (u_y - \hat{\omega}_y)^2} \tag{18}$$

and

$$U_\eta = \hat{\omega}_x - \hat{\omega}_y - \sqrt{(u_x - \hat{\omega}_x)^2 + (\hat{\omega}_y - l_y)^2}, \tag{19}$$

where $\hat{\omega}_x$ and $\hat{\omega}_y$ are denoted in Eq. (5), $l_x$ and $u_x$ are denoted in Eq. (16), and $l_y$ and $u_y$ are denoted in Eq. (17).

Then, the $100(1-\alpha)\%$ two-sides confidence interval for the difference between the coefficients of variation of inverse Gaussian distribution based on the MOVER is given by

$$CI_{MOVER} = (L_{MOVER}, U_{MOVER}) = (L_\eta, U_\eta), \tag{20}$$

where $L_\eta$ and $U_\eta$ are defined in Eqs. (18) and (19), respectively.

# 3 Simulation Studies

A Monte Carlo simulation studies out to evaluate the coverage probabilities and average lengths of the confidence intervals for the difference between the coefficients of variation of inverse Gaussian distributions based on GCI, AGCI, BPCI, and MOVER. The simulations were run by using R statistics programming language. In the simulation, The sample sizes were $(n_x, n_y)$ = (5, 5), (5, 10), (10, 10), (10, 30), (30, 30), (30, 50), (50, 50), (50, 100), and (100, 100); $\mu_x = 0.5$, $\mu_y = 0.5, 1$, $\lambda_x = 10$ and $\lambda_y = 1, 2, 5, 10$. The nominal confidence level is at 0.95. The number of simulation replications for each situation was 10,000 replications, 1,000 bootstrap samples and 5,000 pivotal quantities for GCI and AGCI. The confidence interval which has the coverage probability was greater than or close to the nominal confidence level and the shortest expected lengths are chosen.

The estimated coverage probability and estimated average length for this simulation study are respectively given as:

$$CP = \frac{c(L_{(g)} \leq \eta \leq U_{(g)})}{M}, \qquad AL = \frac{\sum_{i=1}^{m}(U_{(g)} - L_{(g)})}{M}, \tag{21}$$

where $M$ is the number of simulation replications, and $c(L_{(g)} \leq \eta \leq U_{(g)})$ is the numbers of simulation replications for $\eta$ which lies within the confidence interval.

The following algorithm is used to construct the coverage probability for the difference between the coefficients of variation:

Algorithm 3.

Step 1. For a given $M, m, n_1, n_2, \mu_x, \mu_y, \lambda_x$, and $\lambda_y$.

Step 2. For $g = 1$ to $M$.

Step 3. Generate $X_1, X_2, ..., X_n$ and $Y_1, Y_2, ..., Y_n$ from inverse Gaussian distribution.

Step 4. Use Algorithm 1, Eq. (14), Algorithm 2, and Eq. (20) to construct lower and upper limits for GCI, AGCI, BPCI, and MOVER, respectively.

Step 5. If $(L_{(g)} \leq \eta \leq U_{(g)})$, set $P_{(g)} = 1$; else set $P_{(g)} = 0$.

Step 6. Calculate $U_{(g)} - L_{(g)}$.

Step 7. End $g$ loop.

Step 8: Compute the coverage probability and the average length.

The results in Table 1 showed that the difference between the coefficients of variation, the coverage probabilities of GCI method and AGCI method were greater than or equal to the nominal level in all cases. However, the average lengths of the AGCI method were shorter than the GCI method. For small sample sizes, the coverage probabilities of the MOVER were close to the nominal confidence level of 0.95 and the shortest average length.

## 4 An Empirical Study

Herein, we illustrate the methods used to computation of confidence intervals proposed.

**Example 1:** The data were given by Mudholkar and Hutson [25] for the consecutive annual flood discharge rates of the Floyd river at James, Iowa. The data are as follows:

In 1935–1944 : 1460, 4050, 3570, 2060, 1300, 1390, 1720, 6280, 1360, 7440.

In 1945–1954 : 5320, 1400, 3240, 2710, 4520, 4840, 8320, 13900, 71500, 6250.

The summary statistics of data are $n_1 = 10$, $n_2 = 10$, $\hat{\mu}_1 = 3063$, $\hat{\mu}_2 = 12200$, $\hat{\lambda}_1 = 6529.078$, $\hat{\lambda}_2 = 6434.176$, $\hat{\omega}_1 = 0.6849$, $\hat{\omega}_2 = 1.3770$, and the difference between the coefficients of variation $\hat{\eta} = -0.6921$. Based on the 95% two sided confidence interval for the difference between the coefficients of variation using GCI method was $(-0.0051, 0.0148)$ with interval length of 0.0199; AGCI method was $(-0.0045, 0.0144)$ with interval length of 0.0189; BPCI method was $(-1.4848, 0.3342)$ with interval length of 1.8189, and MOVER was $(-1.8489, 0.0182)$ with interval length of 1.8667. Therefore, the results confirm that the simulation results for difference between the coefficients of variation are not different from the results of the previous study.

**Example 2:** The real data were provided by Eilam et al. [26]. The plasma bradykininogen levels were measured in healthy subjects, in patients with active Hodgkin's disease and in patients with inactive Hodgkin's disease. The outcome variable is measured in micrograms of bradykininogen per milliliter of plasma. The data are as follows:

Active Hodgkin's disease: 3.96, 3.04, 5.28, 3.40, 4.10, 3.61, 6.16, 3.22, 7.48, 3.87, 4.27, 4.05, 2.40, 5.81, 4.29, 2.77, 4.40.

Inactive Hodgkin's disease: 5.37, 10.60, 5.02, 14.30, 9.90, 4.27, 5.75, 5.03, 5.74, 7.85, 6.82, 7.90, 8.36, 5.72, 6.00, 4.75, 5.83, 7.30, 7.52, 5.32, 6.05, 5.68, 7.57, 5.68, 8.91, 5.39, 4.40, 7.13.

The summary statistics of data are $n_1 = 17$, $n_2 = 28$, $\hat{\mu}_1 = 4.2418$, $\hat{\mu}_2 = 6.7914$, $\hat{\lambda}_1 = 50.8394$, $\hat{\lambda}_2 = 86.0289$, $\hat{\omega}_1 = 0.2889$, $\hat{\omega}_2 = 0.2810$, and the difference between the coefficients of variation $\hat{\eta} = 0.0079$. Based on the 95% two sided confidence interval for the difference between the coefficients of

**Table 1.** The coverage probability (CP) and average length (AL) of 95% two-sided confidence intervals for the difference between the coefficients of variation of inverse Gaussian distribution: $(\mu_X, \lambda_x) = (0.5, 10)$.

| $n_x$ | $n_y$ | $(\mu_y, \lambda_y)$ | Coverage Probability (Average Length) | | | |
|---|---|---|---|---|---|---|
| | | | $CI_{GCI}$ | $CI_{AGCI}$ | $CI_{BPCI}$ | $CI_{MOVER}$ |
| 5 | 5 | (0.5, 1) | 0.9710 (2.777) | **0.9518 (1.6723)** | 0.6611 (0.6904) | 0.9423 (1.5387) |
| | | (0.5, 2) | 0.9669 (1.7457) | 0.9552 (1.2727) | 0.7599 (0.5161) | **0.9555 (1.1874)** |
| | | (0.5, 5) | 0.9570 (1.0560) | 0.9514 (0.9264) | 0.9055 (0.3696) | **0.9590 (0.8749)** |
| | | (0.5, 10) | 0.9547 (0.8214) | 0.9518 (0.7605) | 0.9654 (0.3021) | **0.9593 (0.7204)** |
| | | (1, 1) | 0.9837 (4.4063) | **0.9551 (2.2717)** | 0.5931 (0.9684) | 0.9310 (2.0541) |
| | | (1, 2) | 0.9744 (2.7700) | **0.9568 (1.6717)** | 0.6578 (0.6907) | 0.9498 (1.5356) |
| | | (1, 5) | 0.9641 (1.5205) | 0.9556 (1.1721) | 0.7907 (0.4728) | **0.9577 (1.1003)** |
| | | (1, 10) | 0.9583 (1.0541) | 0.9537 (0.9252) | 0.9059 (0.3692) | **0.9594 (0.8749)** |
| 5 | 10 | (0.5, 1) | 0.9769 (1.3175) | **0.9585 (0.9814)** | 0.8482 (0.5851) | 0.9460 (0.9528) |
| | | (0.5, 2) | 0.9667 (0.9175) | 0.9553 (0.7973) | 0.9129 (0.4353) | **0.9543 (0.7717)** |
| | | (0.5, 5) | 0.9550 (0.6986) | 0.9504 (0.6475) | 0.9363 (0.3190) | **0.9550 (0.6205)** |
| | | (0.5, 10) | 0.9515 (0.6173) | 0.9484 (0.5787) | 0.8641 (0.2690) | **0.9537 (0.5475)** |
| | | (1, 1) | 0.9874 (2.3690) | **0.9566 (1.2685)** | 0.7904 (0.8328) | 0.9285 (1.2216) |
| | | (1, 2) | 0.9777 (1.3214) | **0.9564 (0.9850)** | 0.8527 (0.5881) | 0.9444 (0.9568) |
| | | (1, 5) | 0.9620 (0.8517) | 0.9511 (0.7577) | 0.9288 (0.4014) | **0.9528 (0.7325)** |
| | | (1, 10) | 0.9513 (0.6978) | 0.9462 (0.6465) | 0.9270 (0.3190) | **0.9515 (0.6190)** |
| 10 | 10 | (0.5, 1) | 0.9758 (1.1737) | **0.9532 (0.8416)** | 0.7937 (0.5659) | 0.9376 (0.8096) |
| | | (0.5, 2) | 0.9682 (0.7317) | 0.9533 (0.6285) | 0.8427 (0.4098) | **0.9506 (0.6115)** |
| | | (0.5, 5) | 0.9601 (0.4817) | 0.9548 (0.4536) | 0.9059 (0.2862) | **0.9554 (0.4450)** |
| | | (0.5, 10) | 0.9568 (0.3863) | 0.9523 (0.3723) | 0.9307 (0.2340) | **0.9559 (0.3662)** |
| | | (1, 1) | 0.9867 (2.2265) | **0.9502 (1.1421)** | 0.7431) (0.8128) | 0.9103 (1.0868) |
| | | (1, 2) | 0.9773 (1.1674) | **0.9505 (0.8379)** | 0.7827 (0.5666) | 0.9341 (0.5081) |
| | | (1, 5) | 0.9670 (0.6529) | 0.9559 (0.5792) | 0.8695 (0.3726) | **0.9534 (0.5649)** |
| | | (1, 10) | 0.9583 (0.4819) | 0.9510 (0.4536) | 0.9054 (0.2868) | **0.9538 (0.4451)** |
| 10 | 30 | (0.5, 1) | 0.9783 (0.5631) | **0.9535 (0.4708)** | 0.9251 (0.3991) | 0.9429 (0.4678) |
| | | (0.5, 2) | 0.9654 (0.4173) | **0.9501 (0.3824)** | 0.9300 (0.2985) | 0.9466 (0.3739) |
| | | (0.5, 5) | 0.9548 (0.3269) | 0.9490 (0.3133) | 0.8996 (0.2240) | **0.9501 (0.3075)** |
| | | (0.5, 10) | 0.9551 (0.2932) | **0.9505 (0.2835)** | 0.8548 (0.1944) | 0.9497 (0.2758) |
| | | (1, 1) | 0.9908 (0.8725) | **0.9550 0.6089)** | 0.8971 (0.5749) | 0.9145 (0.6042) |
| | | (1, 2) | 0.9765 (0.5634) | **0.9518 (0.4709)** | 0.9178 (0.3996) | 0.9352 (0.4686) |
| | | (1, 5) | **0.9599 (0.3864)** | 0.9489 (0.3602) | 0.9293 (0.2747) | 0.9482 (0.3565) |
| | | (1, 10) | 0.9575 (0.3276) | 0.9489 (0.3140) | 0.9075 (0.2243) | **0.9527 (0.3082)** |
| 30 | 30 | (0.5, 1) | 0.9813 (0.5018) | **0.9532 (0.4053)** | 0.8842 (0.3759) | 0.9295 (0.4002) |
| | | (0.5, 2) | 0.9691 (0.3357) | **0.9502 (0.3011)** | 0.9063 (0.2673) | 0.9415 (0.2986) |
| | | (0.5, 5) | 0.9606 (0.2237) | 0.9531 (0.2140) | 0.9247 (0.1848) | **0.9505( 0.2132)** |
| | | (0.5, 10) | 0.9569 (0.1802) | 0.9520 (0.1754) | 0.9344 (0.1499) | **0.9500 (0.1749)** |
| | | (1, 1) | 0.9898 (0.8297) | **0.9514 (0.5579)** | 0.8655 (0.1500) | 0.9493 (0.1749) |
| | | (1, 2) | 0.9801 (0.5023) | **0.9501 (0.4056)** | 0.8872 (0.3750) | 0.9296 (0.4004) |
| | | (1, 5) | 0.9670 (0.2997) | **0.9525 (0.2749)** | 0.9124 (0.2416) | 0.9450 (0.2729) |
| | | (1, 10) | 0.9588 (0.2242) | 0.9529 (0.2144) | 0.9262 (0.1853) | **0.9508 (0.2136)** |

*(continued)*

**Table 1.** (*continued*)

| $n_x$ | $n_y$ | $(\mu_y, \lambda_y)$ | Coverage Probability (Average Length) | | | |
|---|---|---|---|---|---|---|
| | | | $CI_{GCI}$ | $CI_{AGCI}$ | $CI_{BPCI}$ | $CI_{MOVER}$ |
| 30 | 50 | (0.5, 1) | 0.9785 (0.3824) | **0.9534 (0.3156)** | 0.9213 (0.3105) | 0.9265 (0.3141) |
| | | (0.5, 2) | 0.9699 (0.2652) | **0.9537 (0.2409)** | 0.9355 (0.2239) | 0.9469 (0.2402) |
| | | (0.5, 5) | 0.9618 (0.1880) | 0.9519 (0.1806) | 0.9386 (0.1610) | **0.9536 (0.1802)** |
| | | (0.5, 10) | **0.9505 (0.1585)** | 0.9458 (0.1544) | 0.9238 (0.1354) | 0.9438 (0.1537) |
| | | (1, 1) | 0.9911 (0.6119) | **0.9506 (0.4278)** | 0.9050 (0.4548) | 0.9003 (0.4242) |
| | | (1, 2) | 0.9798 (0.3832) | **0.9513 (0.3161)** | 0.9188 (0.3102) | 0.9278 (0.3147) |
| | | (1, 5) | 0.9651 (0.2399) | **0.9510 (0.2224)** | 0.9329 (0.2044) | 0.9446 (0.2221) |
| | | (1, 10) | **0.9553 (0.1871)** | 0.9476 (0.1798) | 0.9328 (0.1607) | 0.9451 (0.1794) |
| 50 | 50 | (0.5, 1) | 0.9803 (0.3729) | **0.9525 (0.3729)** | 0.9088 (0.3019) | 0.9235 (0.3021) |
| | | (0.5, 2) | 0.9700 (0.2499) | **0.9511 (0.2253)** | 0.9193 (0.2136) | 0.9390 (0.2242) |
| | | (0.5, 5) | 0.9601 (0.1671) | **0.9510 (0.1601)** | 0.9358 (0.1457) | 0.9478 (0.1598) |
| | | (0.5, 10) | 0.9570 (0.1344) | **0.9517 (0.1310)** | 0.9401 (0.1199) | 0.9478 (0.1308) |
| | | (1, 1) | 0.9933 (0.6070) | **0.9522 (0.4201)** | 0.9001 (0.4527) | 0.8994 (0.4159) |
| | | (1, 2) | 0.9816 (0.3720) | **0.9508 (0.3040)** | 0.9110 (0.3027) | 0.9235 (0.3017) |
| | | (1, 5) | 0.9681 (0.2241) | **0.9538 (0.2063)** | 0.9255 (0.1938) | 0.9421 (0.2054) |
| | | (1, 10) | 0.9608 (0.1671) | **0.9527 (0.1602)** | 0.9363 (0.1478) | 0.9486 (0.1599) |
| 50 | 100 | (0.5, 1) | 0.9801 (0.2650) | **0.9516 (0.2209)** | 0.9344 (0.2281) | 0.9278 (0.2204) |
| | | (0.5, 2) | **0.9670 (0.1855)** | 0.9478 (0.1695) | 0.9372 (0.1659) | 0.9377 (0.1694) |
| | | (0.5, 5) | 0.9616 (0.1342) | 0.9548 (0.1293) | 0.9433 (0.1219) | **0.9501 (0.1291)** |
| | | (0.5, 10) | **0.9548 (0.1150)** | 0.9488 (0.1121) | 0.9297 (0.1041) | 0.9474 (0.1118) |
| | | (1, 1) | 0.9926 (0.4175) | **0.9552 (0.2979)** | 0.9283 (0.3369) | 0.9009 (0.2971) |
| | | (1, 2) | 0.9811 (0.2648) | **0.9511 (0.2209)** | 0.9377 (0.2281) | 0.9296 (0.2205) |
| | | (1, 5) | 0.9655 (0.1689) | **0.9533 (0.1573)** | 0.9424 (0.1522) | 0.9448 (0.1572) |
| | | (1, 10) | 0.9581 (0.1341) | **0.9507 (0.1292)** | 0.9391 (0.1218) | 0.9481 (0.1290) |
| 100 | 100 | (0.5, 1) | 0.9823 (0.2555) | **0.9514 (0.2102)** | 0.9276 (0.2208) | 0.9249 (0.2094) |
| | | (0.5, 2) | 0.9721 (0.1721) | **0.9551 (0.1556)** | 0.9391 (0.1554) | 0.9441 (0.1552) |
| | | (0.5, 5) | 0.9596 (0.1148) | **0.9518 (0.1102)** | 0.9404 (0.1068) | 0.9463 (0.1101) |
| | | (0.5, 10) | 0.9563 (0.0923) | **0.9510 (0.0900)** | 0.9428 (0.0869) | 0.9487 (0.0899) |
| | | (1, 1) | 0.9931 (0.4113) | **0.9504 (0.2899)** | 0.9182 (0.3302) | 0.8932 (0.2883) |
| | | (1, 2) | 0.9840 (0.2554) | **0.9536 (0.2101)** | 0.9270 (0.2205) | 0.9239 (0.2095) |
| | | (1, 5) | 0.9673 (0.1540) | **0.9523 (0.1421)** | 0.9399 (0.1405) | 0.9425 (0.1418) |
| | | (1, 10) | 0.9599 (0.1190) | **0.9515 (0.1103)** | 0.9410 (0.1068) | 0.9474 (0.1101) |

variation, the results showed that GCI method $CI_{GCI} = (0.0188, 0.1497)$ with interval length of 0.1309, AGCI method $CI_{AGCI} = (0.0213, 0.1447)$ with interval length of 0.1234, BPCI method $CI_{BPCI} = (-0.1245, 0.1274)$ with interval length of 0.2519 and MOVER method $CI_{MOVER} = (0.1697, 0.2873)$ with interval length of 0.2873. Therefore, the results from above examples support our simulation results.

## 5 Conclusions

The new confidence intervals for the difference between the coefficients of variation of inverse Gaussian distributions based on GCI, AGCI, BPCI, and MOVER were presented. The performances of these confidence intervals were assessed in terms of their coverage probabilities and the average lengths. The results obtained from the GCI and the AGCI methods were satisfactory in all cases.

However, the AGCI method was better than the GCI method in terms of the average length. For small sample sizes, MOVER provided the shortest average length. Meanwhile, based on the findings of this study, BPCI is not recommended because the coverage probabilities were under the nominal confidence level in all cases.

**Acknowledgement.** This research was funded by King Mongkut's University of Technology North Bangkok. Contract number:KMUTNB-PHD-63-01.

## References

1. Schrödinger, E.: Zür theorie der fall-und steigversuche an teilchen mit brownscer bewegung. Physikaliche Zeitschrift **16**, 289–295 (1915)
2. Wise, M.E.: Skew distributions in biomedicine including some with negative powers of time. In: Patil, G.P., Kotz, S., Ord, J.K. (eds.) A Modern Course on Statistical Distributions in Scientific Work. NATO Advanced Study Institutes Series (Series C – Mathematical and Physical Sciences), vol. 17, pp. 241–262. Springer, Dordrecht (1975). https://doi.org/10.1007/978-94-010-1845-6_18
3. Liu, X., Li, N., Hu, Y.: Combining inferences on the common mean of several inverse Gaussian distributions based on confidence distribution. Stat. Probab. Lett. **105**, 136–142 (2015)
4. Banerjee, A.K., Bhattacharyya, G.K.: Bayesian result for the inverse Gaussian distribution with an application. Technometrics **21**(2), 247–251 (1979)
5. Lancaster, T.: A stochastic model for the duration of a strike. J. Roy. Stat. Soc. Ser. A (General) **135**(2), 257–271 (1972)
6. Sheppard, C.W.: Basic Principles of the Tracer Method. Wiley, New York (1962)
7. Chhikara, R.S., Folks, J.L.: The Inverse Gaussian Distribution. Marcel Dekker, New York (1989)
8. Krishnamoorthy, K., Lu, Y.: Inference on the common mean of several normal populations based on the generalized variable method. Biometrics **59**(2), 237–247 (2003)
9. Tian, L., Wu, J.: Inferences on the mean response in a log-regression model: the generalized variable approach. Stat. Med. **26**, 5180–5188 (2007)
10. Lin, S.H., Lee, J.C., Wang, R.S.: Generalized inferences on the common mean vector of several multivariate normal populations. J. Stat. Plan. Infer. **137**(7), 2240–2249 (2007)
11. Ye, R.D., Ma, T.F., Wang, S.G.: Inference on the common mean of several inverse Gaussian populations. Comput. Stat. Data Anal. **54**, 906–915 (2010)
12. Wongkhao, A., Niwitpong, S., Niwitpong, S.: Confidence intervals for the ratio of two independent coefficients of variation of normal distribution. Far East J. Math. Sci. **98**(6), 741–757 (2015)
13. Banik, S., Kibria, B.M.G.: Estimating the population coefficient of variation by confidence intervals. Commun. Stat. Simul. Comput. **40**, 1236–1261 (2011)
14. Mahmoudvand, R., Hassani, H.: Two new confidence intervals for the coefficient of variation in a normal distribution. J. Appl. Stat. **36**(4), 429–442 (2009)
15. Hasan, M.S., Krishnamoorthy, K.: Improved confidence intervals for the ratio of coefficients of variation of two lognormal distributions. J. Stat. Theory Appl. **16**(3), 345–353 (2017)

16. Sangnawakij, P., Niwitpong, S.-A., Niwitpong, S.: Confidence intervals for the ratio of coefficients of variation of the gamma distributions. In: Huynh, V.-N., Inuiguchi, M., Denoeux, T. (eds.) IUKM 2015. LNCS (LNAI), vol. 9376, pp. 193–203. Springer, Cham (2015). https://doi.org/10.1007/978-3-319-25135-6_19
17. Thangjai, W., Niwitpong, S.: Confidence intervals for the weighted coefficients of variation of two-parameter exponential distributions. Cogent Math. **4**, 1315880 (2017). https://doi.org/10.1080/23311835.2017.1315880
18. Yosboonruang, N., Niwitpong, S.-A., Niwitpong, S.: Confidence intervals for the coefficient of variation of the delta-lognormal distribution. In: Anh, L.H., Dong, L.S., Kreinovich, V., Thach, N.N. (eds.) ECONVN 2018. SCI, vol. 760, pp. 327–337. Springer, Cham (2018). https://doi.org/10.1007/978-3-319-73150-6_26
19. Tian, L., Wilding, G.E.: Confidence intervals of the ratio of means of two independent inverse gaussian distributions. J. Stat. Plan. Infer. **133**(2), 381–386 (2005)
20. Krishnamoorthy, K., Tian, L.: Inferences on the difference and ratio of the means of two inverse Gaussian distributions. J. Stat. Plan. Infer. **133**, 381–386 (2008)
21. Weerahandi, S.: Generalized confidence intervals. J. Am. Stat. Assoc. **88**(423), 899–905 (1993)
22. Efron, B., Tibshirani, R.J.: An Introduction to Bootstrap. Chapman & Hall/CRC, Boca Raton (1993)
23. Gulhar, M., Golam Kibria, B.M., Albatineh, A.N., Ahmed, N.U.: A comparison of some confidence intervals for estimating the population coefficient of variation: a simulation study. Stat. Oper. Res. Trans. **36**, 45–68 (2012)
24. Donner, A., Zou, G.Y.: Closed-form confidence intervals for functions of the normal mean and standard deviation. Stat. Methods Med. Res. **21**, 347–359 (2010)
25. Mudholkar, G.S., Hutson, A.: The exponentiated Weibull family: some properties and a flood data application. Commun. Stat. Theory Method **25**(12), 3059–3083 (1996)
26. Eilam, N., Johnson, P.K., Creger, W.P.: Bradykininogen levels in Hodgkin's disease. Cancer **22**, 631–634 (1968)

# Methods for Testing the Difference Between Two Signal-to-Noise Ratios of Log-Normal Distributions

Wararit Panichkitkosolkul(✉) and Kamon Budsaba

Department of Mathematics and Statistics, Faculty of Science and Technology, Thammasat University, Pathum Thani 12121, Thailand
wararit@mathstat.sci.tu.ac.th

**Abstract.** This study presents three methods for testing the difference between two signal-to-noise ratios (SNRs) of log-normal distributions. The proposed statistical tests were based on the generalized confidence interval (GCI) approach, the large sample (LS) approach and the method of variance estimates recovery (MOVER) approach. To compare the performance of the proposed statistical tests, a simulation study was conducted under several values of SNRs in log-normal distributions. The performance of the statistical tests was compared based on the empirical size and power of the test. The simulation results showed that the statistical test based on the GCI approach performed better than the statistical tests based on the LS and the MOVER approaches in terms of the attained nominal significance level and empirical power of the test and is thus recommended for researchers. The performance of the proposed statistical tests is also illustrated through a numerical example.

**Keywords:** Coefficient of variation · Type I error rate · Power of the test · Skewed distribution · Simulation study

## 1 Introduction

The coefficient of variation (CV) is a dimensionless measure of variability relative to the mean. It has been more widely used than other measures of dispersion for comparing the variations of several variables obtained by different units [1]. The population CV is defined as the ratio of the population standard deviation ($\sigma$) to the population mean ($\mu$), given by $\mathrm{CV} = \sigma/\mu$. The reciprocal of the CV is called the signal-to-noise ratio (SNR), which is defined by $\mathrm{SNR} = \mu/\sigma$. The SNR is a measure of the signal strength relative to the background noise in analog and digital communication. The SNR has been applied in several fields, such as image processing, quality control, medicine and business. For instance, in image processing, the SNR of an image is usually computed as the ratio of the mean pixel value to the standard deviation of the pixel values over a given neighborhood [2–5]. In quality control, the SNR shows the degree of the predictability of the performance of a product, or process, in the presence of noise factors [6]. In medicine, a simple method for measuring the SNR in magnetic resonance imagining (MRI) was presented


V.-N. Huynh et al. (Eds.): IUKM 2020, LNAI 12482, pp. 384–395, 2020.
https://doi.org/10.1007/978-3-030-62509-2_32

by Kaufman et al. [7]. Later, McGibney and Smith [8] and Firbank et al. [9] studied the methods for measuring the SNR in MRI. In addition, Czanner et al. [10] studied the SNR measurement of a neuron. In business, service quality in the hotel business was improved using collective quality function deployment and the SNR [11]. Liu [12] proposed a data envelopment analysis ranking method based on cross-efficiency intervals and the SNR. However, the literature on testing the SNR and the difference between two SNRs are limited. Recently, Thangjai and Niwitpong [13] presented the confidence intervals for the difference between two SNRs of log-normal distributions. Three proposed confidence intervals were based on the generalized confidence interval (GCI) approach, the large sample (LS) approach and the method of variance estimates recovery (MOVER) approach. Further, Panichkitkosolkul and Tulyanitikul [14] proposed the methods for testing the SNR of a log-normal distribution. Therefore, Thangjai and Niwitpong's [13] confidence intervals for the difference between two SNRs can be used to test the hypothesis for the difference between two SNRs for log-normal distributions.

In this study, an evaluation of the methods for testing the difference between two SNRs based on the GCI approach, the large sample approach and the MOVER approach was conducted using simulated log-normal data from many values of SNRs. The comparisons were made using the empirical type I error rates and the power of the test. Because a theoretical comparison is not possible, a simulation study was conducted to compare the performance of these methods. Based on the simulation results, the test statistic with high power that attained a nominal significance level is recommended for practitioners.

The structure of this paper is as follows. Section 2 proposes statistical tests for the difference between two SNRs of log-normal distributions. In Sect. 3, the simulation study and simulation results are discussed. A numerical example is provided in Sect. 4. Finally, the conclusion is presented in the last section.

## 2 Statistical Tests for the Difference Between Two Signal-to-Noise Ratios of Log-Normal Distributions

The log-normal distribution is widely used in several areas, such as environmental study, survival analysis, biostatistics and other statistical fields. It is a right skewed distribution with a long tail. The probability density function of the log-normal random variable is given by

$$f(x; \mu_x, \sigma_x^2) = \frac{1}{x\sigma_x\sqrt{2\pi}} \exp\left[-\frac{(\ln(x) - \mu_x)^2}{2\sigma_x^2}\right], \quad x > 0.$$

Let $X$ and $Y$ be the independent random variables from log-normal distributions. Define $U = \log(X)$ and $V = \log(Y)$. Then $U$ and $V$ are normally distributed with mean $(\mu_u, \mu_v)$ and variance $(\sigma_u^2, \sigma_v^2)$, respectively. The means of $X$ and $Y$ are, respectively, given by

$$E(X) = \exp(\mu_u + \sigma_u^2/2)$$
$$\text{and } E(Y) = \exp(\mu_v + \sigma_v^2/2).$$

The variances of $X$ and $Y$ are, respectively, given by

$$Var(X) = \left[\exp(\sigma_u^2) - 1\right]\exp(2\mu_u + \sigma_u^2)$$
$$\text{and } Var(Y) = \left[\exp(\sigma_v^2) - 1\right]\exp(2\mu_v + \sigma_v^2).$$

Therefore, the SNRs of $X$ and $Y$ are, respectively, given by

$$\theta_x = \frac{E(X)}{\sqrt{Var(X)}} = \frac{1}{\sqrt{\exp(\sigma_u^2) - 1}} \text{ and } \theta_y = \frac{E(Y)}{\sqrt{Var(Y)}} = \frac{1}{\sqrt{\exp(\sigma_v^2) - 1}}.$$

Let $X_1, \ldots, X_n$ and $Y_1, \ldots, Y_m$ be independent and identically distributed (i.i.d.) random samples from log-normal distributions and let $U_i = \log(X_i)$ and $V_j = \log(Y_j)$ where $i = 1, \ldots, n$ and $j = 1, \ldots, m$ be the log-transformed data. Thus, $U_1, \ldots, U_n$ and $V_1, \ldots, V_m$ are random samples from normal distributions with means $(\mu_u, \mu_v)$ and variances $(\sigma_u^2, \sigma_v^2)$, respectively. Let $(\bar{U}, S_u^2)$ and $(\bar{V}, S_v^2)$ be the sample mean and sample variance of log-transformed data $U_i$ and $V_j$. The estimators of $\theta_x$ and $\theta_y$ are, respectively

$$\hat{\theta}_x = \frac{1}{\sqrt{\exp(S_u^2) - 1}} \text{ and } \hat{\theta}_y = \frac{1}{\sqrt{\exp(S_v^2) - 1}}. \quad (1)$$

The variances of $\hat{\theta}_x$ and $\hat{\theta}_y$ were derived by Thangjai et al. [15]. They have the form

$$Var(\hat{\theta}_x) = \frac{\sigma_u^4 \exp(2\sigma_u^2)}{2(n-1)\left[\exp(\sigma_u^2) - 1\right]^3} \text{ and } Var(\hat{\theta}_y) = \frac{\sigma_v^4 \exp(2\sigma_v^2)}{2(m-1)\left[\exp(\sigma_v^2) - 1\right]^3}.$$

Therefore, the difference between $\hat{\theta}_x$ and $\hat{\theta}_y$ is

$$\hat{\delta} = \hat{\theta}_x - \hat{\theta}_y = \frac{1}{\sqrt{\exp(S_u^2) - 1}} - \frac{1}{\sqrt{\exp(S_v^2) - 1}}.$$

Using the Bienaymé formula [16], the variance of the sum of uncorrelated random variables is the sum of their variances. Moreover, using the linearity of the expectation operator and the assumption that $X$ and $Y$ are independent, the variance of $\hat{\theta}_x - \hat{\theta}_y$ is

$$Var(\hat{\delta}) = Var(\hat{\theta}_x - \hat{\theta}_y) = \frac{\sigma_u^4 \exp(2\sigma_u^2)}{2(n-1)\left[\exp(\sigma_u^2) - 1\right]^3} + \frac{\sigma_v^4 \exp(2\sigma_v^2)}{2(m-1)\left[\exp(\sigma_v^2) - 1\right]^3}.$$

We wanted to test for the difference between two SNRs. The null and alternative hypotheses are defined as follows:

$$H_0 : \theta_x - \theta_y = \delta_0 \text{ versus } H_a : \theta_x - \theta_y \neq \delta_0,$$

where $\delta_0$ is the hypothesized value of the difference between two SNRs.

In this section, we will propose three test statistics for the difference between two SNRs based on the GCI approach, the large sample approach and the MOVER approach.

### 2.1 The GCI Approach for the Difference Between Two SNRs

The generalized confidence interval (GCI) approach was first introduced by Weerahandi [17]. For the application of the GCI approach, we refer our readers to Jose and Thomas [18], Thangjai et al. [19] and Tian and Cappelleri [20]. Let $X_1, \ldots, X_n$ be an i.i.d. random sample of size $n$ having a probability density function $f(x; \lambda, \beta)$, where $\lambda$ is a parameter of interest and $\beta$ is a nuisance parameter. Let $x_1, \ldots, x_n$ be the observed sample of $X_1, \ldots, X_n$. A generalized pivotal quantity $Q(X; x, \lambda, \beta)$ is considered and satisfies the following conditions:

(i) The distribution of $Q(X; x, \lambda, \beta)$ is free of all unknown parameters.
(ii) The observed value of $Q(X; x, \lambda, \beta)$ is the parameter of interest.

Condition (i) is imposed to guarantee that a subset of the sample space of the possible values of $Q(X; x, \lambda, \beta)$ can be found at a given value of the confidence coefficient, with no knowledge of the parameters. Condition (ii) is imposed to ensure that such probability statements, based on a generalized pivotal quantity, lead to confidence regions involving the observed data $x$ only. The GCI for $\lambda$ is computed using the percentiles of the generalized pivotal quantity. Let $[Q(\alpha/2), Q(1-\alpha/2)]$ be a $(1-\alpha)100\%$ GCI for the parameter of interest, where $Q(\alpha/2)$ and $Q(1-\alpha/2)$ are the $100(\alpha/2)$ and the $100(1-\alpha/2)$ percentiles of $Q(X; x, \lambda, \beta)$, respectively.

Suppose that $S_u^2$ and $S_v^2$ are the variances of the log-transformed sample. Let $s_u^2$ and $s_v^2$ be the observed values of $S_u^2$ and $S_v^2$, respectively. Since $s_u^2$ has a chi-squared distribution with $n-1$ degrees of freedom, defined by $s_u^2 \sim \sigma_u^2 \chi_{n-1}^2/(n-1)$, then $\sigma_u^2 = (n-1)s_u^2/\chi_{n-1}^2$. Similarly, $s_v^2$ has a chi-squared distribution with $m-1$ degrees of freedom, defined by $s_v^2 \sim \sigma_v^2 \chi_{m-1}^2/(m-1)$, then $\sigma_v^2 = (m-1)s_v^2/\chi_{m-1}^2$. We define the generalized pivotal quantity for $\sigma_u^2$ and $\sigma_v^2$ as

$$Q_{\sigma_u^2} = \frac{(n-1)s_u^2}{\chi_{n-1}^2} \text{ and } Q_{\sigma_v^2} = \frac{(m-1)s_v^2}{\chi_{m-1}^2}, \tag{2}$$

where $\chi_{n-1}^2$ and $\chi_{m-1}^2$ are random variables from the chi-squared with $n-1$ and $m-1$ degrees of freedom, respectively.

Therefore, the difference between two generalized pivotal quantities $Q_{\theta_x} - Q_{\theta_y}$, based on the generalized pivotal quantities for $\sigma_u^2$ and $\sigma_v^2$ is given by

$$Q_\delta = Q_{\theta_x} - Q_{\theta_y} = \frac{1}{\sqrt{\exp\left(Q_{\sigma_u^2}\right) - 1}} - \frac{1}{\sqrt{\exp\left(Q_{\sigma_v^2}\right) - 1}}. \tag{3}$$

The $(1-\alpha)100\%$ confidence interval for the difference between two SNRs, $\delta$, of log-normal distributions based on the GCI approach is given by

$$CI_{GCI} = \left[Q_\delta(\alpha/2),\ Q_\delta(1-\alpha/2)\right], \tag{4}$$

where $Q_\delta(\alpha/2)$ and $Q_\delta(1-\alpha/2)$ are the $100(\alpha/2)$ and the $100(1-\alpha/2)$ percentiles of $Q_\delta$, respectively. The following algorithm was used to construct the GCI for the difference between two SNRs of log-normal distributions:

*Algorithm: The GCI for the difference between two SNRs*
For a given $s_u^2$ and $s_v^2$
For $i=1$ to $h$
  Generate $\chi_{n-1}^2$ and $\chi_{m-1}^2$ from the chi-squared distribution with $n-1$ and $m-1$ degrees of freedom
  Compute $Q_{\sigma_u^2}$ and $Q_{\sigma_v^2}$ from Eq. (2)
  Compute $Q_\delta$ from Eq. (3)
End $i$ loop
  Compute the $100(\alpha/2)$ percentile of $Q_\delta$ defined by $Q_\delta(\alpha/2)$

We will reject the null hypothesis, $H_0$, if $\delta_0 < Q_\delta(\alpha/2)$ or $\delta_0 > Q_\delta(1-\alpha/2)$, where $Q_\delta(\alpha/2)$ and $Q_\delta(1-\alpha/2)$ are shown in Eq. (4).

### 2.2 The Large Sample Approach for the Difference Between Two SNRs

Using the central limit theorem (CLT), the $(1-\alpha)100\%$ confidence interval for the difference between two SNRs, $\delta$, of log-normal distributions based on the large sample approach is given by

$$CI_{LS} = \left[\hat{\delta} - z_{1-\alpha/2}\sqrt{\widehat{Var}(\hat{\delta})}, \hat{\delta} + z_{1-\alpha/2}\sqrt{\widehat{Var}(\hat{\delta})}\right], \tag{5}$$

where $z_{1-\alpha/2}$ is the $100(1-\alpha/2)$ percentile of a standard normal distribution and

$$\widehat{Var}(\hat{\delta}) = \frac{S_u^4\exp(2S_u^2)}{2(n-1)\left[\exp(S_u^2)-1\right]^3} + \frac{S_v^4\exp(2S_v^2)}{2(m-1)\left[\exp(S_v^2)-1\right]^3}.$$

We will reject the null hypothesis, $H_0$, if $\delta_0 < \hat{\delta} - z_{1-\alpha/2}\sqrt{\widehat{Var}(\hat{\delta})}$ or $\delta_0 > \hat{\delta} + z_{1-\alpha/2}\sqrt{\widehat{Var}(\hat{\delta})}$ where $\hat{\delta} \mp z_{1-\alpha/2}\sqrt{\widehat{Var}(\hat{\delta})}$ are shown in Eq. (5).

### 2.3 The MOVER Approach for the Difference Between Two SNRs

The MOVER approach is a method for estimating the confidence interval for the functions of parameters, i.e., in the form of $\theta_1 + \theta_2$ and $\theta_1/\theta_2$. This method was introduced by Donner and Zou [21], and several researchers have applied the MOVER method for constructing confidence intervals; see Zou et al. [22], Li et al. [23], Newcombe [24], Sangnawakij and Niwitpong [25], and Thangjai and Niwitpong [13] for more details. The idea of this approach is to find the separate confidence intervals for two single parameters, recover the variance estimates from the confidence intervals, and then form the confidence interval for the function of the parameters.

We explain the MOVER approach based on the CLT to find the confidence interval for $\theta_1 + \theta_2$. Therefore, the general form of two-sided confidence interval, under the assumption of independent between the estimators $\hat{\theta}_1$ and $\hat{\theta}_2$, is given by

$$[L, U] = \left[(\hat{\theta}_1 + \hat{\theta}_2) \mp z_{1-\alpha/2}\sqrt{Var(\hat{\theta}_1) + Var(\hat{\theta}_2)}\right],$$

where $Var(\hat{\theta}_1)$ and $Var(\hat{\theta}_2)$ are the unknown variances of $\hat{\theta}_1$ and $\hat{\theta}_2$, respectively. Zou et al. [26] assumed that $[l_i, u_i]$ are the $(1-\alpha)100\%$ confidence intervals for $\theta_i, i = 1, 2$. Furthermore, they pointed out that the value of $l_1 + l_2$ is similar to $L$ and $u_1 + u_2$ is similar to $U$. In order to estimate $Var(\hat{\theta}_i)$, using the CLT and under the conditions $\theta_1 = l_1$ and $\theta_2 = l_2$, the estimated variances recovered from $l_i$ to obtain $L$ can be derived as $\widehat{Var}(\hat{\theta}_i) \approx (\hat{\theta}_i - l_i)^2/z_{1-\alpha/2}^2$. On the other hand, under the conditions $\theta_1 = u_1$ and $\theta_2 = u_2$, the estimated variances recovered from $u_i$ to obtain $U$ can be derived as $\widehat{Var}(\hat{\theta}_i) \approx (u_i - \hat{\theta}_i)^2/z_{1-\alpha/2}^2$. Replacing the corresponding estimated variances into the confidence interval $[L, U]$, we then obtain the $(1-\alpha)100\%$ confidence interval for $\theta_1 + \theta_2$. Similarly, the confidence interval for the difference of parameters can be developed by changing $\theta_1 - \theta_2$ into the form $\theta_1 + (-\theta_2)$ and then recovering the variance estimates by following the above approach.

Following the concept of Donner and Zou [21], the $(1-\alpha)100\%$ confidence interval for the difference between two SNRs, $\delta$, of log-normal distributions based on the MOVER approach is given by

$$CI_{MOVER} = \left[ (\hat{\theta}_x - \hat{\theta}_y) - \sqrt{(\hat{\theta}_x - l_x)^2 + (u_y - \hat{\theta}_y)^2}, (\hat{\theta}_x - \hat{\theta}_y) + \sqrt{(u_x - \hat{\theta}_x)^2 + (\hat{\theta}_y - l_y)^2} \right], \tag{6}$$

where $[l_x, u_x]$ and $\left[l_y, u_y\right]$ are the $(1-\alpha)100\%$ confidence intervals for $\theta_x$ and $\theta_y$, respectively. The confidence intervals $[l_x, u_x]$ and $\left[l_y, u_y\right]$ in Eq. (6) are defined by

$$[l_x, u_x] = \left[ \hat{\theta}_x - z_{1-\alpha/2}\sqrt{\widehat{Var}(\hat{\theta}_x)}, \hat{\theta}_x + z_{1-\alpha/2}\sqrt{\widehat{Var}(\hat{\theta}_x)} \right]$$

$$\text{and } \left[l_y, u_y\right] = \left[ \hat{\theta}_y - z_{1-\alpha/2}\sqrt{\widehat{Var}(\hat{\theta}_y)}, \hat{\theta}_y + z_{1-\alpha/2}\sqrt{\widehat{Var}(\hat{\theta}_y)} \right],$$

where $\hat{\theta}_x$ and $\hat{\theta}_y$ are defined in Eq. (1),

$$\widehat{Var}(\hat{\theta}_x) = \frac{S_u^4 \exp(2S_u^2)}{2(n-1)\left[\exp(S_u^2) - 1\right]^3} \text{ and } \widehat{Var}(\hat{\theta}_y) = \frac{S_v^4 \exp(2S_v^2)}{2(m-1)\left[\exp(S_v^2) - 1\right]^3}.$$

We will reject the null hypothesis, $H_0$, if $\delta_0 < (\hat{\theta}_x - \hat{\theta}_y) - \sqrt{(\hat{\theta}_x - l_x)^2 + (u_y - \hat{\theta}_y)^2}$ or $\delta_0 > (\hat{\theta}_x - \hat{\theta}_y) + \sqrt{(u_x - \hat{\theta}_x)^2 + (\hat{\theta}_y - l_y)^2}$ where $[l_x, u_x]$ and $\left[l_y, u_y\right]$ are shown in Eq. (6).

## 3 Simulation Study

In this study, three methods for testing the difference between two SNRs of log-normal distributions are proposed. Because a theoretical comparison is not possible, a Monte Carlo simulation was conducted using the R version 3.6.1 statistical software [27] to compare the performance of the statistical tests. The methods were compared in terms of their attainment of empirical type I error rates and the power of their performance. For more on simulation studies, we refer our readers to Bhat and Rao [28], Panichkitkosolkul [29], and Niwitpong and Kirdwichai [30], among others. The simulation results are presented only for the significance level $\alpha = 0.05$ because (i) $\alpha = 0.05$ is widely used to compare the power of the test and (ii) similar conclusions were obtained for other values of $\alpha$.

The data were generated from two normal distributions, $U$ and $V$, with means $\mu_u = \mu_v = 1$ and the SNR of population $X$ was $\theta_x = 1, 2, 5$ and 10. Therefore, the population variances were $\sigma_u^2 = \ln\left((1/\theta_x^2) + 1\right)$ and $\sigma_v^2 = \ln\left((1/\theta_y^2) + 1\right)$. To observe the behavior of small, moderate, and large sample sizes, we used $(n, m) = (10, 10)$, (30, 30) and (50, 50). The number of simulations was fixed at 5,000. The null and alternative hypotheses were defined as follows:

$$H_0 : \theta_x - \theta_y = 0 \text{ versus } H_a : \theta_x - \theta_y \neq 0.$$

The simulation results of empirical type I error rates and the power of the test are presented in Table 1. The empirical type I error rates of the statistical tests based on the GCI and the LS approaches were close to the nominal significance level of 0.05 in all situations. In addition, the statistical test based on the MOVER approach provides type I error rates close to the 0.05 nominal value for small sample sizes. Regarding the empirical power comparisons, the test statistic based on the GCI approach performed better than the test statistics based on the LS and the MOVER approaches for almost all situations. The present study confirms previous findings from Thangjai and Niwitpong [13] that the GCI approach was better than the LS and the MOVER approaches. We observed a general pattern; when the sample size increases, the empirical power of the test also increases, and the empirical type I error rate approaches 0.05. Additionally, the empirical power increases as the difference between the values of $\theta_x$ and $\theta_y$. It was observed that for large sample sizes, the performance of all the test statistics did not differ greatly in the terms of empirical power and the attainment of the nominal significance level of the test.

**Table 1.** Empirical type I error rates (bold numeric values) and the power of the test for log-normal distributions.

| $\theta_x$ | $\theta_y$ | $(n, m) = (10, 10)$ | | | $(n, m) = (30, 30)$ | | | $(n, m) = (50, 50)$ | | |
|---|---|---|---|---|---|---|---|---|---|---|
| | | GCI | LS | MOVER | GCI | LS | MOVER | GCI | LS | MOVER |
| 1.0 | 0.2 | 0.588 | 0.601 | 0.470 | 0.983 | 0.984 | 0.980 | 0.999 | 0.999 | 0.999 |
| | 0.4 | 0.320 | 0.324 | 0.206 | 0.789 | 0.792 | 0.765 | 0.953 | 0.955 | 0.949 |
| | 0.6 | 0.150 | 0.154 | 0.082 | 0.394 | 0.393 | 0.362 | 0.628 | 0.624 | 0.604 |
| | 0.8 | 0.078 | 0.077 | 0.033 | 0.125 | 0.123 | 0.103 | 0.189 | 0.188 | 0.176 |
| | 1.0 | **0.050** | **0.047** | **0.021** | **0.051** | **0.051** | **0.040** | **0.051** | **0.051** | **0.045** |
| | 1.2 | 0.066 | 0.063 | 0.028 | 0.104 | 0.105 | 0.090 | 0.162 | 0.159 | 0.148 |
| | 1.4 | 0.111 | 0.106 | 0.052 | 0.279 | 0.278 | 0.247 | 0.433 | 0.431 | 0.412 |

*(continued)*

**Table 1.** *(continued)*

| $\theta_x$ | $\theta_y$ | $(n, m) = (10, 10)$ | | | $(n, m) = (30, 30)$ | | | $(n, m) = (50, 50)$ | | |
|---|---|---|---|---|---|---|---|---|---|---|
| | | GCI | LS | MOVER | GCI | LS | MOVER | GCI | LS | MOVER |
| | 1.6 | 0.175 | 0.166 | 0.089 | 0.498 | 0.494 | 0.457 | 0.736 | 0.732 | 0.716 |
| | 1.8 | 0.262 | 0.250 | 0.143 | 0.700 | 0.696 | 0.664 | 0.907 | 0.907 | 0.898 |
| 2.0 | 1.2 | 0.241 | 0.229 | 0.121 | 0.627 | 0.622 | 0.586 | 0.846 | 0.844 | 0.832 |
| | 1.4 | 0.136 | 0.128 | 0.067 | 0.367 | 0.361 | 0.332 | 0.571 | 0.565 | 0.542 |
| | 1.6 | 0.081 | 0.074 | 0.035 | 0.175 | 0.172 | 0.150 | 0.260 | 0.259 | 0.242 |
| | 1.8 | 0.059 | 0.052 | 0.024 | 0.072 | 0.073 | 0.061 | 0.100 | 0.099 | 0.092 |
| | 2.0 | **0.046** | **0.040** | **0.015** | **0.053** | **0.049** | **0.041** | **0.054** | **0.053** | **0.046** |
| | 2.2 | 0.054 | 0.051 | 0.021 | 0.074 | 0.072 | 0.061 | 0.088 | 0.082 | 0.076 |
| | 2.4 | 0.065 | 0.058 | 0.027 | 0.139 | 0.135 | 0.115 | 0.199 | 0.199 | 0.183 |
| | 2.6 | 0.095 | 0.084 | 0.041 | 0.251 | 0.244 | 0.215 | 0.392 | 0.389 | 0.369 |
| | 2.8 | 0.144 | 0.132 | 0.064 | 0.361 | 0.357 | 0.322 | 0.576 | 0.570 | 0.553 |
| 5.0 | 4.2 | 0.076 | 0.065 | 0.029 | 0.145 | 0.141 | 0.120 | 0.220 | 0.214 | 0.200 |
| | 4.4 | 0.063 | 0.057 | 0.023 | 0.098 | 0.093 | 0.078 | 0.138 | 0.133 | 0.123 |
| | 4.6 | 0.052 | 0.045 | 0.017 | 0.072 | 0.070 | 0.060 | 0.089 | 0.086 | 0.078 |
| | 4.8 | 0.054 | 0.045 | 0.017 | 0.058 | 0.055 | 0.044 | 0.057 | 0.054 | 0.050 |
| | 5.0 | **0.052** | **0.045** | **0.016** | **0.050** | **0.049** | **0.040** | **0.047** | **0.045** | **0.041** |
| | 5.2 | 0.050 | 0.043 | 0.018 | 0.056 | 0.052 | 0.043 | 0.052 | 0.050 | 0.044 |
| | 5.4 | 0.052 | 0.044 | 0.018 | 0.060 | 0.057 | 0.046 | 0.085 | 0.083 | 0.076 |
| | 5.6 | 0.062 | 0.052 | 0.024 | 0.089 | 0.084 | 0.070 | 0.130 | 0.129 | 0.118 |
| | 5.8 | 0.074 | 0.066 | 0.027 | 0.114 | 0.110 | 0.095 | 0.171 | 0.166 | 0.155 |
| 10.0 | 9.2 | 0.057 | 0.048 | 0.018 | 0.073 | 0.069 | 0.056 | 0.084 | 0.082 | 0.075 |
| | 9.4 | 0.053 | 0.046 | 0.019 | 0.069 | 0.065 | 0.054 | 0.070 | 0.067 | 0.059 |
| | 9.6 | 0.052 | 0.043 | 0.016 | 0.062 | 0.057 | 0.047 | 0.065 | 0.063 | 0.057 |
| | 9.8 | 0.052 | 0.044 | 0.019 | 0.049 | 0.047 | 0.038 | 0.048 | 0.048 | 0.044 |
| | 10.0 | **0.048** | **0.040** | **0.015** | **0.048** | **0.045** | **0.035** | **0.050** | **0.049** | **0.042** |
| | 10.2 | 0.050 | 0.042 | 0.016 | 0.050 | 0.046 | 0.038 | 0.047 | 0.045 | 0.039 |
| | 10.4 | 0.048 | 0.040 | 0.017 | 0.055 | 0.053 | 0.042 | 0.059 | 0.058 | 0.051 |
| | 10.6 | 0.056 | 0.049 | 0.019 | 0.064 | 0.061 | 0.050 | 0.074 | 0.071 | 0.064 |
| | 10.8 | 0.057 | 0.050 | 0.020 | 0.068 | 0.064 | 0.054 | 0.085 | 0.084 | 0.075 |

## 4 Numerical Example

To illustrate the application of the methods for testing the difference between two SNRs proposed in the previous section, we used data from the Regenstrief Medical

Record System, as reported in McDonald et al. [31], Zhou et al. [32] and Jafari and Abdollahnezhad [33]. A randomized, parallel group experiment was conducted with 20 subjects to compare a new test formulation ($x$), with a reference formulation ($y$), of a drug product with a long half-life. The data from this study is as follows:

| $x$ | 732.89 | 1371.97 | 614.62 | 557.24 | 821.39 |
|---|---|---|---|---|---|
| | 363.94 | 430.95 | 401.42 | 436.16 | 951.46 |
| $y$ | 1053.63 | 1351.54 | 197.95 | 1204.72 | 447.20 |
| | 3357.66 | 567.36 | 668.48 | 842.19 | 284.86. |

The sample means are $\bar{x} = 668.204$ and $\bar{y} = 997.559$, and the sample standard deviations are $s_x = 314.861$ and $s_y = 913.428$. The sample means for the log-transformed data are $\bar{u} = 6.417$ and $\bar{v} = 6.601$, and the sample standard deviations for the log-transformed data are $s_u = 0.429$ and $s_v = 0.817$. The histogram, density plot, box and whisker plot, and normal quantile-quantile plot of $x$ and $y$ are shown in Fig. 1.

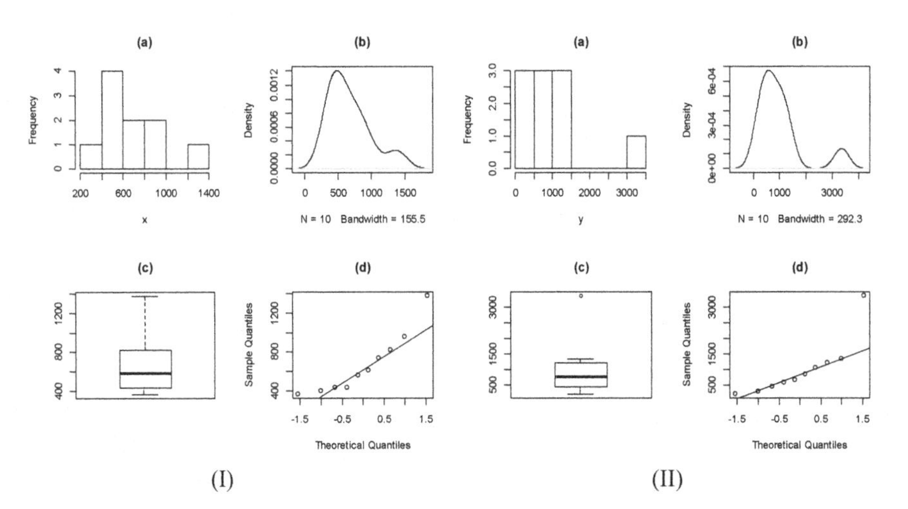


**Fig. 1.** (I) (a) histogram, (b) density plot, (c) box and whisker plot and (d) normal quantile-quantile plot of a new test formulation ($x$) and (II) (a) histogram, (b) density plot, (c) box and whisker plot and (d) normal quantile-quantile plot of a reference formulation ($y$).

Using Minitab program, the probability plots of the data are displayed in Fig. 2. By the Anderson-Darling goodness-of-fit test, the new test formulation ($x$) had a log-normal distribution with a location parameter, $\hat{\mu}_x = 6.417$ and scale parameter, $\hat{\sigma}_x^2 = 0.429$. Similarly, the reference formulation ($y$) had a log-normal distribution with a location parameter, $\hat{\mu}_y = 6.601$ and scale parameter, $\hat{\sigma}_y^2 = 0.817$. The estimators of SNRs are $\hat{\theta}_x = \left(\sqrt{\exp(S_u^2) - 1}\right)^{-1} = 2.222$ and $\hat{\theta}_y = \left(\sqrt{\exp(S_v^2) - 1}\right)^{-1} = 1.026$. Therefore, we were interested in testing the population difference between two SNRs. The null and alternative hypotheses are given as follows:

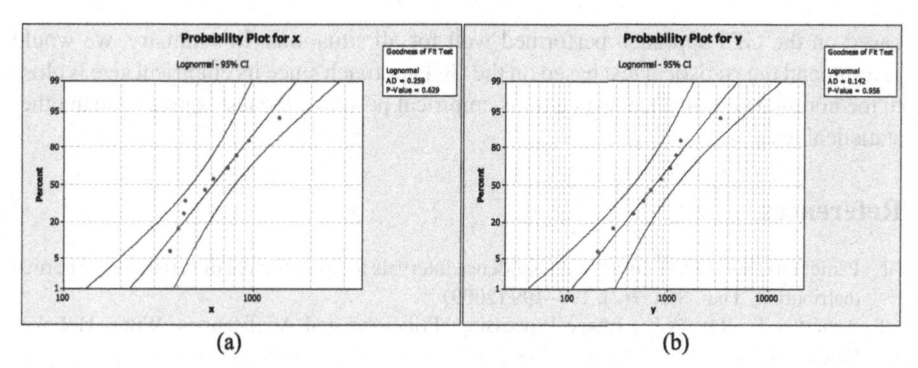


(a) (b)

**Fig. 2.** (a) probability plot of a new test formulation ($x$) and (b) a reference formulation ($y$).

$$H_0 : \theta_x - \theta_y = 0$$

$$H_a : \theta_x - \theta_y \neq 0.$$

The lower and upper critical values of the test statistics based on the GCI, the LS and the MOVER approaches were calculated using Eqs. (4), (5) and (6), respectively. As can be seen in Table 2, the null hypothesis, $H_0$ was not rejected because $-0.0446 < \delta_0 < 2.4161$, $-0.1019 < \delta_0 < 2.4945$ and $-0.3021 < \delta_0 < 2.6947$ using the test statistics based on the GCI, the LS and the MOVER approaches, respectively. We conclude that the difference between two SNRs is not different from zero at the significance level of 0.05.

**Table 2.** Critical values of test statistic based on the GCI, the LS and the MOVER approaches.

| Approach | Critical values | |
|---|---|---|
| | Lower | Upper |
| GCI | −0.0446 | 2.4161 |
| LS | −0.1019 | 2.4945 |
| MOVER | −0.3021 | 2.6947 |

## 5 Conclusion

This study considered three methods for testing the difference between two signal-to-noise ratios of log-normal distributions. Because a theoretical comparison was not possible, we used a simulation study to compare the performance of these methods. In the case of empirical type I error rates, the simulation results indicated that the test statistics based on the GCI and the LS approaches performed well for all situations. In the case of the power of the test, the simulation results suggested that the test statistic based on the GCI approach performed well for all situations. In summary, we would recommend the statistical test based on the GCI approach since its empirical size is close to the nominal significance level and its empirical power of the test is greater than other statistical tests.

## References

1. Panichkitkosolkul, W.: Improved confidence intervals for a coefficient of variation of a normal distribution. Thai. Stat. **7**(2), 193–199 (2009)
2. Acharya, T., Ray, A.K.: Image Processing: Principles and Applications. Wiley, Hoboken (2005)
3. Russ, C.J.: The Image Processing Handbook. CRC Press, Boca Raton (2011)
4. Rafael, C.G., Richard, E.W.: Digital Image Processing. Prentice Hall, Upper Saddle River (2008)
5. Tania, S.: Image Fusion: Algorithms and Applications. Academic Press, San Diego (2008)
6. Kapur, K., Chen, G.: Signal-to-noise ratio development for quality engineering. Qual. Reliab. Eng. Int. **4**(2), 133–141 (1988)
7. Kaufman, L., Kramer, D.M., Crooks, L.E., Ortendahl, D.A.: Measuring signal-to-noise ratios in MR imaging. Radiology **173**(1), 265–267 (1989)
8. McGibney, G., Smith, M.R.: An Unbiased signal-to-noise ratio measure for magnetic resonance images. Med. Phys. **20**(4), 1077–1079 (1993)
9. Firbank, M.J., Coulthard, A., Harrison, R.M., Williams, E.D.: A comparison of two methods for measuring the signal to noise ratio on MR images. Phys. Med. Biol. **44**(12), 261–264 (1999)
10. Czanner, G., et al.: Measuring the signal-to-noise ratio of a neuron. Nat. Acad. Sci. **112**(23), 7141–7146 (2015)
11. Patil, A.N., Hublikar, S.P., Faria, L.S., Khadilkar, S.S.: Improving service quality of hotel business using collective QFD and signal to noise ratio. OmniScience: Multi-Disc. J. **9**(1), 34–41 (2019)
12. Liu, S.T.: A DEA ranking method based on cross-efficiency intervals and signal-to-noise ratio. Ann. Oper. Res. **261**(1), 207–232 (2018)
13. Thangjai, W., Niwitpong, S.-A.: Confidence intervals for the signal-to-noise ratio and difference of signal-to-noise ratios of log-normal distributions. Stats **2**, 164–173 (2019)
14. Panichkitkosolkul, W., Tulyanitikul, B.: Performance of statistical methods for testing the signal-to-noise ratio of a log-normal distribution. In: 2020 IEEE 7th International Conference on Industrial Engineering and Applications (ICIEA), Bangkok, Thailand, pp. 656–661 (2020)
15. Thangjai, W., Niwitpong, S.-A., Niwitpong, S.: Simultaneous fiducial generalized confidence intervals for all differences of coefficients of variation of log-normal distributions. In: Huynh, V.-N., Inuiguchi, M., Le, B., Le, B.N., Denoeux, T. (eds.) IUKM 2016. LNCS (LNAI), vol. 9978, pp. 552–561. Springer, Cham (2016). https://doi.org/10.1007/978-3-319-49046-5_47
16. Loève, M.: Probability Theory I: Graduate Texts in Mathematics. Springer, New York (1977). https://doi.org/10.1007/978-1-4684-9464-8
17. Weerahandi, S.: Generalized confidence intervals. J. Am. Stat. Assoc. **88**(423), 899–906 (1993)
18. Jose, S., Thomas, S.: Interval estimation of the overlapping coefficient of two normal distributions: one way ANOVA with random effects. Thai. Stat. **17**(1), 84–92 (2019)
19. Thangjai, W., Niwitpong, S.-A., Niwitpong, S.: Simultaneous confidence intervals for all differences of means of normal distributions with unknown coefficients of variation. In: Kreinovich, V., Sriboonchitta, S., Chakpitak, N. (eds.) TES 2018. SCI, vol. 753, pp. 670–682. Springer, Cham (2018). https://doi.org/10.1007/978-3-319-70942-0_48

20. Tian, L., Cappelleri, J.C.: A new approach for interval estimation and hypothesis testing of a certain intraclass correlation coefficient: the generalized variable method. Stat. Med. **23**(13), 2125–2135 (2004)
21. Donner, A., Zou, G.Y.: Closed-form confidence intervals for functions of the normal mean and standard deviation. Stat. Methods Med. Res. **21**(4), 347–359 (2012)
22. Zou, G.Y., Taleban, J., Huo, C.Y.: Confidence interval estimation for lognormal data with application to health economics. Comput. Stat. Data Anal. **53**(11), 3755–3764 (2009)
23. Li, H.Q., Tang, M.L., Poon, W.Y., Tang, N.S.: Confidence intervals for difference between two Poisson rates. Commun. Stat.-Simul. Comput. **40**(9), 1478–1493 (2011)
24. Newcombe, R.G.: MOVER-R confidence intervals for ratios and products of two independently estimated quantities. Stat. Methods Med. Res. **25**(5), 1774–1778 (2016)
25. Sangnawakij, P., Niwitpong, S.-A.: Confidence intervals for coefficients of variation in two-parameter exponential distributions. Commun. Stat.-Simul. Comput. **46**(8), 6618–6630 (2017)
26. Zou, G.Y., Huang, W., Zhang, X.: A note on confidence interval estimation for a linear function of binomial proportions. Comput. Stat. Data Anal. **53**(4), 1080–1085 (2009)
27. Ihaka, R., Gentleman, R.: R: a language for data analysis and graphics. J. Comput. Graph. Stat. **5**(3), 299–314 (1996)
28. Bhat, K., Rao, K.A.: On tests for a normal mean with known coefficient of variation. Int. Stat. Rev. **75**(2), 170–182 (2007)
29. Panichkitkosolkul, W.: A unit root test based on the modified least squares estimator. Sains Malaysiana **43**(10), 1623–1633 (2014)
30. Niwitpong, S., Kirdwichai, P.: Adjusted Bonett Confidence interval for standard deviation of non-normal distribution. Thai. Stat. **6**(1), 1–6 (2008)
31. McDonald, C.J., Blevins, L., Tierney, W.M., Martin, D.K.: The Regenstrief medical records. MD Comput. **5**(5), 34–47 (1988)
32. Zhou, X.H., Gao, S., Hui, S.L.: Methods for comparing the means of two independent log-normal samples. Biometrics **53**(3), 1129–1135 (1997)
33. Jafari, A.A., Abdollahnezhad, K.: Inferences on the means of two log-normal distributions: a computational approach test. Commun. Stat.-Simul. Comput. **44**(7), 1659–1672 (2015)

# Generalized Confidence Interval of the Ratio of Coefficients of Variation of Birnbaum-Saunders Distribution

Wisunee Puggard, Sa-Aat Niwitpong(✉), and Suparat Niwitpong

Faculty of Applied Science, Department of Applied Statistics, King Mongkut's University of Technology North Bangkok, Bangkok, Thailand
{sa-aat.n,suparat.n}@sci.kmutnb.ac.th

**Abstract.** The coefficient of variation (CV) is wildly used as the index of reliability of measurement. Hence, the problem of comparing two coefficients of variation (CVs) is interesting. Moreover, Birnbaum-Saunders (BS) distribution is frequently used for analysing the lifetime of materials and equipment. In this paper, the generalized confidence interval (GCI) was proposed to make inference about the ratio of two CVs of the BS distribution. A Monte Carlo simulation study was carried out to compare the performance of GCI with the biased-corrected percentile bootstrap (BCPB) confidence interval and the biased-corrected and accelerated (BCa) confidence interval based on the coverage probability and the average length. The simulation results indicate that GCI is recommended. The fatigue life of 6061-T6 aluminum coupons are applied to illustrate the proposed method.

**Keywords:** Birnbaum-Saunders distribution · Coefficients of variation · Generalized confidence interval · Bootstrap

## 1 Introduction

The Birnbaum-Saunders (BS) distribution, originally introduced by Birnbaum and Saunders [1], is a two parameters family of life time distributions which is derived from a fatigue model caused under cyclic loading. It was assumed that the initiation, growth and ultimate extension of a dominate crack are the causes of fatigue failure. Supposed that a random variable $X$ follows BS distribution with parameter $\alpha$ and $\beta$, denoted by $X \sim BS(\alpha, \beta)$. The cumulative distribution function (cdf) is given by

$$F(x) = \Phi[\frac{1}{\alpha}(\sqrt{\frac{x}{\beta}} - \sqrt{\frac{\beta}{x}})], x > 0, \alpha > 0, \beta > 0, \quad (1)$$

where $\Phi(\cdot)$ is the standard normal cdf, $\alpha$ and $\beta$ are the shape and scale parameters, respectively. The BS distribution has been widely applied in several fields, such as industry, environment, engineering, finance and medicine. For example,


V.-N. Huynh et al. (Eds.): IUKM 2020, LNAI 12482, pp. 396–406, 2020.
https://doi.org/10.1007/978-3-030-62509-2_33

Birnbaum and Saunders [2] fitted this model to some data sets on the fatigue life of 6061-T6 aluminum coupons. Durham and Padgett [4] applied the BS distribution to the data of fatigue of carbon fibers under increasing stress concentrations.

The coefficient of variation (CV) was introduced by Pearson [13]. It is defined as the ratio of the standard deviation to the mean and it is used as a measure of dispersion of data so that a higher value of the CV implies a higher level of dispersion around the mean. The CV has been used rather than the standard deviation because the value of the CV is free of the unit of measurement scale. For example, when comparing variability of two or more populations with different measurement scales or very different mean values, the coefficient of variation can be a useful alternative or complement to the standard deviation. It is widely applied in many fields, such as medicine, economics, science and engineering. In previous studies, confidence intervals are widely considered to make inference about the unknown population CV. A confidence interval of nominal level $100(1-\gamma)\%$ indicate that the population parameter will be within this interval with $100(1-\gamma)\%$ of the time. Tian [16] constructed the confidence interval for the common CV of normal distribution based on generalized confidence interval (GCI). Mahmoudvand and Hassani [12] proposed an approximately unbiased estimator for the population CV and used this estimator and its variance to approximate confidence intervals for for the CV of normal distribution. Sangnawakij and Niwitpong [14] used the method of variance of estimates recovery (MOVER), GCI, and the asymptotic confidence interval (ACI) to construct confidence intervals for the single CV and the difference of coefficients of variation (CVs) in the two parameter exponential distributions. Moreover, the problem of comparing the CVs of two independence populations arises in many practical situation. Confidence intervals of the ratio of the CVs are also considered to make inference about the CVs of two independence populations. For instance, Verrill and Johnson [17] investigated confidence interval for the ratio of CVs in a normal distribution by asymptotic procedure and simulation procedure. Buntao and Niwitpong [3] constructed confidence intervals for the ratio of CVs of a delta-lognormal distribution based on the concept of MOVER based on Wald interval and the generalized pivotal approach (GPA). Hasan and Krishnamoorthy [10] used MOVER approach and fiducial approach to construct confidence intervals for the ratio of CVs of two lognormal distributions.

In the past years, several studies have examined confidence intervals for the functions of parameters, that is, mean, $\alpha$, $\beta$, quantile and reliability function of the Birnbaum-Saunders distribution. For example, Wang [18] proposed GCI for $\alpha$, mean, quantiles and reliability function of the BS distribution. Subsequently, Gua et al. [9] considered hypothesis testing and confidence intervals for the mean of several BS populations based on hybrids between the generalized inference method and the large-sample theory. Recently, Li and Xu [11] examined fiducial inference for the parameters of BS distribution based on the inverse method and Hannig's method. Unfortunately, statistical methods for comparing two CVs based on Birnbaum-Saunders distributions have not been considered and are not available in the previous studies. Therefore, this article proposes confidence intervals to make inference about the ratio of coefficients of variation of Birnbaum-Saunders distribution using the concept of generalized confidence intervals and compares with the biased-corrected percentile bootstrap (BCPB) confidence interval and the biased-corrected and accelerated (BCa) confidence interval.

The paper is organized as follows. Section 2, the details of BS distribution and methods to construct confidence interval for the ratio of CVs are presented. Simulation studies are carried out to examine the performance of the proposed method in Sect. 3. In Sect. 4, the two real data sets are analyzed for illustration. Finally, the conclusion is presented in Sect. 5.

## 2 Confidence Intervals for the Ratio of Coefficients of Variation of Birnbaum-Saunders Distribution

Suppose $X_{ij} = (X_{i1}, X_{i2}, ..., X_{in_i})$, $i = 1, 2$, $j = 1, 2, ..., n_i$ be a vector of random sample from BS distribution with parameter $\alpha_i$ and $\beta_i$ then it can be denoted as $X_{ij} \sim BS(\alpha_i, \beta_i)$. The probability density function (pdf) of BS distribution is given by

$$f(x_{ij}, \alpha_i, \beta_i) = \frac{1}{2\alpha_i\beta_i\sqrt{2\pi}}\left\{\left(\frac{\beta_i}{x_{ij}}\right)^{\frac{1}{2}} + \left(\frac{\beta_i}{x_{ij}}\right)^{\frac{3}{2}}\right\}exp\left[-\frac{1}{2\alpha_i^2}\left(\frac{x_{ij}}{\beta_i} + \frac{\beta_i}{x_{ij}} - 2\right)\right] \quad (2)$$

where $x_{ij} > 0, \alpha_i > 0, \beta_i > 0$. The cdf of BS distribution is given by Eq. (1). If follows BS distribution with parameter $(\alpha_i, \beta_i)$, Then

$$Y_{ij} = \frac{1}{2}\left(\sqrt{\frac{X_{ij}}{\beta_i}} - \sqrt{\frac{\beta_i}{X_{ij}}}\right) \sim N(0, \frac{\alpha_i^2}{4}) \quad (3)$$

where $N(\mu, \sigma^2)$ refers to the normal distribution with mean $\mu$ and varince $\sigma^2$. Hence

$$X_{ij} = \beta_i\left(1 + 2Y_{ij}^2 + 2Y_{ij}\sqrt{1 + Y_{ij}^2}\right) \quad (4)$$

Therefore, the normal random variable based on this relationship can be used to generate BS random variable $X_{ij}$. The BS distribution has several interesting properties. For example, $X_{ij} \sim BS(\alpha_i, \beta_i)$, then

(a) $\alpha_i^{-1}(\sqrt{x_{ij}/\beta_i} - \sqrt{\beta_i/x_{ij}}) \sim N(0, 1)$
(b) $CX_{ij} \sim BS(\alpha_i, C\beta_i)$, where $C$ is a constant,
(c) $X_{ij}^{-1} \sim BS(\alpha_i, \beta_i^{-1})$.

From (a), the expected value and variance of $X_{ij}$ are given by $E(X_{ij}) = \beta_i(1 + \frac{1}{2}\alpha_i^2)$, $Var(X_{ij}) = (\alpha_i\beta_i)^2(1 + \frac{5}{4}\alpha_i^2)$. Hence, the CV, denoted by $\tau_i$, is

$$\tau_i = \frac{\sqrt{Var(X_{ij})}}{E(X_{ij})} = \frac{\alpha_i\sqrt{1 + \frac{5}{4}\alpha_i^2}}{1 + \frac{1}{2}\alpha_i^2}. \quad (5)$$

Since $X_{ij}$ are independent, then the ratio of CVs, denoted by $\eta$ can be written as

$$\eta = \frac{\tau_1}{\tau_2} = \frac{\alpha_1\sqrt{1+\frac{5}{4}\alpha_1^2/(1+\frac{1}{2}\alpha_1^2)}}{\alpha_2\sqrt{1+\frac{5}{4}\alpha_2^2/(1+\frac{1}{2}\alpha_2^2)}}. \tag{6}$$

### 2.1 Generalized Confidence Interval

Generalized confidence interval (GCI) was proposed by Weerahandi [19] based on the concept of generalized pivotal quantity (GPQ) which is a generalization of the usual pivot. The GPQ is allowed to be a function of nuisance parameter, whereas usual pivotal quantities can only be a function of the sample and the parameter of interest. Let $R = R_\theta(X; x, \theta, \delta)$ be a function of $X$, $x$, $\theta$ and $\delta$ where $X = (X_1, X_2, ..., X_n)$ is a random sample from a distribution $F_X(x; \theta, \delta)$, $x = (x_1, x_2, ..., x_n)$ is the observed value of $X$, $\theta$ is an unknown parameter of interest and $\delta$ is a vector of nuisance parameters. Then $R$ is a GPQ if it satisfies the following two conditions:

**GPQ1**: The distribution of $R_\theta(X; x, \theta, \delta)$ is free of unknown parameter
**GPQ2**: The observed pivotal $r_{obs} = R(X; x, \theta, \delta)$ does not depend on $\delta$ .

Therefore, the $100(1-\gamma)\%$ confidence interval for $\theta$ is given by

$$CI_\theta = [L_\theta, U_\theta] = [R(\gamma/2), R(1-\gamma/2)], \tag{7}$$

where $R(\nu)$ denotes the $100\nu\%$ percentile of $R_\theta(X; x, \theta, \delta)$.

Considering the GPQ for the shape parameter $\alpha_i$ and the scale parameter $\beta_i$ of BS distribution. A brief description is given as follows. Suppose $X_{ij} = (X_{i1}, X_{i2}, ..., X_{in_i})$, $i = 1, 2$, $j = 1, 2, ..., n_i$ be a random sample from BS distribution (1) with sample size $n_i$, it is easy to see that

$$Y_{ij} = \left(\sqrt{\frac{X_{ij}}{\beta_i}} - \sqrt{\frac{\beta_i}{X_{ij}}}\right) \sim N(0, \alpha_i^2). \tag{8}$$

Then $\bar{Y}_i = n_i^{-1}\sum_{j=1}^{n_i} Y_{ij}$, $S_i^2 = (n_i - 1)^{-1}\sum_{j=1}^{n_i}(Y_{ij} - \bar{Y}_i)^2$. It is know that $\bar{Y}_i$ and $S_i^2$ are independently distribution with $\bar{Y}_i \sim N(0, \alpha_i^2/n_i)$, $(n_i-1)S_i^2/\alpha_i^2 \sim \chi^2(n_i - 1)$, where $\chi^2(n_i-1)$ denotes the Chi-squared distribution with $n_i - 1$ degrees of freedom and $R_i(X_{ij}; \beta_i) = \sqrt{n_i}(\bar{Y}_i/S_i)$. Then

$$R_i(X_{ij}; \beta_i) = \sqrt{n_i(n_i-1)}\frac{I_i - \beta_i J_i}{\sqrt{K_i - 2n_i(1 - I_i J_i)\beta_i + L_i\beta_i^2}} \sim t(n_i - 1) \tag{9}$$

where $I_i = n_i^{-1}\sum_{j=1}^{n_i}\sqrt{X_{ij}}$, $J_i = n_i^{-1}\sum_{j=1}^{n_i} 1/\sqrt{X_{ij}}$, $K_i = \sum_{j=1}^{n_i}(\sqrt{X_{ij}} - I_i)^2$ and $L_i = \sum_{j=1}^{n_i}(1/\sqrt{X_{ij}} - J_i)^2$ and $t(n_i - 1)$ refers to the $t$ distribution with $n_i - 1$ degrees of freedom. $R_i(X_{ij}; \beta_i)$ is free of the unknown parameter $\beta_i$, so that $R_i(X_{ij}; \beta_i)$ satisfies GPQ1. Moreover, $R_i(X_{ij}; \beta_i)$ is only a function of the parameter $\beta_i$, which does not relate to the other parameter $\alpha_i$. Therefore, $R_i(X_{ij};\beta_i)$ can be used to construct a GPQ for $\beta_i$.

Sun [15] showed that $R_i(X_{ij};\beta_i)$ is a strictly decreasing function of $\beta_i$, then the solution of the equation $R_i(X_{ij};\beta_i)=T_i$ is given by

$$R_{\beta_i} := R_{\beta_i}(x_{ij};T_i) = \begin{cases} max(\beta_{i1},\beta_{i2}), & if\ T_i \leq 0 \\ min(\beta_{i1},\beta_{i2}), & if\ T_i > 0 \end{cases} \tag{10}$$

where $T_i \sim t(n_i-1)$ and $\beta_{i1},\beta_{i2}$ are the solution of the following quadratic equation:

$$\left[(n_i-1)J_i^2-\tfrac{1}{n_i}L_iT_i^2\right]\beta_i^2-2\left[(n_i-1)I_iJ_i-(1-I_iJ_i)T_i^2\right]\beta_i+(n_i-1)I_i^2-\tfrac{1}{n_i}K_iT_i^2=0$$

Therefore, $R_{\beta_i}$ is a GPQ for $\beta_i$.

To construct a GPG for $\alpha_i$, Wang [18] proposed a GPQ for $\alpha_i$ as follows. Let

$$U_i(X_{ij};\beta_i) = \frac{n_i\bar{Y}_i^2/\alpha_i^2}{n_i\bar{Y}_i^2/\alpha_i^2+(n_i-1)S_i^2/\alpha_i^2} = \frac{n_i\bar{Y}_i^2}{\sum_{j=1}^{n_i}Y_{ij}^2} \sim Beta(\frac{1}{2},\frac{n_i-1}{2}) \tag{11}$$

$$V_i(X_{ij};\beta_i,\alpha_i) = \frac{n_i\bar{Y}_i^2}{\alpha_i^2}+\frac{(n_i-1)S_i^2}{\alpha_i^2} = \frac{\sum_{j=1}^{n_i}Y_{ij}^2}{\alpha_i^2} \sim \chi^2(n_i) \tag{12}$$

where $Beta(1/2,(n_i-1)/2)$ denotes the Beta distribution with parameter 1/2 and $(n_i-1)/2)$ and $\chi^2(n_i)$ denotes the Chi-squared distribution with $n_i$ degrees of freedom. It is widely known that $\bar{Y}_i$ and $S_i^2$ are mutually independent, then $U_i(X_{ij};\beta_i)$ and $V_i(X_{ij};\beta_i,\alpha_i)$ are also independent.

Since $U_i(X_{ij};\beta_i) = R_i^2(X_{ij};\beta_i)/(R_i^2(X_{ij};\beta_i)+n_i-1)$, then it can be concluded that $R_i(X_{ij};\beta_i)$ and $V_i(X_{ij};\beta_i,\alpha_i)$ are independent. Hence the unique solution for $V_i(X_{ij};\beta_i,\alpha_i)=v_i$ is obtained by

$$\left[\frac{S_{i2}\beta_i^2-2n_i\beta_i+S_{i1}}{\beta_i v_i}\right]^{1/2} \tag{13}$$

where $S_{i1}=\sum_{j=1}^{n_i}X_{ij}$, $S_{i2}=\sum_{j=1}^{n_i}1/X_{ij}$, $v_i\sim\chi^2(n_i)$ and $\beta_i$ is substituted by $R_{\beta_i}$ in Eq. (10) based on substitution method propose by Weerahandi [19]. Then a GPQ for $\alpha_i$ is defined by

$$R_{\alpha_i} := R_{\alpha_i}(x_{ij};v_i,T_i) = \left[\frac{S_{i2}R_{\beta_i}(x_{ij};T_i)^2-2n_iR_{\beta_i}(x_{ij};T_i)+S_{i1}}{R_{\beta_i}(x_{ij};T_i)v_i}\right] \tag{14}$$

By Eq. (6), the pivotal quantity for $\eta$ can be expressed by

$$R_\eta = \frac{R_{\alpha_1}(x_{1j};v_1,T_1)\sqrt{1+\frac{5}{4}R_{\alpha_1}(x_{1j};v_1,T_1)}/(1+\frac{1}{2}R_{\alpha_1}(x_{1j};v_1,T_1)}{R_{\alpha_2}(x_{2j};v_2,T_2)\sqrt{1+\frac{5}{4}R_{\alpha_2}(x_{2j};v_2,T_2)}/(1+\frac{1}{2}R_{\alpha_2}(x_{2j};v_2,T_2)} \tag{15}$$

Therefore, The $100(1-\gamma)\%$ confidence interval for $\eta$ based on GCI is

$$CI_{GCI} = [L_\eta, U_\eta] = [R_\eta(\gamma/2), R_\eta(1-\gamma/2)], \tag{16}$$

where $R(\nu)$ denotes the $100\nu\%$ percentile of $R_\eta$.

**Algorithm for GCI method**

*Step* 1. Generate $x_{ij} \sim BS(\alpha_i, \beta_i), i = 1, 2, j = 1, 2, ..., n_i$.

*Step* 2. Compute the corresponding value $I_i, J_i, K_i, L_i, S_{i1}$ and $S_{i2}$.

*Step* 3. From $m = 1$, generate $T_i$ from $t$ distribution with $n_i - 1$ degrees of freedom and $\upsilon_i$ from Chi-square distribution with $n_i$ degrees of freedom, independently (If $R_{\beta_i}(x_{ij}; T_i) < 0$, regenerate $T_i \sim t(n_i - 1)$). Using this value, compute the value of $R_{\beta_i}$ and $R_{\alpha_i}$ in Eqs. (10) and (14), respectively. Then, $R_\eta$ can be obtained by replaced $R_{\alpha_i}$ in Eq. (15).

*Step* 4. Repeat step 3 until $m = 5000$.

*Step* 5. Rank $R_{\eta,1}, R_{\eta,2}, ..., R_{\eta,m}$ from the smallest to the largest, then compute $R_\eta(\gamma/2)$ and $R_\eta(1 - \gamma/2)$.

## 2.2 Bootstrap Confidence Interval

The bootstrap method was originally proposed by Efron [5]. It is a re-sampling technique for assigning measures of accuracy to statistical estimate based on a random selection of resamples from the original sample with replacement. There are different type of bootstrap methods for constructing confidence interval, which are, the standard bootstrap method (SB), the percentile bootstrap method (PB), the bias-corrected percentile bootstrap method (BCPB) and the bias-corrected and accelerated bootstrap method (BCa) [6–8]. For independent and identically distributed random variables, the bootstrap procedure can be explained as follows.

Let $x_{ij} = (x_{i1}, x_{i2}, ..., x_{in_i})$, $i = 1, 2$, $j = 1, 2, ..., n_i$ be an original random sample of size $n_i$ drawn from the BS distribution with parameter $\alpha_i$ and $\beta_i$ and $\hat{\eta}$ represents the estimator of $\eta$. Then $\hat{\eta}$ is given by

$$\hat{\eta} = \frac{\hat{\sigma_1}/\hat{\mu_1}}{\hat{\sigma_2}/\hat{\mu_2}} \tag{17}$$

where $\hat{\sigma_i} = \sqrt{\sum_{j=1}^{n_i}(X_{ij} - \bar{X}_i)^2/(n_i - 1)}$ and $\hat{\mu_i} = \sum_{j=1}^{n_i} X_{ij}/n_i$. A bootstrap sample size $n_i$ drawn with replacement from the original sample is denoted by $x_{ij}^* = \left(x_{1j}^*, x_{1j}^*, ..., x_{1n_i}^*\right)$, $i = 1, 2$ and the estimator of $\eta$ for the bootstrap sample is denoted by $\hat{\eta}^*$. Suppose that B bootstrap samples are available, then B bootstrap $\eta$'s can be obtained and can be arranged in ascending order, denoted by $\hat{\eta}^*(1), \hat{\eta}^*(2), ..., \hat{\eta}^*(B)$, which constitutes and empirical bootstrap distribution of $\eta$. In this study, the biased-corrected percentile bootstrap (BCPB) and the biased-corrected and accelerated bootstrap (BCa) are considered to construct confidence interval for $\eta$.

**The Biased-Corrected Percentile Bootstrap Confidence Interval.** The bootstrap distribution for $\hat{\eta}^*$ maybe a biased distribution. Therefore, The BCPB approach has been introduced in order to correct this potential bias, which requires the following steps. First, the probability, denoted by $P_0$, is calculated using the ordered distribution of $\hat{\eta}^*$ as

$$P_0 = Pr(\hat{\eta}^* \leq \hat{\eta}) \tag{18}$$

where $\hat{\eta}$ is the estimated value of $\eta$ from the original sample. Second, compute

$$z_0 = \Phi^{-1}(p_0) \tag{19}$$

$$P_L = \Phi(2z_0 + Z_{\gamma/2}) \tag{20}$$

$$P_U = \Phi(2z_0 + Z_{1-\gamma/2}) \tag{21}$$

where $\Phi(\cdot)$ is the standard normal cumulative distribution function.Finally, the $100(1-\gamma)\%$ BCPB confidence interval is defined by

$$CI_{BCPB} = [L_\eta, U_\eta] = [\hat{\eta}^*_{(P_L \times B)}, \hat{\eta}^*_{(P_U \times B)}] \tag{22}$$

**The Biased-Corrected and Accelerated Confidence Interval.** The BCa confidence interval was proposed by Efron and Tibshirani [8] to improve the performance of BCPB approach based on two number $z_0$ and $a$, where $z_0$ is the biased-correction and $a$ is the acceleration. The biased-correction can be calculate from Eq. (19) and the Jackknife value is used to estimate the acceleration $a$. Then, the acceleration $a$ is calculated as follows,

$$a = \frac{\sum_{i=1}^{n}(\hat{\eta}_{(\cdot)} - \hat{\eta}_{(i)})^3}{6[\sum_{i=1}^{n}(\hat{\eta}_{(\cdot)} - \hat{\eta}_{(i)})^2]^{3/2}} \tag{23}$$

where $\hat{\eta}_{(i)}$ is the bootstrap estimate of $\eta$ calculated from the original sample with $i$-th point deleted, for $i = 1, 2, ..., n$ and $\hat{\eta}_{(\cdot)} = \sum_{i=1}^{n} \hat{\eta}_{(i)}/n$. With the value of $z_0$ and $a$, the value $P_{AL}$ and $P_{AU}$ are computed,

$$P_{AL} = \Phi(z_0 + \frac{z_0 + Z_{\gamma/2}}{1 - a(z_0 + Z_{\gamma/2})}) \tag{24}$$

$$P_{AU} = \Phi(z_0 + \frac{z_0 + Z_{1-\gamma/2}}{1 - a(z_0 + Z_{1-\gamma/2})}) \tag{25}$$

Then, the $100(1-\gamma)\%$ BCa confidence interval is obtained by

$$CI_{BCa} = [L_\eta, U_\eta] = [\hat{\eta}^*_{(P_{AL} \times B)}, \hat{\eta}^*_{(P_{AU} \times B)}] \tag{26}$$

**Algorithm for BCPB and BCa methods**

*Step* 1. Generate $x_{ij} \sim BS(\alpha_i, \beta_i), i = 1, 2, j = 1, 2, ..., n_i$.

*Step* 2. Draw a bootstrap sample $x^*_{1j}, x^*_{1j}, ..., x^*_{1n_i}$ from step 1.

*Step* 3. Compute $\hat{\eta}^*$.

*Step* 4. Repeat step 2 and 3, B times, Yielding estimators $\hat{\eta}^*(1), \hat{\eta}^*(2), ...,$ $\hat{\eta}^*(B)$.

*Step* 5. Compute $z_0$ and $a$ from Eqs. (19) and (23).

*Step* 6. Compute $P_L$, $P_U$, $P_{AL}$ and $P_{AU}$ from Eqs. (20), (21), (24) and (25).

*Step* 7. Rank $\hat{\eta}^*(1), \hat{\eta}^*(2), ..., \hat{\eta}^*(B)$ from the smallest to the largest, then compute $\hat{\eta}^*_{(P_L \times B)}$ and $\hat{\eta}^*_{(P_U \times B)}$ for BCPB and $\hat{\eta}^*_{(P_{AL} \times B)}$ and $\hat{\eta}^*_{(P_{AU} \times B)}$ for BCa.

## 3 Simulation Study

A Monte Carlo simulation study was designed to compare the performance of GCI BCPB and BCa. The simulation study was carried out for different combinations of shape parameter and sample size. For equal sample size ($n_1 = n_2$), the sample sizes ($n_1, n_2$) were set at (10, 10), (20, 20), (30, 30), (50, 50) and (100, 100). For unequal sample size ($n_1 \neq n_2$), the sample sizes ($n_1, n_2$) were set at (10, 20), (30, 20), (30, 50) and (100, 50). The value of shape parameter ($\alpha_1, \alpha_2$) are (0.25, 0.25), (0.25, 0.50), (0.25, 1.00), (0.25, 2.00) and (0.25, 3.00), respectively. Since $\beta$ is the scale parameter, with out loss of generality, $\beta_1 = \beta_2 = 1$ are considered in this simulation study. Simulation results were based on 5000 replications for each different case and 5000 pivotal quantities for GCI and 1000

**Table 1.** The coverage probability and average length of nominal 95% two-sided confidence intervals for the ratio of coefficients of variation of Birnbaum-Saunders distributions: equal sample sizes ($n_1 = n_2$).

| $n_1$ | $n_2$ | $\beta_1 = \beta_2$ | $\alpha_1 : \alpha_2$ | Coverage probability | | | Average length | | |
|---|---|---|---|---|---|---|---|---|---|
| | | | | GCI | BCPB | BCa | GCI | BCPB | BCa |
| 10 | 10 | 1 | 0.25:0.25 | 0.950 | 0.934 | 0.927 | 1.6318 | 1.6373 | 1.6377 |
| | | | 0.25:0.50 | 0.951 | 0.931 | 0.930 | 0.8097 | 0.8282 | 0.8246 |
| | | | 0.25:1.00 | 0.949 | 0.920 | 0.928 | 0.3945 | 0.4434 | 0.4349 |
| | | | 0.25:2.00 | 0.944 | 0.920 | 0.930 | 0.2078 | 0.2825 | 0.2731 |
| | | | 0.25:3.00 | 0.950 | 0.927 | 0.937 | 0.1621 | 0.2425 | 0.2330 |
| 20 | 20 | 1 | 0.25:0.25 | 0.951 | 0.925 | 0.926 | 1.0006 | 0.9667 | 0.9686 |
| | | | 0.25:0.50 | 0.951 | 0.921 | 0.921 | 0.4960 | 0.4957 | 0.4985 |
| | | | 0.25:1.00 | 0.950 | 0.904 | 0.910 | 0.2389 | 0.2692 | 0.2698 |
| | | | 0.25:2.00 | 0.949 | 0.903 | 0.915 | 0.1242 | 0.1736 | 0.1717 |
| | | | 0.25:3.00 | 0.944 | 0.912 | 0.926 | 0.0981 | 0.1517 | 0.1492 |
| 30 | 30 | 1 | 0.25:0.25 | 0.953 | 0.930 | 0.931 | 0.7841 | 0.7622 | 0.7635 |
| | | | 0.25:0.50 | 0.952 | 0.919 | 0.929 | 0.3884 | 0.3920 | 0.3950 |
| | | | 0.25:1.00 | 0.951 | 0.906 | 0.912 | 0.1867 | 0.2133 | 0.2155 |
| | | | 0.25:2.00 | 0.950 | 0.905 | 0.918 | 0.0966 | 0.1384 | 0.1380 |
| | | | 0.25:3.00 | 0.949 | 0.911 | 0.923 | 0.0763 | 0.1219 | 0.1209 |
| 50 | 50 | 1 | 0.25:0.25 | 0.950 | 0.931 | 0.937 | 0.5869 | 0.5862 | 0.5874 |
| | | | 0.25:0.50 | 0.948 | 0.924 | 0.932 | 0.2905 | 0.3034 | 0.3065 |
| | | | 0.25:1.00 | 0.946 | 0.903 | 0.915 | 0.1393 | 0.1658 | 0.1685 |
| | | | 0.25:2.00 | 0.945 | 0.896 | 0.915 | 0.0718 | 0.1073 | 0.1079 |
| | | | 0.25:3.00 | 0.946 | 0.901 | 0.916 | 0.0570 | 0.0946 | 0.0946 |
| 100 | 100 | 1 | 0.25:0.25 | 0.946 | 0.936 | 0.941 | 0.4049 | 0.4151 | 0.4153 |
| | | | 0.25:0.50 | 0.948 | 0.933 | 0.940 | 0.2010 | 0.2169 | 0.2187 |
| | | | 0.25:1.00 | 0.946 | 0.915 | 0.920 | 0.0963 | 0.1196 | 0.1216 |
| | | | 0.25:2.00 | 0.947 | 0.914 | 0.920 | 0.0495 | 0.0775 | 0.0784 |
| | | | 0.25:3.00 | 0.948 | 0.918 | 0.925 | 0.0393 | 0.0684 | 0.0688 |

bootstrap sample for BCPB and BCa. The coverage probability and the average length were considered as the criteria for evaluating the performance of the methods. The confidence interval is the best performing method when the coverage probability is greater than or close to the nominal confidence level 0.95 and the average length is the shortest. The coverage probability and the average length of 95% two-sided confidence interval for the ratio of CVs for equal and unequal sample sizes were presented in Tables 1 and 2, respectively. From Table 1, The results showed that the coverage probabilities of GCI are above or close to the nominal confidence level 0.95 for all cases, whereas the coverage probabilities of BCPB and BCa are less than the nominal confidence level 0.95 for all cases. Hence the BCPB and BCa are not recommended to construct confidence interval for the ratio of the CVs of BS distribution. The average lengths showed that GCI are shorter than other confidence intervals for all cases. Moreover, when the sample size $(n_1, n_2)$ are increases, the average lengths of three methods tend to decrease. These results are similar to Table 1 where we set the unequal sample size (results are shown in Table 2).

**Table 2.** The coverage probability and average length of nominal 95% two-sided confidence intervals for the ratio of coefficients of variation of Birnbaum-Saunders distributions: unequal sample sizes ($n_1 \neq n_2$).

| $n_1$ | $n_2$ | $\beta_1 = \beta_2$ | $\alpha_1 : \alpha_2$ | Coverage probability | | | Average length | | |
|---|---|---|---|---|---|---|---|---|---|
| | | | | GCI | BCPB | BCa | GCI | BCPB | BCa |
| 10 | 20 | 1 | 0.25:0.25 | 0.946 | 0.901 | 0.890 | 1.3718 | 1.1381 | 1.0987 |
| | | | 0.25:0.50 | 0.947 | 0.906 | 0.895 | 0.6801 | 0.5834 | 0.5642 |
| | | | 0.25:1.00 | 0.944 | 0.906 | 0.903 | 0.3357 | 0.3165 | 0.3068 |
| | | | 0.25:2.00 | 0.952 | 0.909 | 0.913 | 0.1868 | 0.2027 | 0.1962 |
| | | | 0.25:3.00 | 0.948 | 0.912 | 0.916 | 0.1528 | 0.1755 | 0.1697 |
| 30 | 20 | 1 | 0.25:0.25 | 0.948 | 0.923 | 0.923 | 0.8874 | 0.8844 | 0.9175 |
| | | | 0.25:0.50 | 0.950 | 0.912 | 0.913 | 0.4399 | 0.4532 | 0.4713 |
| | | | 0.25:1.00 | 0.951 | 0.892 | 0.891 | 0.2090 | 0.2459 | 0.2525 |
| | | | 0.25:2.00 | 0.948 | 0.896 | 0.907 | 0.1033 | 0.1592 | 0.1593 |
| | | | 0.25:3.00 | 0.950 | 0.907 | 0.921 | 0.0786 | 0.1398 | 0.1387 |
| 30 | 50 | 1 | 0.25:0.25 | 0.952 | 0.931 | 0.929 | 0.6932 | 0.6692 | 0.6605 |
| | | | 0.25:0.50 | 0.951 | 0.926 | 0.930 | 0.3433 | 0.3451 | 0.3418 |
| | | | 0.25:1.00 | 0.951 | 0.916 | 0.928 | 0.1675 | 0.1876 | 0.1865 |
| | | | 0.25:2.00 | 0.949 | 0.913 | 0.926 | 0.0914 | 0.1200 | 0.1190 |
| | | | 0.25:3.00 | 0.949 | 0.914 | 0.928 | 0.0744 | 0.1048 | 0.1038 |
| 100 | 50 | 1 | 0.25:0.25 | 0.949 | 0.927 | 0.928 | 0.5006 | 0.5068 | 0.5216 |
| | | | 0.25:0.50 | 0.950 | 0.908 | 0.906 | 0.2481 | 0.2636 | 0.2741 |
| | | | 0.25:1.00 | 0.951 | 0.885 | 0.881 | 0.1163 | 0.1451 | 0.1511 |
| | | | 0.25:2.00 | 0.951 | 0.891 | 0.892 | 0.0548 | 0.0952 | 0.0967 |
| | | | 0.25:3.00 | 0.951 | 0.903 | 0.909 | 0.0410 | 0.0849 | 0.0855 |

# 4 Application

In this section, the real fatigue dataset was applied to illustrate the proposed method. The two data set of the fatigue life of 6061-T6 aluminum coupons cut parallel to the direction of rolling and oscillated at 18 cycles per second were given by Birnbaum and Saunders [2]. Sample (i) and (ii) consist of 101 observations with maximum stress per cycle 31k and 21k, respectively. The sample mean, sample variance and CV of fatigue life are 133.7327, 499.7778, 0.1672 from sample (i); and 1400.911, 153134.5, 0.2793 from sample (ii). Then, the ratio of CVs is 0.5984. The 95% confidence interval based on GCI, BCPB and BCa for the ratio of CVs were presented in Table 3. As a result, the length of GCI is shorter than BCPB and BCa. These results are the same as the results of the simulation study in the previous section when $(n_1, n_2) = (100, 100)$. Furthermore, all confidence intervals in Table 3 shown that the lower and upper confidence levels are not including 1. This evidence can be concluded that the CV of example (i) and (ii) are different.

**Table 3.** Three nominal 95% two-sided confidence intervals and their lengths.

| | Interval | Length |
|---|---|---|
| GCI | [0.4469–0.6618] | 0.2149 |
| BCPB | [0.4800–0.7541] | 0.2741 |
| BCa | [0.4591–0.7194] | 0.2603 |

# 5 Conclusions

This paper proposed confidence interval for the ratio of CVs of BS distribution based on GCI. A Monte Carlo simulation study was used to compare the performance of GCI with BCPB and BCa in term of the coverage probability and the average length. Simulation results and real data set of fatigue life demonstrated that the coverage probabilities of GCI are above or close to the nominal confidence level 0.95 for all cases and its average lengths are the shortest. Therefore, GCI is recommended for constructing confidence interval for the ratio of CVs of BS distribution while BCPB and BCa are not recommended since the coverage probabilities of BCPB and BCa are less than the nominal confidence level 0.95 for all cases.

# References

1. Birnbaum, Z.W., Saunders, S.C.: A new family of life distributions. J. Appl. Probab. **6**(2), 319–327 (1996)
2. Birnbaum, Z.W., Saunders, S.C.: Estimation for a family of life distributions with applications to fatigue. J. Appl. Probab. **6**(2), 328–347 (1996b)

3. Buntao, N., Niwitpong, S.-A.: Confidence intervals for the ratio of coefficients of variation of delta-lognormal distribution. Appl. Math. Sci. **7**(77), 3811–3818 (2013)
4. Durham, S.D., Padgett, W.J.: A cumulative damage model for system failure with ap-plication to carbon fibers and composites. Technometrics **39**(1), 34–44 (1997)
5. Efron, B.: Bootstrap methods: another look at the jackknife. Ann. Stat. **7**(1), 1–26 (1979)
6. Efron, B., Gong, G.: A leisurely look at the bootstrap, the jackknife, and cross-validation. Am. Stat. **37**(1), 36–48 (1983)
7. Efron, B., Tibshirani, R.J.: Bootstrap methods for standard errors, confidence intervals, and other measures of statistical accuracy. Stat. Sci. **1**(1), 54–77 (1986)
8. Efron, B., Tibshirani, R.J.: An Introduction to the Bootstrap. Chapman and Hall CRC, London (1998)
9. Guo, X., Wu, H., Li, G., Li, Q.: Inference for the common mean of several Birnbaum-Saunders populations. J. Appl. Stat. **44**(5), 941–954 (2017)
10. Hasan, M.S., Krishnamoorthy, K.: Improved confidence intervals for the ratio of coefficients of variation of two lognormal distributions. J. Stat. Theory Appl. **16**(3), 345–353 (2017)
11. Li, Y., Xu, A.: Fiducial inference for Birnbaum-Saunders distribution. J. Stat. Comput. Simul. **86**(9), 1673–1685 (2016)
12. Mahmoudvand, R., Hassani, H.: Two new confidence intervals for the coefficient of variation in a normal distribution. J. Appl. Stat. **36**(4), 429–442 (2009)
13. Pearson, K.: Mathematical contributions to the theory of evolution. III. Regression, Heredity, and Panmixia. Philos. Trans. Royal Soc. London. Ser. A, Containing Pap. Math. Phys. Charact. **187**, 253–318 (1896)
14. Sangnawakij, P., Niwitpong, S.-A.: Confidence intervals for coefficients of variation in two-parameter exponential distributions. Commun. Stat. Simul. Comput. **46**(8), 6618–6630 (2017)
15. Sun, Z.L.: The confidence intervals for the scale parameter of the Birnbaum-Saunders fatigue life distribution. Acta Armamentarii **30**(11), 1558–1561 (2009). (in Chinese)
16. Tian, L.: Inferences on the common coefficient of variation. Stat. Med. **24**(14), 2213–2220 (2005)
17. Verrill, S., Johnson, R.A.: Confidence bounds and hypothesis tests for normal distribution coefficients of variation. Commun. Stat. Theory Methods **36**(12), 2187–2206 (2007)
18. Wang, B.X.: Generalized interval estimation for the Birnbaum-Saunders distribution. Comput. Stat. Data Anal. **56**(12), 4320–4326 (2012)
19. Weerahandi, S.: Generalized confidence intervals. J. Am. Stat. Assoc. **88**(423), 899–905 (1993)

# Confidence Interval for Coefficient of Variation of Inverse Gamma Distributions

Theerapong Kaewprasert(✉), Sa-Aat Niwitpong, and Suparat Niwitpong

Department of Applied Statistics, Faculty of Applied Science,
King Mongkut's University of Technology North Bangkok, Bangkok, Thailand
theerapong.kps@gmail.com, sa-aat.n@sci.kmutnb.ac.th,
suparat.n@sci.kmutnb.ac.th

**Abstract.** This paper proposes confidence intervals for a single coefficient of variation (CV) in the inverse gamma distribution, using the score method, the Wald method, and the percentile bootstrap (PB) confidence interval. Mote Carlo simulations were used to investigate the performance of these confidence intervals in terms of coverage probabilities (CP) and expected length (EL). The results showed that the Wald method for the single CVs are better than the other confidence intervals. Finally, these methods applied to determine the confidence interval of a real data from rainfall of Thailand.

**Keywords:** Inverse gamma distribution · Coefficient of variation · Score method · Wald method · Percentile bootstrap confidence interval

## 1 Introduction

The coefficient of variation is defined as the ratio of the standard deviation to the mean. It is a helpful quantity to describe the variation in evaluating results from different population [8]. It is also commonly used in many field, such as engineering, physics, medicine, geology and economics. For example, [11] assess the value of the coefficient of variation in assessing reproducibility of electrocardiogram (ECG) measurements. [3] discusses use of the coefficient of variation in appraising variability in population studies. [10] and [5] explain why the coefficient of variation is the best measure of the variability of population size over time if there are zeros in the data.

In probability and statistic, the inverse gamma distribution is a two-parameter family of continuous probability distributions on the positive real line, which is the distribution of the reciprocal of a variable distributed according to the gamma distribution [1]. The inverse gamma distribution is most often used as a conjugate prior distribution in Bayesian statistics. There are many papers to study the distribution of the inverse gamma. For example, [7] studied inverse gamma distribution as a prior distributions for variance parameters in hierarchical models. [1] studied some issues related with inverted gamma distribution which is the reciprocal of the gamma distribution. [9] introduced five different


V.-N. Huynh et al. (Eds.): IUKM 2020, LNAI 12482, pp. 407–418, 2020.
https://doi.org/10.1007/978-3-030-62509-2_34

algorithms based on method of moments, maximum likelihood and Baysian to estimate the parameters of inverted gamma distribution. [6] studied the inverse gamma as a survival distribution.

Several studies have investigating methods of constructing confidence intervals for single CV, such as [2] proposes the new confidence interval for the single coefficient of variation and the difference between coefficients of variation of inverse Gaussian distribution using the generalized confidence interval (GCI) and the bootstrap percentile confidence interval. [18] present an analysis of the small sample distribution of a class of approximate pivotal quantities for a normal coefficient of variation. [12] provided confidence intervals for coefficient of variation of lognormal distribution. [14] developed the confidence intervals for the coefficient of variation for the case of a gamma distribution. [15] examines confidence intervals for the single coefficient of variation and the difference of coefficient of variation in the two-parameter exponential distribution. [16] proposed the novel approaches to construct confidence intervals for the common coefficient of variation of several normal populations. Unfortunately, no research has studied the methods of constructing confidence intervals for the coefficient of variation in the inverse gamma distribution. Therefore, this paper proposes confidence intervals for single CV in the inverse gamma distribution. The first confidence interval is construction using the score method. Then, the confidence interval is proposed based on the Wald method. Finally, the confidence interval is constructed according to the percentile bootstrap confidence interval.

This paper is organized as follows. The methods and theories for constructing these confidence intervals are presented in Sect. 2. Our simulation results are presented in Sect. 3. Worked example is presented in Sect. 4. Section 5 summarized this paper.

## 2 Methods

Let $\mathbf{X} = (X_1, X_2, ..., X_n)$ be a random sample from the inverse gamma distributions with the shape parameter $\alpha$ and scale parameter $\beta$, denoted as $X \sim IG(\alpha, \beta)$. The probability density function (pdf) of $X$ is given by

$$f(x; \alpha, \beta) = \frac{\beta^{\alpha}}{\Gamma(\alpha)} (x)^{-\alpha-1} \exp\left(-\frac{\beta}{x}\right), x > 0, \alpha > 0, \beta > 0. \quad (1)$$

The population mean and variance of $X$ are defined as

$$E(X) = \frac{\beta}{\alpha - 1} \quad for \quad \alpha > 1 \quad (2)$$

and

$$Var(X) = \frac{\beta^2}{(\alpha-1)^2(\alpha-2)} \quad for \quad \alpha > 2 \quad (3)$$

Then, the coefficient of variation of $X$ can be expressed as

$$CV(X) = \tau = \frac{1}{\sqrt{\alpha - 2}} \quad (4)$$

Since $\alpha$ is the unknown parameter, it is required to be estimated.

The maximum likelihood estimators (MLE) for $\alpha$ and $\beta$. So, the log-likelihood function of $\alpha$ and $\beta$ is given by

$$lnL(\alpha,\beta) = -\Sigma\frac{\beta}{X_i} - (\alpha+1)\,\Sigma lnX_i - nln\Gamma(\alpha) + n\alpha ln\beta. \tag{5}$$

Taking partial derivatives of the above equation with respect to $\alpha$ and $\beta$, respectively. The score function is derived as

$$U(\alpha,\beta) = \begin{bmatrix} -\Sigma lnX_i - nln\alpha + \frac{n}{2\alpha} + nln\beta \\ -\Sigma X_i^{-1} + \frac{n\alpha}{\beta} \end{bmatrix}. \tag{6}$$

Then, the maximum likelihood estimators for $\alpha$ and $\beta$, respectively

$$\hat{\alpha} = \frac{1}{2\left[\frac{\Sigma lnX_i}{n} + ln\frac{\Sigma X_i^{-1}}{n}\right]} \tag{7}$$

and

$$\hat{\beta} = \frac{n\alpha}{\Sigma X_i^{-1}}. \tag{8}$$

Therefore, the sample coefficient of variation for $\tau$ is given by $\hat{\tau} = \frac{1}{\sqrt{\hat{\alpha}-2}}$

The method to construct the confidence intervals for $\tau$ are proposed in the following section.

### 2.1 Confidence Interval Based on the Score Method

Let $\alpha$ and $\beta$ be the parameter of interest and the nuisance parameter, respectively. In general the score statistic is denoted as

$$W_u = U^T\left(\alpha_0,\hat{\beta}_0\right) I^{-1}\left(\alpha_0,\hat{\beta}_0\right) U\left(\alpha_0,\hat{\beta}_0\right), \tag{9}$$

where $\hat{\beta}$ is the maximum likelihood estimator for $\beta$ under the null hypothesis $H_0 : \alpha = \alpha_0$, $U\left(\alpha_0,\hat{\beta}_0\right)$ is the vector of the score function and $I\left(\alpha_0,\hat{\beta}_0\right)$ is the matrix of Fisher information. The score function is

$$U\left(\alpha_0,\hat{\beta}_0\right) = \begin{bmatrix} -\Sigma lnX_i + \frac{n}{2\alpha_0} + nln\left(\frac{n}{\Sigma X_i^{-1}}\right) \\ 0 \end{bmatrix}. \tag{10}$$

The inverse of the Fisher information can be derived as

$$I^{-1}\left(\alpha_0,\hat{\beta}_0\right) = \begin{bmatrix} 2\alpha_0{}^2/n & -2\alpha_0{}^2/\Sigma X_i^{-1} \\ -2\alpha_0{}^2/\Sigma X_i^{-1} & \alpha_0 n\,(2\alpha_0-1)/\left(\Sigma X_i^{-1}\right)^2 \end{bmatrix}. \tag{11}$$

Using the property of the score function, we can see that the pivotal

$$Z_{score} = \sqrt{\frac{2\alpha_0{}^2}{n}}\left[-\Sigma lnX_i + \frac{n}{2\alpha_0} + nln\left(\frac{n}{\Sigma X_i^{-1}}\right)\right] \tag{12}$$

converges in distribution to the standard normal distribution.

Since the variance of $\hat{\alpha}$ is $2{\alpha_0}^2/n$, we approximate it by substituting $\hat{\alpha}$ in its variance, is given as

$$Z_{score} \cong \sqrt{\frac{2\hat{\alpha}^2}{n}}\left[-\Sigma lnX_i + \frac{n}{2\hat{\alpha}} + nln\left(\frac{n}{\Sigma X_i^{-1}}\right)\right]. \tag{13}$$

From the probability statement,

$$1-\gamma = P\left(-Z_{\gamma/2} \le Z_{score} \le Z_{\gamma/2}\right). \tag{14}$$

It can be simply written as

$$1-\gamma = P\left(l_s \le \tau \le u_s\right). \tag{15}$$

Therefore, the $(1-\gamma)\,100\%$ confidence interval for $\tau$ based on the score method is given by

$$CI_S = [l_s, u_s] = \left[\frac{1}{\sqrt{\frac{n}{2\left(z_1 - Z_{\gamma/2}\sqrt{\frac{2\hat{\alpha}^2}{n}}\right)} - 2}}, \frac{1}{\sqrt{\frac{n}{2\left(z_1 + Z_{\gamma/2}\sqrt{\frac{2\hat{\alpha}^2}{n}}\right)} - 2}}\right], \tag{16}$$

where $z_1 = \Sigma lnX_i - nln\left(\frac{n}{\Sigma X_i^{-1}}\right)$ and $Z_{\gamma/2}$ is the $(\gamma/2)\,100$ th percentile of the standard normal distribution.

## 2.2 Confidence Interval Based on the Wald Method

The general form of the Wald statistic under the null hypothesis $H_0 : \alpha = \alpha_0$ is defined as

$$W_e = (\hat{\alpha} - \alpha_0)^T \left[I^{\alpha\alpha}\left(\hat{\alpha}, \hat{\beta}\right)\right]^{-1} (\hat{\alpha} - \alpha_0) \tag{17}$$

where $I^{\alpha\alpha}\left(\hat{\alpha}, \hat{\beta}\right)$ is the estimated variance of $\hat{\alpha}$ obtained from the first row and the first column of $I^{-1}\left(\hat{\alpha}, \hat{\beta}\right)$.

The inverse matrix is given by

$$I^{-1}\left(\hat{\alpha}, \hat{\beta}\right) = \begin{bmatrix} 2\hat{\alpha}^2/n & -2\hat{\alpha}^2/\Sigma X_i^{-1} \\ -2\hat{\alpha}^2/\Sigma X_i^{-1} & \hat{\alpha}n\,(2\hat{\alpha}-1)/\left(\Sigma X_i^{-1}\right)^2 \end{bmatrix}. \tag{18}$$

Therefore under $H_0$ we obtain the Wald statistic

$$Z_{wald} \cong \sqrt{\frac{n}{2\hat{\alpha}^2}}\,(\hat{\alpha} - \alpha). \tag{19}$$

So, the $(1-\gamma)\,100\%$ confidence interval for $\tau$ based on the score method is given by

$$CI_W = [l_w, u_w] = \left[ \frac{1}{\sqrt{\hat{\alpha} - 2 + \left( Z_{\gamma/2}\sqrt{\frac{2\hat{\alpha}^2}{n}} \right)}}, \frac{1}{\sqrt{\hat{\alpha} - 2 - \left( Z_{\gamma/2}\sqrt{\frac{2\hat{\alpha}^2}{n}} \right)}} \right] \quad (20)$$

### 2.3 Confidence Interval Based on the Percentile Bootstrap (PB)

Bootstrap sampling is a method of drawing samples (with replacement) from the underlying probability distribution. [4] introduced a computationally intensive method of estimation called *Bootstrap*, a technique of computer based simulation for estimating the parameters under consideration. We describe the constructions for the confidence interval of the difference between two process capability indices $\tau$ using bootstrap techniques.

Let $\mathbf{X}^*_{k,b}$ be a random sample from inverse gamma distribution, where $1 \leq b \leq B$, be the $b^{th}$ bootstrap samples and let $\mathbf{X}^*_{k,1}, ..., \mathbf{X}^*_{k,B}$ be the $B$ bootstrap samples.

Therefore, the $(1-\gamma)\,100\%$ confidence interval for $\tau$ based on the bootstrap percentile method is given by

$$CI_{PB} = [l_{pb}, u_{pb}] = \left[ \hat{\tau}^*_{(\gamma/2)}, \hat{\tau}^*_{(1-\gamma/2)} \right], \quad (21)$$

where $\hat{\tau}^*_{(r)}$ is the $r^{th}$ ordered value on the list of the $B$ bootstrap estimator of $\tau$.

## 3 Simulation Studies

In this study, to compare the performance of the confidence intervals for $\tau$ by the estimated coverage probability and expected length are defined as

$$CP = \frac{c\,(L \leq \tau \leq U)}{M} \quad and \quad EL = \frac{\Sigma\,(U-L)}{M} \quad (22)$$

respectively, where $c\,(L \leq \tau \leq U)$ is the number of simulation runs for $\tau$. Then, we choose the confidence interval with a coverage probability greater than or close to the nominal confidence level $1-\gamma = 0.95$ and short length interval. The simulations are done using Monte Carlo simulation that R statistical program [13] with $M$ = 15,000 replications in each case and repeated the exercise $B$ = 1,000 times for the PB confidence interval.

The data are generated from $IG(\alpha, \beta)$ with $\beta = 1$ and $\alpha$ is adjusted to get the required coefficient of variation $\tau$. Then, we set $\tau = 0.05, 0.10, 0.15, 0.20, 0.25, 0.30, 0.35, 0.40, 0.45$ and $0.50$. The sample sizes are chosen to be $n = 10, 30, 50, 100$ and $200$.

**Table 1.** The coverage probabilities and expected lengths of 95% confidence intervals for the coefficient of variation in the inverse gamma distribution

| n | $\tau$ | Coverage probabilities | | | Expected lengths | | |
|---|---|---|---|---|---|---|---|
| | | $CI_S$ | $CI_W$ | $CI_{PB}$ | $CI_S$ | $CI_W$ | $CI_{PB}$ |
| 10 | 0.05 | 0.8043 | **0.9713** | 0.7151 | 0.0472 | 0.1001 | **0.0354** |
| | 0.10 | 0.8043 | **0.9722** | 0.7197 | 0.0954 | 0.2172 | **0.0716** |
| | 0.15 | 0.8077 | **0.9725** | 0.7212 | 0.1468 | 0.3999 | **0.1105** |
| | 0.20 | 0.8065 | **0.9991** | 0.7219 | 0.2005 | 0.7585 | **0.1514** |
| | 0.25 | 0.8093 | **0.9980** | 0.7307 | 0.2618 | 1.1592 | **0.1972** |
| | 0.30 | 0.8138 | **0.9971** | 0.7275 | 0.3352 | 1.5023 | **0.2538** |
| | 0.35 | 0.8150 | **0.9950** | 0.7287 | 0.4320 | 1.7131 | **0.3207** |
| | 0.40 | 0.8131 | **0.9912** | 0.7373 | 0.5688 | 1.8331 | **0.4248** |
| | 0.45 | 0.8127 | **0.9893** | 0.7417 | 0.6862 | 1.9473 | **0.5169** |
| | 0.50 | 0.8051 | **0.9813** | 0.7411 | 0.8278 | 2.0490 | **0.6749** |
| 30 | 0.05 | 0.8955 | **0.9586** | 0.8604 | 0.0257 | 0.0298 | **0.0233** |
| | 0.10 | 0.8977 | **0.9605** | 0.8603 | 0.0520 | 0.0609 | **0.0471** |
| | 0.15 | 0.8961 | **0.9595** | 0.8639 | 0.0804 | 0.0954 | **0.0731** |
| | 0.20 | 0.8962 | **0.9560** | 0.8743 | 0.1102 | 0.1332 | **0.1006** |
| | 0.25 | 0.8994 | **0.9574** | 0.8744 | 0.1439 | 0.1785 | **0.1312** |
| | 0.30 | 0.9024 | **0.9591** | 0.8717 | 0.1832 | 0.2359 | **0.1674** |
| | 0.35 | 0.9035 | **0.9552** | 0.8779 | 0.2295 | 0.3119 | **0.2098** |
| | 0.40 | 0.9096 | **0.9562** | 0.8827 | 0.2881 | 0.4315 | **0.2629** |
| | 0.45 | 0.9101 | **0.9600** | 0.8854 | 0.3389 | 0.5698 | **0.3118** |
| | 0.50 | 0.9097 | **0.9864** | 0.8813 | 0.4358 | 0.8125 | **0.3952** |
| 50 | 0.05 | 0.9151 | **0.9549** | 0.9020 | 0.0198 | 0.0215 | **0.0186** |
| | 0.10 | 0.9186 | **0.9556** | 0.8974 | 0.0402 | 0.0439 | **0.0378** |
| | 0.15 | 0.9185 | **0.9550** | 0.9011 | 0.0621 | 0.0683 | **0.0586** |
| | 0.20 | 0.9222 | **0.9549** | 0.9003 | 0.0852 | 0.0947 | **0.0804** |
| | 0.25 | 0.9227 | **0.9533** | 0.9053 | 0.1112 | 0.1252 | **0.1053** |
| | 0.30 | 0.9269 | **0.9531** | 0.9047 | 0.1416 | 0.1622 | **0.1336** |
| | 0.35 | 0.9265 | **0.9535** | 0.9088 | 0.1770 | 0.2074 | **0.1673** |
| | 0.40 | 0.9310 | **0.9509** | 0.9106 | 0.2209 | 0.2673 | **0.2084** |
| | 0.45 | 0.9332 | **0.9502** | 0.9154 | 0.2577 | 0.3215 | **0.2441** |
| | 0.50 | 0.9331 | **0.9455** | 0.9147 | 0.3173 | 0.4219 | **0.3023** |
| 100 | 0.05 | 0.9311 | **0.9487** | 0.9221 | 0.0140 | 0.0145 | **0.0135** |
| | 0.10 | 0.9346 | **0.9514** | 0.9277 | 0.0284 | 0.0296 | **0.0275** |
| | 0.15 | 0.9329 | **0.9522** | 0.9238 | 0.0439 | 0.0459 | **0.0424** |
| | 0.20 | 0.9357 | **0.9512** | 0.9266 | 0.0602 | 0.0633 | **0.0583** |
| | 0.25 | 0.9385 | **0.9495** | 0.9305 | 0.0786 | 0.0831 | **0.0761** |
| | 0.30 | 0.9363 | **0.9477** | 0.9287 | 0.1000 | 0.1065 | **0.0969** |
| | 0.35 | 0.9421 | **0.9445** | 0.9319 | 0.1248 | 0.1342 | **0.1214** |
| | 0.40 | 0.9446 | **0.9451** | 0.9313 | 0.1551 | 0.1689 | **0.1509** |
| | 0.45 | **0.9467** | 0.9441 | 0.9378 | 0.1801 | 0.1982 | **0.1752** |
| | 0.50 | **0.9487** | 0.9419 | 0.9333 | 0.2192 | 0.2454 | **0.2132** |
| 200 | 0.05 | 0.9383 | **0.9499** | 0.9361 | 0.0099 | 0.0101 | **0.0097** |
| | 0.10 | 0.9429 | **0.9528** | 0.9386 | 0.0200 | 0.0205 | **0.0197** |
| | 0.15 | 0.9441 | **0.9484** | 0.9404 | 0.0310 | 0.0317 | **0.0305** |
| | 0.20 | 0.9441 | **0.9492** | 0.9380 | 0.0425 | 0.0436 | **0.0419** |
| | 0.25 | 0.9445 | **0.9493** | 0.9390 | 0.0555 | 0.0571 | **0.0547** |
| | 0.30 | 0.9495 | **0.9495** | 0.9438 | 0.0706 | 0.0728 | **0.0696** |
| | 0.35 | **0.9509** | 0.9447 | 0.9425 | 0.0881 | 0.0912 | **0.0868** |
| | 0.40 | **0.9517** | 0.9431 | 0.9441 | 0.1093 | 0.1138 | **0.1079** |
| | 0.45 | **0.9490** | 0.9390 | 0.9437 | 0.1266 | 0.1325 | **0.1249** |
| | 0.50 | **0.9521** | 0.9397 | 0.9443 | 0.1536 | 0.1619 | **0.1518** |

The coverage probability and expected length of the 95% confidence intervals for $\tau$ are evaluated and shown in Table 1. We consider coverage probabilities close to 0.95 or not. Then consider the expected lengths as the next step. The confidence interval based on the Wald method, $CI_W$, provides coverage probabilities greater than or close to the nominal coverage level at 0.95 in almost all cases. Regarding the score method, $CI_S$ and PB confidence interval, $CI_{PB}$, respectively provide coverage probabilities less than 0.95. For some cases, the score method performed better than other methods when the coefficient of variation increases in a large sample sizes ($n = 100, 200$). It was found that the expected lengths of $CI_{PB}$ are shorter than those of $CI_W$ and $CI_S$ in all cases. In addition, the expected lengths of confidence intervals tend to increases when $\tau$ increases in all sample sizes. Although expected lengths of $CI_{PB}$ and $CI_S$ are shorter than $CI_W$ but, coverage probabilities of $CI_{PB}$ and $CI_S$ are less than 0.95. While coverage probabilities of $CI_W$ greater than $CI_S$ and $CI_{PB}$. So, $CI_W$ method is better than others.

The results from Table 1, we plots graph of coverage probability and expected length for $n = 10, 30, 50, 100$ and 200 are presented in Fig. 1, Fig. 2, Fig. 3, Fig. 4 and Fig. 5 respectively.

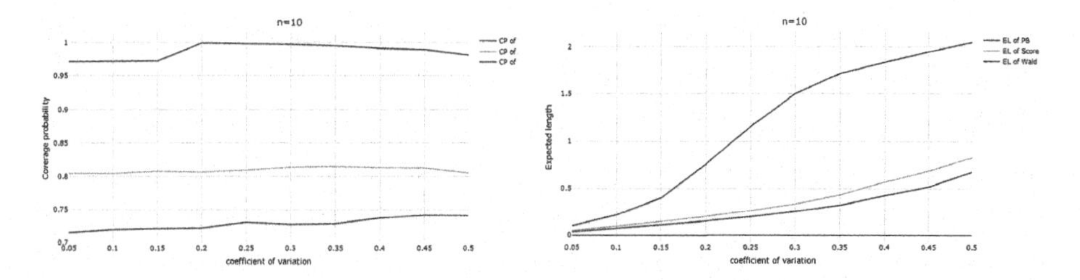


**Fig. 1.** Plots of coverage probability and expected length for $n = 10$

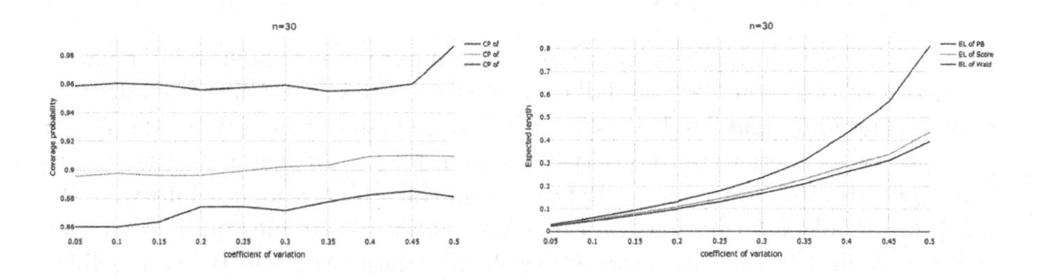


**Fig. 2.** Plots of coverage probability and expected length for $n = 30$

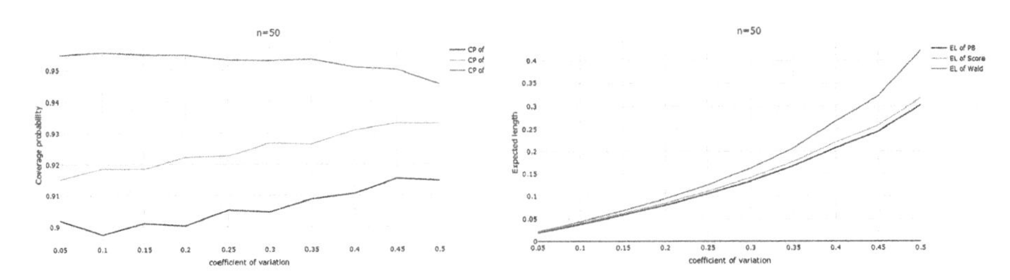


**Fig. 3.** Plots of coverage probability and expected length for $n = 50$

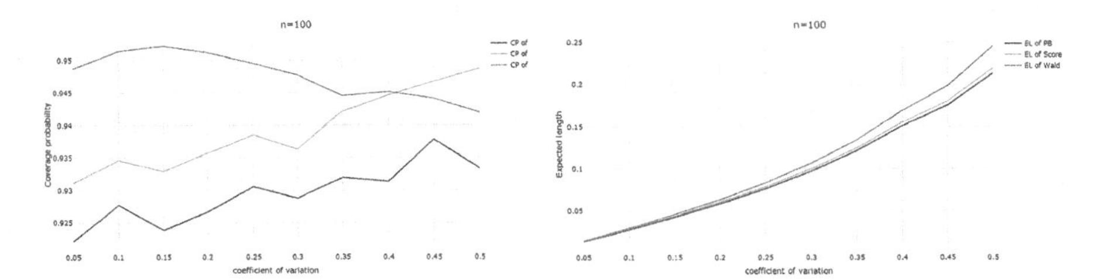


**Fig. 4.** Plots of coverage probability and expected length for $n = 100$

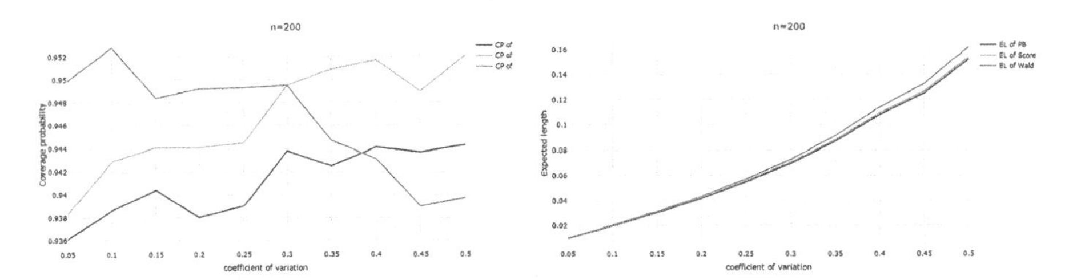


**Fig. 5.** Plots of coverage probability and expected length for $n = 200$

# 4 An Empirical Application

The data of yearly rainfall(mm) are used to compute the confidence intervals for coefficient of variation. The rainfall data were collected from Chae Hom district, Lampang province and San Kamphaeng, Chiang Mai province, Thailand [17]. For this data series, there were 24 observations from 1996 to 2019 and 18 observations from 2002 to 2019 respectively. Before computing the confidence intervals, the minimum Akaike information criterion (AIC) and Bayesian information criterion (BIC) were first to test the distribution of these data. AIC and BIC are defined as

$$AIC = -2lnL + 2k \tag{23}$$

and

$$BIC = -2lnL + 2kln(n) \tag{24}$$

where $L$ be the likelihood function, $k$ be the number of parameters, and $n$ be the number of recorded measurements.It was found that the data of rainfall fit the inverse gamma distribution, because the AIC value in Table 2 and Table 4 of the inverse gamma distribution was smallest. The inverse gamma Q-Q plot of these data are presented in Fig. 6 and Fig. 7.

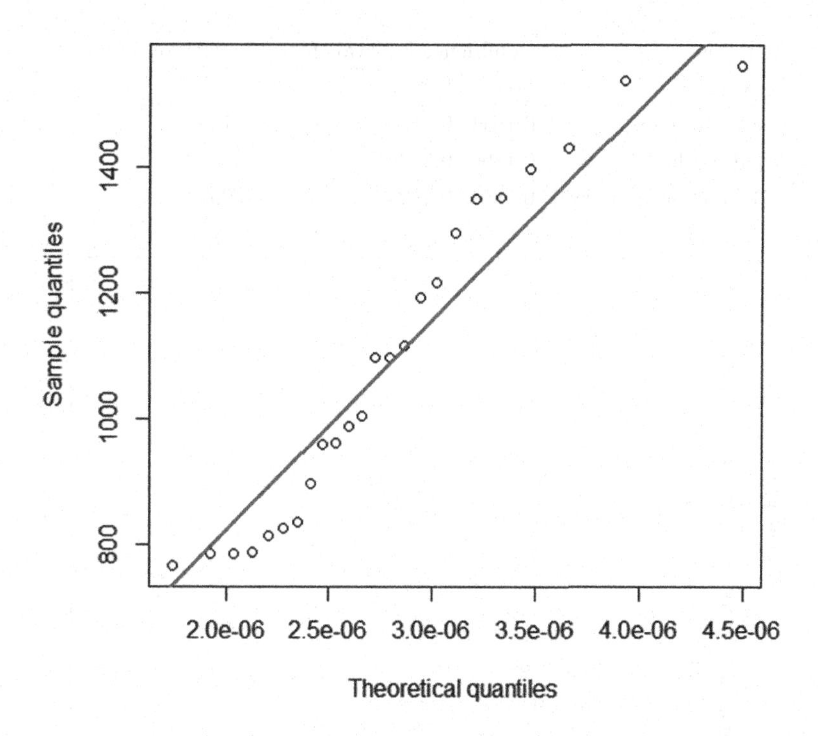


**Fig. 6.** The inverse gamma Q-Q plot for a rainfall data from Lampang province

**Table 2.** AIC and BIC results to check the distribution of rainfall data from Lampang province

| Densities | Normal | Cauchy | Exponential | Weibull | Gamma | Inverse Gamma |
|---|---|---|---|---|---|---|
| AIC | 337.5703 | 350.8302 | 385.5251 | 338.2257 | 336.2078 | 332.7863 |
| BIC | 339.9264 | 353.1863 | 386.7032 | 340.5818 | 338.5639 | 330.4301 |

Next, summary statistics were computed: rainfall data from Lampang have $n = 24$, $\hat{\alpha} = 19.1518$ and the maximum likelihood estimator for $\tau$ is $\hat{\tau} = 0.2415$, and rainfall data from Chiang Mai have $n = 18$, $\hat{\alpha} = 14.1233$ and $\hat{\tau} = 0.2872$ . The 95% confidence intervals for $\tau$ were calculated, as reported in Table 3 and Table 5 respectively. Like the simulation in the previous section, the confidence intervals of $CI_W$ provides coverage probabilities greater than or close to the nominal confidence level 0.95 in a small sample sizes and the length of $CI_{PB}$ is the shortest, which is consistent with the simulation.

**Table 3.** The 95% confidence intervals for a single coefficient of variation of rainfall data from Lampang province

| Methods | Confidence intervals for $\tau$ | | Length of intervals |
|---|---|---|---|
| | Lower | Upper | |
| Score method | 0.1541 | 0.3126 | 0.1585 |
| Wald method | 0.1890 | 0.3979 | 0.2089 |
| PB confidence interval | 0.1851 | 0.2824 | 0.0933 |

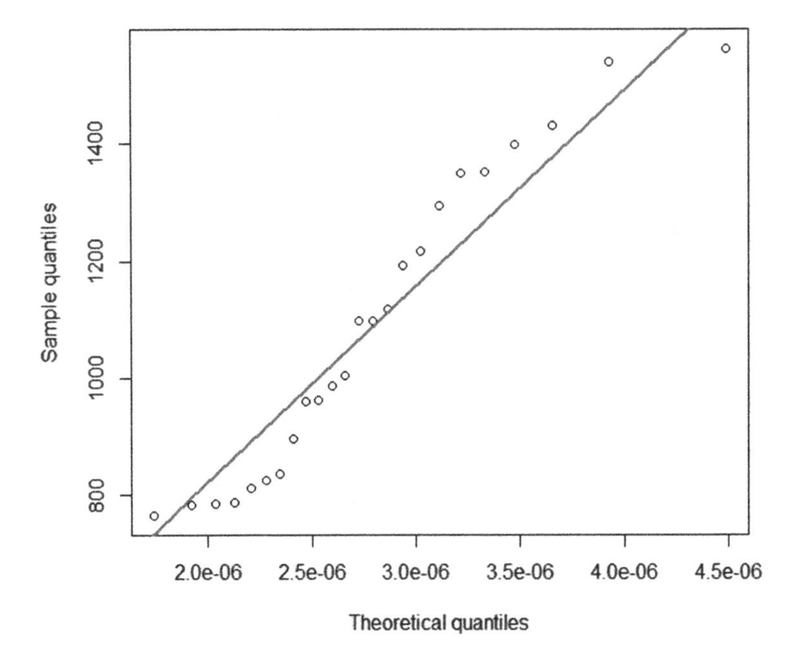


**Fig. 7.** The inverse gamma Q-Q plot for a rainfall data from Chiang Mai province

**Table 4.** AIC and BIC results to check the distribution of rainfall data from Chiang Mai province

| Densities | Normal | Cauchy | Exponential | Weibull | Gamma | Inverse Gamma |
|---|---|---|---|---|---|---|
| AIC | 261.4749 | 268.4911 | 292.5031 | 261.7443 | 260.6007 | 245.6159 |
| BIC | 263.2556 | 270.2719 | 293.3934 | 263.5250 | 262.3815 | 243.8352 |

**Table 5.** The 95% confidence intervals for a single coefficient of variation of rainfall data from Chiang Mai province

| Methods | Confidence intervals for $\tau$ | | Length of intervals |
|---|---|---|---|
| | Lower | Upper | |
| Score method | 0.1607 | 0.3910 | 0.2303 |
| Wald method | 0.2164 | 0.5876 | 0.3712 |
| PB confidence interval | 0.1978 | 0.3635 | 0.1718 |

## 5 Conclusions

This paper proposes confidence intervals for a single coefficient of variation of a inverse gamma distribution. The first confidence interval is constructed based on the score method. Then, the confidence interval is proposed based on the Wald method. Finally, the PB confidence interval is constructed. The performance of the confidence intervals was evaluated using the coverage probability and expected length through Monte Carlo simulations. The simulation studies showed that the confidence interval based on the Wald method, $CI_W$, is recommended to construct the confidence interval for a single coefficient of variation for the inverse gamma distribution. Furthermore, $CI_{PB}$ is better than the other methods in terms of the expected length but the coverage probability of $CI_{PB}$ less than 0.95.

**Acknowledgements.** The authors would to thank the referees for their constructive comments and suggestions. The first author is grateful for receiving the financial support from Science Achievement Scholarship of Thailand.

## References

1. Abid, S.H., Al-Hassany, S.A.: On the inverted gamma distribution. Int. J. Syst. Sci. Appl. Math. **1**(3), 16–22 (2016)
2. Chankham, W., et al.: Confidence intervals for coefficient of variation of inverse gaussian distribution. In: Proceedings of the 3rd International Conference on Vision, Image and Signal Processing, pp. 1–6 (2019). https://doi.org/10.1145/3387168.3387254
3. Eberhardt, L.L.: Appraising variability in population studies. J. Wildl. Manag. **42**(2), 207–238 (1978)
4. Efon, B.: The jackknif, the bootstrap and other re-sampling plans. In: 38th CMBS-NSF Regional Conference Series in Applied Mathematics. SIAM, Philadelphia (1982)
5. Gaston, K.J., McArdle, B.: The temporal variability of animal abundances: measures, methods and patterns. Philos. Trans. R. Soc. Lond. **345**(1314), 335–358 (1994)

6. Glen, A.G., Leemis, L.M. (eds.): Computational Probability Applications. ISORMS, vol. 247. Springer, Cham (2017). https://doi.org/10.1007/978-3-319-43317-2
7. Gelman, A.: Prior distributions for variance parameters in hierarchical models. Bayesian Anal. **1**(3), 515–533 (2006)
8. Liu, S.: Confidence interval estimation for coefficient of variation. Thesis. Georgia State University, Georgia (2012)
9. Llera, A., Beckmann, C.F.: Estimating an inverse gamma distribution. Technical report, Radboud University Nijmegen, Donders Institude for Brain Cognition and Behaviour. arXiv: 1605.01019v2 (2016)
10. McArdle, B., et al.: Variation in the size of animal populations: patterns, problems and artefacts. J. Anim. Ecol. **59**(2), 439–454 (1990)
11. McLaughlin, S.C., et al.: The value of the coefficient of variation in assessing repeat variation in ECG measurements. Eur. Heart J. **19**(2), 342–351 (1998)
12. Niwitpong, S.: Confidence intervals for the coefficients of variation for lognormal distributions with restricted parameter space. Appl. Math. Sci. **7**(77), 3805–3810 (2013)
13. R Core Team, An introduction to R: Notes on R, A programming environment for data analysis and graphics. http://cran.r-project.org/doc/manuals/R-intro.pdf. Accessed 5 Dec 2019
14. Sangnawakij, P., Niwitpong, S.: Confidence intervals for functions of coefficients of variation with bounded parameter spaces in two gamma distributions. Songklanakarin J. Sci. Technol. **39**(1), 27–39 (2015)
15. Sangnawakij, P., Niwitpong, S.: Confidence intervals for coefficients of variation in two-parameter exponential distributions. Commun. Stat. Simul. Comput. **46**(8), 6618–6630 (2016). https://doi.org/10.1080/03610918.2016.1208236
16. Thangjai, W., et al.: Confidence intervals for mean and difference between means of normal distributions with unknown coefficients of variation. Mathematics **5**(3), 39 (2016). https://doi.org/10.3390/math5030039
17. Upper Northern Region Irrigation Hydrology Center. http://hydro-1.rid.go.th/. Accessed 1 Mar 2020
18. Vangel, M.G.: Confidence intervals for a normal coefficient of variation. Am. Stat. **50**(1), 21–26 (1996)

# The Bayesian Confidence Interval for the Mean of the Zero-Inflated Poisson Distribution

Sunisa Junnumtuam, Sa-Aat Niwitpong(✉), and Suparat Niwitpong

Department of Applied Statistics, King Mongkut's University of Technology North Bangkok, Bangkok 10800, Thailand
sa-aat.n@sci.kmutnb.ac.th

**Abstract.** In this paper, the aim is to propose the confidence interval for the mean of Zero-inflated Poisson distribution. The two methods namely the Markov chain Monte Carlo (MCMC) and the highest posterior density (HPD) are applied to avoid the complex variance of mean of Zero-inflated Poisson distribution. Both the simulation study and the real-life data of the number of new daily COVID-19 cases in Laos are considered. The results show that Markov chain Monte Carlo method perform better than the highest posterior density method.

**Keywords:** Markov chain Monte Carlo · Highest posterior density · Count data

## 1 Introduction

Count data is usually analyzed by Poisson distribution. The probability mass function (pmf) of Poisson distribution is given by

$$f(x;\lambda) = \frac{e^{-\lambda}\lambda^{x}}{x!}; x = 0, 1, 2, ... \tag{1}$$

where $\lambda$ is the rate parameter and $\lambda > 0$. Since the Poisson distribution has property; *equidispersion* [1]; which means the mean and the variance are equal. When data is excess zeros such as the number of new daily COVID-19 cases in Laos. The data from 25/03/2020 to 07/07/2020 of new daily cases in Laos are used to study. The report shows that there are 95 days with no new cases, 1 new case for 5 days, 2 cases for 2 days, 3 cases for 2 days, and 4 cases for 1 day. This clearly that data has variance exceeding mean and this led to the overdispersion. Thus, the Poisson distribution cannot apply to these data. Thus some models are proposed to depart from this standard count model such as zero-inflated (ZI) models and hurdle models. Both the zero-inflated models and hurdle models provide an alternative way to describe count data with excess zeros. Lambert [2]


V.-N. Huynh et al. (Eds.): IUKM 2020, LNAI 12482, pp. 419–430, 2020.
https://doi.org/10.1007/978-3-030-62509-2_35

proposed a zero-inflated Poisson (ZIP) model where the probability of zeros is logit and the base count density is Poisson. The pmf is given by

$$f(x;\lambda,\omega) = \begin{cases} \omega + (1-\omega)e^{-\lambda} & ; x = 0 \\ \frac{(1-\omega)e^{-\lambda}\lambda^x}{x!} & ; x = 1,2,3,... \end{cases} \quad (2)$$

with mean equals $(1-\omega)\lambda$ and the variance is $(1-\omega)\lambda + \frac{\omega}{1-\omega}((1-\omega)\lambda)^2; \lambda > 0$ when $\omega$ is the parameter of the proportion of zeros. The ZIP distribution has been studied in various field, such as biomedical applications, Bohning *et al.* [3] have used dental epidemiology to study the amount of the decayed, teeth extraction, and teeth filling in both deciduous and succedaneous tooth of children with DMFT (The decayed, missing and filled teeth) index. The results have clearly shown that the ZIP model is useful to describe the DMFT index. In financial insurance, Kusuma and Purwono [4] adjusted the Poisson regression and the ZIP regression to the claim frequency data from the health insurance company PT.XYZ. The result of this study has shown that the ZIP regression model is more suitable than the ordinary Poisson regression model for modeling the frequency data of claims. In natural resources, Lee and Kim [5] have followed the number of torrential rainfall occurrences at the Daegu and the Busan rain gauges in South Korea in order to apply and compare the developed models such as the Poisson distribution, the Generalized Poisson distribution (GPD), ZIP, the Zero-inflated Generalized Poisson (ZIGP), and the Bayesian ZIGP model. The result of this study showed that the Bayesian ZIGP model provided the most accurate results and the ZIP model is recommended to be an alternative from a practical viewpoint. Beckett *et al.* [6] showed that datasets from some natural disasters such as earthquake, wildfire, hurricane, tornado, and lightning dataset can be modeled by the ZIP distribution. In manufacturing, Xie *et al.* [7] analyzed the defect count data of the read-write errors in a computer hard disk from a manufacturing process by using the ZIP model and Poisson model. The result showed that the ZIP model is more suitable than the Poisson model.

Moreover, Bayesian approach is applied to analyze the ZIP distribution in many researches for example Jose [8] studied the ZIP distribution by using the Bayesian method with the noninformative priors to estimate the number of roots produced by 270 shoots of the apple cultivar *Trajan* and the results showed the great fitting of the ZIP distribution at the point zero. Xu *et al.* [9] investigated the noninformative priors for the ZIP distribution with two parameters: the probability of zeros and the mean of the Poisson part and presented the point estimation and the 95% CI for the two parameters. Unhapipat *et al.* [10] applied the Bayesian predictive inference with many types of prior distribution such as generalized noninformative prior, Jeffrey's noninformative prior, and Beta-Gamma prior for the ZIP distribution and this research also demonstrated the real-life data in public health, natural calamities, and vehicle accidents.

Since the Poisson mean is investigated by many researchers in various ways for example Yip [11] compared the standard likelihood approach and the conditional likelihood approach of estimating the mean of a Poisson distribution in a nuisance parameter. Tanusit [12] compared seven method confidence intervals for the Poisson means such as Wald CI, Score CI, Score continuity correction CI, Agresti and Coull CI, Bayes Wald CI, Bayes Score CI, and Bayes Score continuity correction CI. Patil and Kulkarni [13] studied and compared nineteen methods of the confidence interval for Poisson mean. And, Bityukov *et al.* [14] investigated the numerical procedure of constructing confidence intervals for parameter of the Poisson distribution. In this study, the confidence interval for the zero-inflated Poisson mean was studied based on Bayesian analysis. Since the marginal posterior distribution of Poisson part ($\lambda$) is no closed form so, the approximately Bayesian method namely the Markov chain Monte Carlo (MCMC) was used to sampling parameters and lead to construct the confidence interval also the highest posterior density (HPD) interval.

Currently, the coronavirus disease of 2019 (also known as COVID-19) is caused by severe acute respiratory syndrome coronavirus 2 (SARS-CoV-2). The COVID-19 outbreak was first identified in Wuhan (Hubei, China) in December of 2019 and it has since been declared a pandemic by the World Health Organization in March of 2020 [15]. In this pandemic, Laos is one of the countries that has quickly virus response, so almost of the number of new daily COVID-19 cases in Laos is zero. Thus, in this study, the data of new daily COVID-19 in Laos are investigated with the confidence interval for the zero-inflated Poisson mean.

## 2 Bayesian Analysis of ZIP-Distribution

Suppose that $X = (X_1, ..., X_n)$ is a vector of the independent random variables generated from the ZIP distribution. Let $\hat{\theta}$ represent the estimator of $E(X)$. The sample estimate of the mean or $E(X)$ is given by

$$\widehat{E(X)} = (1 - \hat{\omega})\hat{\lambda}. \tag{3}$$

Let $A = x_i : x_i = 0, i = 1, ..., n$ and $m$ is the number of $A$, then the likelihood function is [8]

$$L[\lambda, \omega] = [\omega + (1 - \omega)p(0|\lambda)]^m (1 - \omega)^{n-m} \prod_{x_i \notin A} p(x_i|\lambda). \tag{4}$$

Since the elements of the set $A$ can be generated from two different parts: (1) the real zeros part and (2) the Poisson distribution. Then an unobserved latent allocation variable can defined as

$$I_i = \begin{cases} 1 \ ; \ p(\lambda, \omega) \\ 0 \ ; \ 1 - p(\lambda, \omega) \end{cases}$$

where $i = 1, ..., m$ and

$$p(\lambda, \omega) = \frac{\omega}{\omega + (1 - \omega)p(0|\lambda)}. \tag{5}$$

Thus, the likelihood function based on the augmented data $D = \{X, I\}$, where $I = (I_1, ..., I_m)$ [16] is

$$\begin{aligned} L[\lambda, \omega|D] &= L[\omega, \lambda] \prod_{i=1}^{m} p(\lambda, \omega)^{I_i} (1 - p(\lambda, \omega))^{1-I_i} \\ &= \omega^S (1-\omega)^{n-S} p(0|\lambda)^{m-S} \prod_{x_i \notin A} p(x_i|\lambda) \end{aligned} \tag{6}$$

where $S = \sum_{i=1}^{m} I_i \sim Bin[m, p(\lambda, \omega)]$. Thus, the likelihood function based on the augmented data is [8]

$$L[\lambda, \omega] \propto \omega^S (1-\omega)^{n-S} \lambda^{\sum_{x_i \notin A} x_i} e^{-(n-S)\lambda} \tag{7}$$

and

$$p(\lambda, \omega) = \frac{\omega}{\omega + (1-\omega)e^{-\lambda}}. \tag{8}$$

The likelihood function suggests the following independent priors:

$$\pi(\lambda) \sim Gamma[a, b] \tag{9}$$

$$\pi(\omega) \sim Beta[c, d]. \tag{10}$$

So, the joint posterior distribution for $(\lambda, \omega)$, given $D$

$$\pi(\lambda, \omega|D) \propto \omega^{S+c-1} (1-\omega)^{n-S+d-1} \lambda^{\sum_{x_i \notin A} x_i + a - 1} e^{-(n-S+b)\lambda}. \tag{11}$$

Since $\omega$ and $\lambda$ are independent given $D$, thus the marginal posterior distribution of $\omega$ is a Beta distribution; that is,

$$\pi(\omega|D) = Beta(S + c, n - S + d), \tag{12}$$

and the marginal posterior distribution of $\lambda$ is

$$\pi(\lambda|D) \propto \lambda^{\sum_{x_i \notin A} x_i + a - 1} e^{-(n-S+b)\lambda}. \tag{13}$$

### 2.1 Approximately Bayesian Interval by MCMC Algorithm

**Algorithm 1**
Given a,c,d = 0.5 and b = 0 for noninformative prior for $X \sim ZIP(\lambda^{(0)}, \omega^{(0)})$ and $t = 1, ..., 10$.

1. Calculate $p(\lambda^{(0)}, \omega^{(0)}) = \frac{\omega^{(0)}}{\omega^{(0)} + (1-\omega^{(0)})e^{-\lambda^{(0)}}}$.
2. Generate $S^{(t)}$ from $Bin(m, p(\lambda^{(t-1)}, \omega^{(t-1)}))$.
3. Generate $\omega^{(t)}$ from $Beta(S^{(t)} + c, n - S^{(t)} + d)$
4. Generate $\lambda^{(t)}$ from $Gamma(\sum_{i=1}^{n} x_i + a, n - S^{(t)} + b)$.
5. Repeat step 2–4 for $t$ times to update the sample.
6. Collect $\omega^{(t)}$ and $\lambda^{(t)}$ for 5,000 samples.

7. Burn-in 1,000 samples and calculate the estimator of $\theta$ in Eq. 3.

Then, the $100(1-\alpha)\%$ Approximately Bayesian confidence interval of $\theta$ is calculated by

$$CI_{MCMC} = (L.CI, U.CI), \tag{14}$$

where $L.CI = quantile(\hat{\theta}, \frac{\alpha}{2})$ and $U.CI = quantile(\hat{\theta}, 1-\frac{\alpha}{2})$ respectively.

### 2.2 Bayesian-Based HPD Interval

The Bayesian credible or the highest posterior density (HPD) interval is the shortest interval containing $100(1-\alpha)\%$ of the posterior probability. So that the density within the interval has a higher probability than outside of interval. The credible intervals can be constructed by using a Markov chain Monte Carlo (MCMC) method [17].

The two major properties of the highest posterior density interval were explained by Box and Tiao [18] as follows:
(a) The density for every point inside the interval is greater than that for every point outside the interval.
(b) For a given probability content (say, $1-\alpha$) the interval is of the shortest length. In this paper, the MCMC method is used to estimate the HPD intervals for the mean of the ZIP distribution. This approach requires only a MCMC sample generated from the marginal posterior distribution of two parameters: $\lambda$ and $\omega$. In simulation and computation, the HPD intervals are computed by using the package *HDInterval* version 0.2.0 from RStudio.

**Algorithm 2**
Given a,c,d = 0.5 and b = 0 for noninformative prior for $X \sim ZIP(\lambda^{(0)}, \omega^{(0)})$ and $t = 1, ..., 10$.

1. Calculate $p(\lambda^{(0)}, \omega^{(0)}) = \frac{\omega^{(0)}}{\omega^{(0)}+(1-\omega^{(0)})e^{-\lambda^{(0)}}}$.
2. Generate $S^{(t)}$ from $Bin(m, p(\lambda^{(t-1)}, \omega^{(t-1)}))$.
3. Generate $\omega^{(t)}$ from $Beta(S^{(t)} + c, n - S^{(t)} + d)$
4. Generate $\lambda^{(t)}$ from $Gamma(\sum_{i=1}^{n} x_i + a, n - S^{(t)} + b)$.
5. Repeat step 2–4 for $t$ times to update the samples.
6. Collect $\omega^{(t)}$ and $\lambda^{(t)}$ for 5,000 samples.
7. Burn-in 1,000 samples.
8. Compute the HPD intervals $100(1-\alpha)\%$ for $\theta$.

# 3 Simulation Results

In the simulation study, sample size $n$ was set as 50, 100 and 200; $\lambda$ = 1, 5, 10, 15, 20 and 25 and $\omega = 0.1(0.1)0.9$. The simulation data were generated form Package *gamlss* version 5.1-6 from RStudio, https://rstudio.com, and the number of replications were set as 1,000 repetitions. The nominal confidence level was 0.95. The simulation results, in Table 1, present the coverage probability (CP) and the average lengths (AL) of the confidence interval only $n$ is 100. Looking at the Table 1, for $n = 100$ and $\lambda = 1$, the MCMC methods performed

**Table 1.** The coverage probability (CP) and the average lengths (AL) of the confidence interval at the confidence coefficient of 0.95

| $n$ | $\lambda$ | $\omega$ | $E(X)$ | Coverage probabilities (Average lengths) | |
|---|---|---|---|---|---|
| | | | | MCMC | HPD |
| 100 | 1 | 0.1 | 0.9 | 0.9500 (0.3929) | 0.9470 (0.3901) |
| | | 0.2 | 0.8 | 0.9480 (0.3862) | 0.9410 (0.3831) |
| | | 0.3 | 0.7 | 0.9400 (0.3741) | 0.9350 (0.3708) |
| | | 0.4 | 0.6 | 0.9450 (0.3585) | 0.9420 (0.3546) |
| | | 0.5 | 0.5 | 0.9620 (0.3389) | 0.9550 (0.3346) |
| | | 0.6 | 0.4 | 0.9450 (0.3164) | 0.9360 (0.3113) |
| | | 0.7 | 0.3 | 0.9450 (0.2803) | 0.9440 (0.2741) |
| | | 0.8 | 0.2 | 0.9510 (0.2396) | 0.9480 (0.2319) |
| | | 0.9 | 0.1 | 0.9340 (0.1805) | 0.9190 (0.1689) |
| | 5 | 0.1 | 4.5 | 0.9380 (1.0158) | 0.9350 (1.0113) |
| | | 0.2 | 4.0 | 0.9420 (1.1041) | 0.9370 (1.0991) |
| | | 0.3 | 3.5 | 0.9410 (1.1473) | 0.9380 (1.1419) |
| | | 0.4 | 3.0 | 0.9460 (1.1634) | 0.9450 (1.1570) |
| | | 0.5 | 2.5 | 0.9450 (1.1476) | 0.9420 (1.1408) |
| | | 0.6 | 2.0 | 0.9540 (1.0968) | 0.9420 (1.0885) |
| | | 0.7 | 1.5 | 0.9380 (1.0099) | 0.9330 (0.9999) |
| | | 0.8 | 1.0 | 0.9480 (0.8711) | 0.9420 (0.8572) |
| | | 0.9 | 0.5 | 0.9450 (0.6583) | 0.9420 (0.6374) |
| | 10 | 0.1 | 9.0 | 0.9460 (1.6555) | 0.9420 (1.6474) |
| | | 0.2 | 8.0 | 0.9470 (1.8999) | 0.9440 (1.8907) |
| | | 0.3 | 7.0 | 0.9550 (2.0535) | 0.9480 (2.0442) |
| | | 0.4 | 6.0 | 0.9580 (2.1214) | 0.9510 (2.1115) |
| | | 0.5 | 5.0 | 0.9410 (2.1177) | 0.9360 (2.1067) |
| | | 0.6 | 4.0 | 0.9560 (2.0531) | 0.9500 (2.0411) |
| | | 0.7 | 3.0 | 0.9540 (1.9010) | 0.9510 (1.8852) |
| | | 0.8 | 2.0 | 0.9430 (1.6493) | 0.9340 (1.6285) |
| | | 0.9 | 1.0 | 0.9520 (1.2388) | 0.9410 (1.2052) |
| | 15 | 0.1 | 13.5 | 0.9490 (2.2771) | 0.9520 (2.2639) |
| | | 0.2 | 12.0 | 0.9400 (2.6917) | 0.9390 (2.6778) |
| | | 0.3 | 10.5 | 0.9460 (2.9412) | 0.9450 (2.9281) |

*(continued)*

**Table 1.** (*continued*)

| $n$ | $\lambda$ | $\omega$ | $E(X)$ | Coverage probabilities (Average lengths) | |
|---|---|---|---|---|---|
| | | | | MCMC | HPD |
| 100 | 15 | 0.4 | 9.0 | 0.9470 (3.0640) | 0.9500 (3.0505) |
| | | 0.5 | 7.5 | 0.9520 (3.0842) | 0.9480 (3.0694) |
| | | 0.6 | 6.0 | 0.9500 (2.9922) | 0.9430 (2.9749) |
| | | 0.7 | 4.5 | 0.9400 (2.7764) | 0.9360 (2.7559) |
| | | 0.8 | 3.0 | 0.9430 (2.4084) | 0.9350 (2.3803) |
| | | 0.9 | 1.5 | 0.9490 (1.8126) | 0.9300 (1.7672) |
| | 20 | 0.1 | 18.0 | 0.9390 (2.8821) | 0.9350 (2.8618) |
| | | 0.2 | 16.0 | 0.9570 (3.4622) | 0.9580 (3.4417) |
| | | 0.3 | 14.0 | 0.9450 (3.8332) | 0.9440 (3.8149) |
| | | 0.4 | 12.0 | 0.9580 (4.0226) | 0.9520 (4.0044) |
| | | 0.5 | 10.0 | 0.9500 (4.0475) | 0.9510 (4.0277) |
| | | 0.6 | 8.0 | 0.9430 (3.9236) | 0.9390 (3.9018) |
| | | 0.7 | 6.0 | 0.9580 (3.6727) | 0.9540 (3.6464) |
| | | 0.8 | 4.0 | 0.9540 (3.1756) | 0.9460 (3.1382) |
| | | 0.9 | 2.0 | 0.9320 (2.3993) | 0.9280 (2.3425) |
| | 25 | 0.1 | 22.5 | 0.9490 (3.4499) | 0.9440 (3.4235) |
| | | 0.2 | 20.0 | 0.9450 (4.2520) | 0.9420 (4.2254) |
| | | 0.3 | 17.5 | 0.9450 (4.7216) | 0.9410 (4.6953) |
| | | 0.4 | 15.0 | 0.9670 (4.9628) | 0.9610 (4.9394) |
| | | 0.5 | 12.5 | 0.9470 (5.0140) | 0.9410 (4.9912) |
| | | 0.6 | 10.0 | 0.9490 (4.8883) | 0.9430 (4.8624) |
| | | 0.7 | 7.5 | 0.9510 (4.5519) | 0.9480 (4.5185) |
| | | 0.8 | 5.0 | 0.9410 (3.9605) | 0.9410 (3.9177) |
| | | 0.9 | 2.5 | 0.9550 (2.9658) | 0.9410 (2.8959) |

well for all most cases except $\omega = 0.9$ and the HPD method performed well for some cases. Furthermore, when $\omega$ is increased, the average lengths trend to decrease. When $\lambda$ is increased to 5, the performance of the MCMC and HPD are slightly dropped except case $\omega = 0.9$. When $\lambda$ is 10, the CPs of the MCMC and HPD method are slightly increased except $\omega = 0.5$. and the CPs of two methods are not changed much when the $\lambda$ is increased to 15. However, when the $\lambda$ is 20, the CPs of two methods are decreased for $\omega$ 0.1 and 0.9. And when $\lambda$ is increased to 25, the CPs of the MCMC method is approximately 0.95 for all cases while, the CPs of the HPD method is less than the MCMC method for all cases. Moreover, the coverage probabilities (CP) for the ZIP mean can be seen clearly in Fig. 1. The coverage probabilities for two methods are slightly different and approach to 0.95. Furthermore, the MCMC method provides the CPs more than the HPD method even though the $\lambda$ and $\omega$ are changed. In the Fig. 2, the average lengths (AL) are considered by level of $\lambda$. The average lengths of two methods are not different. In addition, when the $\lambda$ is increased, the average lengths are also increased.

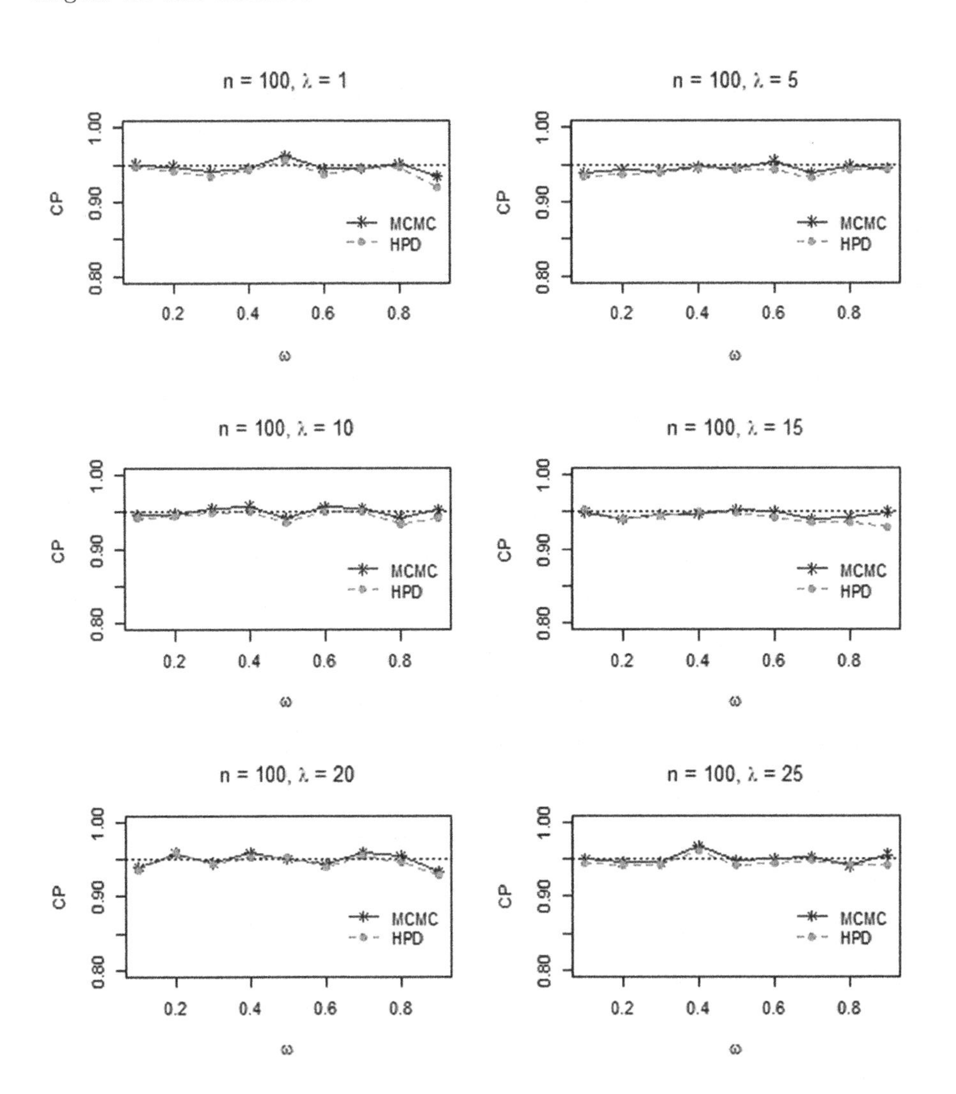


**Fig. 1.** The coverage probabilities for n = 100

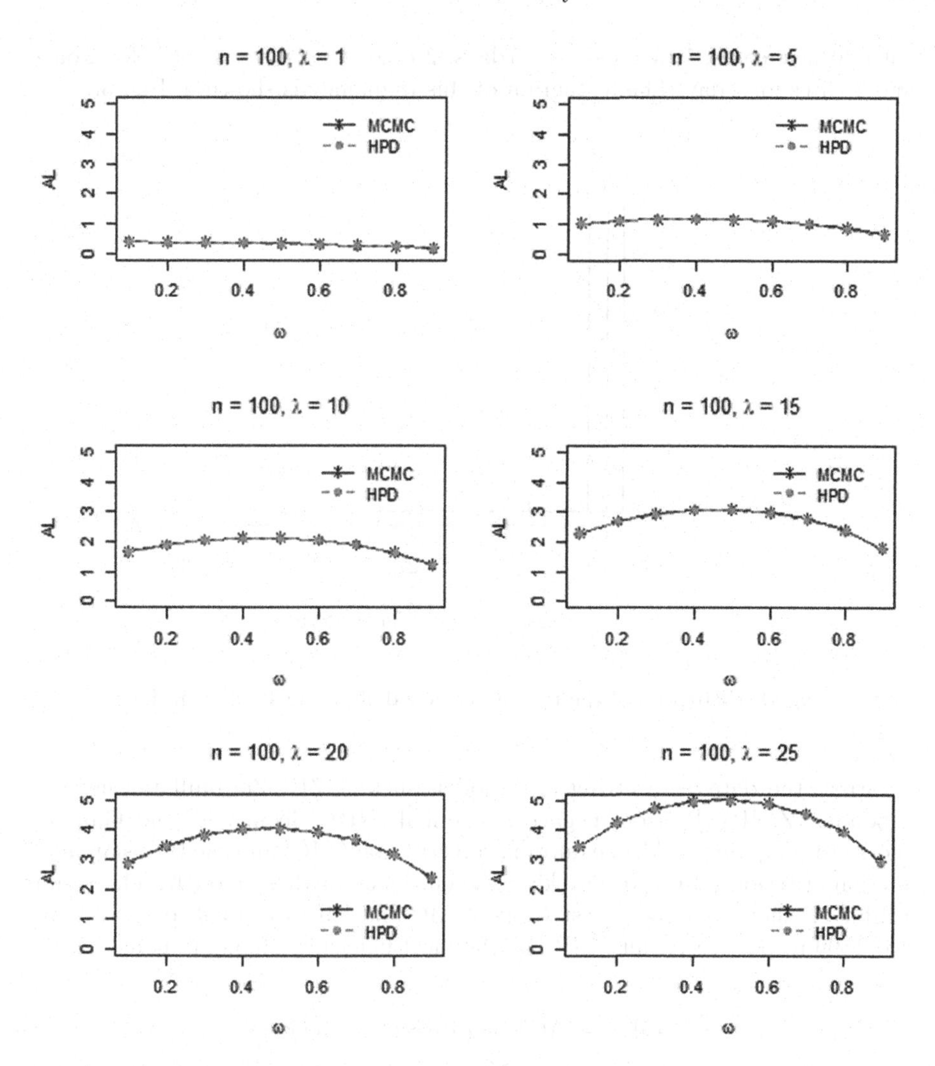


**Fig. 2.** The average lengths for n = 100

## 4 The Empirical Study

The database of daily covid-19 cases in Laos is considered for ZIP distribution. This data is retrieved from the R package 'utils' version 3.6.3 in data sheet 'https://opendata.ecdc.europa.eu/covid19/casedistribution/csv'. The downloadable data file is updated daily and contains the latest available public data on COVID-19. Each row/entry contains the number of new cases reported daily for each country with ECDC's copyright policy. In this study, data from 25/03/2020 to 07/07/2020 of new daily cases in Laos are used to analyze. There are 105 days in total of this observation. The report shows that there are 95 days with no new cases, 1 new case for 5 days, 2 cases for 2 days, 3 cases for 2 days, and 4 cases for 1 day. The histogram of this frequency is shown in Fig. 3.

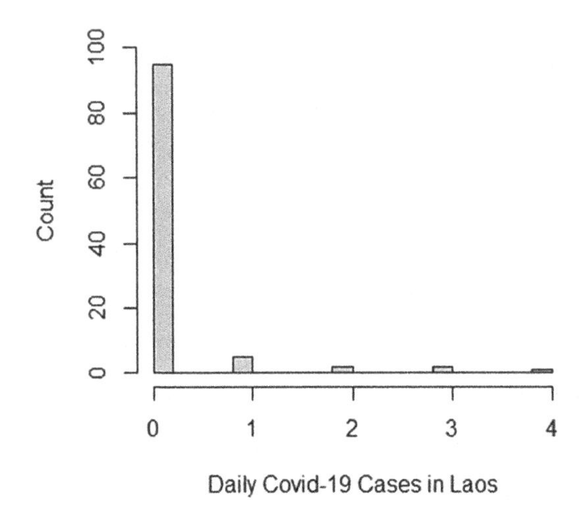


**Fig. 3.** Histogram of the number of new daily covid-19 cases in Laos

First, the date were fitting in 6 models such as ZIP, Zero-inflated negative binomial (ZINB), Poisson, negative binomial (NB), Geometric, and Gaussian model to compare the Akaike information criterion (AIC) and the Bayesian information criterion (BIC) for checking the efficiency of those models. The results (Table 2.) show that the lowest value of AIC and BIC are equal to 95.8730 and 101.1809 respectively; thus, ZIP has the most efficiency among 6 models.

**Table 2.** AICs and BICs of six models

| Models | ZIP | ZINB | Poisson | NB | Geometric | Gaussian |
|---|---|---|---|---|---|---|
| **AIC** | **95.8730** | 97.8673 | 121.2575 | 97.0780 | 108.2095 | 209.6005 |
| **BIC** | **101.1809** | 105.8292 | 123.9115 | 102.3859 | 110.8634 | 214.9084 |

In this study, we require package *pscl* to construct model of ZIP. The model provided count model coefficients for Poisson distribution with log link is equal to 0.3769, then $\hat{\lambda} = e^{0.3769} = 1.4578$ and Zero-inflation model coefficients for binomial distribution with logit link is equal to 1.954, then $\hat{\omega} = e^{1.954}/(1 + e^{1.954}) = 0.8759$. Thus the estimator of mean is $(1 - \omega) * \lambda = 0.1809$ and the confidence interval for mean is given by Table 3.

As the results in Table 3, the 95% confidence interval for mean from two methods are cover the point estimator. According to the simulation results in

**Table 3.** The estimation of the number of new daily covid-19 cases in Laos

| Methods | Estimation of mean | |
|---|---|---|
| | 95% CI | Length of CI |
| MCMC | (2.1749 3.1820) | 1.0072 |
| HPD | (2.1215 3.1165) | 0.9950 |

case of $n = 100$, $\lambda = 1$, and $\omega$ around 0.8–0.9, the MCMC method is the best way to construct the confidence interval for mean because the coverage probability is close to 0.95 more than the HPD method.

## 5 Discussion and Conclusion

Since the highest posterior density method provides the coverage probabilities less than 0.95 in many cases. Because the complex form of the ZIP model with two sources of zero: (1) the zero count form the Bernoulli part and (2) the zero count form the Poisson part. This complexity could cause some errors of estimation and conclusion. However, when the overdispersion occurs such as the data are excessive number of zeros, the ZIP distribution is the great way to analyze data. The results from simulation study recommend the approximately Bayesian by Markov chain Monte Carlo method with the noninformative prior for constructing the 95% confidence interval for the ZIP mean. Because this method provides the coverage probability approximately 0.95 also performs better than the highest posterior density method.

In the empirical study shows that when count data excess zeros, the zero-inflated Poisson distribution is one of the models that can describe and construct the confidence interval for the ZIP mean. By the result, the approximately Bayesian by Markov chain Monte Carlo method with the noninformative prior is recommended to construct the 95% confidence interval for the mean of the ZIP distribution since it provides the coverage probability approximately 0.95. However, the Bayesian method with noninformative prior does not perform well in many cases, so that in the future research, the other priors are considered to compare the efficiency with the noninformative prior.

## References


1. Cameron, A.C., Trivedi, P.K.: Regression Analysis of Count Data, 4th edn. Cambridge University Press, New York (2013)
2. Lambert, D.: Zero-inflated poisson regression with an application to defects in manufacturing. Technometrics **34**, 1–14 (1992)
3. Böhning, D., Dietz, E., Schlattmann, P., Mendonça, L., Kirchner, U.: The zero-inflated Poisson model and the decayed, missing and filled teeth index in dental epidemiology. J. Roy. Stat. Soc. Ser. A (Stat. Soc.) **162**(2), 195–209 (1999)

4. Kusuma, R.D., Purwono, Y.: Zero-inflated poisson regression analysis on frequency of health insurance claim PT. XYZ. In: Proceedings of the 12th International Conference on Business and Management Research (ICBMR 2018) (2019)
5. Lee, C., Kim, S.: Applicability of zero-inflated models to fit the torrential rainfall count data with extra zeros in South Korea. Water **9**(2), 123 (2017)
6. Beckett, S., et al.: Zero-Inflated Poisson (ZIP) distribution: parameter estimation and applications to model data from natural calamities. Involve J. Math. **7**(6), 751–767 (2014)
7. Xie, M., He, B., Goh, T.N.: Zero-inflated Poisson model in statistical process control. Comput. Stat. Data Anal. **38**, 191–201 (2001)
8. Rodrigues, J.: Bayesian analysis of zero-inflated distributions. Commun. Stat. Theo. Meth. **32**(2), 281–289 (2003)
9. Xu, H.-Y., Xie, M., Goh, T.N.: Objective Bayes analysis of zero-inflated Poisson distribution with application to healthcare data. IIE Trans. **46**(8), 843–852 (2014)
10. Unhapipat, S., Tiensuwan, M., Pal, N.: Bayesian predictive inference for zero-inflated poisson (ZIP) distribution with applications. Am. J. Math. Manage. Sci. **37**(1), 66–79 (2018)
11. Yip, P.: Inference about the mean of a Poisson distribution in the presence of a nuisance parameter. Austral. J. Statist. **30**(3), 299–306 (1988)
12. Tanusit, M.: Two-side confidence intervals for the Poisson means. Int. J. Model. Optimization **2**(5), 589–591 (2012)
13. Patil, V.V., Kulkarni, H.V.: Comparison of confidence intervals for the poisson mean: some new aspects. REVSTAT-Stat. J. **10**(2), 211–227 (2012)
14. Bityukov, S.I., Krasnikov, N.V., & Taperechkina, V.A., Confidence intervals for poisson distribution parameter. https://arxiv.org/abs/hep-ex/0108020v1 (2001)
15. Valencia, D.N.: Brief Review on COVID-19: The 2020 Pandemic Caused by SARS-CoV-2. Cureus **12**(3) (2020)
16. Tanner, M.A., Wong, W.H.: The calculation of posterior distributions by data augmentation. J. Am. Stat. Assoc. **82**(398), 528–540 (1987)
17. Chen, M.-H., Shao, Q.-M.: Monte Carlo estimation of Bayesian credible and HPD intervals. J. Comput. Graph. Stat. **8**(1), 69–92 (1999)
18. Box, G.E.P., Tiao, G.C.: Bayesian Inference in Statistical Analysis. Wiley, New York (1992)

# Author Index

Printed in the United States
By Bookmasters